Management of Fungal Plant Pathogens

Management of Fungal Plant Pathogens

Edited by

Arun Arya

Professor and Head, Department of Botany and
Coordinator Environment Science Programme, Faculty of Science
The Maharaja Sayajirao University of Baroda,
Vadodara, India

and

Analía Edith Perelló

Assistant Professor and Research Scientist, CONICET - CIDEFI,
and Coordinator MSc Vegetal Protection Programme, Plant Pathology,
Facultad de Ciencias Agrarias y Forestales, Universidad Nacional
de La Plata, Provincia de Buenos Aires, Argentina

www.cabi.org

CABI is a trading name of CAB International

CABI Head Office
Nosworthy Way
Wallingford
Oxfordshire OX10 8DE
UK

Tel: +44 (0)1491 832111
Fax: +44 (0)1491 833508
E-mail: cabi@cabi.org
Website: www.cabi.org

CABI North American Office
875 Massachusetts Avenue
7th Floor
Cambridge, MA 02139
USA

Tel: +1 617 395 4056
Fax: +1 617 354 6875
E-mail: cabi-nao@cabi.org

A catalogue record for this book is available from the British Library, London, UK

Library of Congress Cataloging-in-Publication Data

Management of fungal plant pathogens / edited by Arun Arya, Analía Edith Perelló.
p. cm.
Includes bibliographical references and index.
ISBN 978-1-84593-603-7 (alk. paper)
1. Fungal diseases of plants. 2. Phytopathogenic fungi–Control. 3. Plant-pathogen relationships. I. Arya, Arun. II. Perelló, Analía Edith. III. title.

SB733.M36 2010
632'.4–dc22

2009023395

ISBN-13: 978 1 84593 603 7

Typeset by AMA Dataset, Preston, UK.
Printed and bound in the UK by the MPG Books Group.

Contents

The colour plate section can be found following page 52.

Contributors

Acharya, S.N., *Agriculture and Agri-Food Canada Research Centre, Lethbridge, AB, Canada T1J 4B1*

Adiver, S.S., *Oilseeds Scheme, Main Agricultural Research Station, University of Agricultural Sciences, Dharwad 580 005, Karnataka, India (shivaputra_adiver@rediffmail.com)*

Arya, Arun, *Department of Botany, Faculty of Science, The Maharaja Sayajirao University of Baroda, Vadodara 390002, India (aryaarunarya@rediffmail.com)*

Arya, Chitra, *Department of Botany, Faculty of Science, The Maharaja Sayajirao University of Baroda, Vadodara 390002, India (caarya@yahoo.co.in)*

Astiz Gassó, Marta M., *Instituto Fitotécnico Santa Catalina (IFSC), Facultad de Ciencias Agrarias y Forestales, Universidad Nacional de La Plata, CC 4, 1836 Llavallol, Buenos Aires, Argentina (astizgasso@yahoo.com.ar)*

Awasthi, Pallavi, *Mycology and Plant Pathology Division, Department of Botany, University of Lucknow, Lucknow 226007, India*

Basu, S.K., *Department of Biological Sciences, University of Lethbridge, Lethbridge, AB, Canada T1K 3M4 (saikat.basu@uleth.ca)*

Bhowmik, P.K., *Bioproducts and Bioprocesses, Lethbridge Research Center, Agriculture and Agri-Food Canada, Lethbridge, AB Canada T1J 4B1*

Cabello, Marta N., *Comisión de Investigaciones Científicas de la Provincia de Buenos Aires (CICBA) – Instituto de Botánica Spegazzini, Calle 53 N° 577, 1900 La Plata, Argentina (mcabello@museo.fcnym.unlp.edu.ar)*

Cordo, Cristina A., *Comisión de Investigaciones Científicas de la Provincia de Buenos Aires, Centro de Investigaciones de Fitopatología (CIDEFI) – Facultad de Ciencias Agrarias y Forestales, 60 y 119, (1900) La Plata, Argentina (cristcordo@hotmail.com)*

Dal Bello, Gustavo, *Centro de Investigaciones de Fitopatología (CIDEFI), Facultad de Ciencias Agrarias y Forestales, Universidad Nacional de La Plata, 60 y 119, CC 31, 1900 La Plata, Argentina (dalbello@speedy.com.ar)*

Dubey, N.K., *Centre of Advanced Study in Botany, Banaras Hindu University, Varanasi 221005, India (nkdubey2@rediffmail.com)*

Esh, Ayman M.H., *Biotechnology and Tissue Culture Laboratories, Sugar Crops Research Institute, Agricultural Research Center, Giza, Egypt (aymanesh@gmail.com)*

Gond, S.K., *Mycopathology and Microbial Technology Laboratory, Centre of Advanced Study in Botany, Banaras Hindu University, Varanasi 221005, India*

Gonthier, Paolo, *Department of Exploitation and Protection of Agricultural and Forestry Resources (DIVAPRA), Plant and Forest Pathology, University of Torino, Via L. da Vinci, 44, I-10095 Grugliasco (TO), Italy (paolo.gonthier@unito.it)*

Goyal, Aakash, *Agriculture and Agri-Food Canada, Lethbridge Research Center, Lethbridge AB-T1J4B1, Canada (akgroyal@gmail.com)*

Gupta, Anuja, *Indian Agricultural Research Institute, Regional Station, Karnal – 132 001, Haryana, India (agupta_2005@yahoo.com)*

Khan, M.A., *Department of Weed Science, NWFP Agricultural University, Peshawar, NWFP, Pakistan 25130*

Kharwar, R.N., *Mycopathology and Microbial Technology Laboratory, Centre of Advanced Study in Botany, Banaras Hindu University, Varanasi 221005, India (rnkharwar@yahoo.com)*

Kumar, Ashok, *Centre of Advanced Study in Botany, Banaras Hindu University, Varanasi 221005, India*

Kumari, *Oilseeds Scheme, Main Agricultural Research Station, University of Agricultural Sciences, Dharwad 580 005, Karnataka, India*

Larrán, Silvina, *Centro de Investigaciones de Fitopatología (CIDEFI), Facultad de Ciencias Agrarias y Forestales, Universidad Nacional de La Plata, 60 y 119, CC 31, 1900 La Plata, Argentina*

Mandal N., Bidhan Chandra Krishi Vishavidalay, Nadia, WB, India 741252

Mishra, A., *Mycopathology and Microbial Technology Laboratory, Centre of Advanced Study in Botany, Banaras Hindu University, Varanasi 221005, India*

Misra, Renu, *Department of Botany, Faculty of Science, The Maharaja Sayajirao University of Baroda, Vadodara 390002, India*

Molina, María del C., *Consejo de Investigaciones Científicas y Técnicas (CONICET), Facultad de Ciencias Agrarias y Forestales, Universidad Nacional de La Plata, CC 4, 1836 Llavallol, Buenos Aires, Argentina*

Mónaco, Cecilia, *Centro de Investigaciones de Fitopatología (CIDEFI), Facultad de Ciencias Agrarias y Forestales, Universidad Nacional de La Plata, 60 y 119, CC 31, 1900 La Plata, Argentina (cecilia.monaco7@gmail.com)*

Moreno, M.V., *CONICET – Facultad de Agronomía de Azul, Universidad Nacional del Centro de la Provincia de Buenos Aires, República de Italia No. 780, Azul CP 7300, Buenos Aires, Argentina (morevir@yahoo.com.ar)*

Nandy, S., *Bioproducts and Bioprocesses, Lethbridge Research Center, Agriculture and Agri-Food Canada, Lethbridge, AB Canada T1J 4B1*

Noelting, M.C.I., *Instituto Fitotécnico de Santa Catalina, Facultad de Ciencias Agrarias y Forestales, Universidad Nacional de La Plata, Garibaldi 3400, Llavallol 1836 CC 4 Buenos Aires, Argentina (mcnoelting@hotmail.com)*

Ogaraku, A.O., *Plant Science and Biotechnology Unit, Department of Biological Sciences, Nasarawa State University, PMB 1022, Keffi, Nasarawa State, Nigeria (ogara006@yahoo.com)*

Patil, Anjali, *Department of Botany, Rajaram College, Kolhapur 416004 (M.S.), India (dhirajanj@gmail.com)*

Patil, M.S., *Department of Botany, Shivaji University, Kolhapur (M.S.), India*

Perelló, Analía Edith, *CIDEFI (Centro de Investigaciones de Fitopatología) – CONICET (Consejo Nacional de Investigaciones Científicas y Técnicas), Facultad de Ciencias Agrarias y Forestales de la Universidad Nacional de La Plata, La Plata, Provincia de Buenos Aires, Argentina (anaperello@yahoo.com.ar)*

Prasad, Rajib, *Agriculture and Agri-Food Canada, Lethbridge Research Center, Lethbridge AB-T1J4B1, Canada*

Rogers, W.J., *Laboratorio de Biología Funcional y Biotecnología (BIOLAB), Facultad de Agronomía, Universidad Nacional del Centro de la Provincia de Buenos Aires (UNICEN),*

Av. República de Italia # 780 (CC 47), (7300) Azul, Buenos Aires, Argentina; FIBA – Consejo Nacional de Investigaciones Científicas y Técnicas (CONICET), Argentina

Sandoval, M.C., *Facultad de Ciencias Agrarias, UNLZ, Ruta 4 Km 2 Llavallol, Buenos Aires, Argentina*

Schalamuk, Santiago, *CONICET – Centro de Investigaciones de Fitopatología (CIDEFI) y Cerealicultura, Facultad de Ciencias Agrarias y Forestales, Universidad Nacional de La Plata, 60 y 119, CC 31, 1900 La Plata, Argentina (sschala@yahoo.com.ar)*

Sharma, Neeta, *Mycology and Plant Pathology Division, Department of Botany, University of Lucknow, Lucknow 226007, India (dr_neeta_sharma2003@yahoo.com)*

Shukla, A.K., *Department of Botany, Rajiv Gandhi University, Rono Hills, Itanagar 791 112, India*

Singh, Priyanka, *Centre of Advanced Study in Botany, Banaras Hindu University, Varanasi 221005, India*

Simón, María R., *Cerealicultura, Facultad de Ciencias Agrarias y Forestales, Universidad Nacional de La Plata, 60 y 119, CC 31, 1900 La Plata, Argentina (mrsimon@agro.unlp.edu.ar)*

Sisterna, Marina, *Centro de Investigaciones de Fitopatología (CIDEFI), Facultad de Ciencias Agrarias y Forestales, Universidad Nacional de La Plata, 60 y 119, CC 31, 1900 La Plata, Argentina (mnsisterna@infovia.com.ar)*

Stenglein, S.A., *Laboratorio de Biología Funcional y Biotecnología (BIOLAB), Facultad de Agronomía, Universidad Nacional del Centro de la Provincia de Buenos Aires (UNICEN), Av. República de Italia # 780 (CC 47), (7300) Azul, Buenos Aires, Argentina; FIBA – Consejo Nacional de Investigaciones Científicas y Técnicas (CONICET), Argentina (stenglein@faa.com.unicen.edu.ar)*

Thomas, J.E., *Department of Biological Sciences, University of Lethbridge, Lethbridge, AB, Canada T1K 3M4*

Tripathi, Abhishek, *Department of Bioscience and Biotechnology, Banasthali University, PO Banasthali Vidyapith, 304022 Rajasthan, India (abhitri77@yahoo.com)*

Tripathi, Nidhi, *Department of Bioscience and Biotechnology, Banasthali University, PO Banasthali Vidyapith, 304022 Rajasthan, India*

Tripathi, Pramila, *Department of Botany, D.A.V.-P.G. College, Kanpur 208001 (U.P.), India (pramilatripathi_bhu@rediffmail.com)*

Verma, V.C., *Mycopathology and Microbial Technology Laboratory, Centre of Advanced Study in Botany, Banaras Hindu University, Varanasi 221005, India*

Xue, Allen G., *Eastern Cereal and Oilseed Research Centre, Agriculture and Agri-Food Canada, 960 Carling Ave., Ottawa, Ontario, Canada, K1A 0C6 (axue@agr.gc.ca)*

Zhang, Shuzhen, *Soybean Research Institute, Key Laboratory of Soybean Biology of Chinese Education Ministry, Northeast Agricultural University, Harbin, Heilongjiang, China, 150030*

Preface

Oldest life forms have been reported from the North Pole Dome area of Western Australia, which dates back 3556 million years. Non-septate mycelium remains of *Eomycetopsis robusta* were recovered from late Precambrian chert of Australia. Having appeared first on planet Earth, microbes have immense potential to influence all other life forms. Plant diseases have caused epidemics and have had a profound influence on wars, famine and the changing economy. Microbes including fungi need no introduction to common man; they are progressive, ever changing and evolving in their own way, so they are capable of adapting to every condition of life. The French biochemist, Louis Pasteur, once said, 'The role of the infinitely small is infinitely large.'

Potentially immortal fungi spread their tentacles in 1845, when potato late blight fungus caused havoc in Ireland. Soon after, *Plasmopara viticola* threatened the wine industry in France. First reported in 1819 in Sweden, apple scab disease caused by *Venturia inaequalis* threatened apple cultivation in the Kashmir Valley in India in 1973. Panama disease of banana, wilt diseases of pigeon pea, castor and guava and smut and rust of cereals are some other serious fungal diseases. The chance discovery of Bordeaux mixture by P.A. Millardet in France paved the way to the chemical control of plant diseases. Phytopathologists are confronted by a volley of challenges in the wake of a resurgence of new diseases and the obligation to fulfil international trade agreements. We have to protect the environment and at the same time ensure the safety and security of farmers in the field by making a concentrated effort to minimize crop losses due to fungi and other microbes.

This book provides an overview of our current knowledge of some plant–pathogen interactions in economically important crops, emphasizing the importance of pathogenic fungi on fruits, cereals, postharvest crops and the establishment of plant diseases and drawing together fundamental new information on their management strategies based on conventional and eco-friendly methods, with an emphasis on the use of microorganisms and various biotechnological aspects of agriculture, which could lead to sustainability in modern agriculture.

The book examines the role of microbes in growth promotion, as bioprotectors and bioremediators, and presents practical strategies for using microbes in sustainable agriculture. In addition, the use of botanicals vis-à-vis chemical pesticides has also been reviewed. Contributions on new research fields such as mycorrhizae and endophytes have been included. The book also examines in different chapters host–pathogen interactions in the light of the new tools and techniques of molecular biology and genetics.

Dr Arya expresses his deep sense of indebtedness and admiration to the late Dr S.N. Bhargava and to Professor Bihari Lal, ex Head of the Department of Botany, University of Allahabad, who taught him his first lessons in plant pathology at the University of Allahabad. He is grateful to his father, the late Shri O.P. Arya, for inspiring him to write about the management of plant diseases and pests, which has proved most useful to plant growers. He honours his grandfather, Baba Shankaranand, who fed him with sweet mangoes during his childhood and who motivated him to love plants and to learn how to nurture them and research into new and improved varieties.

We are grateful to the entire staff of our institutions and the cooperation and collaborative efforts of the plant pathology experts of Argentina (Universidad Nacional de La Plata, Universidad Nacional de Lomas de Zamora, Universidad Nacional del Centro) and India (Botany Department, The Maharaja Sayajirao University of Baroda), who made this book possible.

We thank all those who have contributed their valuable articles to this volume and are sure that the present work, which consists of 27 different chapters written by learned experts in the field, will be immensely useful to postgraduate students, researchers, academics, progressive farmers and practising horticulturists, as well as those involved in the various agro-industries. We are hopeful that the available knowledge in the field, newer technologies and disease-resistant varieties will be used in different parts of the world and that ultimately the plant disease scenario will change. All appreciations and good wishes are extended to the members of the CABI team, particularly Ms. Sarah Mellor, for helpful discussions and skilled assistance in the reviewing of the manuscripts, and also for helping us in various ways to accomplish this project satisfactorily in the stipulated time. And also for the cooperation and collaborative effort of the Plant Pathology experts that made this book possible.

Arun Arya
Analia Edith Perelló

Part I

Botanicals in Fungal Pest Management

1 Recent Advances in the Management of Fungal Pathogens of Fruit Crops

Arun Arya
Department of Botany, Faculty of Science, The Maharaja Sayajirao University of Baroda, Vadodara, India

Abstract

Fruits constitute a rich source of sugars, vitamins, minerals and medicinally important compounds like flavonoids, which prevent cancer and cardiovascular diseases. These are eaten as a dessert or processed into jams, jellies, ice creams and drinks; grapes are dried to make raisins. The science of protecting fruit crops began with the discovery of Bordeaux mixture by P.A. Millardet in France. But still we have yet to find many new techniques and fungicide formulations to control diseases; such as bunch rot of grapes (*Botrytis cinerea*), apple scab (*Venturia inaequalis*), wilt of guava (*Fusarium solani*), Panama wilt of banana (*F. cubense*), mango malformation (*F. moniliforme*), blue mould of citrus (*Penicillium citrinum*) and anthracnose of papaya (*Colletotrichum papayae*), etc. Losses from postharvest fruit diseases range from 1 to 20% in the USA and from 10 to 40% in India. The pathogens have developed resistance against various fungicides and the postharvest phase is minimized. Alternative strategies like the use of biocontrol methods and the application of botanicals have been tried. A large number of plants are screened for the presence of effective secondary metabolites. Integrated pest management, using improved cultural practices (pruning methods to control *Botrytis* bunch rot in grapes), the use of solarization (in strawberries), the application of growth hormone (NAA in the case of mango malformation), along with minimum dosage of fungicides, are recommended to control various fruit diseases.

The world fruit market is expanding; we are more concerned about human nutrition now, but at the same time serious enough to protect the environment from pollution. The economics of a success story will have to revolve around the use of various cutting-edge technologies and, at the same time, the use of simpler and more effective methods acceptable to fruit growers. Biotechnologists have tried to enhance the activity of biocontrol agents; at the same time, efforts are being made for genetic transformation involving molecular breeding. This technology involves intimate knowledge of the gene, regulatory components and gene functional environment (i.e. the domain where the gene is located). Once an understanding of the molecular basis of genes involved in resistance has been achieved, we will be able to isolate the alleles of those genes and their inclusion will lead to transformed, disease-free plants.

Introduction

Fruits constitute an important component of our daily diet. The use of dates, fig, mango and grapes is mentioned in ancient texts. Taken either as a dessert or processed, the nutritional value of fruits depends chiefly on the quality and concentration of sugars, vitamins and other essential minerals. Plants suffer with a number of diseases and pests

(eds A. Arya and A.E. Perelló)

during their growth phase. Fungi not only blemish, disfigure or cause rot to a number of fruits but also reduce their market value (Arya, 2004). Realizing the importance of postharvest diseases, Stevens and Stevens (1952) mentioned, 'of all losses caused by plant disease those that occur after harvest are the most costly, whether measured in monetary terms or in man hours'. India is the second most important fruit producing country of the world. It produces the highest quantity of mango, while the productivity of grapes in India at 56 t/ha is a world record. The export of mango increased in the nineties from 25,000 to 44,000 t (a 25% share of world trade) (Neelam, 1993). Fruit production is 49 Mt (Arya, 2004).

The fruit growing industry has developed a lot, overcoming the hurdles of biotic and abiotic stresses. The industry needs a comprehensive strategy to face the challenges and opportunities of a global economy. Lewis (1985) stated that 'a few key discoveries have led to a breakthrough in our understanding of the biological genome and our ability to alter it, which may equal in significance the development of nuclear energy in the Physical Sciences.'

Fungal Infection of Fruits and Fruit Trees

Brown rot of citrus fruit is caused in the orchard by *Phytophthora citrophthora*. Many fungi that penetrate the host in the field cause quiescent infection. The grey mould in strawberries is caused by *B. cinerea*. The *Botrytis* spores with which the strawberry field is filled during bloom can germinate in a drop of water on the petal or other parts of the flower and later penetrate the senescenced parts of the flower into the edge of the receptacle of the strawberry, where they develop a dormant mycelium. During ripening and storage, as the resistance of the fruit to the pathogen decreases, the preliminary mycelium enters an active stage and decay develops (Powelson, 1960; Jarvis, 1962). Postharvest pathogens can be divided, according to the timing of their penetration of the host, into those that penetrate the fruits while still in the field, but develop in their tissues only after harvest, during storage or marketing, and those that initiate penetration during or after harvest. Symptoms in stylar end rot of guava caused by *Phomopsis psidii* become more prominent during storage (Arya, 1983). Verhoeff (1974) describes how quiescent infection is established in young fruits:

1. Shortage of adequate substances in young fruits.
2. The incapability of the pathogen to produce cell wall degrading enzymes in the young fruit.
3. The presence of antifungal compounds.
4. The accumulation of phytoalexins (Swinburne, 1983).

The first theory claims that young unripe fruit does not provide the pathogen with the nutrition and energy required for its development. The artificial increase of the sugar level in apple was achieved by the use of a chemical such as 2,4-dinitrophenol on the fruit. It accelerated the decay caused by *Botryosphaeria ribis* (Sitterly and Shay, 1960). It has been found that antifungal compounds become toxic in the presence of sugars.

The second theory suggests that the unripe fruit does not supply the pathogen with compounds that induce activity in cell wall degrading pectolytic enzymes.

The third and fourth theories point to a relation between the formation of antifungal compounds in the young tissues and the creation of quiescent infections. Chemicals such as 3,4-dihydroxy benzaldehyde have proven fungistatic activity in the green banana fruit. In unripe avocado fruit, a link has been established between the presence of a diene and monoene antifungal compounds in the fruit rind and the quiescent infection of *C. gloeosporioides* in such a fruit. The reduction in the concentration of the diene probably results from lipoxygenase enzymatic activity that increases as ripening progresses and the fruit softens (Prusky *et al.*, 1982, 1985). The dormant state of *Alternaria alternata* in young mango fruits has been attributed to the presence of two antifungal resorcinols in the unripe fruit rind (Droby *et al.*, 1986).

Recent Advances in the Management of Fungal Pathogens

Cultural practices

Initial infection of most temperate fruits is carried from the orchard; therefore, preharvest cultural practices, if adopted, considerably reduce postharvest diseases during transit and storage. Strict orchard hygiene and maintenance of tree vigour is recommended to reduce losses from *Botryosphaeria* rot of apple. *Pezicula malicorticis* and *Nectria galligena* infection in apple start from cankered portions. The removal of dead and senile plant parts and canker portions helps to reduce the incidence of many postharvest diseases. The incidence of many rots may also be reduced if the rotted fruits are frequently collected and dumped in a deep trench and later covered with a thick layer of soil to prevent the dissemination of their spores. If such rotted fruits are destroyed by burning some distance away from the orchard, this also helps to reduce the incidence of many rots in temperate fruits. Proper pruning can prevent *Botrytis* rot of grapes (Philips *et al.*, 1990).

The influence of N, P, K, Ca and Mg nutrients on storage rots of apple and pear has been studied extensively (Sharples, 1980). Susceptibility to *Gloeosporium* rot was correlated negatively with fruit Ca, but correlated positively with K/Ca ratios. Higher doses of nitrogen increase the incidence of *G. album* (Montgomery and Wilkinson, 1962). Calcium sprays to control bitter pit in apples also confer resistance to *P. expansum*.

Fumigation

Safe fumigating agents that disappear after a short time, such as the use of ozone and sulphur dioxide and acetic acid, can be recommended to reduce dependence on conventional fungicides. Ozone application to grapes (0.1 mg/g grapes) during 20 min exposure reduced decay caused by *Rhizopus stolonifer* and prolonged shelf life. This treatment was as effective as sulphur dioxide (Sarig *et al.*, 1996). The tolerance of grapes to sulphur dioxide is unique among fresh fruits. It eradicates most of the postharvest pathogens. However, the benefits of sulphur dioxide disappear after a short period of time. Hence, sodium bisulphate in packing cases reacts with the moisture in the air in grape containers. This treatment is used exclusively for the long distance transportation of grapes (Hedberg, 1977). Fumigation with acetic acid is effective in controlling *M. fructicola*, *R. stolonifer* and *Alternaria* species on peaches, nectarines, apricot and cherries (Sholberg and Gaunce, 1996). Relatively few fumigation treatments have been developed for pome and stone fruits.

Heat treatments

Heat treatments may be applied by hot water dips or hot vapour exposure. Hot water is useful in controlling fungal infections, while exposure to hot vapour controls insects. Postharvest decay of strawberries caused by *B. cinerea* and *R. stolonifer* has been controlled by exposing the fruits to humid air at 44°C for 40–60 min (Couey and Follstad, 1966). Akamine and Arisumi (1953) have reported hot water treatments for fruit rot of papaya (48°C for 20 min). Two methods have been suggested: one involves a short-term heat treatment above 40°C (usually 44–55°C) for a few minutes to 1 h and in the other, the fruits are exposed to 38–46°C but for a longer duration (12 h to 4 days) (Fallik *et al.*, 1996). The LD_{50} temperature for sporangiospores of *R. stolonifer* exposed to hot water for 4 min was 49°C, whereas that for germinating spores was only 39°C (Eckert and Sommer, 1967).

Ionizing radiation and UV illumination

Ionizing radiation may harm the genetic material of the living cell directly, leading to mutagenesis and eventually to cell death. Most studies are carried out with Co^{60} gamma rays. It has been seen that multicellular conidia of *Alternaria* and *Stemphylium* or bicellular spores such as *Cladosporium* and *Diplodia* are more resistant to gamma radiation than the unicellular spores of other fungal species (Sommer *et al.*, 1964). Since radiation can penetrate fruit tissues, it has a therapeutic effect. Plant tissues can produce phytoalexins (defence chemicals) in response to radiation effect. Low doses of UV-C light (wavelength 190–280 nm) can induce resistance in a wide range of fruit and vegetables (Barkai-Golan, 2001). UV light has a germicidal effect and, at the same time, it induces activity of PAL and peroxidase enzymes (Droby *et al.*, 1993).

Chemically impregnated wrappers

Wrapping grape clusters in tissue paper impregnated with sodium orthophenyl butyrate and sodium metabisulphate reduces postharvest decay. Volatile fungal inhibitors also provide effective control of grapes against *A. niger* and *P. canescens* (Sharma and Vir, 1984). Potassium iodide wraps provide effective control of *G. roseum* on apples (Sharma and Kaul, 1988). Development of *Botryodiplodia* rot of apples was retarded by wrapping them in papers dipped in culture filtrate of *Streptomyces thermoflavus* (Gupta and Gupta, 1983).

Fruit skin coatings

Skin coatings can improve the keeping quality of fruits by decreasing water loss and retarding ripening and rotting by various pathogens. Coating is generally done with oils, waxes and colloidal solutions of carboxymethyl cellulose. Apples coated with mustard oil, paraffin and castor oil checked the infection of a large number of pathogens (Sumbali and Mehrotra, 1980; Kaul and Munjal, 1982; Sharma and Kaul, 1988). Application of hydrogenated groundnut oil provided effective control of *Alternaria* rot of apple (Tak *et al.*, 1985). Skin coating with neem oil completely checked blue mould rot in apples (Kerni *et al.*, 1983).

Another treatment for extending the postharvest life of apple, pear and plum is by coating the skin with a product called 'Prolong', a mixture of sucrose esters of fatty acids and polysaccharide (Banks and Harper, 1981). It alters the permeability of fruits to gases in such a way that oxygen permeability is reduced considerably, while carbon dioxide permeability is little affected. This coating had little effect on grapes and strawberries.

Search for the antagonists: criteria of selection

Various strains of antagonist must be compared for effectiveness in controlling fruit decay and for phenotypic characteristics that are useful in determining their commercial potential; for example, the differentiation criteria for decay control on apple includes the biological control efficacy of the strains, spectrum of activity (pathogens to be tested, cultivar range, fruit maturity stages), ability to colonize wounded and sound fruit surfaces under various conditions, utilization of substrates occurring in fruits, or growth at cold storage temperatures and at 37°C.

In addition, these antagonists must meet strict regulations for safety as they are being applied to consumable commodities, i.e. fruits. Thus, in developing biocontrol systems for postharvest disease management of fruits, the key requirements for successful commercialization of an antagonist must be well defined and strain searches should continue until adequate strains are found that meet all the safety requirements.

Enhancement in biocontrol activity of antagonists

Postharvest environments are better defined than field conditions, wherein abiotic and biotic factors can be determined with relative ease and manipulated to the antagonist's advantage, although the mechanism(s) of biocontrol have not yet been fully explained and, to date, there have been only a few attempts to exploit these mechanisms to improve postharvest biocontrol (Janisiewiez *et al.*, 1992). The reports available on the mechanism of the biocontrol of postharvested commodities suggest that competition for nutrients and space plays a major role in most cases (Wisniewski *et al.*, 1991; Calvente *et al.*, 1999). In most of the systems where microbial communities are involved, interactions are density dependent and often more than one type of interaction occurs at a specific time which is dependent on the growth phase of different microorganisms, population density and species diversity. Basically, three different types of interactions, namely competition for nutrients, competition for space and inhibition by secondary metabolites, have been observed in preharvest sprays of *B. subtilis* to control *C. gloeosporioides* on avocado (Korsten *et al.*, 1997). The main approaches used to improve biological control in postharvest systems are: (i) manipulation of the environment; (ii) use of mixed cultures of antagonists; (iii) physiological and genetic manipulation of antagonists; (iv) combining field and postharvest applications; (v) manipulation of formulations; and (vi) integration with other methods.

In the case of the development of Bio-Save, the effectiveness of the antagonist, a saprophytic strain of *P. syringae* L-59-66, in reducing blue mould and grey mould decay on apples and pears in a laboratory setting was demonstrated to EcoScience Corp (Orlando, Florida, USA). The commercial setting of the test, the involvement of industry in conducting those tests and the encouraging results were the key factors in obtaining a commitment to develop the antagonist for commercial use. EcoScience Corp then investigated the potential for registration and formulation of the antagonist before making this commitment. Mass production by fermentation and the biomass yield of *P. syringae* strain L-59-66 was determined before scale-up experiments (Janisiewiez, 1998). Extensive technical support and quality control have been instrumental in the success of this product. Similar support and testing need to be conducted for the development of many more biocontrol agents for postharvest fruit rots.

Biocontrol: an integrated approach

Recently there has been an increased interest in enhancing the efficacy of biocontrol agents by adding some synthetic chemicals like calcium chloride or nitrogenous compounds or sugar analogues. For example, a mixture of *Cryptococcus laurentis* and thiabendazole has been observed to reduce 95% of *P. expansum* infection in pear (Sugar *et al.*, 1994). Enhancement of biocontrol activity of antagonists by the addition of nitrogenous (L-asparagine, and L-proline) and carbohydrate (2-deoxy-D-glucose) compound has been reported in apple and pear fruit (Janisiewiez, 1994). Similarly, a combination of 2-deoxy-D-glucose and *Candida saitoana* is reported to be useful in reducing postharvest diseases (Wilson and El-Ghaouth, 1997). Recently, a bioactive coating having a combination of *C. saitoana* and 0.2% glycolchitosan has been found more effective in controlling rot development caused by *B. cinerea*, *P. digitatum* and *P. expansum* in several cultivars of apples, oranges and lemon (El-Ghaouth *et al.*, 2000a,b). The same group of researchers showed that the application of *C. saitoana* with 0.2% 2-deoxy-D-glucose, before inoculation of pathogens, was more effective in controlling the decay of apple, orange and lemon caused by *B. cinerea*, *P. expansum* and *P. digitatum* than either *C. saitoana* or 0.2% 2-deoxy-D-glucose alone.

For the postharvest treatment of fruits, stock of biocontrol agent is usually made in lyophilized cultures, agar slant or spore suspensions and is maintained at low temperature and at the same osmotic concentration in culture medium (Churchill, 1982).

Botanicals as Antifungal Agents in Postharvest Disease Control of Fruits

Fruits and vegetables have a number of constituents and inducible volatile aromatic and flavour compounds (Tripathi, 2007).

These aromatic and flavour components are produced generally by fruits during ripening and provide resistance to the fruits at the postharvest stage. The flavour compounds are secondary metabolites having unique properties of volatility and low water solubility. As potential fungicides, their natural occurrence as part of the diet, their ephemeral nature and their biodegradability suggest low toxic residue problems. Such compounds could be extracted and applied to other harvested perishables. Some of the volatile aromatic components, namely acetaldehyde, six carbon (C_6) aldehydes, benzaldehyde, hexenel and hexanal, are of significant importance.

Vapours of acetaldehyde have been used to control *B. cinerea* (Prasad and Stadelbacher, 1973). Avissar and Pesis (1991) reported acetaldehyde to be active against *B. cinerea* and *R. stolonifer* causing rot to strawberry fruits. The effect of *trans*-2-hexenel on the control of blue mould disease (*P. expansum*) in the reduction of patulin content and on fruit quality improvement of 'Conference' pears was evaluated and greater reduction of decay was obtained by treatment at 12.5 μl/l at 20°C for 24 or 48 h after inoculation (Neri *et al.*, 2006).

Jasmonates are naturally occurring plant growth regulators that are widely distributed in the plant kingdom and are known to regulate various aspects of plant development and responses to environmental stresses. The antifungal activity of six glucosinolates has been tested on several postharvest pathogens, namely *B. cinerea*, *R. stolonifer*, *Monilinia laxa*, *Mucor piriformis* and *P. expansum*, both *in vitro* and *in vivo* (Mari *et al.*, 1996).

Fumigation of apples with acetaldehyde, a natural volatile compound produced by various plant organs, inhibits *P. expansum* development in the fruit (Stadelbacher and Prasad, 1974), while fumigation of strawberries with acetaldehyde considerably reduces decay caused by *R. stolonifer* and *B. cinerea*. Evaluation of 15 volatile odour compounds, released from raspberries and strawberries during ripening, for their ability to inhibit postharvest decay fungi showed that 5 of them inhibited the growth of *A. alternata*, *B. cinerea* and *C. gloeosporioides* directly on the fruit at 0.4 μl ml (Vaughn *et al.*, 1993). Among the five compounds, benzaldehyde was the most toxic to the fungi.

Plant extracts

Fungitoxic activity of plant extracts can be tested by the poisoned food technique (Grover and Moore, 1962). Tripathi (2005) tested 24 taxa belonging to 12 different families for their antifungal activity against *P. italicum*. Most of the plants showed either poor or moderate (50–100%) activity. Leaf extracts of seven plants, namely *Acacia nilotica* (ethyl alcohol), *Citrus aurantifolia* (ethyl acetate), *Murraya koenigii* (ethyl acetate), *Nerium indicum* (ethyl acetate), *Ocimum gratissimum* (benzene, ethyl acetate), *O. sanctum* (petroleum ether), *Prunus persica* (ethyl acetate) and bark extract of *A. farnesiana* and *A. nilotica* (ethyl acetate extract) showed 100% activity against test fungus. The leaves of *Achyranthes aspera* and *Hyptia suaveolens* showed poor activity.

Arya (1988) tried leaf extracts of *Aegle marmelos*, *O. sanctum*, *Azadirachta indica*, *Crataeva nurvala*, *Ephedra foliata* (shoot), *Eucalyptus occidentalis*, *Lawsonia inermis* and *Strichnos nux vomica* in three different concentrations on two fruit rot pathogens, *P. psidii* and *P. viticola*. Extracts obtained from *Ephedra* and *Eucalyptus* were most effective at 25% concentration in the case of *P. viticola*, while a higher concentration (75%) leaf extract of 'neem' (*A. indica*) was most effective, causing 82.3% spore inhibition. Tulsi caused 76.4% inhibition. The fungicidal nature of 'neem' and 'tulsi' was reported earlier by Pandey *et al.* (1983) against *Pestalotia psidii*.

Essential oils

Volatile oils are sweet-smelling lipids synthesized and stored in various plant parts. These oils are essentially mixtures of two classes of terpenoids, i.e. the monoterpenes and the sesquiterpenes, the former predominating in most cases. Among the 49 essential oils tested, those of palmrosa (*Cymbopogon martini*) and red thyme (*Thymus zygis*) showed the greatest inhibitory effect on *B. cinerea* spore germination at the lowest concentration. The next best inhibitors were essential oils of clove buds (*Eugenia caryophyllata*) and cinnamon leaf (*Cinnamomum zeylanicum*). The most frequently occurring constituents in essential oils showing high antifungal activity were: D-limonene, cineole, α-pinene, β-pinene, β-myrcene and camphor. The fungicidal activity of the individual components, singly and in combination, is being studied (Wilson *et al.*, 1997). Essential oil derived from another species of *Thymus*, *T. capitatus*, reduced the development of *B. cinerea* markedly in inoculated mandarin fruits when applied as a vapour. Scanning electron microscopic observations indicated a direct damaging effect of the thyme oil on fungal hyphae (Arras and Piga, 1994).

Gel and latex

Gel derived from *Aloe vera* has been found to have antifungal activity against four common postharvest pathogens, *P. digitatum*, *P. expansum*, *B. cinerea* and *A. alternata*. The natural gel suppressed both germination and mycelial growth. Latex present in some fruits is another natural fungicide which is effective against diseases of banana, papaya and other fruits (Adikaram *et al.*, 1996). Papaya latex contains proteases, glucosidases, chitinases and lipases, while a cysteinrich protein, hevien, was isolated from the latex of rubber tree (*Hevea brasiliensis*). It showed a strong antifungal activity *in vitro* against *B. cinerea* and species of *Fusarium* and *Trichoderma* (van Parijs *et al.*, 1991).

Use of Plantibodies for Disease Control

Drawing a clue from the potential antibodies in combating human diseases, plant scientists are now geared to extend this remarkable technology to plant disease control. Antibodies are produced in response to invasion of an antigen. The remarkable potential of recombinant DNA technology has made it possible for plants to express antibodies against pathogen proteins, which in turn enable them to defend against the target pathogen. The expression of pathogen-specific antibody in plants is termed 'plantibody' (Smith, 1996; Gibbs, 1997). The plantibodies produced in the cell cytosol are expected to interact with their targets, rendering them inactive (Zhang and Wu, 1998).

Induced Resistance

Induced resistance is a new concept proposed by the American phytopathologist, Joseph Kuc (1995). According to Kuc, resistance in plant tissues can be enhanced by modulating their natural defence mechanisms. Various physical, chemical and biological elicitors can enhance resistance in plants. Use of chitosan, a deacetylated derivative of chitin, and salicylic acid can be made to offer a possible alternative to synthetic pesticides. ASM (acibenzolar-*s*-methyle) is the first commercially available product that activates a systemic acquired resistance (SAR) in plants like other biological inducers.

Host Defence Through Gene Silencing

Scientists working on *Eutypa* dieback disease of grapevine in Switzerland (2008) found the involvement of glutathion-*s*-transferase in the detoxification of toxins, of the jasmonic acid signalling path way, and of several effector genes underlying a more general response where the toxins could be recognized as an elicitor for the trunk pathogens. Grapevines were tested for infiltration of double standard RNA into leaves for easy testing of genes. dsRNA were functional in *Puccinia striiformis* to suppress recognition by host plants (Newton, 2002). Genes that encode for post-transcriptional gene silencing have been characterized in plants and fungi (Dalmay *et al.*, 2000).

A variety of gene silencing phenomena that have been discovered are: (i) the duplicated DNA sequence is inactivated by mutation in the meiotic phase, a process known as repeat induced mutation (RIP) (Selker *et al.*, 1987); (ii) the duplicated DNA sequence during the meiotic phase is inactivated by methylation, methylation induced premeiotically (MIP) (Goyon and Faugeron, 1989); (iii) multiple copies of transgenes in the vegetative phase are irreversely inactivated and silencing is called 'quelling' (Romano and Macino, 1992); and (iv) silencing is maintained even in the absence of transgenes (van West *et al.*, 1999) or another process called MSUD (Shiu *et al.*, 2001).

Disease-resistant Transgenic Plants

Newly developed techniques in plant breeding such as restriction fragment length polymorphism techniques and gene transfer methods can be used to develop these cultivars. In contrast to conventional breeding, this later technology allows the transfer of traits from one species into the genomes of plants of other species with the preservation of the intrinsic properties of the acceptor plant (Cornelissen and Melchers, 1993).

A transgenic plant contains, within its genome, a foreign DNA that has been introduced artificially via genetic engineering. The creation of such plants involves the introduction of genes for resistance from unrelated plant species. Desirable target genes are isolated from plant viruses, bacteria, fungi or other plants and introduced in the plants. Genes have been transferred by scientists in India from *Amaranthus* to potato for improving protein quality and quantity, and from mangroves to annual crops for imparting tolerance to salinity. Powell *et al.* (1994) reported that transgenic tomato fruits expressing the gene of fungal PG-inhibiting glycoproteins of plants were more resistant to *B. cinerea* than the control fruits. Scientists have tried to prevent ethylene production by plant tissue using an antisense gene. The fruits would not ripen here until treated exogenously with ethylene. PR protein genes appear to be a very potential source for candidate genes providing fungal resistance. These proteins may play a direct role in defence by attacking and degrading pathogen cell wall components.

The first specific fungal-resistant gene, Hm1, has been isolated from maize, conferring resistance to race 1 of the fungus *Helminthosporium carbonum* (Johal and Briggs, 1992). After fungal-resistance genes have been isolated, they can be transferred to provide resistance to a specific race of fungal pathogens. Woloshuk *et al.* (1991) identified in tobacco a salt stress-inducible vacuolar protein with an inhibitory effect on the growth of *P. infestans in vitro.* It was suggested that this protein, described as Osmotin, inhibited growth by interfering with the fungal membrane, hence disturbing cellular function. As with class I hydrolyses, the protein could be arrested extracellularly by modification of the corresponding gene (Melchers *et al.*, 1993).

References

Adikaram, N.K.B., Indrakeerthi, S.R.P., Charmalie, A., Menike, P.R. and Ajani, K. (1996) Antifungal activity in fruit and postharvest disease. In: *Proceedings of the Australian Postharvest Horticulture Conference, Science and Technology for the Fresh Food Revolution*, Melbourne, Australia, pp. 381–385.

Akamine, E.K. and Arisumi, T. (1953) Control of postharvest storage decay of fruits of papaya (*Carica papaya* L) with special reference to the effect of hot water. *Proceedings of the American Society for Horticultural Science* 61, 270–274.

Arras, G. and Piga, A. (1994) *Thymus capitatus* essential oil reducing citrus fruit decay. In: Ait-Oubahou, A. and El-Otmani, M. (eds) *Postharvest Physiology, Pathology and Technology for Horticultural Commodities, Recent Advances.* Proceedings of International Symposium, Agadir, Moracco, pp. 426–428.

Arya, A. (1983) Cultural and pathological studies of certain fungi-imperfecti. DPhil thesis, Allahabad University, Allahabad, India, 246 pp.

Arya, A. (1988) Control of *Phomopsis* fruit rot by leaf extract of certain medicinal plants. In: Kaushik, P. (ed.) *Indigenous Medicinal Plants.* Symposium. Today and Tomorrows Printers and Publishers, New Delhi, pp. 41–46.

Arya, A. (2004) *Tropical Fruit Diseases and Pests.* Kalyani Publications, Ludhiana, India, 217 pp.

Avissar, I. and Pesis, E. (1991) The control of postharvest decay in table grapes using acetaldehyde vapours. *Annals of Applied Biology* 18, 229–237.

Banks, N.H. and Harper, G.R. (1981) Gaseous exchange, fruit surface and costing. In: *Proceeding of 150th Anniversary of British Association for the Advancement of Science*, Annual meeting, 3 August to 4 September 1981. Sec. D. No. 81.

Barkai-Golan, R. (2001) *Postharvest Diseases of Fruits and Vegetables – Development and Control.* Elsevier Science B.V., Netherlands, 418 pp.

Calvente, V., Benuzzi, D. and de Tosetti, M.I.S. (1999) Antagonistic action of siderophores from *Rhodotorula glutinis* upon the postharvest pathogen *Penicillium expansum. International Biodeterioration and Biodegradation* 43, 167–172.

Churchill, W.B. (1982) *Mass Production of Microorganism for Biological Control of Weeds with Plant Pathogens* (Charudattan, R. and Walker, H.L., eds). John Wiley and Sons, New York, 139 pp.

Cornelissen, B.J.C. and Melchers, L.S. (1993) Strategies for control of fungal diseases with transgenic plants. *Plant Physiology* 101, 709–712.

Couey, H.M. and Follstad, M.N. (1966) Heat pasteurization for control of postharvest decay in fresh strawberries. *Phytopathology* 56, 1345–1347.

Dalmay, T., Hamilton, A., Rudd, S., Angel, S. and Daulcombe, D.C. (2000) An RNA-dependent RNA polymerase gene in Arabidopsis is required for post-transcriptional gene silencing mediated by a transgene but not by a virus. *Cell* 101, 453–553.

Droby, S., Prusky, D., Jacoby, B. and Goldman, A. (1986) Presence of antifungal compounds in the peel of mango fruits and their relation to latent infection of *Alternaria alternata. Physiological and Molecular Plant Pathology* 9, 173–183.

Droby, S., Chalutz, E., Horv, B., Cohen, L., Gabai, V., Wilson, C.L. and Wisniewski, M.E. (1993) Factors affecting UV-induced resistance in grapefruit against the green mould decay caused by *Penicillium digitatum. Plant Pathology* 2, 418–424.

Eckert, J.W. and Sommer, N.F. (1967) Control of diseases of fruits and vegetable by postharvest treatment. *Annual Review of Phytopathology* 5, 391–432.

El-Ghaouth, A., Smilanick, J.L., Brown, G.E., Ippolito, A., Wisniewski, M. and Wilson, C.L. (2000a) Application of *Candida saitoana* and Glycolchitosan for the control of post-harvest diseases of apple and citrus fruit under semi-commercial conditions. *Plant Disease* 84, 243–248.

El-Ghaouth, A., Smilanick, J.L., Wisniewski, M. and Wilson, C.L. (2000b) Improved control of apple and citrus fruit decay with a combination of *Candida saitoana* and 2-deoxy-D-glucose. *Plant Disease* 84, 249–253.

Fallik, E., Grinberg, S., Alkaline, S. and Lurie, S. (1996) The effectiveness of post-harvest hot water dipping on the control of gray and black moulds in sweet red pepper (*Capsicum annuum*). *Plant Pathology* 45, 644–699.

Gibbs, W.W. (1997) Biotechnology – plantybodies. *Scientific American* 277, 44.

Goyon, C. and Faugeron, G. (1989) Targeted transformation of *Ascobolus immerses* and de novo methylation of duplicated DNA sequences. *Molecular and Cellular Biology* 9, 2818–2827.

Grover, R.K. and Moore, J.D. (1962) Toximetric study of fungicides against brown rot organism *Sclerotinia fructicola* and *S. laxa. Phytopathology* 52, 876–880.

Gupta, V.P. and Gupta, M.N. (1983) Study on the use of antibiotic secretion of *Streptomyces thermoflavus* against Botryodiplodia fruit rot during storage. *Progressive Horticulture* 15, 232–235.

Hedberg, P.R. (1977) Techniques for long-term storage of grapes. *Australian Journal of Experimental Agriculture and Animal Husbandry* 17, 866–870 (http://www.sbf.admin.ch/htm/dokumentation/publikationen/international/cost/cd2008/cos, accessed 7 March 2007).

Janisiewiez, W.J. (1994) Enhancement of biocontrol of blue mold with nutrient analog 2-deoxy-D-glucose on apples and pears. *Applied and Environmental Microbiology* 60, 2671–2676.

Janisiewiez, W.J. (1998) Biological control of postharvest disease of temperate fruits: challenges and opportunities. In: Boland, G.J. and Kuykendal, L.D. (eds) *Plant–Microbes Interaction and Biological Control.* Marcel Dekker, Inc, New York, pp. 171–198.

Janisiewiez, W.J, Usall, J. and Bors, B. (1992) Nutritional enhancement of biocontrol of blue mold on apples. *Phytopathology* 82,1364–1370.
Jarvis, W.R. (1962) The infection of strawberry and raspberry fruits by *Botrytis cinerea* Fr. *Annals of Applied Biology* 50, 569–575.
Johal, G.S. and Briggs, S.P. (1992) The HMI disease resistance gene in maize encodes a reductase activity. *Science* 258, 985–987.
Kaul, J.L. and Munjal, R.L. (1982) Apple losses in Himachal Pradesh due to postharvest fungal pathogens. *Indian Journal of Mycology and Plant Pathology* 18, 137–139.
Kerni, P.N., Shant, P.S. and Singh, D. (1983) Effect of various vegetable oils in controlling blue mould (*Penicillium expansum*) rot of apple. *Progressive Horticulture* 15, 129–131.
Korsten, L., De Villiers, E.E., Wehner, F.C. and Kotze, J.M. (1997) Field sprays of *Bacillus subtilis* and fungicides for control of preharvest fruit diseases of avocado in South Africa. *Plant Disease* 81, 455–459.
Kuc, J. (1995) Phytoalexins, stress metabolism and disease resistance in plants. *Annual Review of Phytopathology* 33, 275–297.
Lewis, L.N. (1985) Genetic engineering one leg of three-legged stool. *California Agriculture* 39(1–2), 2.
Mari, M., Leoni, O., Lori, R. and Marchi, A. (1996) Bioassay of glucosinolate derived isothyocyanates against postharvest pear pathogens. *Plant Pathology* 45, 753–760.
Melchers, L.S., Sela-Buurlage, M.B., Vloemans, S.A., Woloshuk, C.P., van Roekel, J.S.C., Pen, J., van den Elzen, P.J.M. and Cornelissen, B.J.C. (1993) Extracellular targeting of the vacuolar tobacco proteins AP24, chitinase and β-1,3-glucase in transgenic plants. *Plant Molecular Biology* 21, 20–23.
Montgomery, H.B.S. and Wilkinson, B.G. (1962) Storage experiments with Cox's Orange Pippin apples from a manurial trial. *Journal of Horticultural Science* 37, 150–158.
Neelam (1993) Is it flexing the export muscle or catching up with times? *Indian Horticulture* 38(2), 1.
Neri, F, Mari, M., Meniti, A.M. and Brigati, S. (2006) Activity of trans-2-hexenal against *Penicillium expansum* in 'Conference' pears. *Journal of Applied Microbiology* 100, 1186–1193.
Newton, A.C. (2002) Gene silencing mutation in stability and ds RNA micro viruses in rust and other fungal pathogens (http://www.crpmb.org/2002/1108newton/, accessed 12 January 2009).
Pandey, R.S., Bhargava S.N., Shukla, D.N. and Dwivedi, D.K. (1983) Control of Pestalotia fruit rot of guava by leaf extract of two medicinal plants. *Revista Mexicana de Fitopatologia* 2, 15–16.
Parijs, J. van, Brockaert, W.F. and Peumans, W.J. (1991) Hevein: an antifungal protein from rubber tree (*Hevea brasiliensis*) latex. *Plant* 183, 258–264.
Philips, P.A., Foott, J.H. and Righetti, L. (1990) Grape pruning methods can affect Botrytis bunch rot. *California Agriculture* 44(3), 9–10.
Powell, A.L.T., Hallewin, D.G., Hall, B.D., Stotz, H., Labavitch, J.M. and Bennett, A.B. (1994) Glycoprotein inhibitors of fugual polygalacturonases: expression of pear PGIP improves resistance in transgenic tomatoes. *Plant Physiology* 105, 159.
Powelson, R.L. (1960) Initiation of strawberry fruit rot caused by *Botrytis cinerea*. *Phytopathology* 50, 491–494.
Prasad, K. and Stadelbacher, G.J (1973) Control of postharvest decay of fresh raspberries by acetaldehyde vapor. *Plant Disease Reporter* 57, 795–797.
Prusky, D., Keem, N.T., Simus, J.J. and Midland, S.L. (1982) Possible involvement of an antifungal diene in the latency of *Colletotrichum gloeosporioides* on unripe avocado fruit. *Phytopathology* 72, 1578–1582.
Prusky, D., Kobiler, I., Jacoby, B., Sims, J.J. and Midland, S.L. (1985) Inhibitors of avocado lipoxygenase: their possible relationship with the latency of *Colletotrichum gloeosporioides*. *Physiological Plant Pathology* 27, 269–279.
Romano, M. and Macino, G. (1992) Quelling: transient inactivation of gene expression in *Neurospora crassa* by transformation with homologous sequences. *Molecular Microbiology* 6, 3343–3353.
Sarig, P., Zahavi, T., Zutkhi, Y., Yannai, S., Liskar, N. and Ben-Arie, R. (1996) Ozone for the control of postharvest decay of table grapes caused by *Rhizopus stolonifer*. *Physiological and Molecular Plant Pathology* 48, 403–415.
Selker, E.U., Camberari, E.B., Jensen, B.C. and Haack, K.R. (1987) Rearrangement of duplicated DNA in specializer cells of *Neurospora*. *Cell* 51, 741–752.
Sharma, R.C. and Vir, D. (1984) Efficacy of fungicides XXII. Evaluation of benzimidazoles, an antibiotic and other fungicides against postharvest spoilage of grapes. *International Journal of Tropical Plant Diseases* 2, 5–7.
Sharma, R.L. and Kaul, J.L. (1988) Efficacy of fruit wrappers and skin coatings against brown rot (*Monilinia* spp.) in stored apples. *Plant Disease Research* 3, 247–250.

Sharples, R.O. (1980) The influence of orchard nutrition on the storage quality of apples and pears grown in United Kingdom. In: Atkinson, D., Jackson, J.E. and Sharples, R.O. (eds) *Mineral Nutrition of Fruit Trees*. Butterworths, London, pp. 17–28.

Shiu, P.K.T., Raju, N.B., Zickler, D. and Metzenberg, R.L. (2001) Meiotic silencing by unpaired DNA. *Cell* 107, 905–916.

Sholberg, P.L. and Gaunce, A.P. (1996) Fumigation of stone fruit with acetic acid to control postharvest decay. *Crop Protection* 15, 681–686.

Sitterly, W.R and Shay, J.R. (1960) Physiological factors affecting the onset of susceptibility of apple fruit to rotting by fungus pathogens. *Phytopathology* 50, 91–93.

Smith, M.D. (1996) Antibody production in plant. *Biotechnology Advances* 14, 267–281.

Sommer, N.F., Maxie, E.C. and Fortlage, R.J. (1964) Quantitative dose response of *Prunus* fruit decay to gamma irradiation. *Radiation Botany* 4, 309–316.

Stadelbacher, G.J and Prasad, K. (1974) Postharvest decay control of apple by acetaldehyde vapor. *Journal of American Society of Horticultural Science* 99, 364–368.

Stevens, R.B. and Stevens, N.E. (1952) *Diseases in Plants*. Chronica Botanica Watham, Massachusetts.

Sugar, D., Roberts, R.G., Hitton, R.J., Righetti, T.L. and Sanchez, E.E. (1994) Integration of cultural methods with yeast treatment for control of post-harvest fruit decay in pear. *Plant Disease* 78, 791–795.

Sumbali, G. and Mehrotra, R.S. (1980) Evaluation of some fixed oils for the control of certain temperate fruit rot fungi. *Indian Phytopathology* 33, 517.

Swinburne, T.R. (1983) Quiescent infections in postharvest diseases. In: Dennis, C. (ed.) *Postharvest Pathology of Fruits and Vegetables*. Academic Press, London, pp.1–21.

Tak, S.K., Verma, O.P., Gupta, A.K. and Pathak, V.N. (1985) Control of *Alternaria* rot of apple fruits by postharvest application of chemicals. *Indian Phytopathology* 38, 471–474.

Tripathi, P. (2005) *Botanical Pesticides in the Management of Postharvest Fruit Diseases*. Daya Publishing House, New Delhi, 174 pp.

Tripathi, P. (2007) Biologicals and biorationals in the management of agricultural insect pests: an ecofriendly approach. In: Arya, A. and Monaco, C. (eds) *Seed Borne Diseases: Ecofriendly Management*. Scientific Publishers, Jodhpur, India, pp. 171–189.

Vaughn, S.F., Spencer, G.F. and Shasha, B.S. (1993) Volatile compounds from raspberry fruit inhibit postharvest decay fungi. *Journal of Food Science* 58, 793–796.

Verhoeff, K. (1974) Latent infections by fungi. *Annual Review of Phytopathology* 12, 99–110.

West, P. van, Kamoun, S., Klooter, J.W. van't and Grovers, F. (1999) Internuclear gene silencing in *Phytophthora infestans*. *Molecular Cell* 3, 339–348.

Wilson, C.L. and El-Ghaouth, A. (1997) Biological control of post-harvest diseases by combining a sugar analog with antagonistic yeast. Patent Serial No. 08/0951552.

Wilson, C.L., Solar, J.M., El-Ghaouth, A. and Wisniewski, M.E. (1997) Rapid evaluation of plant extracts and essential oils for antifungal activity against *Botrytis cinerea*. *Plant Disease* 81, 204–210.

Wisniewski, M., Biles, C., Droby, S., McLaughlin, R., Wilson, C. and Chalutz, E. (1991) Mode of action of the postharvest biocontrol yeast, *Pichia guilliermondii*. I. Characterization of attachment to *Botrytis cinerea*. *Physiological and Molecular Plant Pathology* 39, 245–258.

Woloshuk, C.P., Meulenhoff, E.J.S, Sela-Buurlage, M., Vander Elzen, P.Z.M. and Cornelissen, B.J.C. (1991) Pathogen induced proteins with inhibitory activity towards *Phytophthora infestans*. *Plant Cell* 3, 619–628.

Zhang, Z.H. and Wu, L.P. (1998) Research and development of expressing antibodies in plants. *Progress in Biochemistry and Biophysics* 25, 136–139.

2 Botanicals in Agricultural Pest Management

Ashok Kumar, Priyanka Singh and N.K. Dubey
Centre of Advanced Study in Botany, Banaras Hindu University, Varanasi, India

Abstract

The overzealous and indiscriminate use of most of the synthetic fungicides has created different types of environmental and toxicological problems. The ultimate aim of recent research in this area has been the development of alternative control strategies to reduce dependency on synthetic fungicides. Recently, in different parts of the world, attention has been paid to the exploitation of higher plant products as novel chemotherapeutants in plant protection because of their non-phytotoxicity, systemicity and easy biodegradability. The exploitation of natural products to control fungal infestation and prolong storage life of food commodities has received more attention. Biologically active natural products have the potential to replace synthetic fungicides. Currently, different plant products have been formulated for large-scale application as botanical pesticides in the eco-friendly management of plant pests and are being used as alternatives to synthetic pesticides in crop protection. This chapter deals with the current status and future prospects of botanical pesticides in eco-friendly management of different plant pests.

Introduction

The constant growth of the world's population requires substantial resources for the production of food. One of the greatest challenges of the world is to produce enough food for the growing population. Production as well as protection of food commodities is necessary to nourish the ever-growing population. The situation is particularly critical in developing countries, where the rate of net food production is slowing down in relation to population growth. The world food situation is aggravated by the fact that in spite of the use of all available means of plant protection, a major proportion of the yearly production of food commodities of the world is destroyed by various pests, including bacteria, fungi, viruses, insects, rodents, nematodes, etc. Losses at times are so severe as to lead to famine in large areas of the world that are densely populated.

Considerable attention has been given to losses in the field caused by different pests, but research into postharvest losses of food commodities is still required. So, priority should be given to postharvest studies, particularly in hot and humid tropical climates where at least half of the foodstuffs may be lost between harvest and consumption. Considerable postharvest losses of food commodities are brought about due to fungi, insects and rodents. International agencies that monitor world food resources have acknowledged that one of the most feasible options for meeting future food needs is the reduction of postharvest losses (Tripathi and Dubey, 2004).

Fungi are significant destroyers of foodstuffs during storage, rendering them unfit for human consumption by retarding their nutritive value. Many agricultural commodities are vulnerable to attack by a group of fungi that are able to produce toxic metabolites called mycotoxins. Production of mycotoxins by several fungi has added a new dimension to the gravity of the problem. Fungal toxins are low molecular weight chemical compounds which are not detected by the body's antigens. Their effect is more often chronic rather than acute; hence, they produce no obvious symptoms. Thus, mycotoxins are insidious poisons (Pitt, 2002). Cereals and grains are major mycotoxin vectors because they are consumed by both humans and animals. According to FAO estimates, 25% of the world food crops are affected by mycotoxins each year. These toxins can develop during production, harvesting, or storage of grains, nuts and other crops. Mycotoxins are among the most potent mutagenic and carcinogenic substances known. They pose chronic health risks: prolonged exposure through diet has been linked to cancer and kidney, liver and immune system disease (Srivastava *et al.*, 2008). Among mycotoxins, aflatoxins chiefly produced by strains of *Aspergillus flavus* are the most dangerous and about 4.5 billion people in underdeveloped countries are at risk of chronic exposure to aflatoxicosis through contaminated foods (Williams *et al.*, 2004; Srivastava *et al.*, 2008). In most of the developing countries, total permissible aflatoxin content in food has been set around 20 ppb (Mishra and Das, 2003). Aflatoxins are potent toxic, carcinogenic, mutagenic, immunosuppressive agents, produced as secondary metabolites by the fungus *Aspergillus*, *A. parasiticus* and *A. nomius* on a variety of food products. In addition, aflatoxin inhibits seed germination, seedling growth, root elongation, chlorophyll and carotenoid synthesis, as well as protein, nucleic acid and some enzyme synthesis in seeds.

Climatic conditions in India are most conducive to mould invasion and elaboration of mycotoxins. Unseasonal rains and flash floods are very common in India, which enhances the moisture content of the grains, making them more vulnerable to fungal attack (Srivastava, 1987). Fungi can grow on simple and complex food products and produce various metabolites (Khosravi *et al.*, 2007). Up to now, more than 100,000 fungal species are considered as natural contaminants of agricultural and food products (Kacaniova, 2003). The quality and safety of food is of importance so that markets are not compromised by the sale of low quality or unsafe food.

Control of Fungal Infestation During Storage

Attempts to control postharvest diseases have been carried out by different physical and chemical treatments.

Physical methods

Several techniques are used for the preservation of food and feeds. Drying, freeze-drying, cold storage, modified atmosphere storage and heat treatments are all physical methods of food preservation (Farkas, 2001) (Table 2.1).

Cold storage

Low temperature inhibits the germination of spore/conidia and pathogenicity significantly (Tian, 2001). It reduces the metabolic activities of various microbes associated with foodstuffs, which would be helpful in enhancing the shelf life of edibles. However, cold storage has its limitations, such as unavailability in most developing countries and an inability to check psychrophilic microorganisms.

Heat treatment

High temperature plays a significant role in controlling the metabolic activities of organisms because it affects the enzymatic activities in all organisms adversely (Lagunas and Castaigne, 2008; Moatsou *et al.*, 2008). Heat treatment can check microbial

Table 2.1. Some physical and chemical methods used in the prevention of fungal contamination and mycotoxin production.

Methods	Fungi/mycotoxins	References
Physical:		
Sunlight	*Aspergillus flavus*/aflatoxin	Shantha and Sreenivasamurthy (1977)
Solar irradiation	*A. parasiticus*/aflatoxin B_1	Samarajeewa *et al.* (1985)
Electric light	*A. flavus*/aflatoxin B_1	Chourasia and Roy (1991)
UV light	*A. flavus*/aflatoxin	Shantha and Sreenivasamurthy (1977)
UV-C radiation	*Colletotrichum gloeosporioides*	Cia *et al.* (2007)
Infrared light	*Penicillium citrinum*	Qing *et al.* (2002)
γ Radiation	*Cryptococcus neoformans*	Dadachova *et al.* (2004)
α Radiation	*C. neoformans*	Martinez *et al.* (2006)
Autoclaving	All types of moulds	Coomes *et al.* (1966)
Cooking	Food-spoiling moulds	Rehana and Basappa (1990)
Roasting	Food-spoiling moulds	Ogunsanwo *et al.* (2004)
Dry heat	*Fusarium graminearum*	Clear *et al.* (2002)
Low temperature/ refrigeration	Some soil fungi	Janna *et al.* (2005)
Chemical:		
H_2O_2	*A. flavus*/aflatoxin	Sreenivasamurthy *et al.* (1967)
Na-hypochlorite	*A. flavus*/aflatoxin	Shantha *et al.* (1986)
Azoxystrobin	*C. lupini*	Thomas *et al.* (2008)
Chlorothalonil	*C. lupini*	Thomas *et al.* (2008)
Copper oxychloride	*C. lupini*	Thomas *et al.* (2008)
Carbendazim	*A. carbonarius*/ochratoxin	Medina *et al.* (2007)
Mancozeb	*Penicillium* sp., *Trichoderma* sp.	Magarey *et al.* (1997)
Maneb	*F. graminearum*/ZEN	D'Mello *et al.* (1998)
Nitroimidazole	*Sclerophoma pityophila*	Olender *et al.* (2008)
Organotin	*C. gloeosporioides*	Rehman *et al.* (2008)
Blitox	*Aspergillus* spp.	Satish *et al.* (2008)
Captan	*Aspergillus* spp.	Satish *et al.* (2008)
Dithane M-45	*Aspergillus* spp.	Satish *et al.* (2008)
Thiram	*Aspergillus* spp.	Satish *et al.* (2008)
SAAF	*A. flavus*	Kumar *et al.* (2008)
Bavistin	*A. flavus*	Kumar *et al.* (2008)
Wettasul-80	*A. flavus*	Kumar *et al.* (2008)
Ceresan	*A. flavus*	Kumar *et al.* (2008)
Diphenylamine	*A. flavus*	Kumar *et al.* (2008)

growth efficiently but the technique is not suitable for long-term storage.

Radiation

Sun drying of food commodities (grains and pulses) before storage is preferable in most underdeveloped countries but the technique is unsuitable in the case of vegetable crops. High-energy radiation like γ rays (Petushkova *et al.*, 1988), UV rays (Oteiza *et al.*, 2005), infrared (Qing *et al.*, 2002), etc., is also efficient in checking microbial growth and proliferation, as well as mycotoxin production. The irradiation of food commodities during storage is unattainable in developing countries.

Chemical methods

In order to minimize the losses caused by moulds in the field and also during storage, many synthetic fungicides have been introduced (Table 2.1). The discovery of Bordeaux mixture is significant in the history of the chemical control of plant diseases. In the past few decades, various synthetic chemicals have played a significant role in the management of such losses. Several chemical additives also function as preservatives, even though the exact mechanisms or targets are often not known (Davidson, 2001. The organic acids, acetic, lactic, propionic, sorbic and benzoic acids, are used as food preservatives (Brul and Coote, 1999). Both sorbic and benzoic acid have a broad spectrum of activity (Nielsen and De Boer, 2000; Davidson, 2001). Benzoic acid and sodium benzoate are used primarily as antifungal agents (Davidson, 2001). Recently, some technology like TiO_2 photocatalytic ozonation has been found to be efficient in controlling postharvest spoilage of kiwifruit (Hur *et al.*, 2005).

The indiscriminate application of synthetic chemicals as antimicrobials has contributed greatly to the management of losses caused by fungi, but these chemicals have led to a number of ecological and health problems due to their residual toxicity (Knežević and Serdar, 2008), carcinogenicity, teratogenicity, hormonal imbalance, spermatotoxicity, etc. (Pandey, 2003; Kumar *et al.*, 2007). History also shows that overzealous use of synthetic pesticides has led to numerous problems unforeseen at the time of their introduction. Different types of ecological problems have been reported from time to time by these xenobiotics, such as acute and chronic poisoning of applicators, farm workers, and even consumers, extensive groundwater contamination, resistance development in pests (Wilson *et al.*, 1997), effect on non-target organisms (Wiktelius *et al.*, 1999), ozone layer depletion by methyl bromide (Lee *et al.*, 2001), etc.

Biocontrol Agents in Pest Management

Considerable attention has also been given to the potential of biological control of postharvest diseases of fruits, vegetables and other edibles as a viable alternative to the use of present day synthetic fungicides (Wilson *et al.*, 1999; Pang *et al.*, 2002). Microbial antagonists have been reported to protect a variety of harvested perishable commodities against a number of postharvest pathogens (Wisniewski *et al.*, 2001). However, decreasing efficacy and lack of consistency when applied as stand-alone treatments under commercial conditions (Droby *et al.*, 2001) are limiting their use. Hence, these drawbacks in alternative methods have increased interest in developing further alternative control methods, particularly those which are environmentally sound and biodegradable.

Botanicals as Fungitoxicants

Recently, in different parts of the world, attention has been drawn towards the exploitation of higher plant products as novel chemotherapeutants in plant protection. Because of non-phytotoxicity, systemicity, easy biodegradability and the stimulatory nature of host metabolism, plant products possess the potential to be of value in pest management (Mishra and Dubey, 1994). Higher plants contain a wide spectrum of secondary metabolites such as phenols, flavonoids, quinones, tannins, essential oils, alkaloids, saponins and sterols. Such plant-derived chemicals may be exploited for their different biological properties (Tripathi *et al.*, 2004). Terrestrial plants produce a spectrum of natural products, namely terpenoids, phenolics and alkaloids. Many of these are thought to have an ecological function for the plants producing them, serving to defend the plants from herbivores and pathogens (Isman and Akhtar, 2007). Such defensive chemistry is thought to be extremely widespread among the plant kingdom.

The body of scientific literature documenting the bioactivity of plant derivatives to different pests continues to expand; yet only a handful of botanicals are currently used in agriculture in the industrialized world. In the context of agricultural pest management, botanical pesticides are well suited for use in industrialized countries and can play a much greater role in the postharvest protection of food commodities in developing countries (Isman, 2006).

Among the different plant products, the application of essential oils is a very attractive method for controlling postharvest losses (Table 2.2). Production of essential oils by plants is believed to be predominantly a defence mechanism against pathogens and pests (Oxenham, 2003). Essential oils and their components are gaining increasing interest because of their relatively safe status, wide acceptance by consumers and their exploitation for potential multi-purpose use (Sawamura, 2000; Ormancey *et al.*, 2001; Feng and Zheng, 2007). The problem of the development of resistant strains of fungi and other organisms may be solved by the use of essential oils of higher plants as fumigants in the management of storage pests because of synergism between different components of the oils (Varma and Dubey, 1999; Dubey *et al.*, 2006).

The antifungal activity of essential oils is well documented and characterized with their bioactivity in vapour phase. The pesticidal activities of essential oils are due to the presence of some aroma compounds. Fumigation with such aroma compounds greatly reduces postharvest decay without causing any toxicity (Chu *et al.*, 2001; Liu *et al.*, 2002). Recently, some monoterpenes isolated from essential oils exhibited fungicidal activity and have been shown to inhibit fungal rotting of vegetables without altering taste and quality (Hartmans *et al.*, 1995; Oosterhaven, 1995). The fungitoxic properties of essential oils from higher plants are well documented but little attention has been paid towards the bioactivity of essential oil constituents. The fungitoxic activity of some essential oil components is listed in Table 2.3 and Fig. 2.1. However, more work on the bioactivity of plant products including essential oil and constituents in *in vitro* and *in vivo* conditions is required. The literature is also silent on the mode of action of the essential oils and components when used as postharvest fungitoxicants.

Conclusions

Plants are a virtually untapped reservoir of different valuable chemicals that can be used directly or as templates for the formulation of pesticides. Numerous factors have increased the interest of the pesticide industry and the pesticide market in this source of natural products as pesticides. Pesticides based on plant essential oils or their constituents have demonstrated their efficacy against a range of fungal pests responsible for pre- and postharvest diseases, as well as mycotoxin production. Encouraging results on the use of natural products to control postharvest fungal spoilage indicate that we should be able to develop natural pesticides that could be as effective as synthetic fungicides and presumably safer for man and the environment. Biological compounds, because of their natural origin, are comparatively biodegradable and most of them are almost non-residual in nature (Beye, 1978).

During recent years, products of some pesticidal plants have received global attention for the protection of several food commodities because of their antimicrobial properties (Kumar *et al.*, 2007). Such plant products have been formulated for large-scale application as botanical pesticides, which are used as alternatives to synthetic pesticides in crop protection. A consolidated and continuous search of natural products may yield safer alternative control measures comparable to azadirachtin and pyrethroids, which are being used in different parts of the world as ideal natural fungicides. The number of options that must be considered in the discovery and development of a natural product as a pesticide is larger than for a synthetic pesticide. However, current advances in plant chemistry and biotechnology, combined with increasing need and environmental pressure, are greatly increasing the interest in plant products as pesticides. Products from higher plants are a safe and economical option in the management of agricultural pests and will be in high demand in the global pesticide market.

Table 2.2. Efficacy of some higher plant products in checking fungal growth and mycotoxin production.

Plants	Products	Fungi/mycotoxins	References
Hypericum linarioides	EO/PEE/ ME/ChlE	6 *Fusarium* spp.	Cakir *et al.* (2005)
Calocedrus macrolepis	EO	*Colletotrichum gloeosporioides*, *Rhizoctonia solani*, *F. oxysporum*	Chang *et al.* (2008)
Silene armeria	EO/ME/ HexE	*F. oxysporum*, *C. capsici*, *Botrytis cinerea*	Bajpai *et al.* (2008)
Origanum acutidens	EO	17 pathogenic fungi	Kordali *et al.* (2008)
Cinnamomum osmophloeum	EO	*Trametes versicolor*, *Lenzites betulina*, *Laetiporus sulphureus*	Cheng *et al.* (2006)
Thymus numidicus	EO	*Candida albicans*	Giordani *et al.* (2008)
Lantana camara	EO	*Aspergillus niger*, *A. parasiticus*	Deena and Thoppil (2000)
O. glandulosum	EO	*F. oxysporum*, *Cladosporium herbarum*, *A. flavus*	Bendahou *et al.* (2008)
Tarchonanthus camphoratus	EO	*C. albicans*	Matasyoh *et al.* (2007)
Syzygium aromaticum	AqE/EO	*A. flavus*/aflatoxin B_1, *A. flavus*, *Penicillium verrucosum*/ochratoxin A	Omidbeygi *et al.* (2007); Aldred *et al.* (2008); Reddy *et al.* (2008)
Curcuma longa	AqE	*A. flavus*/aflatoxin B_1	Reddy *et al.* (2008)
Allium sativum	AqE	*A. flavus*/aflatoxin B_1	Reddy *et al.* (2008)
Lippia rugosa	EO	*A. flavus*/aflatoxin B_1	Tatsadjieu *et al.* (2009)
Citrus sp.	EO	*A. flavus*, *P. chrysogenum*, *P. verrucosum*	Viuda-Martos *et al.* (2008)
Bidens pilosa	EO/AqE	*Corticium rolfsii*, *F. solani*	Deba *et al.* (2008)
Satureja hortensis	EO/ME	*A. flavus*, *A. parasiticus*/aflatoxin	Omidbeygi *et al.* (2007); Abyaneh *et al.* (2008); Dikbas *et al.* (2008)
T. eriocalyx	EO	*A. niger*	Rasooli *et al.* (2006)
T. x-porlock	EO	*A. parasiticus*/aflatoxin	Rasooli and Abyaneh (2004)
Ocimum basilicum	EO	*A. parasiticus*/aflatoxin	Atanda *et al.* (2007)
Pimpinella anisum	EO	*A. niger*, *P. chrysogenum*	Matan and Matan (2008)
Salvia officinalis	EO	*C. albicans*, *Trichophyton rubrum*, *A. flavus*	Pinto *et al.* (2007)
T. vulgaris	EO	*A. flavus*/aflatoxin B_1	Kumar *et al.* (2008)
Cympopogon citratus	EO	*B. cinerea*, *C. herbarum*, *A. niger*	Tzortzakis and Economakis (2007)
Rosmarinus officinalis	EO	*A. parasiticus*/aflatoxin	Rasooli *et al.* (2008)
Trachyspermum copticum	EO	*A. parasiticus*/aflatoxin	Rasooli *et al.* (2008)
Cordia curassavica	EO/HexE/ ChlE/ME	*R. solani*, *T. mentagrophytes*	Hernandez *et al.* (2007)
Sesuvium portulacastrum	EO	*A. niger*, *A. flavus*, *P. notatum*	Magwa *et al.* (2006)
Calamintha officinalis	EO	*B. cinerea*	Bouchra *et al.* (2003)
Olea europaea	AE/ME	*Alternaria alternata*, *A. flavus*, *F. oxysporum*	Korukluoglu *et al.* (2008)
Citrus sinensis	EO	*A. niger*	Sharma and Tripathi (2008)
Azadirachta indica	AqE	*P. citrinum*/Citrinin	Aparecida *et al.* (2008)
Agave asperrima	ME/AqE	*A. flavus*, *A. parasiticus*/ aflatoxin B_1	Sánchez *et al.* (2005)
Adenocalymma alliaceum	AqE	*A. flavus*/aflatoxin B_1	Shukla *et al.* (2008)
Lupinus albus	AqE	*A. flavus*/aflatoxin B_1	Mahmoud (1999)

Note: EO, essential oil; ME, methanolic extract; AqE, aqueous extract; PEE, petroleum ether extract; AE, acetone extract; ChlE, chloroformic extract; HexE, hexane extract.

Table 2.3. Efficacy of some essential oil components in checking fungal growth.

Compounds of plant origin	Fungi	References
Ajoene	*Aspergillus niger*, *Candida albicans*, *Saccharomyces cerevisiae*	Yoshida *et al.* (1987) Naganawa *et al.* (1996)
Allicin	*C. albicans*	Ankri and Mirelman (1999)
Myrcene	*Rhizoctonia solani*, *Fusarium oxysporum*	Chang *et al.* (2008)
Limonene	*Colletotrichum gloeosporioides*, *Botryosphaeria parva*, *F. verticillioides*	Regnier *et al.* (2008) Dambolena *et al.* (2008)
p-Cymene	*Fusarium* sp.	Kordali *et al.* (2008)
α-Pinene	*C. albicans*, *S. cerevisiae*, *A. niger*	Yousefzadi *et al.* (2008) Sonboli *et al.* (2006)
Caryophyllene	*R. solani*, *F. oxysporum*	Chang *et al.* (2008)
Citral	*A. niger*, *F. oxysporum*, *Penicillium digitatum*, *C. albicans*	Moleyar and Narasimham (1986) Da Silva *et al.* (2008)
Cinnamaldehyde	*Lenzites betulina*, *Laetiporus sulphureus*	Cheng *et al.* (2006)
Camphor	*Fusarium* sp., *R. solani*	Pitarokili *et al.* (2003)
Carvone	*C. gloeosporioides*, *B. parva*	Regnier *et al.* (2008)
Pulegone	*C. albicans*	Duru *et al.* (2004)
Menthone	*F. verticillioides*	Dambolena *et al.* (2008)
Thujone	*Phytophthora capsici*	Shafi *et al.* (2004)
Linalool	*C. camelliae*	Zhang *et al.* (2006)
Geraniol	*C. camelliae*	Zhang *et al.* (2006)
Citronellol	*Rhizopus stolonifer*	Moleyar and Narasimham (1986)
Terpine-4-ol	*A. flavus*, *R. solani*, *P. commune*, *F. oxysporum*	Barra *et al.* (2007)
Menthol	*F. verticillioides*, *R. stolonifer*, *Penicillium* sp., *Monilia* sp.	Dambolena *et al.* (2008) Moleyar and Narasimham (1986) Serrano *et al.* (2005)
Thymol	*C. albicans*, *Fusarium* sp., *F. verticillioides*	Braga *et al.* (2008) Kordali *et al.* (2008) Dambolena *et al.* (2008)
Eugenol	*L. betulina*, *L. sulphureus*, *T. mentagrophytes*, *C. albicans*	Cheng *et al.* (2006) Gayoso *et al.* (2005)
Fenchone	*P. capsici*	Shafi *et al.* (2004)
1,8 Cineole	*C. gloeosporioides*, *B. parva*	Regnier *et al.* (2008)
Asarone	*R. solani*, *P. infestans*, *Cladosporium cucumerinum*, *Pythium ultimum*	Lee (2007) Lee *et al.* (2004)
Zingiberene	*R. solani*	Agarwal *et al.* (2001)
Curcumene	*R. solani*	Agarwal *et al.* (2001)
Verbenone	*Colletotrichum* sp.	Meepagala *et al.* (2003)
Verbenol	*Colletotrichum* sp.	Meepagala *et al.* (2003)
Carvacrol	*Fusarium* sp., *Botrytis cinerea*	Kordali *et al.* (2008) Romero *et al.* (2007)
α-Cadinol	*R. solani*, *F. oxysporum*	Chang *et al.* (2008)
T-muurolol	*R. solani*, *F. oxysporum*	Chang *et al.* (2008)

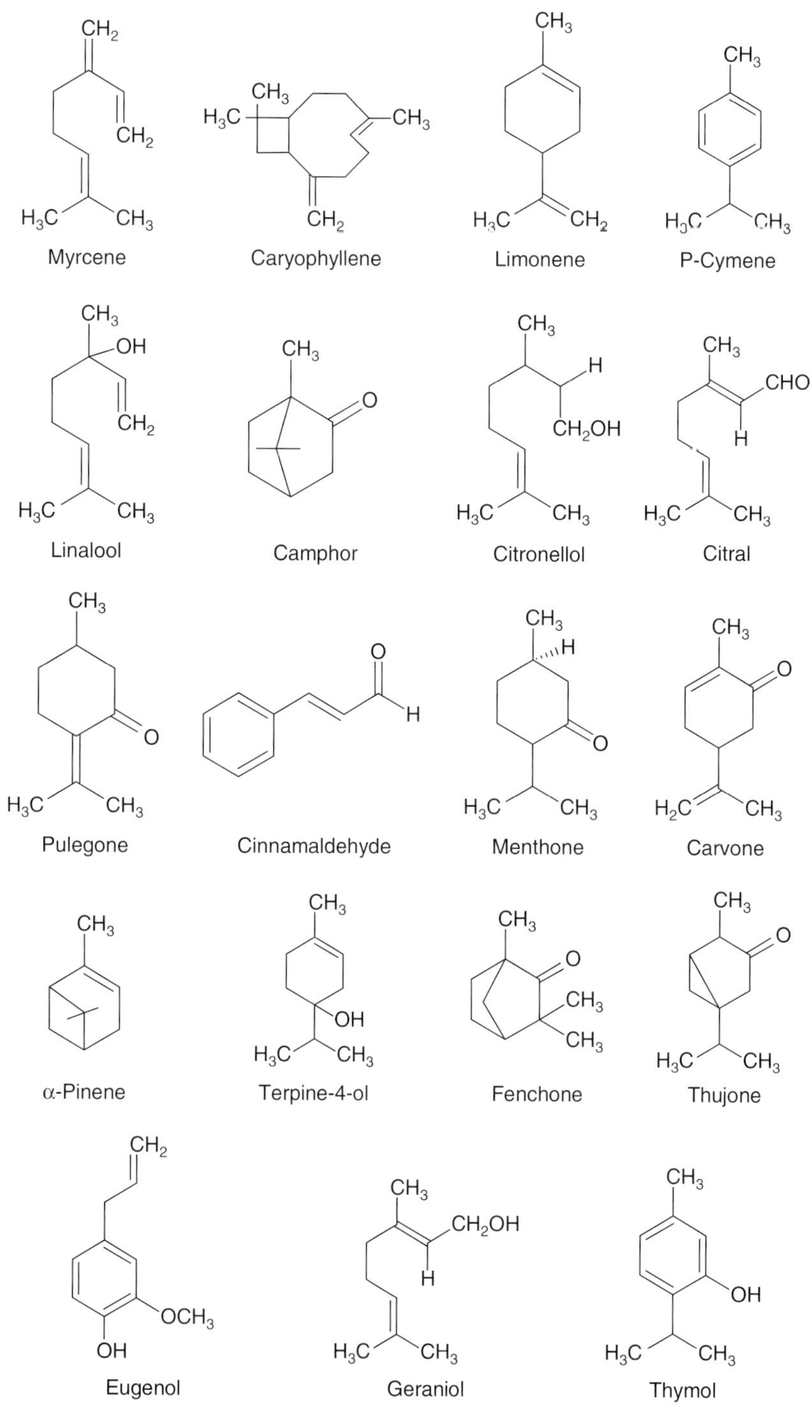

Fig. 2.1. Chemical structures of some bioactive essential oil constituents.

continued

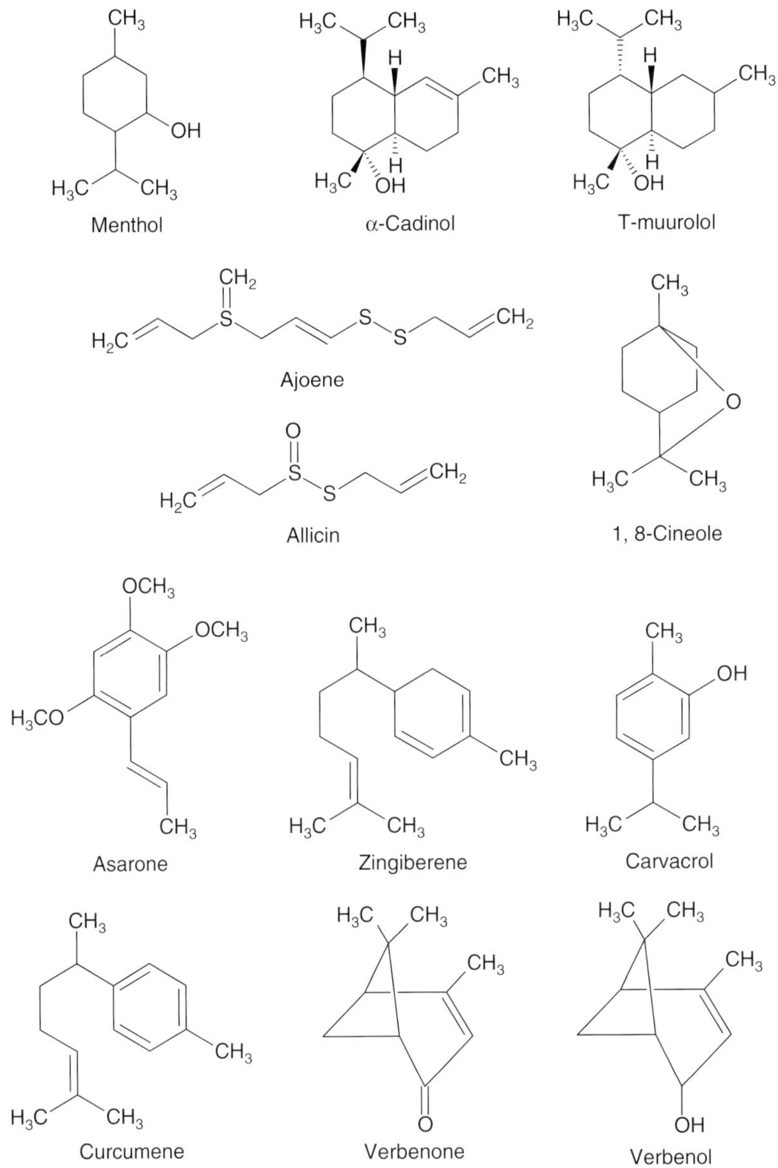

Fig. 2.1. *continued.*

References

Abyaneh, M.R., Ghahfarokhi, M.S., Yoshinari, T., Rezaee, M.B., Jaimand, K., Nagasawa, H. and Sakuda, S. (2008) Inhibitory effects of *Satureja hortensis* L. essential oil on growth and aflatoxin production by *Aspergillus parasiticus*. *International Journal of Food Microbiology* 123, 228–233.

Agarwal, M., Walia, S., Dhingra S. and Khambay, B.P. (2001) Insect growth inhibition, antifeedant and antifungal activity of compounds isolated/derived from *Zingiber officinale* Roscoe (ginger) rhizomes. *Pest Management Science* 57, 289–300.

Aldred, D., Fuller, V.C. and Magan, N. (2008) Environmental factors affect efficacy of some essential oils and resveratrol to control growth and ochratoxin A production by *Penicillium verrucosum* and *Aspergillus westerdijkiae* on wheat grain. *Journal of Stored Products Research* 44, 341–346.

Ankri, S. and Mirelman, D. (1999) Antimicrobial properties of allicin from garlic. *Microbes and Infection* 2, 125–129.

Aparecida, S., Mossini, G. and Kemmelmeier, C. (2008) Inhibition of Citrinin production in *Penicillium citrinum* cultures by Neem [*Azadirachta indica* A. Juss. (Meliaceae)]. *International Journal of Molecular Sciences* 9, 1676–1684.

Atanda, O.O., Akpan, I. and Oluwafemi, F. (2007) The potential of some spice essential oils in the control of *A. parasiticus* CFR 223 and aflatoxin production. *Food Control* 18, 601–607.

Bajpai, V.K., Shukla, S. and Kang, S.C. (2008) Chemical composition and antifungal activity of essential oil and various extract of *Silene armeria* L. *Bioresource Technology* 99, 8903–8908.

Barra, A., Coroneo, V., Dessi, S., Cabras, P. and Angioni, A. (2007) Characterization of the volatile constituents in the essential oil of *Pistacia lentiscus* L. from different origins and its antifungal and antioxidant activity. *Journal of Agricultural Food Chemistry* 55, 7093–7098.

Bendahou, M., Muselli, A., Grignon-Dubois, M., Benyoucef, M., Desjobert, J.M.B., Bernardini, A.F. and Costa, J. (2008) Antimicrobial activity and chemical composition of *Origanum glandulosum* Desf. essential oil and extract obtained by microwave extraction: comparison with hydrodistillation. *Food Chemistry* 106, 132–139.

Beye, F. (1978) Insecticides from vegetable kingdom. *Plant Research and Development* 7, 13–31.

Bouchra, C., Achouri, M., Hassani, L.M.I. and Hmamouchi, M (2003) Chemical composition and antifungal activity of essential oils of seven Moroccan Labiatae against *Botrytis cinerea* Pers: Fr. *Journal of Ethnopharmacology* 89, 165–169.

Braga, P.C., Culici, M., Alfieri, M. and Sasso, M.D. (2008) Thymol inhibits *Candida albicans* biofilm formation and mature biofilm. *International Journal of Antimicrobial Agents* 31, 472–477.

Brul, S. and Coote, P. (1999) Preservative agents in foods. Mode of action and microbial resistance mechanisms. *International Journal of Food Microbiology* 50, 1–17.

Cakir, A., Kordalib, S., Kilicc, H. and Kayad, E. (2005) Antifungal properties of essential oil and crude extracts of *Hypericum linarioides* Bosse. *Biochemical Systematics and Ecology* 33, 245–256.

Chang, H.T., Cheng, Y.H., Wua, C.L., Chang, S.T., Chang, T.T. and Su, Y.C. (2008) Antifungal activity of essential oil and its constituents from *Calocedrus macrolepis* var. *formosana* Florin leaf against plant pathogenic fungi. *Bioresource Technology* 99, 6266–6270.

Cheng, S.S., Liu, J.Y., Hsui, Y.R. and Chang, S.T. (2006) Chemical polymorphism and antifungal activity of essential oils from leaves of different provenances of indigenous cinnamon (*Cinnamomum osmophloeum*). *Bioresource Technology* 97, 306–312.

Chourasia, H.K. and Roy, A.K. (1991) Effect of temperature, relative humidity and light on aflatoxin B_1 production in neem and datura seeds. *Pharmaceutical Biology* 29, 197–202.

Chu, C.L., Liu, W.T. and Zhou, T. (2001) Fumigation of sweet cherries with thymol and acetic acid to reduce post harvest brown rot and blue mold rot. *Fruits* 56, 123–130.

Cia, P., Pascholati, S.F., Benato, E.A., Camili, E.C. and Santos, C.A. (2007) Effects of gamma and UV-C irradiation on the postharvest control of papaya anthracnose. *Postharvest Biology and Technology* 43, 366–373.

Clear, R.M., Patrick, S.K., Turkington, T.K. and Wallis, R. (2002) Effect of dry heat treatment on seedborne *Fusarium graminearum* and other cereal pathogens. *Canadian Journal of Plant Pathology* 24, 489–498.

Coomes, T.J., Crowther, P.C., Feuell, A.J. and Francis, B.J. (1966) Experimental detoxification of groundnut meal containing aflatoxin. *Nature* 209, 406–408.

D'Mello, J.P.F., Macdonald, A.M.C., Postel, D., Dijksma, W.T.P., Dujardin, A. and Placinta, C.M. (1998) Pesticide use and mycotoxin production in *Fusarium* and *Aspergillus* phytopathogens. *European Journal of Plant Pathology* 104, 741–751.

Dadachova, E., Howell, R.W., Bryan, R.A., Frenkel, A., Nosanchuk, J.D. and Casadevall, A. (2004) Susceptibility of the human pathogenic fungi *Cryptococcus neoformans* and *Histoplasma capsulatum* to γ-radiation versus radioimmunotherapy with α- and β-emitting radioisotopes. *The Journal of Nuclear Medicine* 45, 313–320.

Dambolena, J.S., Lopez, A.G., Cánepa, M.C., Theumerc, M.G., Zygadloa, J.A. and Rubinstein, H.R. (2008) Inhibitory effect of cyclic terpenes (limonene, menthol, menthone and thymol) on *Fusarium verticillioides* MRC 826 growth and fumonisin B1 biosynthesis. *Toxicon* 51, 37–44.

Da Silva, C.B., Guterres, S.S., Weisheimer, V. and Schapoval, E.E.S. (2008) Antifungal activity of the lemongrass oil and citral against *Candida* spp. *The Brazilian Journal of Infectious Diseases* 12, 63–66.

Davidson, M.P. (2001) Chemical preservatives and natural antimicrobial compounds. In: Doyle, M.P., Beuchat, L.R. and Montville, T.J. (eds) *Food Microbiology: Fundamentals and Frontiers.* ASM Press, Washington, DC, pp. 593–627.

Deba, F., Xuan, T.D., Yasuda, M. and Tawata, S. (2008) Chemical composition and antioxidant, antibacterial and antifungal activities of the essential oils from *Bidens pilosa* Linn. var. *radiata. Food Control* 19, 346–352.

Deena, M.J. and Thoppil, J.E. (2000) Antimicrobial activity of the essential oil of *Lantana camara. Fitoterapia* 71, 453–455.

Dikbas, N., Kotan, R., Dadasoglu, F. and Sahin, F. (2008) Control of *Aspergillus flavus* with essential oil and methanol extract of *Satureja hortensis. International Journal of Food Microbiology* 124, 179–182.

Droby, S., Cohen, L., Wiess, B., Dans, A. and Wisniewski, M. (2001) Microbial control of postharvest diseases of fruits and vegetables – current status and future outlook. *Acta Horticulture* 553, 371–376.

Dubey, S.C., Suresh, M. and Singh, B. (2006) Evaluation of *Trichoderma* species against *Fusarium oxysporum* f. sp. *ciceris* for integrated management of chickpea. *Biological Control* 40, 118–127.

Duru, M.E., Oztürk, M., Uğur, A. and Ceylan, O. (2004) The constituents of essential oil and *in vitro* antimicrobial activity of *Micromeria cilicica* from Turkey. *Journal of Ethnopharmacology* 94, 43–48.

Farkas, J. (2001) Physical methods for food preservation. In: Doyle, M.P., Beuchat, L.R., Montville and T.J. (eds) *Food Microbiology: Fundamentals and Frontiers.* ASM Press, Washington. DC, pp. 567–592.

Feng, W. and Zheng, X. (2007) Essential oil to control *Alternaria alternata in vitro* and *in vivo. Food Control* 18, 1126–1130.

Gayoso, C.W., Lima, E.O., Olivera, V.T., Pereira, F.O., Souza, E.L., Lima, E.L. and Navarro, D.F. (2005) Sensitivity of fungi isolated from onichomicosis to *Eugenia caryophyllata* essential oil and eugenol. *Fitoterapia* 76, 247–249.

Giordani, R., Hadef, Y. and Kaloustian, J. (2008) Compositions and antifungal activities of essential oils of some Algerian aromatic plants. *Fitoterapia* 79, 199–203.

Hartmans, K.J., Diepenhorst, P., Bakker, W. and Gorris, L.G.M. (1995) The use of carvone in agriculture, sprout suppression of potatoes and antifungal activity against potato tuber and other plant diseases. *Industrial Crops and Products* 4, 3–13.

Hernandez, T., Canales, M., Teran, B., Avila, O., Duran, A., Garcia, A.M., Hernandez, H., Lopez, O.A., Araiza, M.F. and Avila, G. (2007) Antimicrobial activity of the essential oil and extracts of *Cordia curassavica* (Boraginaceae). *Journal of Ethnopharmacology* 111, 137–141.

Hur, J.S., Oh, S.O., Lim, K.M., Jung, J.S., Kim, J.W. and Koh, Y.J. (2005) Novel effects of TiO_2 photocatalytic ozonation on control of postharvest fungal spoilage of kiwifruit. *Postharvest Biology and Technology* 35, 109–113.

Isman, M.B. (2006) Botanical insecticides, deterents, and repellents in modern agriculture and an increasingly regulated world. *Annual Review of Entomology* 51, 45–66.

Isman, M.B. and Akhtar, Y. (2007) Plant natural products as a source for developing environmentally acceptable insecticides. In: Ishaaya, I., Nauen, R. and Horowitz, A.R. (eds) *Insecticides Design Using Advanced Technologies.* Springer-Verlag, Berlin, Heidelberg, pp. 235–248.

Janna, B., Harald, M., Oydis, U. and Magni, M. (2005) Longitudinal study of taste identification of sensory panellists: effect of ageing, experience and exposure. *Food Quality and Preference* 18, 230–241.

Kacaniova, M. (2003) Feeding soybean colonization by microscopic fungi. *Trakya University Journal of Science* 4, 165–168.

Khosravi, A.R., Shokri, H. and Ziglari, T. (2007) Evaluation of fungal flora in some important nut products (pistachio, peanut, hazelnut and almond) in Tehran, Iran. *Pakistan Journal of Nutrition* 6, 460–462.

Kneževi, Z. and Serdar, M. (2008) Screening of fresh fruit and vegetables for pesticide residues on Croatian market. *Food Control* 20, 419–422.

Kordali, S., Cakir, A., Ozer, H., Cakmakci, R., Kesdek, M. and Mete, E. (2008) Antifungal, phytotoxic and insecticidal properties of essential oil isolated from Turkish *Origanum acutidens* and its three components, carvacrol, thymol and p-cymene. *Bioresource Technology* 99, 8788–8795.

Korukluoglu, M., Sahan, Y. and Yigit, A. (2008) Antifungal properties of olive leaf extracts and their phenolic compounds. *Journal of Food Safety* 28, 76–87.

Kumar, A., Shukla, R., Singh, P., Prasad, C.S. and Dubey, N.K. (2008) Assessment of *Thymus vulgaris* L. essential oil as a safe botanical preservative against post harvest fungal infestation of food commodities. *Innovative Food Science and Emerging Technologies* 9, 575–580.

Kumar, R., Mishra, A.K., Dubey, N.K. and Tripathi, Y.B. (2007) Evaluation of *Chenopodium ambrosioides* oil as a potential source of antifungal, antiaflatoxigenic and antioxidant activity. *International Journal of Food Microbiology* 115, 159–164.

Lagunas, L.L.M. and Castaigne, F. (2008) Effect of temperature cycling on allinase activity in garlic. *Food Chemistry* 111, 56–60.

Lee, B., Choi, W., Lee, S. and Park, B. (2001) Fumigant toxicity of essential oils and their constituent towards the rice weevil, *Sitophilus oryzae* (L.). *Crop Protection* 20, 317–320.

Lee, H.S. (2007) Fungicidal property of active component derived from *Acorus gramineus* rhizome against phytopathogenic fungi. *Bioresource Technology* 98, 1324–1328.

Lee, J.Y., Yun, B.S. and Hwang, B.K. (2004) Antifungal activity of β-asarone from rhizome of *Acorus gramineus. Journal of Agricultural Food Chemistry* 52, 776–780.

Liu, W.T., Chu, C.L. and Zhou, T. (2002) Thymol and acetic acid vapors reduce post harvest brown rot of apricot and plums. *HortScience* 37, 151–156.

Magarey, R.C., Yip, H.Y., Bull, J.I. and Johnson, E.J. (1997) Effect of the fungicide mancozeb on fungi associated with sugarcane yield decline in Queensland. *Mycological Research* 101, 858–862.

Magwa, M.L., Gundidza, M., Gwerua, N. and Humphrey, G. (2006) Chemical composition and biological activities of essential oil from the leaves of *Sesuvium portulacastrum. Journal of Ethnopharmacology* 103, 85–89.

Mahmoud, A.L. (1999) Inhibition of growth and aflatoxin biosynthesis of *Aspergillus flavus* by extracts of some Egyptian plants. *Letters in Applied Microbiology* 29, 334–336.

Martinez, L.R., Bryan, R.A., Apostolidis, C., Morgenstern, A., Casadevall, A. and Dadachova, E. (2006) Antibody-guided alpha radiation effectively damages fungal biofilms. *Antimicrobial Agents and Chemotherapy* 50, 2132–2136.

Matan, N. and Matan, N. (2008) Antifungal activities of anise oil, lime oil, and tangerine oil against molds on rubberwood (*Hevea brasiliensis*). *International Biodeterioration and Biodegradation* 62, 75–78.

Matasyoh, J.C., Kiplimo J.J., Karubiu N.M. and Hailstorks T.P. (2007). Chemical composition and antimicrobial activity of essential oil of *Tarchonanthus camphorates. Food Chemistry* 101: 1183–1187.

Medina, A., Mateo, R., Valle-Algarra, F.M., Mateo, E.M. and Jiménez, M. (2007) Effect of carbendazim and physicochemical factors on the growth and ochratoxin A production of *Aspergillus carbonarius* isolated from grapes. *International Journal of Food Microbiology* 119, 230–235.

Meepagala, K.M., Kuhajek, J.M., Sturtz, G.D. and Wedge, D.E. (2003) Vulgarone B, the antifungal constituent in the steam-distilled fraction of *Artemisia douglasiana. Journal of Chemical Ecology* 29, 1771–1780.

Mishra, A.K. and Dubey, N.K. (1994) Evaluation of some essential oils for their toxicity against fungi causing deterioration of stored food commodities. *Applied and Environmental Microbiology* 60, 1101–1105.

Mishra, H.N. and Das, C. (2003) A review on biological control and metabolism of aflatoxin. *Critical Reviews in Food Science and Nutrition* 43, 245–264.

Moatsou, G., Katsaros, G., Bakopanos, C., Kandarakis, I., Taoukis, P. and Politis, I. (2008) Effect of high-pressure treatment at various temperatures on activity of indigenous proteolytic enzymes and denaturation of whey proteins in ovine milk. *International Dairy Journal* 18, 1119–1125.

Moleyar, V. and Narasimham, P. (1986) Antifungal activity of some essential oil components. *Food Microbiology* 3, 331–336.

Naganawa, R., Iwata, N., Ishikawa, K., Fukuda, H., Fujino, T. and Suzuki, A. (1996) Inhibition of microbial growth by ajoene, a sulfur-containing compound derived from garlic. *Applied and Environmental Microbiology* 62, 4238–4242.

Nielsen, P.V. and De Boer, E. (2000) Food preservatives against fungi. In: Samson, R.A., Hoekstra, E.S., Frisvad, J.C. and Filtenborg, O. (eds) *Introduction to Food- and Airborne Fungi.* Centraal Bureau voor Schimmelcultures, Utrecht, Netherlands, pp. 357–363.

Ogunsanwo, B.M., Faboya, O.O.P., Idowu, O.R., Lawal, O.S. and Bankole, S.A. (2004) Effect of roasting on the aflatoxin contents of Nigerian peanut seeds. *African Journal of Biotechnology* 3, 451–455.

Olender, D., Żwawiak, J., Lukianchuk, V., Lesyk, R., Kropacz, A., Fojutowski, A. and Zaprutko, L. (2008) Synthesis of some N-substituted nitroimidazole derivatives as potential antioxidant and antifungal agents. *European Journal of Medicinal Chemistry* 44, 645–652.

Omidbeygi, M., Barzegar, M., Hamidi, Z. and Naghdibadi, H. (2007) Antifungal activity of thyme, summer savory and clove essential oils against *Aspergillus flavus* in liquid medium and tomato paste. *Food Control* 18, 1518–1523.

Oosterhaven, J. (1995) Different aspects of *S*-carvone – a natural potato sprout growth inhibitor. Thesis, Landbouwuniversiteit, Wageningen, Cip-data Konin Klije Bibliotheek Den Haag.

Ormancey, X., Sisalli, S. and Coutiere, P. (2001) Formulation of essential oils in functional perfumery. *Parfums, Cosmetiques, Actualites* 157, 30–40.

Oteiza, J.M., Peltzer, M., Gannuzzi, L. and Zaritzky, N. (2005) Antimicrobial efficacy of UV radiation on *Escherichia coli* O157:H7 (EDL 933) in fruit juices of different absorptivities. *Journal of Food Protection* 68, 49–58.

Oxenham, S.K. (2003) Classification of an *Ocimum basilicum* germplasm collection and examination of the antifungal effect of the essential oil of the basil. PhD thesis, University of Glasgow, Glasgow, UK.

Pandey, R. (2003) Pesticides and sterility. *Everyman's Science* 38, 84–86.

Pang, X., Zhang, Z.Q.Q. and Xue, M.H. (2002) Biological control of postharvest diseases of fruits and vegetables. *Journal of Tropical and Subtropical Botany* 10, 186–192.

Petushkova, I.P., Lyalikova, N.N. and Nichiporov, F.G. (1988) Effect of ionizing radiation on monument deteriorating microorganisms. *Journal of Radioanalytical and Nuclear Chemistry* 125, 367–371.

Pinto, E., Salgueiro, L.R., Cavaleiro, C., Palmeira, A. and Gonçalves, M. (2007) *In vitro* susceptibility of some species of yeasts and filamentous fungi to essential oils of *Salvia officinalis*. *Industrial Crops and Products* 26, 135–141.

Pitarokili, D., Tzakou, O., Loukis, A. and Harvala, C. (2003) Volatile metabolites from *Salvia fruticosa* as antifungal agents in soilborne pathogens. *Journal of Agricultural Food Chemistry* 51, 3294–3301.

Pitt, J.I. (2002) An introduction to mycotoxins. In: Semple, R.L., Frio, A.S., Hicks, P.A. and Lozare, J.V. (eds) A collaborative publication of the UNDP/FAO Regional Network Inter-country Cooperation on Preharvest Technology and Quality Control of Food Grains (REGNET) and the ASEAN Grain Postharvest Programme. Bangkok, Thailand.

Qing, X., Wen, C., Feng, Z. and Run-zhang, Y. (2002) A novel infrared radiant glaze exhibiting antibacterial and antifungal functions. *Journal of Wuhan University of Technology* 17, 10–13.

Rasooli, I. and Abyaneh, M.R. (2004) Inhibitory effects of thyme oils on growth and aflatoxin production by *Aspergillus parasiticus*. *Food Control* 15, 479–483.

Rasooli, I., Rezaei, M.B. and Allameh, A. (2006) Growth inhibition and morphological alterations of *Aspergillus niger* by essential oils from *Thymus eriocalyx* and *Thymus x-porlock*. *Food Control* 17, 359–364.

Rasooli, I., Fakoor, M.H., Yadegarinia, D., Gachkar, L., Allameh, A. and Rezaei, M.B. (2008) Antimycotoxigenic characteristics of *Rosmarinus officinalis* and *Trachyspermum copticum* L. essential oils. *International Journal of Food Microbiology* 122, 135–139.

Reddy, K.R.N., Reddy, C.S. and Muralidharan, K. (2008) Potential of botanicals and biocontrol agents on growth and aflatoxin production by *Aspergillus flavus* infecting rice grains. *Food Control* 20, 173–178.

Regnier, T., Plooy, W., Combrinck, S. and Botha, B. (2008) Fungitoxicity of *Lippia scaberrima* essential oil and selected terpenoid components on two mango postharvest spoilage pathogens. *Postharvest Biology and Technology* 48, 254–258.

Rehana, F. and Basappa, S.C. (1990) Detoxification of aflatoxin B1 in maize by different cooking methods. *Journal of Food Science and Technology* 27, 379–399.

Rehman, W., Badshah, A., Baloch, M.K. and Muhammad, B. (2008) Synthesis, spectroscopic characterization, and *in vitro* fungicidal activity of some organotin (IV) complexes. *Russian Journal of Coordination Chemistry* 34, 256–258.

Romero, D.M., Guillén, F., Valverde, J.M., Bailén, G., Zapata, P., Serrano, M., Castillo, S. and Valero, D. (2007) Influence of carvacrol on survival of *Botrytis cinerea* inoculated in table grapes. *International Journal of Food Microbiology* 115, 144–148.

Samarajeewa, U., Jayatilake, C.L.V., Ranjithan, A., Gamage, T.V. and Arseculeratne, S.N. (1985) A pilot plant for detoxification of aflatoxin B_1 contaminated coconut oil by solar irradiation. *MIRCEN Journal of Applied Microbiology and Biotechnology* 1, 333–343.

Sánchez, E., Heredia, N. and García, S (2005) Inhibition of growth and mycotoxin production of *Aspergillus flavus* and *Aspergillus parasiticus* by extracts of *Agave* species. *International Journal of Food Microbiology* 98, 271–279.

Satish, S., Raghavendra, M.P., Mohana, D.C. and Raveesha, K.A. (2008) Antifungal activity of a known medicinal plant *Mimusops elengi* L. against grain moulds. *Journal of Agricultural Technology* 4, 151–165.

Sawamura, M. (2000) Aroma and functional properties of Japanese yuzu (*Citrus junos* Tanaka) essential oil. *Aroma Research* 1, 14–19.

Serrano, M., Romero, D.M., Castillo, S., Guillén, F. and Valero, D. (2005) The use of natural antifungal compounds improves the beneficial effect of MAP in sweet cherry storage. *Innovative Food Science and Emerging Technologies* 6, 115–123.

Shafi, P.M., Nambiar, M.K.G., Clery, R.A., Sarma, Y.R. and Veena, S.S. (2004) Composition and antifungal activity of the oil of *Artemisia nilagirica* (Clarke) Pamp. *Journal of Essential Oil Research* 16, 377–379.

Shantha, T. and Sreenivasamurthy, V. (1977) Photodestruction of aflatoxin in groundnut oil. *Indian Journal of Technology* 15, 453–454.
Shantha, T., Sreenivasamurthy, V., Rati, E.R. and Prema, V. (1986) Detoxification of groundnut seeds by urea and sunlight. *Journal of Food Safety* 7, 225–231.
Sharma, N. and Tripathi, A. (2008) Effects of *Citrus sinensis* (L.) Osbeck epicarp essential oil on growth and morphogenesis of *Aspergillus niger* (L.) van Tieg. *Microbiological Research* 163, 337–344.
Shukla, R., Kumar, A., Prasad, C.S., Srivastava, B. and Dubey, N.K. (2008) Antimycotic and antiaflatoxigenic potency of *Adenocalymma alliaceum* Miers. on fungi causing biodeterioration of food commodities and raw herbal drugs. *International Biodeterioration and Biodegradation* 62, 348–351.
Sonboli, A., Babakhani, B. and Mehrabian, A.R. (2006) Antimicrobial activity of six constituents of essential oil from *Salvia. Zeitschrift für Naturforschung C* 61, 160–164.
Sreenivasamurthy, V., Parpia, A.B., Srikanta, S. and Shankar, A. (1967) Detoxification of aflatoxin in peanut meal by H_2O_2. *Journal of the Association of Official Analytical Chemists* 50, 350–354.
Srivastava, B., Singh, P., Shukla, R. and Dubey, N.K. (2008) A novel combination of the essential oils of *Cinnamomum camphora* and *Alpinia galanga* in checking aflatoxin B_1 production by a toxigenic strain of *Aspergillus flavus. World Journal of Microbiology and Biotechnology* 24, 693–697.
Srivastava, J.L. (1987) Mycotoxin problems in food in India. Second International Conference on Mycotoxins held at Bangkok, FAO/WHOIUNDP.
Tatsadjieu, N.L., Dongmo, P.M.J., Ngassoum, M.B., Etoa, F.X. and Mbofung, C.M.F. (2009) Investigations on the essential oil of *Lippia rugosa* from Cameroon for its potential use as antifungal agent against *Aspergillus flavus* Link ex. Fries. *Food Control* 20, 161–166.
Thomas, G.J., Sweetingham, M.W. and Adcock, K.G. (2008) Application of fungicides to reduce yield loss in anthracnose-infected lupins. *Crop Protection* 27, 1071–1077.
Tian, S.P. (2001) Effects of low temperature on mycelial growth and spore germination of *Botrytis cinerea in vitro* and on its infectivity to stored chicory. *Acta Phytopathologica Sinica* 31, 56–62.
Tripathi, P. and Dubey, N.K. (2004) Exploitation of natural products as an alternative strategy to control post harvest fungal rotting of fruits and vegetables. *Postharvest Biology Technology* 32, 235–245.
Tripathi, P., Dubey, N.K., Banerji, R. and Chansouria, J.P.N. (2004) Evaluation of some essential oils as botanical fungitoxicants in management of post-harvest rotting of *Citrus* fruits. *World Journal of Microbiology and Biotechnology* 20, 317–321.
Tzortzakis, N.G. and Economakis, C.D. (2007) Antifungal activity of lemongrass (*Cympopogon citratus* L.) essential oil against key postharvest pathogens. *Innovative Food Science and Emerging Technologies* 8, 253–258.
Varma, J. and Dubey, N.K. (1999) Prospectives of botanical and microbial products as pesticides of tomorrow. *Current Science* 76, 172–179.
Viuda-Martos, M., Ruiz-Navajas, Y., Lopez, J.F. and Pérez-Alvarez, J. (2008) Antifungal activity of lemon (*Citrus lemon* L.), mandarin (*Citrus reticulata* L.), grapefruit (*Citrus paradisi* L.) and orange (*Citrus sinensis* L.) essential oils. *Food Control* 19, 1130–1138.
Wiktelius, S., Chiverton, P.A., Meguenm, H., Bennaceur, M., Ghezal, F., Umeh, E.D.N., Egwuatu, R.I., Minja, E., Makusi, R., Tukahirwa, E., Tinzaara, W. and Deedat, Y. (1999) Effects of insecticides on non-target organisms in African agroecosystems: a case for establishing regional testing programmes. *Agriculture, Ecosystems and Environment* 75, 121–131.
Williams, H.J., Phillips, T.D., Jolly, E.P., Stiles, K.J., Jolly, M.C. and Aggarwal, D. (2004) Human aflatoxicosis in developing countries: a review of toxicology, exposure, potential health consequences and interventions. *American Journal of Clinical Nutrition* 80, 1106–1122.
Wilson, C.L., Solar, J.M., El Ghaouth, A. and Wisniewski, M.E. (1997) Rapid evaluation of plant extracts and essential oils for antifungal activity against *Botrytis cinerea. Plant Disease* 81, 204–210.
Wilson, C.L., El-Ghaouth, A. and Wisniewski, M.E. (1999) Prospecting in nature's storehouse for biopesticides. *Conferencia Magistral Revista Maxicana de Fitopatologia* 17, 49–53.
Wisniewski, M., Wilson, C., El-Ghaouth, A. and Droby, S. (2001) Non-chemical approaches to postharvest disease control. *Acta Horticulturae* 553, 407–412.
Yoshida, S., Kasuga, S., Hayashi, N., Ushiroguchi, T., Matsuura, H. and Nakagawa, S. (1987) Antifungal activity of ajoene derived from garlic. *Applied and Environmental Microbiology* 53, 615–617.
Yousefzadi, M., Sonboli, A., Ebrahimi, S.N. and Hashemi, S.H. (2008) Antimicrobial activity of essential oil and major constituents of *Salvia chloroleuca. Zeitschrift für Naturforschung C* 63, 337–340.
Zhang, Z.Z., Li, Y.B., Qi, L. and Wan, X.C. (2006) Antifungal activities of major tea leaf volatile constituents toward *Colletorichum camelliae* Massea. *Journal of Agricultural Food Chemistry* 54, 3936–3940.

3 Deleterious Effects of Fungi on Postharvest Crops and Their Management Strategies

A.O. Ogaraku
Plant Science and Biotechnology Unit, Department of Biological Sciences, Nasarawa State University, Keffi, Nigeria

Abstract
Fungi influence our lives in many ways. The parasitic forms cause serious diseases in crop plants and pose hazards to the lives of animals and humans whenever they infect consumable crops. Most consumable crops are susceptible to fungal infection. The most prominent types of fungi attacking commodities are species of *Aspergillus*, *Penicillium* and *Rhizopus*, etc. Types of crop deterioration caused by fungi include discoloration, flavours and odour, rotting and caking, destruction of viability and production of mycotoxins on food before infestation. Conditions that favour the development of fungi on harvested and stored crops include moisture, preharvest infection and lapses in the processing method. Method of control involves drying of produce to a safe moisture level, non-mixing of new produce with old ones, avoidance of pre-storage damage and use of chemicals, fungicides and medicinal plants in treating the produce.

Introduction

Fungi are one of the most important groups of organisms on the planet. They are microscopic, achlorophyllous and non-vascular plants. They cause deterioration of postharvest crops (Ogundana *et al.*, 1970).

Deterioration means that something is made to be of less value or worse in quality (Adebayo *et al.*, 1994). It is a common phenomenon in agricultural crops, either on the farm, at harvest or during storage. Fungi are known to cause various types of deterioration and pose a hazard to humans and animals whenever they infect crops. Fungal deterioration can be defined as any change resulting from the activities of fungi which renders a product unsuitable for its intended use or reduces the economic value of the materials (Opadokun *et al.*, 1979).

It is also noteworthy to mention some other factors that have been identified as causing damage to crops, namely:

- insects and mites
- microorganisms, such as bacteria, actinomycetes, yeasts and virus
- rodents and birds
- physical factors, such as temperature and relative humidity of the storage environment
- harvesting, handling and transportation (Clarke, 1968).

Before the 17th century, scientists concentrated on damage caused by insects on stored products. This was because damage by insects was usually conspicuous, easy to quantify and these insects were visible to the naked eye. But, awareness of the losses caused by fungi, also referred to as 'moulds', came with the discovery of a toxic metabolite called aflatoxin in 1968 caused by a fungus called *Aspergillus flavus*, which killed over 100,000 turkeys in Britain when fed with groundnut cakes that were infected by this organism.

Studies in Nigeria have revealed the presence of aflatoxin in Nigerian groundnuts and livestock feed maize; hence, there is a need to take extremely good care of these products during storage (Akano and Atanda, 1989). Some crops in which fungal deterioration can take place are as follows: maize, sorghum, millet, cowpea, beans, groundnut, cocoa beans, palm kernels and tubers.

Deleterious Effects of Fungi on Postharvest Crops

Fungi occur everywhere and have a profound effect on their environment. Like other microorganisms, fungi may be good or harmful, depending on the species involved. The deterioration of postharvest crops by fungi can be either by destruction of the produce itself or by presenting a potential hazard to animals or humans. Some of the deleterious effects of fungi on postharvest crops are as follows:

Discoloration

Fungi come in various colours, i.e. green, brown, white, grey, black, etc. They impart these colours on postharvest crops, thereby changing the original appearance. Discoloured produce is often disliked by consumers and manufacturers in that the colours affect the end products from such produce.

Cocoa beans, melon seeds, palm kernels, groundnuts, maize, yam and cassava are examples of produce in which deterioration is accomplished by marked discoloration. For instance, *Lasiodiplodia theobromae* is responsible for the disease which discolours cocoa, widely known as 'concealed damage' in Alabama, while *Macrophomina phaseoli* causes 'black mars' in Gambian groundnuts. In fruits and seeds, the micropyle is the common place for infections to begin, but fungi, bacteria and actinomycetes can develop in any other region of the seed or fruit, causing abnormal colouring, either localized or generalized (Clarke, 1968). However, it is not all discoloration on produce that is caused by fungi; sometimes, it may be due to genetic mutations.

Flavour and odour

The flavour and odour of produce caused by moulds usually affect the taste of the end products and are not acceptable to consumers. The change in the flavour and odour is usually as a result of the biochemical change which takes place in the stored produce.

For example, mouldy groundnuts have a very unpleasant and sour taste when consumed and these are usually spat out from the mouth as soon as they are chewed. Undesirable flavour is easily noticed in mouldy cocoa beans, as it can be detected by tasting a sample of chocolate which has passed through all the normal manufacturing processes. Banana and plantain affected by mould also have a detectable flavour and odour. Mouldy produce can also have an odour, ranging from the musty odour of mouldy grains to the foul smell of rotten grains (Ogundana *et al.*, 1970).

Biochemical effects

The development of moulds leads to a great modification in the chemical composition of the infected produce. One such effect is an increase in the free fatty acid (FAA) content of the produce. This acid is one of the intermediate products of spoilage in materials containing fats and oils and its formation results in rancidity. Many of the mould species infecting our crops are known to produce lipases, which can hydrolyse fats into fatty acids by a process called lipolysis, thereby increasing the free fatty acid content of the produce and resulting in a decrease in oil content and a low protein content.

Kuku (1972) isolated a number of moulds from palm oil and showed that many of these increased the FAA of palm oil in pure culture studies. Coursey *et al.* (1963) isolated a number of lipolytic fungi from Nigerian palm kernels. These included *A. chevalieri*, *A. fumigatus*, *Paecilomyces variotii* and *P. steckii*.

The development of mould on produce causes other modifications; generally, an increase in reducing sugars and a loss in protein, which may lead to flours unsuitable for bread making. Moreover, mouldy rice grain breaks easily during polishing. If we preserve damp grain in an anaerobic environment, fermentation results in the release of carbon dioxide, alcohol and other volatile substances. The compounds formed give a bad taste, which remains even after drying in the open air.

Rotting and caking

Extensive mould activities usually result in rotting and caking. Rotting and caking renders produce unsightly, decreases milling yield and quality. Studies carried out by several workers, including Adeniyi (1970) and Ogundana *et al.* (1970), revealed that a number of fungal species, for example *A. niger*, *Fusarium moniliforme*, *P. exalicum*, etc., caused rotting in Nigerian yams. Oyeniran (1970) and other workers carried out studies that showed that over 30 mould species could be associated with the deterioration of maize in Nigeria.

Weight loss

Fungi growing on plant parts or produce use them as food substrate. They produce a variety of enzymes, i.e. amylases, cellulases, pectinases and lipases, which hydrolyse the food substances into soluble forms. The food components easily absorbed and utilized are carbohydrates, proteins, fats and oil. This breakdown invariably leads to absolute weight loss. Scientists have reported up to 10% weight loss in rotting yam tubers during storage (Ogundana *et al.*, 1970). Some of the fungi that can cause weight loss in maize are *A. flavus*, *A. niger*, *A. candidus*, *Mucor racemosus* and *P. pallitans*.

Destruction of viability

Fungi reduce the viability of seeds by infecting and destroying their embryo. This in turn affects the germinability of the seeds during planting. Broadbent (1967) found samples of mouldy maize from government farms in southern Nigeria had only 7–14% germination, while the mould-free samples had 100% germination.

Heating

If crops with high water content are piled up together in one place, heat is generated and decay sets in. Heating or production of hot spots is one of the characteristics that results in rapid fungal development in moist stored produce, mostly grains and tubers. Heating in bulk storage is evidence of spoilage in progress or spoilage already completed.

Growth abnormalities

Groundnut and maize contaminated with *A. flavus* produce deformed plants. The infected young groundnut or maize plant will have a greatly decreased growth. The follicles develop poorly and are elongated in form. During growth, a large number of sick plants die. Others are continually abnormal in appearance, while some evolve into normal plants.

Preparation of the Material for Attack by Other Agents

It is sometimes difficult in many crops to separate deterioration or spoilage due to insects from that caused by fungi, but that the two are interrelated is in no doubt. What is in doubt, however, is the exact sequence of events and the relative damage caused by the two agents.

Invasion of stored produce by fungi prepare such commodities for attack by other agents of deterioration, especially bacteria, insects and mites. In fact, some insects are known to feed on fungi and in this way they help to spread the spores. These storage insects can live, develop and reproduce entirely on certain fungi and thus undoubtedly play an important part as carriers in the spread of the fungi. An example of such an insect is *Adhasverus advena*.

Production of Toxic Metabolites (Mycotoxins)

Toxic metabolite production is the most serious effect of microbiological deterioration of stored products because of its poisoning nature. There are two kinds of poisoning by fungi, mycetism and mycotoxicosis. In mycetism, the toxic substances are constituents of the fungi, large enough to be eaten alone. In mycotoxicosis, the fungus is a contaminant of and has produced toxic product in some food. The effects of mycetism include diarrhoea and jaundice, while mycotoxicoses were defined by Clarke (1968) as diseases of animals and humans caused by ingesting poisonous metabolite fungi that have grown in the food previously before ingestion. Some notable examples of mycotoxicoses are:

1. Ergotism – diseases of cattle in central Europe caused by the fungi, *Claviceps purpurea*.
2. Yellow rice disease of humans in Japan caused by the fungi, *P. citrinum*.
3. Alimentary toxic aleukia (ATA) of humans and cattle caused by *F. sporotrichioides*.
4. Importantly, aflatoxicosis of poultry and livestock caused by *A. flavus*.

This last mentioned toxin disease, aflatoxicosis, has been receiving worldwide attention since 1960, when it was reported to have caused the death of about 100,000 turkeys in Britain when they were fed with groundnut cakes which was infected with *A. flavus*. The toxic substance was therefore called 'aflatoxin'. Different mycotoxins affect different sites of the body. Aflatoxins produced by *A. flavus* are the commonest of all the toxins and affect the liver, causing aflatoxicosis or liver poisoning. High levels of aflatoxin have also been reported to cause infertility (abnormality in the spermatozoa) in samples of semen from men fed on diets contaminated with *A. flavus* (Ibeh *et al.*, 1994). The production of aflatoxins on maize grains and other consumable foods in Nigeria has been reported by many researchers, including Broadbent (1967), Oyeniran (1970), Opadokun *et al.* (1979) and Akano and Atanda (1989). Other common mycotoxins are:

- Fumonism – this causes oesophageal cancer in horses and humans. It is produced by *F. graminearum* on maize.
- Ochratoxin – produced by *A. ochareus*, which causes serious nephropathy in pigs and humans. It is commonly found in milk and cereals (processed or raw).

Some examples of mould species and the toxins they produce are shown in Table 3.1.

Conditions that Favour Development of Fungi on Harvested and Stored Crops

Fungi, like other living organisms, require certain conditions for growth and development. These conditions are as follows:

Moisture

It is not the moisture content as such that is the controlling factor in biological deterioration; it is the relative humidity of the air in and around the crop. Although relative humidity is the controlling factor, attention is usually focused on the moisture content because relative humidity of produce is difficult to measure, while moisture content

Table 3.1. Examples of mould species and the toxins they produce.

Mould species	Toxin produced
Aspergillus flavus	Aflatoxin
A. ochareus	Ochratoxin
A. chevalieri	Xanthocillin
A. nidulans	Sterigmatocystin
A. ruber	Rubratoxin
A. niger	Oxalic acid
Penicillium islandicum	Islanditoxin
P. notatum	Xanthocillin
P. rubrum	Rubratosin
P. citrinum	Citrinin
P. patulinium	Patulin
Fusarium graminearum	Zearalenone

is not. Moisture in stored produce is divisible into two main types: chemically bound water, which is the part of the intrinsic composition, and physically bound water, some of which is held loosely on the commodity.

Moisture in terms of water is necessary for mould spores to germinate, and it also helps in the process of dissolution of food materials. The moisture level of a stored product therefore determines the development rate of the storage fungi.

Mould species vary in their water requirement; for instance, there are those that thrive at low moisture levels and are said to be xerophytic. Examples are *A. flavus*, *A. chevalieri* and *A. repens*. Others require high levels of moisture before they can survive and are said to be hydrophilic, e.g. *Penicillium* species.

Temperature

All living things have a minimum and maximum temperature for growth. Fungi are a co-exception. Most fungi will grow at temperatures between 5°C and 35°C. These are the mesophilic species. There are those that thrive at 35°C and above and are said to be thermophilic. Some thrive at very cold temperatures and are said to be psychrophilic. This means that fungi thrive well in a very wide temperature range, which gives room for existence in postharvest crops.

Oxygen

Most fungi are aerobic; they require oxygen to survive, like other living organisms. Any device which cuts off oxygen from the storage environment will reduce, if not totally eliminate, fungi. This is why storage at an inert temperature has been effective.

Nutrients

All biological systems, from microorganisms to humans, share a set of nutritional requirements with regards to the chemicals necessary for their growth and normal functioning. The great diversity of nutritional types required are energy, carbon, nitrogen, sulphur and phosphorus, metallic elements and vitamins. All these nutritional requirements are present in food substances, such as carbohydrates, proteins, fats and oil, which fungi need in soluble forms for metabolic processes.

Fungi produce a variety of enzymes which break down complex food substances. Some of these enzymes are as follows:

- Cellulases – break down cellulose in plant materials.
- Amylases – hydrolyse carbohydrates.
- Lipases – hydrolyse fats to fatty acids and glycerol.
- Proteases – hydrolyse proteins.
- Pectinases – hydrolyse the pectic materials of plant tissues.

Heating

If crops with a high water content are allowed to overlap or are piled together in one place, yam for example, heat is generated under moist conditions and decay will set in.

Insufficient drying

Some crops grow mouldy if insufficiently dried. Fungi can creep in to destroy the crops.

Preharvest infection

Produce destined for storage is sometimes infected by moulds before harvest. Most fungi species also invade, especially following natural or artificial wounds. Some examples are attack of cocoa beans by *Lasiodiplodia theobromae* and other moulds, attack of groundnut by *Macrophomina phaseoli* and attack of maize by *F. moniliforme* and *P. citrinum*.

Attack during preparation

During the process of preparation, mould attacks some produce as a result of lapses in cultural practices; for example, during cocoa fermentation mould could infect and penetrate the beans if the fermenting mass of beans is not stirred or mixed thoroughly at intervals. In palm produce, mould can attack the fruits and sometimes the kernels when they are heaped on the ground just before de-husking. In groundnut, the crop has to be lifted at certain times to avoid mould contamination.

Types of stores

If, for instance, through economy a store is poorly constructed and the roof is holding water, it is possible to cause leakage and water will drip on to the commodity and thereby cause deterioration.

There are other factors which contribute to the development of fungi in crops apart from those mentioned above and they are:

1. The degree to which the grain has already been invaded by storage fungi before it arrives at a given site.
2. The amount of foreign material present in the grain.
3. The activity of mites and insects. Bored holes serve as an entry for mould spores. Some insects, such as *A. advena*, and mites feed on mould spores and therefore help to spread the fungi, as well as increase their activities in storage.

Economic loss

1. There is a monetary loss because of inaccessibility to foreign trade due to the poor quality of the produce. There is also a monetary loss because of the poor health of animals fed with inferior feeds.
2. Some fungi, for example *Fusarium* species, can grow on stored animal feeds, generating products that are highly toxic to swine and other animals.
3. Infections leading to disease of crops are extremely important because of the famine, malnutrition and dietary deficiency they may cause.
4. Some plant pathogens cause food intoxication when eaten by humans or animals; for example, the fungus, *C. purpurea*, which grows on cereal grains and some grasses, replaces the feed kernels with compact masses of hardened fungus called sclerotia. These contain alkaloids that act on the nervous system of humans and other animals, causing gangrene, convulsions and death.

Control of Fungal Deterioration in Postharvest Crops

If left uncontrolled, these fungi will cause deterioration of food products and many other articles of commerce and industry. For this purpose, a distinction can be made between postharvest produce that is stored dry, such as grains, cocoa, groundnuts, etc., and those which are stored with a high water content, such as yams and other tubers.

However, some of the measures or suggestions listed below will certainly apply to both types:

1. Proper drying of produce to a safe moisture level, either by retaining maize, millet or guinea corn on the cob and storing in a condition where gradual drying by heat or aeration takes place, or otherwise by providing artificial drying.
2. Prevention of damage or wounds on produce so as to forestall a source of entry for moulds.
3. Any produce to be stored must be wholesome and healthy. Bruised yam tubers,

cassava, oranges and fruits should never be stored.

4. Avoid drying the produce on a bare floor because of infestation by soil fungi.

5. Hot produce should not be stored. After drying, allow produce to cool before storage.

6. New produce should not be mixed with an old consignment, to avoid cross-infestation.

7. Bagged produce should not be placed on the ground but on raised plank platforms.

8. Overfermentation should be avoided in produce like cocoa, cassava, etc.

9. The store or warehouse should be leak-proof to prevent moisture reabsorption by the already dried produce.

10. Prevent pockets of heavy insect activity by proper application of insect control measure to avoid localized moisture increases and mould growth in the bulk of the grain.

11. In the case of fruits, harvesting should be done promptly as very old fruits are highly susceptible to fungi infection.

12. If possible, dried produce should be stored in airtight conditions to keep away from fluctuating atmospheric relative humidity, which could lead to an increase in moisture content; for example, store in polythene bags or polythene-lined sacks.

Other methods of controlling deterioration of dry produce are:

13. Use of fungicides – in the case of grains not desired for immediate consumption or use, some fungicides such as captan, benomyl, thiobendazole, borax, etc., have been used to control fungal attack, but their use has been limited because of their toxicity.

14. Use of plant materials – parts or roots with medicinal properties can also be used to suppress mould growth in stored crops. Williams and Akano (1985) reported on the efficacy of dogonyaro (neem) as a filtrate in suppressing rotting fungi growth in stored yam tubers.

15. Addition of chemical preservative agents – the addition of antiseptics to foodstuffs allows for better preservation under certain conditions. The use of these products is subject to regulations in most countries. Examples of such chemical preservatives are propionic acid, ascorbic acid, glycerol, sulphur dioxide and benzoic acid. Their use in many instances has been limited to livestock feeds.

16. Other technical methods of control – other methods by which fungal development in stored products can be controlled are refrigeration, irradiation (for yam) and storage in airtight containers and inert atmosphere for grains.

Ogundana *et al.* (1970) found benomyl and thiabendozole effective in reducing the activities of fungi in causing yam rot during storage, but these chemicals are rather toxic. Research is currently in progress at the Nigerian Stored Products Research Institute on the use of safer fungistatic chemicals to preserve yams against microbiological rot during storage. Adesuyi (1973) stored yams successfully for up to 6 months by using a curing method, cutting off sprouts from healthy undamaged tubers and using low temperature and irradiation techniques.

17. Precautions in mycotoxicoses – it is very important to have a control measure in harvesting produce in order to eliminate the fungi causing mycotoxicoses diseases because of their devastating effect on humans and animals that consume such an infected crop. Standard safe limits should be determined and enforced levels of aflatoxin and other toxins in food and feed. Different countries have a wide variety of tolerance level of mycotoxin between 5 and 50 μg/kg (Hansen, 1993). In the USA, the Food and Drug Association has established an aflatoxin limit of 20 μg/kg for food and feed ingredients.

A regular monitoring programme should be arranged for commodities that are susceptible to aflaxtoxin contamination. Processing, packaging, transportation and storage practices should be well managed to eliminate or reduce infestation by moulds, especially the toxigenic strains. Decontamination procedures are to be designed to remove or inactivate the toxins in feed and food. Mycotoxins can be removed from food by detoxification using chemical agents.

Conclusions

The role of fungi in the deterioration of post-harvest crops is enumerated. The contribution of some workers in providing an insight into the deleterious effects of fungi on harvested and stored crops, economic loss, control of fungal deterioration in postharvest crops and precautions in mycotoxicoses diseases is also highlighted. Not forgotten is the most important aspect, the precautions that need to be taken to control or eliminate the fungi causing mycotoxicoses in humans and animals.

It is pertinent to say that knowledge is far from complete and experts should still endeavour to find total solutions to the various aspects of these problems as the struggle of humans against the menace of fungi continues.

References

Adebayo, L.O., Idowu A. and Adesanya, O.O. (1994) Mycoflora and mycotoxins production in Nigeria corn based snacks. *Mycopathologia* 126, 183–192.

Adeniyi, M.O. (1970) Fungi associated with storage decay of yam in Nigeria. *Phytopathology* 60, 590–592.

Adesuyi, D.A.A. (1973) Curing techniques for reducing incidence of rot in yams. *Nigerian Stored Products Research Institute Technical Report* No. 12, 57–63.

Akano, D.A. and Atanda, O.O. (1989) The present level of aflatoxin in Nigeria groundnut cake. *Letters in Applied Microbiology* 10, 187–189.

Broadbent, J.A. (1967) The micoflora germination and seeding vigour of some maize seeds. *Nigerian Stored Products Research Institute Technical Report* No. 15, 113–114.

Clarke, J.H. (1968) Fungi in stored produce. *Tropical Stored Product Institute Technical Report* 15, 2–14.

Coursey, D.G., Summons, E.A. and Sheridan, A. (1963) Studies on the quality of Nigerian palm kernels. *African Science Association* 8,18–28.

Hansen, T.J. (1993) Quantitative testing for mycotoxins in cereal foods. *World* 38, 346–348.

Ibeh, I.N., Urath and Ogonar, J.I. (1994) Dietary exposure to aflatoxin in human male infertility in Benin City, Nigeria. *International Journal of Fertility and Menopausal Studies* 39, 208–214.

Kuku, F.O. (1972) Some mould induced changes in palm kernels. *Nigerian Stored Product Research Institute, Technical Report* No. 9, 69–72.

Ogundana, S.K., Haviq, S.H. and Ekundayo, J.A. (1970) Fungi associated with soft rot of yams (*Dioscorea* spp) in storage. *Nigerian Stored Product Research Institute Technical Report* No. 10, 41–45.

Opadokun, J.S., Ikeorah, J.N. and Afolabi, E. (1979) The aflatoxin contents of locally consumed food stuffs. *Nigerian Stored Product Research Institute Technical Report* No. 12, 105–108.

Oyeniran, J.O. (1970) Microbiological studies on maize used as poultry and livestock feeds at the research Farms in Kandan, Western State. *Nigerian Stored Product Research Institute Technical Report* No. 6, 47–49.

Williams, J.O. and Akano, D.A. (1985) An assessment of wood ash for yam tuber (*Dioscorea rotundata*) in storage. *Nigerian Stored Product Research Institute Report* No. 2, 31–34.

4 Exploitation of Botanicals in the Management of Phytopathogenic and Storage Fungi

Pramila Tripathi[1] and A.K. Shukla[2]
[1]*Department of Botany, D.A.V.-P.G. College, Kanpur, India;* [2]*Department of Botany, Rajiv Gandhi University, Rono Hills, Itanagar, India*

Abstract
Plants are known to contain a number of secondary substances like phenols, flavonoids, quinines, essential oils, alkaloids, saponins, steroids, etc. Some of these plant-based metabolites have antimicrobial properties and are toxic to phytopathogens. They are also repellant to insects and have fumigant toxicity against pests. Currently, synthetic pesticides are the primary means of controlling pathogens. The adverse effects of synthetic pesticides on human health and from the food safety point of view has enunciated interest in finding an alternative means of controlling phytopathogens and pests. To reduce dependency on synthetic pesticides, the use of plant-based antimicrobial substances (essential oils, volatile aromatic compounds, glucosinolates, jasmonates and acetaldehydes) may help in the management of phytopathogens and pests as an alternative method for sustainable agriculture. Use of botanicals is still on a small scale compared to synthetic chemicals; therefore, it is timely to exploit and formulate low-cost, effective, free of human hazard and eco-friendly plant-based products for the management of pests and pathogens.

Introduction

To control fungal diseases, synthetic fungicides are usually applied as effective, dependable and economical control measures. However, the indiscriminate use of chemical fungicides has resulted in several problems, such as toxic residues in food, water and soil and disruption of the ecosystem, leading to the fear that their regular use may harm the environment further. Hardly 0.1% of the agrochemicals used in crop protection reach the target pest, leaving the remaining 99.9% to enter the environment to cause a hazard to non-target organisms, including humans (Pimentel and Levitan, 1986). According to WHO estimates, approximately 0.75 million people are becoming ill every year with pesticide poisoning. Further, the resistance of pathogens to fungicides has rendered certain fungicides ineffective, giving rise to a new physiological race of pathogens. Basic research for over more than 40 years in biology and biochemistry has made it possible to envisage not only how new pesticides may be synthesized but also has generated a completely new approach to the protection of plants using secondary plant products which may be toxic to a specific pest yet harmless to humans. Pesticidal plants have been in nature and its compounds for millions of years without having any ill or adverse effects on the ecosystem and, because of their renewability, they have a distinct advantage in the management of disease-causing pests. Plants have a natural potential to withstand the aggressiveness of pathogenic species.

Plants synthesize a dazzling array of structural variety, which inhibits an almost equally dazzling array of biological activities. A wide spectrum of secondary substances is contained in higher plants, namely phenols, flavonoids, quinines, tannins, essential oils, alkaloids, saponins and steroids. The total number of plant chemicals may exceed 4000 and of these, 1000 are secondary metabolites. These secondary metabolites have a major defensive role for plants (Swain, 1977). The search for botanicals from plant species is one of the important areas where Indian scientists can take a lead and capture the global market. India enjoys the benefits of a varied climate, from an alpine climate in Himalaya to a tropical one in the south and an arid one in Rajasthan to a highly humid climate in Assam and Bengal. This is consequently reflected in the rich and diversified flora, which is often quite distinct, thanks to the natural barriers that India has all along its frontiers. It is estimated that India has about 17,000 species of angiosperms. There is a need for extensive screening programmes at different regional centres of the country so that knowledge on the various types of biological properties of angiospermic flora may be gathered. This type of scientific testing would definitely be helpful in the conservation of plant resources and in proving our sovereign right over our plant biodiversity. Under these conditions, in any meaningful search for better and cheaper substitutes, plant resources for India are a natural choice. Hopefully, this will lead to new information on plant application and a new perspective on the potential use of these natural products. This chapter explores the potential to use a variety of botanicals in the form of plant extracts and essential oils to control various fungal phytopathogens and fungi related to the storage of grains and the postharvest pathogens of perishables.

Essential Oils

Essential oils from different plant species are known to exhibit various kinds of biological activities. The volatility, ephemeral nature and biodegradability of such volatile components of angiosperms will be especially advantageous if they are developed as pesticides (French, 1985). Essential oils are a complex natural mixture of volatile secondary metabolites isolated from plants by hydro or steam distillation and by expression. The main constituents of essential oils are mono- and sesquiterpenes, along with carbohydrates, alcohols, ethers, aldehydes and ketones, polyphenolic compounds, oxides, nitrogen and sulphur compounds and organic acids, etc. The chemical composition of essential oils is extremely complex and varies with the geographical as well as the environmental conditions where the plants are grown (Bhaskara *et al.*, 1998; Vanneste *et al.*, 2002). The essential oils are extracted from various parts of plants such as flowers, fruits, leaves and wood. They are normally formed in special cells or groups of cells or as glandular hairs. Oils occur as a globule or globules in the cell and may also be excreted from cells lining the schizogenous ducts or canals. They may be present in glandular regions such as leaves, bark or fruit and, when occurring in various organs in one plant, may possess different individual chemical compounds (Bonner, 1991; Hili *et al.*, 1997). The general antifungal activity of essential oils is well documented (Tripathi *et al.*, 2007, 2008). These essential oils are thought to play a role in plant defence mechanisms against phytopathogenic microorganisms (Mihaliak *et al.*, 1991). The emerging picture is that certain specific oils and their chemical constituents have traditionally been used to protect stored grains and to repel flying insects in the home and have demonstrable contact and fumigant toxicity to a number of economically important insects and mite pests, as well as to pathogenic fungi. The essential oils or their major constituents could be effective fumigants and also could be integrated with other pest management programmes. Natural pesticides based on plant essential oils could represent alternative crop protectants. The essential oils produced by different plant species are, in many cases, biologically active and have antimicrobial, allelopathic, antioxidant and bioregulatory properties (Caccioni and Guizzardi, 1994; Vaughan and Spencer, 1994). Sometimes, the chemicals in the oil, as well as the oil itself, are registered as pesticide active ingredients. It is also fairly common for two or more oils to be used in the same commercial product. Since the essential oils as such are a mixture of different major and minor components which act synergistically in the biological efficacy of the oil, there would be less chance of the development of physiological races of the target pathogens if the oils as such were formulated as botanical pesticides and fumigants. Essential oils as botanical pesticides may be produced easily, even by small-scale industries, as there is no sophisticated procedure for their distillation and most aromatic plants are available locally. They thus constitute a friendly, natural alternative in pest control.

Essential Oils Against Phytopathogenic Fungi

The antifungal activity of essential oils has been studied by a number of workers (Apablaza *et al.*, 2004; Harish *et al.*, 2004; Muller-Ribeau *et al.*, 1995). Singh *et al.* (1980) found that essential oils from *Cymbopogon* spp. and *Trachyspermum ammi* L. exhibited strong antifungal activity against *Bipolaris oryzae*. Carvone, a monoterpene isolated from the essential oil of *Carum carvi*, was found to inhibit the sprouting of potatoes during storage.

Carvone was also found to have fungicidal activity that helped to protect potato tubers from fungal rotting without exhibiting mammalian toxicity (Hartmans *et al.*, 1995). It has been introduced in the Netherlands under the trade name TALENT. Besides, the essential oils of *Salvia officinalis* have also shown practical potency in enhancing the storage life of some vegetables by protecting them from fungal rotting (Bang, 1997). Powdery mildew of *Cucurbita maxima* is caused by *Sphaerotheca fuliginea*. *Reynoutria* extracts and olive oil were found to be effective in controlling the disease (Cheah and Cox, 1995). Since olive oil is used in cooking, food additives and medicines, it does not cause any human health or environmental problems. Recent studies in Ghana confirm that *Ocimum gratissimum* and *Syzigium aromaticum* are very effective in preventing fungal growth (FAO, 1999).

Essential Oils Against Fungal Pathogens of Seeds

The fungicidal effect of essential oils against pathogens of cereal grains has been tested successfully. It is especially significant in the case of stored rice, where currently fungicides are not used to control fungal pests. Peppermint (*Mentha piperata*), thyme (*Thymus copitatus*) and caraway (*C. carvi*) oils have demonstrated effective control against fungal pathogens like *Fusarium* sp., *Macrophomina phaseolina* and *Colletotrichum dematium* (Abdelmonem *et al.*, 2001). Essential oils from oregano (*Origanum vulgare*) and thyme were applied as fumigants against the mycelia and spores of *Aspergillus flavus, A. niger* and *A. ochraceus* infesting wheat grains. Only oregano essential oil exhibited fungicidal activity (Paster *et al.*, 1995). The antifungal activity of the essential and fixed oils of thyme, clove, peppermint, soybean and groundnut were tested against *A. flavus, A. niger, F. oxysporum, F. equiseti* and *Penicillium chrysogenum in vitro* on the cowpea (*Vigna unguiculata*) (Kritzinger *et al.*, 2002). Thyme and clove oils inhibited growth of all the fungi significantly at concentrations of 500 and 1000 ppm. Peppermint oil inhibited growth of the above-mentioned fungi successfully at 2000 ppm (Kritzinger *et al.*, 2002). In blackgram (*V. mungo*), essential oil extracted from wood chips of cedar (*Cedrus deodara*) and that from seeds of *T. ammi* exhibited antifungal activity, inhibiting the mycelial growth of *A. niger* and *Curvularia ovoidea*, two storage fungi found on seeds (Singh and Tripathi, 1999). *A. flavus* was also found infesting seeds of guar (*Cyamopsis tetragonoloba*), a native plant of India which has main commercial value due to its seed gum (galactomannan gum). In this case, *A. flavus* was controlled by cumin (*Cuminum cyminum* L.) oil extracted from its seeds (Dwivedi *et al.*, 1991). Chemical studies indicated that the greater part of this antimicrobial activity might be attributed to the cuminaldehyde that is present in the dried fruit of this plant (De *et al.*, 2003). The essential oils of *Cassulia allaris* and *M. arvens* have been reported as botanical fumigants for management of the biodeterioration of wheat from *A. flavus* (Varma and Dubey, 2001).

Essential Oils Against Aflatoxicogenic and Mycotoxicogenic Fungi

The aflatoxins are well known for their carcinogenic, mutagenic and teratogenic effects on humans and domestic animals (Wyllie and Morehouse, 1978). A natural fungicide against aflatoxigenic fungi to protect stored rice using the essential oil of lemongrass (*C. citrates*) was developed by Paranagama *et al.* (2003). Lemongrass oil was tested against *A. flavus* and the test oil was fungistatic and fungicidal against the test pathogen at 0.6 and 1.0 mg/ml, respectively. Aflatoxin production was inhibited completely at 0.1 mg/ml. Citral has been found as a fungicidal compound in lemongrass oil. During the fumigant toxicity assay of lemongrass oil, the sporulation and mycelial growth of the test pathogen were inhibited at a concentration of 2.80 and 3.46 mg/ml, respectively. Lemongrass oil could be used to manage aflatoxin production and to inhibit the fungal growth of *A. flavus* in stored rice.

Putative mycotoxicogenic fungi were partially or completely sensitive to different essential oils extracted from different medicinal plants (Soliman and Badeaa, 2002). Seed treated with cinnamon, palmarosa and lemongrass oils at 500 mg/kg showed antimycotoxigenic ability against fumonisin B1 accumulation produced by *F. vesticillioides* and *F. proliferatum* (Marin *et al.*, 2003). Velutti *et al.* (2004) reported antimycotoxicogenic activity of the essential oils against *F. graminearum* infested seeds. The essential oils of oregano, cinnamon, lemongrass, clove and palmarosa effect the growth rate of *F. graminearum* and mycotoxin Zearalenone (ZEA) and Deoxynivalenol (DON) production at two concentrations (500 and 1000 mg/kg).

Plant Extracts Against Phytopathogenic Fungi

The preservative nature of some plant extracts has been known for centuries and there has been renewed interest in the antimicrobial properties of extracts from aromatic plants. The application of the extracts of higher plants to control plant diseases was first attempted by Democritus as early as 470 BC. Plant extracts have assumed special significance nowadays as an eco-friendly method for plant disease management. Plants contain alkaloids, tannins, quinines, coumarins, phenolic compounds, phytoalexins and ipomeamarone in the extract, which are known for their antifungal property (Datar, 1999). Use of plant extracts for seed treatment is one of the alternative methods of preventing pathogen problems of agricultural crops. Plant materials as such can be used as soil amendments that can serve as both a nutrient as well as an antifungal agent. Plant extracts have also been reported to stimulate the growth of targeted plant species. This is probably due to some hormones and allied substances like IAA, IBA, etc.

However, the active principles of some plants have been isolated phytochemically and have shown a strong inhibitory action against a number of fungi. Antifungal activity of plant extracts against a wide range of fungi has been reported by a number of workers (Grange and Ahmed, 1988; Davidson and Parish, 1989). Bhargava *et al.* (1981) screened extracts of some plant species and found *O. canum* to be most effective against *A. flavus* and *A. versiolor*. Pandey *et al.*

(1982) evaluated the seed extract of 30 plants and found soybean, *Leonotis nepetaefolis*, *Parpalum* and *Peltophorum* to exhibit an inhibitory effect against the fungi, *Alternaria alternata* and *A. niger*. Ark and Thompson (1959) found the leaf extract of *Allium sativum* to be effective against various plant pathogens. *Acacia nilotica* (leaf and bark) and *A. farnasiana* (bark) of *Mimosaceae* showed high activity, while *A. catechu* of the same family did not show activity either from leaf or from bark (Tripathi, 2005). Four compounds, i.e. iritin A, iritin B, flavonone-dehydroulogonin and sesquiterpene pygmol, were isolated with dichloro-methane extract of the aerial parts of *Chenopodium procerum*. These compounds have been found to inhibit the growth of the plant pathogenic fungi, *Cladosporium cacumerinum* (Bergeron *et al.*, 1995). Kim *et al.* (2004) evaluated *Achyranthus japonica* and *Rumex crispus* for activity against various plant pathogenic fungi and control of powdery mildew. Methanol extract of the fresh material of 183 plants was screened *in vivo* for antifungal activity against *Magnaporthe grisea, Corticium sasaki, Botrytis cinerea, Phytophthora infestans, Puccinia recondita* and *Erisiphe graminis*. Among them, 33 plant extracts showed disease control efficacy. The methanol extract of *Achranthes japonica* (whole plant) and *R. crispus* (roots) at a concentration greater than 11 g fresh weight of plant tissue per litre aqueous Tween 20 solution controlled the development of barley powder mildew caused by *E. graminis* effectively in an *in vivo* assay using plant seedlings. Some fungi like *F. solani* and *Verticillium alboatrum* have been shown to be susceptible to tannins extracted from the bark of various trees, including chestnut and wattle (Lewis and Papavizas, 1967). The effects of aqueous and methanol, petroleum ether, chloroform and ethyl acetate extracts of *Cyprus rotundus* were tested on spore germination of *F. solani*. Ethyl acetate extract exhibited an inhibitory effect on spore germination at 1000 μg/ml (Singh and Tripathi, 1999). In the field, reduction of disease incidence has been recorded as a result of plant seed treatment with extract, and an increase in yield was also noted.

Plant Extracts in the Management of Fungal Seed Diseases

Cereal seeds carry a wide range of fungi that are known to play a significant role in spoilage and probably rank second only to insects as a cause of deterioration and loss in all kinds of field and storage crops throughout the world (Christensen and Kaufman, 1974). The information on fungal association with important cereal grains is relevant in assessing the potential risk of mycotoxin contamination. In recent years, the use of plant extracts for controlling fungal seed disease has also been of renewed interest. Carvone (monoterpene compound) completely inhibited *F. oxysporum* and *A. pisi*. African yam bean, *Sphenostylis stenocarpa,* is an important grain legume in most tropical African countries (Nwachukwu and Umechuruba, 2001). Major pathogenic fungi associated with this crop are *A. niger, A. flavus, Lasiodiplodia theobromae* and *F. moniliforme*. Associated fungi could be controlled by using crude and aqueous extract of basil (*O. basilicum*), bitter leaf (*Vernonia amydalina*), neem and pawpaw (*Carica papaya*). Parimelazhagan and Francis (1999) reported reduction in the radial growth of *Curvularia lunata* associated with rice seeds when treated with leaf extract of *Clerodendrum viscosum*, which also increased seed germination, root and shoot length of the rice. The same results were observed by using plant extracts to control *B. oryzae* on rice seeds, which have a high natural infection of the fungus (Alice and Rao, 1986). In Bangladesh, use of the extract of *Polygonum hydropiper, A. cepa, A. sativum* and *A. jidia* demonstrated to be effective against *B. oryzae* at higher concentrations. Among them, neem and garlic were the most effective at 1:1 dilution and inhibited the occurrence of the pathogen by 91 and 83%, respectively (Ahmed *et al.*, 2002).

Alternaria padwickii, another important seedborne pathogen of rice, was also inhibited by aqueous extract of *Strychnos nux-vomica*, garlic bulbs, ginger rhizome, basil leaves and fruits of *A. indica* (Shetty *et al.*, 1989). The ability of natural plant extracts to prevent the growth of fungi naturally infesting grains was also studied. Before sowing, wheat seeds were soaked in an aqueous plant extract of *O. gratissimum* and disease transmission was evaluated. The rate of infection decreased with the extract at concentrations higher than 10% (Rodrigues *et al.*, 2001). Leaf extracts of *Delonix regia, Pongamia glabra* and *A. nilotica* significantly inhibit spore germination, mycelial growth and spore production of *A. helianthi, M. phaseolina* and *F. solani* from sunflower seeds (Tribuhavanaamala and Narsimhan, 1998). Melon seeds are very important as as condiment and constitute a very valuable source of oil and protein for many people of West Africa (Oyolu, 1977). After 6 months of incubation, all the melon seeds treated with leaf extract showed no infection except *M. phaseolina*. Ahmad and Prasad (1995) evaluated that post-infection treatment of sponge-gourd fruits with the extracts of *Azadirachta indica, Lantana camara, Murraya exotica, O. sanctum, Datura fistulosa* and *Catharanthus roseus* almost fully inhibited the spread of disease caused by *Helminthosporium spiciferum* and *F. scirpi*.

Application of Botanicals in Seed Storage

Quality seed should have higher vigour and viability and these two characteristics cannot be maintained in storage because they deteriorate rapidly under storage conditions and suffer quantitative and qualitative losses due to pests and diseases. Therefore, treating seeds with synthetic chemicals is vital for successful storage. However, these chemicals are hazardous to humans. Therefore, use of natural plant products for long-term seed storage has multi-purpose benefits as eco-friendly protection against the ageing process, prevention of insects and fungi and for their cost effectiveness (Vanangamudi *et al.*, 2007). During storage, the enzymatic activity (amylase, catalase, peroxidase, superoxide dismutase and dehydrogenase) responsible for maintenance of seed quality and the antioxidant (ascorbic acid) contents are maintained at optimum level in botanically treated seeds (Umarani, 1999). Common botanicals, arrapu (*Abizia amaru*), neem (*A. indica*), notchi (*Vitex negundo*), *Prosopis* sp., pungam (*Pongamia glabra*), moringa and tamarind, contain an auxin-like substance which regulates seedling growth and initial establishment. In botanicals, a gibberellin-like substance is also present in addition to saponin and other nutrients, which interact with amino acids, tryptophane to form the indole acetic acid (IAA), which leads to release of plant hormones that are responsible in cell elongation and vegetative growth. In botanical seed pelleting, the leaf powder acts as a water pad by absorbing/regulating soil moisture availability, which enhances a better seed–soil relationship (Narasimha, 1994). Seeds are stored by pelleting them with botanical products. The aim of botanical pelleting in seed storage is to extend storage potential, besides maintaining its ability to produce normal seedlings. Jegathambal (1996) found that sorghum seeds hardened and pelleted with arappu leaf powder could be stored for up to 2 weeks with higher germinability. Papaya seeds pelleted with botanicals or presoaked with botanicals gave improved germination, vigour index and field emergence when compared to the control or water socking (Ananthakalaiselvi, 1995). Dry dressing of seeds with botanicals prolongs the storability of the seeds in many crops, especially in pulses, and acts as a dual-purpose technologically for seed storage by preventing biotic organisms attacking the seeds during storage. Sabir (1989) reported that soybean seeds treated with sambangi (*Polianthes tuberose*) seed powder at a ratio of 1:100 maintained a higher germination rate (70%), even up to 8 months after storage. Pea seeds dried and mixed with notchi (*V. negundo*) powder or sambangi seed powder at a ratio of 1:100 maintained a higher germination rate after up to 8 months in storage (Paramasivam, 1990). Umarani (1999) reported that dressing dried *Casuarina* seeds with neem leaf powder extended the storability of the seeds for up to 9 months.

Biocide Formulation of Essential Oils

The formulation of plant metabolites must be introduced to overcome their degradation and to be used practically during handling and application as biocides. Such formulation could be used easily and diluted with water to form the appropriate concentrations in different applications. Study should be continued to evaluate the pesticidal activity of the produced formulated biocides against some plant pathogenic microorganisms. Narsimhan *et al.* (1988) demonstrated that neem oil (*A. indica*) and pungam oil (*P. pinnata*) emulsifiable concentrate formulation prevented sheath rot (*Sarocladium oryzae*) of rice. Gascon *et al.* (1999) showed that the essential oils of rosemary, jarilla, mendocina, tomillo mendocina, origanum, tarragon, lavandins and eucalyptus were emulsified with different formulations of water suspensions of wall support systems using both a hand-held propeller blender and a high pressure, double effect homogenizer. Also, Bowers and Locke (2000) report that several commercial formulations of botanical extracts and essential oils have been investigated as possible alternatives for soil fumigation to control *Fusarium* wilt disease. Essential oils of fennel, peppermint and caraway have been formulated in the form of stable emulsifiable concentrates.

Botanicals in the Management of Postharvest Diseases of Perishables

Botanicals as antifungal agents in postharvest disease control of fruits

Fruits and vegetables have a number of constituents and inducible volatile aromatic and flavour compounds (Tripathi, 2007). These aromatic and flavour components are generally produced by fruits during ripening and provide resistance to the fruits at the postharvest stage. The flavour compounds are secondary metabolites having unique properties of volatility and low water solubility. As potential fungicides, their natural occurrence as part of the diet, their ephemeral nature and their biodegradability suggest low toxic residue problems. Such compounds could be extracted and applied to other harvested perishables. Some of the volatile aromatic components, namely acetaldehyde, 6-carbon (C_6) aldehydes, benzaldehyde, hexenel and hexanal, are of significant importance.

Aldehydes

Vapours of acetaldehyde have been used to control *B. cinerea* (Prasad and Stadelbacher, 1973). Avissar and Pesis (1991) reported acetaldehyde to be active against *B. cinerea* and *Rhizopus stolonifer* causing rot to strawberry fruits. Benzaldehyde has been used in the laboratory to fumigate peaches and to protect them against *Rhizopus* rot. It inhibits spore germination of *B. cinerea* totally at 25 μl/l and germination of *Monilinia fructicola* at 125 μl/l (Wilson *et al.*, 1987). The aldehydes, benzaldehyde, acetaldehyde and cinnamaldehyde, ethanol and benzyl alcohol were found to be the strongest growth inhibitors and the most lethal to fungal spores and mycelia of fruit and vegetable pathogens like *P. digitatum*, *R. stolonifer* and *Colletotrichum* during *in vitro* trials.

Hexenal and hexanal

(*E*)-2-Hexenal and hexanal are two different volatile flavour compounds. Hexenal vapours have a number of attributes that may be important in consumer demand for more natural measures to combat fruit diseases with fewer toxic residues. Hexenal vapour inhibited hyphal growth of apple slices (Song *et al.*, 1996). Archbold *et al.* (1999) showed (*E*)-2-hexenal to be an efficient fumigant in controlling mould on 'Crimson Seedless' table grapes. (*E*)-2-Hexenal has been found to be strongly antifungal in nature and its *in vitro* and *in vivo* activity against *B. cinerea* has been reported by a number of workers (Hamilton-Kemp *et al.*, 1992; Fallik *et al.*, 1998). The effect of *trans*-2-hexenal on the control of blue mould disease (*P. expansum*) in reducing patulin content and on improving the fruit quality of 'Conference' pears has been evaluated and greater reduction of decay was obtained by treatment at 12.5 μl/l at 20°C for 24 or 48 h after inoculation (Neri *et al.*, 2006).

Acetic acid

Acetic acid is a metabolic intermediate that occurs naturally in many fruits (Nursten, 1970). There are several advantages in using acetic acid fumigation. It is a natural compound found throughout the biosphere, posing little or no residual hazard. Low concentrations, i.e. 2.0 or 4.0 mg/l, of acetic acid in air have been found to be extremely effective for controlling *B. cinerea* conidia on apple ('Red Delicious') fruit (Sholberg and Gaunce, 1995). Acetic acid has been shown to be an effective fumigant for commercial use on apricot and plums (Liu *et al.*, 2002), grapes (Sholberg *et al.*, 1996) and sweet cherries (Sholberg, 1998; Chu *et al.*, 1999, 2001). The use of acetic aid and vinegar is the better choice in most cases because it does not have an objectionable odour and has a long history of use on food (Sholberg *et al.*, 2000).

Jasmonates

The term 'jasmonates' includes jasmonic acid (JA) and methyl jasmonate (MJ). These are naturally occurring plant growth regulators that are widely distributed in the plant kingdom and are known to regulate various aspects of plant development and responses to environmental stresses (Sembdner and Parthier, 1993; Creelman and Mullet, 1995, 1997). Droby *et al.* (1999) found that postharvest application of jasmonates reduced decay caused by grey mould, *P. digitatum,* either after natural or artificial inoculation of 'Marsh Seedless' grapefruit. When applied at low concentrations, jasmonates are potential postharvest treatments to enhance natural resistance and to reduce decay in fruit. Since they are naturally occurring compounds and are given in low doses, jasmonates may provide a more environmentally friendly means of reducing the current chemical usage.

Glucosinolates

Among natural substances with potential antimicrobial activity are the glucosinolates, a large class of approximately 100 compounds produced by members of the family *Cruciferеae*, with well-documented activity (Fenwick *et al.*, 1983). Hydrolysis of glucosinolates produces D-glucose, sulphate ion and a series of compounds such as sothiocyanate (ITC), thiocyanate and nitril. The antifungal activity of six glucosinolates has been tested on several postharvest pathogens, namely *B. cinerea*, *R. stolonifer*, *M. laxa*, *Mucor piriformis* and *P. expansum*, both *in vitro* (Mari *et al.*, 1993) and *in vivo* (Mari *et al.*, 1996). Allyl-isothiocyanate (AITC), a naturally occurring flavour compound in mustard and horseradish, has a well-documented antimicrobial activity. Exposure of pear fruit to an AITC-enriched atmosphere resulted in good control of blue mould, including a TBZ resistant strain on pears (Mari *et al.*, 2002). The use of AITC, produced from purified sinigrin or from *Brassica juncea*, against *P. expansum* appears very promising as an economically viable alternative with moderately low impact on the environment.

Essential oils

The antimicrobial effects of essential oils (EOs) or their constituents on postharvest pathogens have been studied quite extensively (Bishop and Thornton, 1997; Tripathi *et al.*, 2007). The advantage of EOs is their bioactivity in the vapour phase, a characteristic that makes them attractive as possible fumigants for stored product protection. Control of the storage pathogen, *B. cinerea,* on Dutch white cabbage (*B. oleracea* var. *capitata)* by the EOs of *Melaleuca alternifolia* in *in vitro* conditions has been investigated (Bishop and Reagon, 1998). Tripathi *et al.* (2008) evaluated some EOs against moulds of grapes caused by *B. cinerea*. The effect of *C. nardus* EO on the growth and morphogenesis of *A. niger* has been tested (Bellerbeck *et al.*, 2001). The potential of using EOs by spraying or dipping to control postharvest decay has been examined in fruits, namely cherries, citrus fruits, apple, peaches and cabbage (Tiwari *et al.*, 1988; Smid *et al.*, 1994; Dixit *et al.*, 1995). Thymol is an EO component from thyme (*T. capitatus*). Fumigation of sweet cherries with thymol was effective in controlling postharvest grey mould rot caused by *B. cinerea* (Chu *et al.*, 1999) and brown rot caused by *M. fructicola* (Chu *et al.*, 2001). The shelf life and safety of some perishable foods treated with EOs have been improved remarkably (Ponce *et al.*, 2004; Holley and Patel, 2005). The EO of *S. officinalis* has also shown practical potency in enhancing the storage life of some vegetables by protecting them from fungal rot (Bang, 1995). Treatment of oranges by fumigation with the EOs of *M. arvensis* (100 μl/l), *O. canum* (200 μl/l) and *Zingiber officinale* (200 μl/l) has been found to control blue mould, thereby enhancing shelf life (Tripathi *et al.*, 2004). Plaza *et al.* (2004) evaluated the potential of thyme, oregano, clove and cinnamon EOs against *P. digitatum* and *P. italicum* on citrus fruits. The postharvest quality of strawberry and tomato fruit was evaluated after treatment with *Eucalyptus* and cinnamon volatile EO vapours (Tzortzakis, 2007).

Plant extracts

Some plants extracted in different organic solvents have shown inhibitory action against different storage fungi (Singh *et al.*, 1993; Hiremath *et al.*, 1996; Rana *et al.*, 1999; Okigbo and Pandalai, 2005). The inhibitory effect of water-soluble extracts of garlic bulbs, green garlic, green onions, hot peppers, ginger, Chinese parsley and basil on the growth of *A. niger* and *A. flavus* was examined. Garlic bulbs, green garlic and green onions showed an inhibitory effect against these two fungi (Yin and Cheng, 1998). Treatment of pineapple fruits infested with *C. paradoxa* by *X. strumarium* extract reduced the severity of the disease (Damayanti *et al.*, 1996). The phytochemical investigation of a methanolic extract of *A. nilotica* resulted in isolation of kaempferol. It has shown antifungal activity against *P. italicum* at 500 μg/l (Tripathi *et al.*, 2002). *In vitro* inhibition of *B. theobromae* causing Java black rot in sweet potato was induced by phenolic compounds, chlorogenic acid giving the highest *in vitro* inhibition, followed by pyrogallol, pyrocatechol, phenol and resorcinol. Low concentrations of phenols are required by the fungus during normal metabolism, but higher concentrations are inhibitory to growth (Mohapotra *et al.*, 2000). The phytochemical investigations of most plants have resulted in the isolation of active principles. These compounds when tested against postharvest fungi have shown pronounced antifungal activity. A naturally occurring compound isolated from the flavedo tissue of 'Star Ruby' grapefruit (*Citrus paradise*) identified as 7-geranoxy coumarins exhibited antifungal activity against *P. italicum* and *P. digitatum* during *in vitro* and *in vivo* tests (Agnioni *et al.*, 1998). Arya (1988) controlled fruit rots by leaf extracts of medicinal plants.

Mode of Action of Essential Oils

The mechanism of action of EOs and other bioactive phytocompounds against microorganisms is a complex process and has not yet been fully explained. It is generally recognized that the antimicrobial action of essential oils depends on their hydrophobic or lipophilic character. Terpenoids may serve as an example of lipid-soluble agent that affects the activities of membrane catalysed enzymes; for example, their action on respiratory pathways. Compounds of EOs either affect the physiological function of microorganisms or cause structural changes of hyphae and spores (Thompson, 1986; Arras *et al.*, 1993; Zambonelli *et al.*, 2004). For instance, the effect of thyme oils and thymol on the hyphae cytomorphology of *F. solani*, *R. solani* and *C. lindemuthianum* increased vacuolization of the cytoplasm and accumulation of lipid bodies, undulation of the plasmalemma and alteration of the mitochondrial and endoplasmic reticulum (Zambonelli *et al.*, 2004). However, variations in the fungicidal action of the compounds seem to depend on solubility, as well as on the capacity to interact with cytoplasmic membrane.

Conclusions

Sustainable agriculture in the 21st centaury will rely increasingly on alternative interventions for pest management that are environmentally friendly and reduce the amount of human contact with chemical pesticides. The use of botanicals in crop protection has now gained popular ground in the world of agriculture as an alternative to the use of toxic, persistent and synthetic compounds. Several factors are now responsible for making the use of alternative methods more attractive. A number of studies have been conducted on the use of botanicals and several plants with promising biocidal properties have been identified. Most of these plants have also been used *in vitro* and *in vivo* in the control of various plant diseases. Certain plant EOs and plant extracts have a broad spectrum of activity against plant pathogenic and other fungi. They have considerable potential as crop protectants. Current information indicates that they are safe to the user and the environment, with few qualifications. With the modern techniques now available and the attention being given to this area, we look forward to intensifying development of the biological activity of botanicals so as to exploit them as fungicides. A consolidated and continuous search for natural products may yield safer alternative control measures like azadirachtin and pyrethroids, which are being used. However, in order to consider the use of any plant material seriously, further information is required. The use of locally available plants avoids the need to establish complex mechanisms for pesticide distribution; the community can collect or grow the plants itself. Effective antifungal plant compounds that can fill the void of phased-out chemicals will require some advances in the study of regional aromatic plants, their production, formulation and possible beneficial mechanism to prevent or control fungal attack, better understanding of how they will fit into integrated systems and thoir interaction with the environment and other IPM components and identification of the optimum concentration of EOs that can control seedborne fungal pathogens without affecting seed germination and seedling growth. The information on the active principles present in various botanicals on germination and seedling vigour is to be elucidated. A proper study of the mode of action and structure activity relationship will bring about a new class of interesting compounds for future pest control. Despite the common belief that phytocompounds are safe, they all have inherent risk, just like synthetic compounds. Thus, it is within the scope of phytoscientists to elucidate the side effects and appropriate doses and identify bioactive phytocompounds and ways of extraction and conservation. As a cautionary note, the EOs that are the most efficacious against pests are often the most phytotoxic. This phytotoxicity requires serious attention when formulating products for agricultural use. Also, selectivity among invertebrates is not well documented. Honeybees appear somewhat susceptible (Lindberg *et al.*, 2000). The susceptibility of various natural enemies has yet to be reported, although the lack of persistence of EOs under field conditions could provide some information on temporal selectivity favouring non-target species. Finally, we should maintain our efforts in considering and valorizing our natural patrimony, as well as conducting more scientific research on aromatic plants for chemical analysis and biological, toxicological and pharmacological investigation of therapeutic aspects. It is important to remember that just because a pesticide is derived from a plant does not mean that it is safe for humans and other mammals, or that it cannot kill a wide variety of other life. Some botanical pesticides can be quite toxic to humans and should not be used on plants for human consumption. For example, methyl salicylate (oil of wintergreen) is commonly used as food flavouring, but it can be quite toxic in large doses (Jonathan and Davis, 2007). Few systematic studies have been conducted to determine how farmers use plant protectants, their effectiveness and method of application. The introduction of rapid rural appraisal (RRA) and participatory rural appraisal (PRA) techniques will facilitate the collection of this type of information. For utilization of botanicals on an industrial scale, it may be necessary to obtain such secondary metabolites from tissue culture-derived materials. There are many advantages to this method of production, including immediate response to an increase in demand irrespective of season, freedom from climatic stresses, pests and diseases and product formation in a clear, sterile environment.

References

Abdelmonem, A.M., Rasmy, M.R. and Hanafy, S.M. (2001) Non-chemical trials to minimize seedborne infection of some seed crops. *Proceedings of 26th ISTA Congress*, Angers, France, 18–20 June, p. 58.

Agnioni, A., Cabras, P., Dhallewin, G., Pirisi, F.M., Reniero, F. and Schirra, M. (1998) Synthesis and inhibitory activity of 7-geranoxy coumarin against *Penicillium* species in citrus fruits. *Phytochemistry* 47, 1521–1525.

Ahmad, S.K. and Prasad, J.S. (1995) Efficacy of foliar extracts against pre- and post-harvest diseases of sponge-gourd fruits. *Letters in Applied Microbiology* 2, 373–375.

Ahmed, M.F., Khalequzzaman, K.M., Islam, M.D., Anam, M.K. and Islam, M.T. (2002) Effect of plant extract against *Bipolaris oryzae* of rice under in vitro conditions. *Pakistan Journal of Biological* Sciences 5, 442–445.

Alice, D. and Rao, A.V. (1986) Management of seedborne *Drechslera orzae* of rice with plant extracts. *International Rice Research Newsletter* 11, 19–24.

Ananthakalaiselvi, A. (1995) Seed technological studies in cowpea (*Vigna unguiculata* L. Walp) cv. CO 5. MSc (Ag.) thesis, Tamil Nadu Agricultural University, Coimbatore, India.

Apablaza, G.E., Moyay, E.A. and Martin, R.M.S. (2004) Observacion microscopica del efecto de control de un extracto de quillay sobre oidio de las Cucurbitaceas. *Fitopatologia* 39,144–149.

Archbold, D.D., Hamilton-Kemp, T.R., Clements, A.M. and Collins, R.W. (1999) Fumigating 'Crimson Seedless' table grapes with (*E*)-2-hexenal reduces mold during long-term post harvest storage. *HortScience* 34, 705–707.

Ark, P.A. and Thompson, J.P. (1959) Control of plants with antibiotics from garlic (*Allium sativum*). *Plant Disease Reporter* 43, 276–282.

Arras, G., Piga, A. and Dhallewin, G. (1993) The use of Thymus capitatus essential oil under vacuum conditions to control *Penicillium digitatum* development on citrus fruit. *Acta Horticulturae* 344, 147–153.

Arya, A. (1988) Control of Phomopsis fruit rot by leaf extracts of certain medicinal plants. In: Kausik, P. (ed.) *Indigenous Medicinal Plant Symposium.* Today and Tomorrow Print and Publishers, New Delhi, pp. 41–46.

Avissar, I. and Pesis, E. (1991) The control of postharvest decay in table grapes using acetaldehyde vapours. *Annals of Applied Biology* 18, 229–227.

Bang, U. (1995) Essential oils as fungicides and sprout inhibitors in potatoes. In: Dowley, L.J., Bannon, E., Cooke, L.R., Keane, T. and O'Sullivan, E. (eds) *Phytophthora infestans* 150. Boole Press Ltd, Dublin, pp. 319–324.

Bang, U. (1997) Control of storage diseases of potato sprout inhibition by plant volatiles. In: Baser, K.S.H. and Kimmer, N. (eds) *Progress in Essential Oil Research. Proceedings of the 25th International Symposium on Essential Oils,* 1–3 September, 1997, Ekiselir, Turkey.

Bellerbeck, V.G., De Roques, C.G., Bessiere, J.M., Fonvieille, J.L. and Dargent, R. (2001) Effect of *Cymbopogon nardus* (L) W. Watson essential oil on the growth and morphogenesis of *Aspergillus niger. Canadian Journal of Microbiology* 47, 9–17.

Bergeron, C., Marston, A., Hakizamungu, E. and Hostettmann, K. (1995) Antifungal constituents of *Chenopodium procerum. International Journal of Pharmacognosy* 33, 115–119.

Bhargava, K.S., Dixit, S.N., Dubey, N.K. and Tripathi, R.D. (1981) Fungitoxic properties of *Ocimum canum. Journal of Indian Botanical Society* 60, 24–27.

Bhaskara, R.M.V., Angers, P., Gosselin, A. and Arul, J. (1998) Characterization and use of essential oil from *Thymus vulgaris* against *Botrytis cinerea* and *Rhizopus stolonifer* in strawberry fruits. *Phytochemistry* 47, 1515–1520.

Bishop, C.D. and Reagon, J. (1998) Control of the storage pathogen *Botrytis cinerea* on Dutch white cabbage (*Brassica oleracea* var. *capitata*) by the essential oil of *Melaleuca alternifolia*. *Journal of Essential Oil Research* 10, 57–60.

Bishop, C.D. and Thornton, I.B. (1997) Evaluation of the antifungal activity of the essential oils of *Monarda citriodora* var. *citriodora* and *Melaleuca alternifolia* on the postharvest pathogens. *Journal of Essential Oil Research* 9, 77–82.

Bonner, J. (1991) Biogenesis of Natural Compounds (Bernfield, P. [ed]). Pergamon Press, London, pp. 241.

Bowers, J.H. and Locke, L.C. (2000) Effect of botanical extracts on the population density of *Fusarium oxysporum* in soil and control of Fusarium wilt in the greenhouse. *Plant Disease* 84, 300–305.

Caccioni, D.R.L. and Guizzardi, M. (1994) Inhibition of germination and growth of fruit and vegetable postharvest pathogen fungi by essential oil components. *Journal of Essential Oil Research* 8, 399–404.

Cheah, L.H. and Cox, J.K. (1995) Screening of plant extract for control of powdery mildew in squash. *Proceedings of the 48th New Zealand Plant Protection Conference.* N.Z. Institute for Crop and Food Research, Levin. New Zealand Plant Protection Society, pp. 340–342.

Christensen, C.M. and Kaufman, H.H. (1974) *Storage of Cereal Grains and Their Products.* American Association of Cereal Chemistry, St Paul, Minnesota, 549 pp.

Chu, C.L., Liu, W.T., Zhou, T. and Tsao, R. (1999) Control of postharvest grey mold rot of modified atmosphere packaged sweet cherries by fumigation with thymol and acetic acid. *Canadian Journal of Plant Science* 79, 686–689.

Chu, C.L., Liu, W.T. and Zhou, T. (2001) Fumigation of sweet cherries with thymol and acetic acid to reduce postharvest brown rot and blue mold rot. *Fruits* 56, 123–130.

Creelman, R.A. and Mullet, J.E. (1995) Jasmonic acid distribution in plants: regulation during development and responses to biotic and abiotic stress. *Proceedings of the National Academy of Sciences of the United States of America* 92, 4114–4119.

Creelman, R.A. and Mullet, J.E. (1997) Biosynthesis and action of jasmonates in plants. *Annual Review of Plant Physiology and Plant Molecular Biology* 48, 355–381.

Damayanti, M., Sucheela, K. and Sharma, G.J. (1996) Effect of plant extracts and systemic fungicides on the pineapple fruit rotting fungus *Ceratocystis paradoxa*. *Cytobios* 86, 155–165.

Datar, V.V. (1999) Bioefficacy of plant extracts against *Macrophomina phaseolina* (Tansi) Goid. The incitant of charcoal rot of sorghum. *Journal of Mycology and Plant Pathology* 29, 251–253.

Davidson, P.M. and Parish, M.E. (1989) Methods for testing the efficacy of food antimicrobials. *Food Technology* 43, 148–155.

De, M., De, A.K., Mukhopadhyay, R., Banerjee, A.B. and Miro, Y.M. (2003) Actividad antimicrobiana de Cuminum cyminum L. *Ars Pharmeceutica* 44, 257–269.

Dixit, S.N., Chandra, H., Tiwari, R. and Dixit, V. (1995) Development of botanical fungicide against blue mold of mandarins. *Journal of Stored Product Research* 31, 165–172.

Droby, S., Porat, R., Cohen, L., Weiss, B., Shapira, B., Philosoph-Hadas, S. and Meir, S. (1999) Suppressing green mold decay in grapefruit with postharvest jasmonates application. *Journal of the American Society for Horticultural Science* 124, 184–188.

Dwivedi, S.K., Pandey, V.N. and Dubey, N.K. (1991) Effect of essential oils of some higher plants on *Aspergillus flavus* Link. infesting stored seeds of gaur (*Cyamopsis tetragonoloba* L.). *Flavour Fragrance Journal* 6, 295–297.

Fallik, E., Arhbold, D.D., Hamilton-Kemp, T.R., Clements, A.M., Collins, R.W. and Barth, M.E. (1998) (*E*)-2-Hexenal can stimulate *Botrytis cinerea* growth *in vitro* and on strawberry fruit *in vivo* during storage. *Journal of the American Society for Horticultural Science* 123, 875–881.

FAO (1999) The use of spices and medicinals as bioactive protectants for grains. *FAO Agricultural Services Bulletin* 137, 249–254.

Fenwick, G.R., Heaney, R.K. and Mullin, W.J. (1983) Glucosinolates and their breakdown products in food and food plants. *CRC Critical Reviews in Food Science and Nutrition* 18, 123–201.

French, R.C. (1985) The bioregulatory action of flavour compounds on fungal spores and other propagules. *Annual Review of Phytopathology* 23, 173–199.

Gascon, A.D., Zuritz, C.A., Bustamantc, J.A., Borbon, L. and Oberti, G. (1999) A study of different formulations of wall support systems for micro-capsulation of antioxidante essential oils. *Acta Horticulture* 503, 53–58.

Grange, M. and Ahmed, S. (1988) *Handbook of Plants with Pest Control Properties*. John Wiley and Sons, New York.

Hamilton-Kemp, T.R., McCraken, C.T. Jr, Loughrin, J.H., Anderson, R.A. and Hildebrand, D.F. (1992) Effect of some natural volatile compounds on the pathogenic fungi *Alternaria alternata* and *Botrytis cinerea*. *Journal of Chemical Ecology* 18, 1083–1091.

Harish, S., Saravanan, T., Radjacommare, R., Ebenezar, E.G. and Seetharaman, K. (2004) Mycotoxic effect of seed extracts against *Helminthosporium oryzae* Breda de Hann, the incitant of rice brown spot. *Journal of Biological Sciences* 4, 366–369.

Hartmans, K.J., Diepenhorst, P., Bakker, W. and Gorris, L.G.M. (1995) The use of carvone in agriculture, sprout suppression of potatoes and antifungal activity against potato tuber and other plant diseases. *Industrial Crops and Products* 4, 3–13.

Hili, P., Evans, C.S. and Veness, R.G. (1997) Antimicrobial action of essential oils: the effect of dimethylsulphoxide on the activity of cinnamon oil. *Letters in Applied Microbiology* 24, 269–275.

Hiremath, S.P., Swamy, H.K.S., Badami, S. and Meena, S. (1996) Antibacterial and antifungal activity of *Striga densiflora* and *Striga orabanchioides*. *Indian Journal of Pharmaceutical Sciences* 58, 174–176.

Holley, A.H. and Patel, H. (2005) Improvement in shelf life and safety of perishable foods by plant essential oils and smoke antimicrobials. *International Journal of Food Microbiology* 22, 273–292.

Jegathambal, R. (1996) Pre-sowing and treatment to augment productivity of sorghum cv. CO 26 under rainfed agriculture. Ph D thesis, Tamil Nadu Agricultural University, Coimbatore, India.

Jonathan, E. and Davis, M.D. (2007) Are one or two dangerous? Methyl salicylate exposure in toddlers. *Journal of Emergency Medicine* 32, 63–69.

Kim, J.C., Choi, G.J., Lee, S.W., Kim, J.S., Chun, K.Y. and Chao, K.Y. (2004) Screening extracts of *Achranthus japonica* and *Rumex crispus* for activity against various plant pathogenic fungi and control of powder mildew. *Post Management Science* 60, 803–808.

Kritzinger, Q., Aveling, T.A.S. and Marasas, W.F.O. (2002) Effect of essential plant oils on storage fungi germination and emergence of cowpea seeds. *Seed Science and Technology* 30, 609–619.

Lewis, J.A. and Papavizas, G.C. (1967) Effects of tannins on spore germination and growth of Fusarium solani f. phaseoli and verticillium alboatrum. *Canadian Journal of Microbiology* 13, 1655–1661.

Lindberg, C., Melathopoulusy, A. and Winston, M. (2000) Laboratory evaluation of miticides to control *Varroa jacobsoni*, a honeybee parasite. *Journal of Economic Entomology* 93, 189–198.

Liu, W.T., Chu, C. and Zhou, T. (2002) Thymol and acetic acid vapors reduce postharvest brown rot of apricot and plums. *HortScience* 37, 151–156.

Mari, M., Lori, R., Leoni, O. and Marchi, A. (1993) *In vitro* activity of glucosinolate derived isothiocyanates against postharvest pear pathogens. *Annals of Applied Biology* 123, 155–164.

Mari, M., Leoni, O., Lori, R. and Marchi, A. (1996) Bioassay of glucosinolate derived isothyocyanates against postharvest pear pathogens. *Plant Pathology* 45, 753–760.

Mari, M., Leoni, O., Lori, R. and Cembali, T. (2002) Antifungal vapour-phase activity of allyl isothyocyanates against *Penicillium expansum* on pears. *Plant Pathology* 51, 231–236.

Marin, S., Velluti, A., Monoz, A., Ramos, A.J. and Sanchis, V. (2003) Control of fumonisin B1 accumulation in naturally contaminated maize inoculated with *Fusarium verticilliodes* and *Fusarium proliferatum* by cinnamon, clove, lemongrass, oregano and palmarosa essential oils. *European Food Research and Technology* 217, 332–337.

Mihaliak, C.A., Gershenzo, J. and Croteau, R. (1991) Lack of rapid monoterpene turnover in rooted plants, implications for theories of plant chemical defense. *Oecologia* 87, 373–376.

Mohapotra, N.P., Pati, S.P. and Ray, R.C. (2000) *In vitro* inhibition of *Botryodiplodia theobromae* (Pat.) causing Java black rot in sweet potato by phenoloic compounds. *Annals of Plant Protection Sciences* 8, 106–109.

Muller-Ribeau, F., Berger, B. and Yegen, O. (1995) Chemical composition and fungitoxic properties to phytopathogenic fungi of essential oils of selected aromatic plants growing wild in Turkey. *Journal of Agricultural and Food Chemistry* 43, 2262–2266.

Narsimha, P.K. (1994) Studies on certain aspects of seed management in soybean (*Glycine max* L.). PhD thesis, Tamil Nadu Agricultural University, Coimbatore, India.

Narsimhan, V., Rajappan, K., Ushamalini, C. and Kareem, A.A. (1988) Efficacy of new EC formulations of neem oil and pungam oil for the management of sheath root disease of rice. *Phytoparasitica* 26, 301–306.

Neri, F., Mari, M., Meniti, A.M. and Brigati, S. (2006) Activity of *trans*-2-hexenal against *Penicillium expansum* in 'Conference' pears. *Journal of Applied Microbiology* 100, 1186–1193.

Nursten, H.E. (1970) Volatile compounds. The aroma of fruits. In: Hulme, A.C. (ed.) *The Biochemistry of Fruits and their Products*. Academic Press, New York, pp. 239–268.
Nwachukwu, E.O. and Umechuruba, C.I. (2001) Antifungal activities of some leaf extract on seedborne fungi of African yam bean seeds, seed germination and seedling emergence. *Journal of Applied Sciences and Environmental Management* 5, 29–32.
Okigbo, R.N. and Pandalai, S.G. (2005) Control of postharvest diseases of tropical fruits in storage. *Recent Research Developments in Biotechnology and Bioengineering* 7, 69–85.
Oyolu, O. (1977) A qualitative and quantitative study of seed types in 'egusi' (*Colocynthis citrullus* L.). *Tropical Science* 19, 51–61.
Pandey, D.K., Tripathi, N.N., Tripathi, R.D. and Dixit, S.N. (1982) Fungitoxic and phytotoxic properties of essential oil of Hyptis suaveolens. *Zeitschrift fur Pflanzenkrankheiten und Pflanzenschutz* 89, 344–349.
Paramasivam, V. (1990) Studies on development, maturation, quality and storability of pea (*Pisum sativum* L.) seeds. MSc (Ag.) thesis, Tamil Nadu Agricultural University, Coimbatore, India.
Paranagama, P.A., Abeysekera, K.H., Abeywichrama, K. and Nugaliyadde, L. (2003) Fungicidal and anti-aflatoxigenic effects of the essential oils of *Cymbopogon citrates* (DC) Stapf. (lemongrass) against *Aspergillus flavus* Link. isolated from stored rice. *Letters in Applied Microbiology* 37, 86–90.
Parimelazhagan, T. and Francis, K. (1999) Antifungal activity of *Clerodendrum viscosum* against *Curvularia lunata* in rice seeds. *Journal of Mycology and Plant Pathology* 29, 139–141.
Paster, N., Menasherov, M., David, U. and Juven, B. (1995) Antifungal activity of oregano and thyme essential oils applied as fumigants against fungi attacking stored grains. *Journal of Food Protection* 58, 81–85.
Pimentel, D. and Levitan, L. (1986) Pesticides: amounts applied and amounts reaching pests. *BioScience* 36, 86–91.
Plaza, P., Torres, R., Vsall, J., Lamara, N. and Vinsa, I.C. (2004) Evaluation of the potential of commercial post-harvest application of essential oils to control citrus decay. *Journal of Horticultural Science and Biotechnology* 76, 935–940.
Ponce, A.G., Valle, C. and del Roura, S.I. (2004) Shelf life of leafy vegetables treated with natural essential oils. *Journal of Food Science* 69, 50–56.
Prasad, K. and Stadelbacher, G.J. (1973) Control of post harvest decay of fresh rasp-berries by acetaldehyde vapor. *Plant Disease Report* 57, 795–797.
Rana, B.K., Taneja, V. and Singh, U.P. (1999) Antifungal activity of an aqueous extract of leaves of garlic creeper (*Adenocalymna alliaceum* Miers.). *Pharmaceutical Biology* 37, 13–16.
Rodrigues, E.A., Schwan, K.R.F., Stangarlin, J.R. and Soares, R.N. (2001) Tratamento de sementes de trigo em extracto bruto de *Ocimum* gratissimum para o controle de *Bipolaris sorokiniana*. (Abstr) *Summa Phytopathologica* 27, 123.
Sabir, A.A. (1989) Studies on the production of quality seed and storage in soybean (*Glycine max* L. Merril). MSc (Ag.) thesis, Tamil Nadu Agricultural University, Coimbatore, India.
Sembdner, G. and Parthier, B. (1993) The biochemistry and the physiological and molecular actions of jasmonates. *Annual Review of Plant Physiology and Plant Molecular Biology* 44, 569–589.
Shetty, S.A., Prakash, H.S. and Shetty, H.S. (1989) Efficacy of certain plant extracts against seedborne infection of *Trchoconiella padwickii* in paddy (*Oryza sativa*). *Canadian Journal of Botany* 57, 1956–1958.
Sholberg, P.L. (1998) Fumigation of fruits with short chain organic acids to reduce the potential of postharvest decay. *Plant Disease* 82, 689–693.
Sholberg, P.L. and Gaunce, A.P. (1995) Fumigation of fruit with acetic acid to prevent postharvest decay. *HortScience* 30, 1271–1275.
Sholberg, P.L., Reynolds, A.G. and Gaunce, A.P. (1996) Fumigation of table grapes with acetic acid to prevent postharvest decay. *Plant Disease* 80, 1425–1428.
Sholberg, P.L., Haag, P., Hocking, R. and Bedford, K. (2000) The use of vinegar vapor to reduce postharvest decay of harvested fruit. *HortScience* 35, 898–903.
Singh, A.K., Dixit, A., Sharma, M. and Dixit, S.N. (1980) Fungitoxic activity of some essential oils. *Economic Botany* 34, 186–190.
Singh, H.N.P., Prasad, M.M. and Sinha, K.K. (1993) Evaluation of medicinal plant extracts against banana rot. *Journal of the Indian Botanical Society* 72, 163–164.
Singh, J. and Tripathi, N.N. (1999) Inhibition of storage fungi of black gram (*Vigna mungo* L.) by some essential oils. *Flavour Fragrance Journal* 14, 1–4.
Smid, E.J., Witte, Y., Vrees, O. de and Gorris, L.M.G. (1994) Use of secondary plant metabolites for the control of postharvest fungal diseases on flower bulbs. *Acta Horticulturae* 368, 523–530.

Soliman, K.M. and Badeaa, R.I. (2002) Effect of oil extracted from some medicinal plants on different mycotoxicogenic fungi. *Food Chemistry and Toxicology* 40, 1669–1675.

Song, J., Leepipattanawit, R., Deng, W. and Beaudry, R.M. (1996) Hexenal vapor is a natural, metabolizable fungicide: inhibition of fungal activity and enhancement of aroma biosynthesis in apple slices. *Journal of the American Society for Horticulture Science* 121, 937–942.

Swain, T. (1977) Secondary compounds as protective agents. *Annual Review of Plant Physiology* 28, 479–501.

Thompson, D.P. (1986) Effect of essential oils on spore germination of *Rhizopus, Mucor* and *Aspergillus* species. *Mycologia* 78, 482–485.

Tiwari, R., Mishra, D.N. and Upadhyay, P.S. (1988) Efficacy of some plant volatiles for the control of black mould of onion caused by *Aspergillus niger* Van Tiegh during storage. *National Academy of Sciences Letters* 11, 345–347.

Tribhuvanamala, G. and Narsimhan, V. (1998) Efficacy of plant extracts on seedborne pathogens of sunflower. *Madras Agricultural Journal* 85, 227–230.

Tripathi, P. (2005) *Botanical Pesticides in the Management of Post Harvest Fruit Diseases.* Daya Publishing House, New Delhi, 174 pp.

Tripathi, P. (2007) Biologicals and biorationals in the management of agricultural insect pests: an ecofriendly approach. In: Arya, A. and Monaco, C. (eds) *Seed Borne Diseases: Ecofriendly Management.* Scientific Publishers, Jodhpur, India, pp. 171–189.

Tripathi, P., Dubey, N.K. and Pandey, V.B. (2002) Kaempferol: the antifungal principle of *Acacia nilotica* Linn. Del. *Journal of the Indian Botanical Society* 81, 51–54.

Tripathi, P., Dubey, N.K., Banergi, R. and Chansuria, J.P.N. (2004) Evaluation of some essential oils as botanical fungitoxicants in management of postharvest rotting of citrus fruits. *World Journal of Microbiology and Biotechnology* 20, 317–321.

Tripathi, P., Dubey, N.K. and Shukla, A.K. (2007) Emerging non-conventional technologies for control of postharvest diseases of perishables. *Fresh Produce* 1, 111–120.

Tripathi, P., Dubey, N.K. and Shukla, A.K. (2008) Use of some essential oils as post-harvest botanical fungicides in the management of grey moulds of grapes caused by *Botrytis cinerea. World Journal of Microbiology and Biotechnology* 24, 39–46.

Tzortzakis, N.G. (2007) Maintaining postharvest quality of fresh produce with volatile compounds. *Innovative Food Science and Emerging Technologies* 8, 111–116.

Umarani, R. (1999) Studies on the physiological and biochemical basis of seed germination and deterioration in Csuarina equisetifolia. PhD thesis, Tamil Nadu Agricultural University, Coimbatore, India.

Vanangamudi, K., Bharathi, A., Parmeswari, K. and Ravichandran, G. (2007) Use of botanicals in seed storage. In: Arya, A. and Monaco, C. (eds) *Seed Borne Diseases: Ecofriendly Management.* Scientific Publishers, Jodhpur, India, pp. 71–92.

Vanneste, J.L., Hill, R.A., Kay, J.S., Farrel, R.L. and Holland, P.T. (2002) Biological control of sapstain fungi with natural products and biological control agents: a review of the work carried out in New Zealand. *Mycological Research* 106, 228–232.

Varma, J. and Dubey, N.K. (2001) Efficacy of essential oils of *Caesulia axillans* and *Mentha arvensis* against some storage pests causing biodeterioration of food commodities. *International Journal of Food Microbiology* 68, 207–210.

Vaughan, S.F. and Spencer, G.F. (1994) Antifungal activity of natural compounds against thiabendazole resistant *Fusarium sambucinum* strain. *Journal of Agriculture and Food Chemistry* 42, 200–203.

Velluti, A., Sanchis, V., Ramos, A.J., Turon, C. and Marin, C. (2004) Impact of essential oils on growth rate, zearalenone and deoxynivalenol production by *Furarium graminearum* under different temperature and water activity conditions in maize grain. *Journal of Applied Microbiology* 96, 716–724.

Wilson, C.L., Franklin, J.D. and Otto, B.E. (1987) Fruit volatiles inhibitory to *Monilinia fructicola* and *Botrytis cinerea. Plant Disease* 71, 316–319.

Wyllie, T.D. and Morehouse, L.G. (1978) *Mycotoxic Fungi, Mycotoxins, Mycotoxicoses: An Encyclopedic Handbook.* Vol. 3. Marcel Dekker, Inc., New York.

Yin, M.C. and Cheng, W.S. (1998) Inhibition of *Aspergillus niger* and *Aspergillus flavus* by some herbs and spices. *Journal of Food Protection* 61, 123–125.

Zambonelli, A., Daulerio, A.Z., Severi, A., Benvenuti, S., Maggi, L. and Bianchi, A. (2004) Chemical composition and fungicidal activity of commercial essential oils of *Thymus vulgaris* L. *Journal of Essential Oil Research* 16, 69–74.

5 Use of Plant Extracts as Natural Fungicides in the Management of Seedborne Diseases

Gustavo Dal Bello and Marina Sisterna

Comisión de Investigaciones Científicas de la Provincia de Buenos Aires, Centro de Investigaciones de Fitopatología (CIDEFI) – Facultad de Ciencias Agrarias y Forestales, Universidad Nacional de La Plata, La Plata, Argentina

Abstract

Seedborne fungi can cause substantial losses to grains, rendering them unfit for human consumption and sowing. Several methods have been used for the control of seedborne diseases and among them chemical control has been the most widely adopted over many decades. The use of most of these fungicides has been restricted because of high and acute toxicity, long degradation periods and bad effects on human health, plants and animals, which is harmful to our environment. Moreover, recent increases in the production and sale of organic seed has heightened the scrutiny of organic seed quality and in particular brought attention to concerns of seedborne disease contamination. In order to meet the demands of consumers and growers alike, exploration of alternative methods for managing fungal diseases is under way. One such eco-friendly approach of controlling seed fungal diseases is the use of natural products, specifically plant-derived compounds. They have played a significant role in reducing the incidence of seedborne pathogens and in the improvement of seed quality and the emergence of plant seeds in the field. It has long been recognized that several plant compounds, such as essential oils, have antifungal activity against both pathogens and spoilage fungi. As a rich source of bioactive chemicals, plants may provide potential alternatives to synthetic fungicides for seed treatment to protect them against seedborne pathogens. Therefore, this chapter discusses the current status of the use of plant extracts to control seedborne fungi.

Introduction

Almost 90% of all the world's food crops are grown from seeds (Schwinn, 1994), which are widely distributed in national and international trade. Many plant pathogens can be seed transmitted and seed distribution is a very efficient means of introducing plant pathogens into new areas, as well as a means of survival of the pathogen between growing seasons. Disease-causing organisms may be carried with, on or in seeds and, in suitable environmental conditions, may be transmitted to cause diseases in developing seedlings or plants. With some diseases, the pathogen attacks the germinating seedling, which affects seedling establishment and hence plant populations; with others, disease symptoms are not seen until a later stage of growth (Rennie and Cockerell, 2006). Furthermore, seedborne pathogens such as bacteria, fungi, viruses and nematodes have the potential to spread disease to the subsequent crop. Seedborne infection of fungal pathogens is important not only for its association with the seeds but also contamination of the soil by permanently establishing its inocula. Additionally, fungi are significant destroyers of foodstuffs and grains during storage, rendering them unfit for human consumption by retarding their nutritive value and often by producing mycotoxins (Satish *et al.*, 2007).

It is, therefore, necessary to search for control measures that are economical, ecologically sound and environmentally safe to eliminate or reduce the incidence of these important pathogens so as to increase seed germination and obtain healthy and vigorous plants with better yield (Hasan *et al.*, 2005).

Seed treatment is the oldest practice in plant protection. Its origin can be traced to the 18th century with the use of brine to control cereal smuts (Neergaard, 1979). The modern era of seed treatments began with the introduction of organomercury fungicides in 1912, which were widely used for several decades. The post-World War II period saw the development of new fungicide chemistry and the first use of seed treatment for insect control. Today, the most widely used application of seed treatment is the traditional one of protecting the germinating seedling against seed- and soilborne fungi in the period immediately after planting (McGee, 1995). Chemical fungicides can control plant diseases but they have bad effects on human health, plants and animals, which is harmful to our environment. Besides, using conventional seed treatment with synthetic fungicides to kill pathogens is a practice not allowed in organic production. Additionally, resistance by pathogens to fungicides has rendered certain fungicides ineffective.

Worldwide ecological awareness requires more natural foods and products, which has influenced the improvement and utilization of integrated pest management. In this kind of control, alternative methods are used to protect seeds to decrease the use of chemical products. Moreover, recent increases in the production and sale of organic seed has heightened the scrutiny of organic seed quality, and in particular brought attention to concerns of seedborne disease contamination. The number of alternative crop production systems has increased in the past decade in response to growing concerns about agricultural concentration and interest in a more ecological, farm-based agriculture. In these low-input systems, some non-chemical substances, such as plant extracts, may be used successfully as a contact fungicide seed treatment for organic crops. As a rich source of bioactive chemicals, plants may provide potential alternatives to be used as pathogen-control agents.

Hamburger and Hostettmann (1991) report that the total number of plant chemicals may exceed 400,000 and of this, more than 10,000 are secondary metabolites whose major role in plants is defensive in nature. Thus, plant-based secondary metabolites that have a defensive role may be exploited for the management of diseases and pests. However, most species of higher plants have never been surveyed. Their chemical or biologically active constituents that have the potential to be used as new sources of commercially valuable pesticides remain to be discovered. This is due mainly to the lack of information on the screening/evaluation of diverse plants for their antifungal potential (Satish *et al.*, 2007). Nevertheless, several higher plants and their constituents have shown success in plant disease control and have proved to be harmless and non-phytotoxic, unlike chemical fungicides.

Essential Oils to Reduce Seedborne Fungi

Plant extracts have played a significant role in reducing the incidence of seedborne pathogens and in the improvement of seed quality and the emergence of plant seeds in the field (Hasan *et al.*, 2005). In recent years, much attention has been paid to essential oils, a group of plant-derived compounds, for seed treatment to protect them

1

2

3

4

5

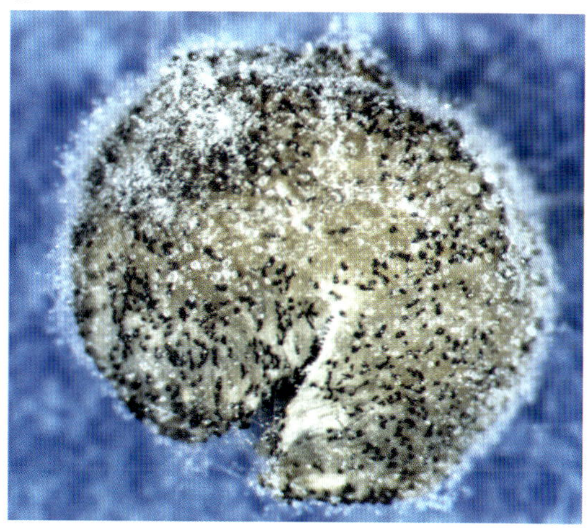

Plate 1. (a) Perithecia of *Gibberella zeae* (anamorph *Fusarium graminearum*) on infected seed of triticale. (b) Cross-section of a perithecium of *G. zeae* showing the ostiole and asci bearing ascospores. (Reprinted with permission from F. Trail and R. Common (2000). Perithecial development by *Gibberella zeae*: a light microscopy study. *Mycologia* 92,130-138. © Mycological Society of America)

Plate 2. *Diaporthe phaseolorum* (anamorph *Phomopsis sojae*) causing seed rot on soybean seeds. (Courtesy M. C. Rollán)

Plate 3. *Fusarium* sp. Infecting soybean seeds. (Courtesy M. C. Rollán)

Plate 4. Germinating onion seed affected by *Botrytis allii*. (Courtesy L. du Toit, Diseases in vegetable seed crops: Identification, biology, and management [Online]. Available at: http://www.seedalliance.org/uploads/pdf/VegSeed-Diseases.pdf)

Plate 5. Seedborne wilt of spinach by *Verticillium dahliae*. (Courtesy L. du Toit)

6

7

8

9

10

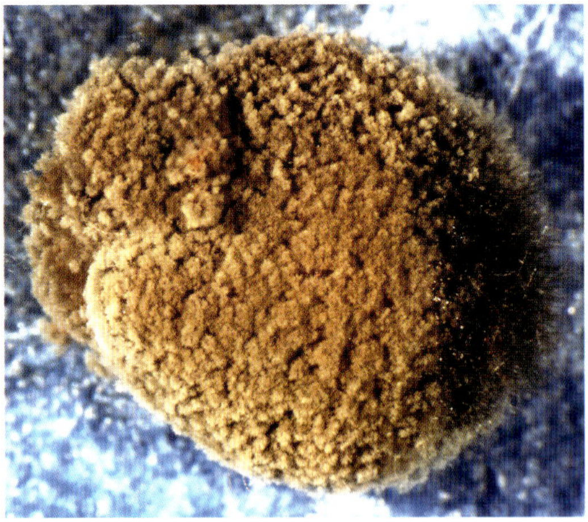

Plate 6. Spinach seed showing stromatisation due to pseudothecia of *Pleospora herbarum* (anamorph *Stemphylium botryosum*). (Courtesy L. du Toit)

Plate 7. Rice seed discoloration caused by a fungi complex.

Plate 8. Wheat seed discoloration caused by a fungi complex.

Plate 9. Open pod of soybean showing purple discoloration caused by *Cercospora kikuchii*. (Courtesy M. C. Rollán).

Plate 10. Conidiophores and conidia of *Cladosporium variabile* on spinach seed. (Courtesy L. du Toit)

against seedborne fungi (Sisterna and Dal Bello, 2007).

The essential oils arise from a secondary metabolism of the plant, normally formed in special cells or groups of cells as glandular hairs, found on many leaves and stems. Oils occur as a globule or globules in the cell and may also be secreted from cells lining the schizogenous ducts or canals. Plant volatile oils are generally isolated from non-woody plant material by several methods, usually distillation, and are a variable mixture of principally terpenoids, specifically monoterpenes [C_{10}] and sesquiterpenes [C_{15}], although diterpenes [C_{20}] may also be present. A variety of other molecules can also occur, such as aliphatic hydrocarbons, acids, alcohols, aldehydes, acyclic esters or lactones and, exceptionally, nitrogen- and sulphur-containing compounds, coumarins and homologues of phenylpropanoids (Dorman and Deans, 2000). Faleiro *et al.* (2003) have shown that the antimicrobial action is determined by more than one component. In such cases, the major component is responsible not only for the antimicrobial activity, but also the synergistic effect that may take place. The mixtures are extremely complex and vary with environmental and genetic factors (Asplund, 1968; Cabo *et al.*, 1986; Arras, 1988; Bhaskara *et al.*, 1998; Vanneste *et al.*, 2002). Moreover, the composition of essential oils from a particular species of plant can differ between harvesting seasons and between geographical sources (Di Pasqua *et al.*, 2005; Di Pasqua, 2006).

Major active compounds from essential oils are known for their broad-spectrum antifungal activity against both human and plant pathogens. These constituents can either affect the physiological functions of microorganisms or cause structural changes of hyphae and spores (Arras *et al.*, 1993; Zambonelli *et al.*, 2004; Kishore *et al.*, 2007), and different fungi appear to react differently to these components (Szczerbanik *et al.*, 2007). The antifungal essential oils reduce hyphal growth and also induce lysis and cytoplasmic evacuation in fungi. Growth inhibition by essential oils often involves induction of changes in cell wall composition (Ghfir *et al.*, 1997), plasma membrane disruption, mitochondrial structure disorganization (de Billerbeck *et al.*, 2001) and interference with enzymatic reactions of the mitochondrial membrane, such as respiratory electron transport, proton transport and coupled phosphorylation steps (Knobloch *et al.*, 1989).

The active components vary between oils. For example, the main component is l-carvone in spearmint (*Mentha spicata* L.), terpinen-4-ol in tea tree (*Melaleuca alternifolia* (Maiden. & Betche.) Cheel.) oil and α-terpineol in pine (*Pinus* spp.) (Knobloch *et al.*, 1989). The essential oils of *Cinnamomum zeylanicum* Blume (cinnamon) and *Syzygium aromaticum* (L.) Merr. & Perry (syn. *Eugenia cariophyllata* Thunb.), consisting of cinnamaldehyde and eugenol, respectively, as major components (Paranagama, 1991), are known to be potent antifungal materials (Beg and Ahmad, 2002; Ranasinghe *et al.*, 2002). Citral and geraniol are the major components in essential oils of *Cymbopogon citratus* (DC.) Stapf (lemongrass) and *C. martinii* (Roxb.) Stapf var. motia (palmarosa), respectively, which are antifungal compounds (Paranagama *et al.*, 2003; Velluti *et al.*, 2004). Thymol was identified as the active ingredient of *Ocimum gratissimum* L. (wild basil) and has been found to suppress fungal growth (Adekunle and Uma, 2005). Linalool is a major component in the essential oil of *Thymus mastichina* L. subsp. *mastichina*, with antimicrobial activity (Faleiro *et al.*, 2003), and both limonene and linalool are the minor components in the essential oils derived from different plants. The majority of these essential oils and their components have proved valuable in protection against postharvest fungal diseases which cause build-up of toxic fungal metabolites in stored foods (Kishore *et al.*, 2007). Therefore, essential oils might substitute agrochemicals or contribute to the development of new agents to inhibit both fungal growth and the production of mycotoxins affecting grain and seed crops.

This chapter discusses the current status of plant extracts and the potential use of essential oils as natural antifungal agents to control the main seedborne pathogens and spoilage fungi.

Symptoms on Seeds Caused by Fungi

Seedborne mycoflora comprise a large number of saprobes and pathogenic fungal species. Pathogenic fungi grown on seeds can cause heavy damage and reduce yields of seed, both quantitatively and qualitatively (Neergaard, 1979). Other fungi, including saprophytes and very weak parasites (Sisterna and Lori, 2005), may lower the quality of seeds by causing discoloration, which may seriously depreciate the commercial value of seeds, particularly of grain when graded for consumption.

Disease and disorder

The following types of disease and disorder are encountered, often in combination (Neergaard, 1979):

Seed abortion

The most prominent examples of fungi producing abortion are the smut fungi, which infect cereals and grasses systemically, and the ergot fungi. The floral parts of the hosts are replaced by the fructifications of the parasites. Other examples are different species of *Fusarium* (in wheat, maize and rice); *Ascochyta rabiei* in chickpea may kill the young seeds; *Drechslera verticillata* causes death of seed primordia in brome grass and in wheat.

Shrunken seeds, reduced in size

Examples of more or less heavy reduction of seed size are: *Alternaria brassicicola* and *Phoma lingam* in crucifers, *Septoria linicola* in flax, *D. teres* in barley, *F. graminearum* and *S. nodorum* in wheat.

Seed rot

Many seedborne fungi produce seed rot either in the crop or during germination. Examples are *F. avenaceum*, *F. graminearum* (Plate 1), *F. moniliforme*, *Bipolaris sorokiniana*, *B. maydis* and *B. oryzae* in cereals; *Colletotrichum graminicola*, *Diaporthe phaseolorum* (Plate 2) and *Fusarium* spp. in soybean (Plate 3); *Botrytis allii* on onion (Plate 4); *Verticillium dahliae* on spinach (Plate 5) and *B. cinerea* in the seeds of many hosts, including forest trees.

Sclerotization and stromatization

Transformation of floral organs or seed into sclerotia or stromata is an important disease condition in certain categories of fungi and host. Ergots produced by *Claviceps purpurea* and other species of *Claviceps* in cereals and grasses exemplify sclerotia of this type. Another example is *Phomopsis viterbensis* in chestnut, *Pleospora herbarum* in spinach (Plate 6) and *Ciboria* spp. in the seeds of forest trees and grasses.

Seed necroses

Many seed-rotting fungi produce superficial necroses in the seed; other fungi never penetrate deeply into the tissues, most seedborne fungi usually not beyond the protective layers, the seed coat or pericarp. Anthracnose fungi, *Colletotrichum* spp. as well as *Ascochyta* spp., often penetrate into the fleshy cotyledons, producing conspicuous necrotic lesions in the seeds of bean, soybean, pea, cowpea and other hosts.

Seed discoloration

Discoloration of seeds is a very important degrading factor, both for consumption (grain) or for industrial purposes (oil seed). It may be a general indication of poor quality (Plates 7 and 8). Well-known examples are the effects of *A. pisi* in pea; *C. lindemuthianum* in bean; *B. sorokiniana* in wheat, *B. oryzae* in rice, *Cercospora kikuchii* (Plate 9) in soybean, etc.

Reduction or elimination of germination capacity, lowered viability

Obviously, necroses or more deeply penetrating rots in seeds reduce the viability of the seeds, their longevity in storage and their emergence in the field.

Physiological alterations or effects in seed

Metabolic products of seedborne microorganisms may affect the seed itself or may have other, sometimes serious consequences such as toxicity to animals and humans (*Aspergillus* spp., *Penicillium* spp., *Fusarium* spp.).

Moreover, seed fungi are classified as field and storage fungi (Christensen and Kaufmann, 1965). Genera such as *Alternaria*, *Cladosporium* (Plate 10), *Fusarium* and *Bipolaris* invade seeds as they are developing on the plants in the field or after they have matured, but before they are harvested, and for this reason, they have been designated 'field fungi'. These fungi require moisture content in equilibrium with a relative humidity of more than 90% to grow and usually do not continue to grow in grains after harvest, since grains and seeds are stored with moisture contents below those required by the field fungi.

The storage fungi consist mainly of several species of *Aspergillus*. Species of *Penicillium* are encountered at times, usually in lots of grain stored at low temperatures and with moisture contents above 16%. The storage fungi do not invade grains to any appreciable degree or extent before harvest.

Fungicidal Effects of Plant Extracts Against Seed Fungi

Numerous studies have described the use of botanicals with a view to exploiting their potential as natural fungicides against seedborne fungi. The following section discusses this alternative method, with particular emphasis on the main seedborne fungal pathogens.

Alternaria

A. padwickii, an important seedborne pathogen of rice (*Oryza sativa* L.), was inhibited by aqueous extracts of *Strychnos nux-vomica* L. (strychnine tree), garlic (*Allium sativum* L.) bulbs, ginger (*Zingiber officinale* Roscoe) rhizomes, basil (*O. basilicum* L.) leaves, and fruits of *Azadirachta indica* A. Juss. (neem) (Shetty *et al.*, 1989).

Positive effects have been recorded on the same fungus with essential oils of *C. citratus*, *O. gratissimum* L. and *Thymus vulgaris* L. (thyme) (Nguefack *et al.*, 2004). The researchers investigated the ability to control seedborne infection and seed–seedling transmission in naturally infected seeds. The essential oils increased the germination capacity of the treated seeds.

Bipolaris

Hasan *et al.* (2005) demonstrated that plant extracts, namely *Z. officinale*, *A. sativum*, *A. cepa* L. (onion), *Adhatoda vasica* Nees (vasaka), *Achyranthes aspera* L. (devil's horsewhip), *A. indica*, *Lawsonia alba* Lamarck (henna), *Cuscuta reflexa* Roxb. (giant dodder), *Vinca rosea* L. and *Nigella sativa* L. (black cumin), significantly reduced seed infection of wheat by *B. sorokiniana* (*Triticum aestivum* L.). Alcoholic extracts of neem and garlic inhibited the presence of *B. sorokiniana* completely, whereas the highest percentage of the fungus was recorded from untreated seeds (control). Water extract of all tested plants had the ability to control seedborne fungi of wheat var. Kanchan, which showed 100% inhibition of *B. sorokiniana* with the application of extracts from *Z. officinale*, *A. sativum*, *A. cepa*, *A. indica*, *C. reflexa* and *N. sativa*, whereas the highest fungal incidence (11.67%) was observed on untreated seed. After treatment with the water extract of *L. alba* and *A. aspera*, only 4.84% and 7.16% incidence of the pathogen, respectively, was recorded. Seeds of wheat treated with *A. vasica* and *V. rosea* gave statistically identical results (5.83% and 5.90% incidence of *B. sorokiniana*).

Alice and Rao (1986) reported good results using plant extracts to control *B. oryzae* on rice seeds which have high natural infection of the fungus. After soaking in the filtrates of different extracts, *A. sativum* and *M. piperita* (peppermint) reduced seed infection by 68%. In Bangladesh, use of extracts of *Polygonum hydropiper* L. (water-pepper), *A. cepa*, *A. sativum* and *A. indica* was demonstrated to be effective against *B. oryzae* at higher concentrations. Among them, neem and garlic were the most effective at 1:1 dilution and inhibited the occurrence of the pathogen by 91 and 83%, respectively (Ahmed *et al.*, 2002).

Neem and pungam (*Pongamia pinnata* (L.) Pierre) oil-based emulsifiable concentrate (EC) formulations were evaluated for their efficacy to inhibit the mycelial growth of the fungus *Helminthosporium oryzae* (syn. *B. oryzae*) causing grain discoloration of rice under *in vitro* conditions. All three formulations, namely neem oil 60 EC (acetic acid), neem oil 60 EC (citric acid) and neem oil + pungam oil 60 EC (citric acid), inhibited mycelial growth of the pathogen; they were effective even after 9 months of storage. These formulations controlled the grain discoloration on rice effectively (Rajappan *et al.*, 2001). The efficacy of essential oils such as clove, ginger, lemongrass, basil, peppermint, anise (*Pimpinella anisum* L.) and cinnamon at different concentrations on growth inhibition of *B. oryzae* was examined by Palaoud (2006). Treatments with clove, anise, ginger and cinnamon oils at 500 ppm provided the best results in controlling the fungus and, after storage for 4 months, seed viability was as high as 97–98%. Also, the extracts of *C. citratus*, *O. gratissimum* and *T. vulgaris* applied to rice seeds infected with *B. oryzae* controlled fungal growth and seedling transmission of the pathogen (Nguefack *et al.*, 2004).

Colletotrichum

Abdelmonem *et al.* (2001) screened oils of *M. piperita*, *T. capitatus* (L.) Hoffmans. and Link and *Carum carvi* L. (caraway) against various seedborne fungi of soybean (*Glycine max* (L.) Merr.) and lentil (*Lens culinaris* Medik.) and found all plant extracts to be highly effective in controlling *C. dematium*.

Among the fibre-producing species, a study on garlic bulb extract reported a fungicidal effect of the oil against numerous seedborne fungal pathogens of white jute (*Corchorus capsularis* L.), one of the most important crops from Bangladesh, India and China. The essential oil produced inhibition in both mycelial growth and spore germination of fungi, including *C. corchori* (Ahmed and Shultana, 1984), which was also strongly inhibited in *in vitro* tests by using crude leaf extracts from *Eupatorium triplinerve* Vehl. (yapana) (Rahman and Junaid, 2008).

Several studies carried out in Burkina Faso underlined the antifungal properties of extracts from some *Cymbopogon* spp. against *C. graminicola*, the causal agent of anthracnose on sorghum (*Sorghum bicolor* (L.) Moench and pearl millet (*Pennisetum glaucum* (L.) R. Br.). Somda *et al.* (2007) demonstrated that the essential oil of *C. citratus* at a concentration of 6% was effective in controlling seedborne infection and seed–seedling transmission of *C. graminicola* without affecting seedling development. Similarly, the essential oils extracted from *C. giganteus* (Hochst.) Chiov., *C. nardus* (L.) Rendle and *C. schoenanthus* Spreng. reduced sorghum seed infection by the pathogen significantly. The lowest rates of infected seeds were recorded on seeds treated with 10 μl and 15 μl of *C. nardus* oil/g seeds. These doses were more efficient than chemical control (Elisabeth *et al.*, 2008).

Curvularia

In blackgram (*Vigna mungo* L.), essential oils extracted from wood chips of cedar (*Cedrus deodara* (Roxb. ex Lamb) G. Don) and that from seeds of *Trachyspermum ammi* (L.) Sprague ex Turrill (ajowan) exhibited absolute toxicity, inhibiting the mycelial growth of *C. ovoidea*, storage fungi found on seeds (Singh and Tripathi, 1999).

Parimelazhagan and Francis (1999) reported reduction in the radial growth of *C. lunata* associated with rice seeds when treated with leaf extracts of *Clerodendrum viscosum* Vent. (glory tree), which also increased seed germination and root and shoot lengths of rice. Considerable research activity has occurred in the Asian-Pacific region on the potential for plant extracts to control seedborne fungi including maize. The oils of cassia (*C. cassia* Blume) and clove inhibited the growth of established seedborne infections of *C. pallescens* (Chatterjee, 1990).

Fusarium

The essential oils and their constituents have been found effective as antifungal agents against the main species of *Fusarium*. Among several plant extracts, Sitara *et al.* (2008) found that essential oils from seed of neem, black cumin and asafoetida (*Ferula asafoetida* L.) showed fungicidal activity of varying degree against *F. oxysporum*, *F. moniliforme* (syn. *F. verticillioides*), *F. nivale* and *F. semitectum*. Of those oils, asafoetida oil at 0.1% and 0.15% inhibited the growth of all test fungi significantly. A variety of wild plants from Mexico were evaluated against several cereal seedborne fungi in *in vitro* tests (Tequida-Meneses *et al.*, 2002). Extracts from leaves and stems of *Larrea tridentata* (Sessé & Moc. ex DC.), Coville (creosote bush) and *Datura discolor* Bernh. (desert thorn apple) in methanol or ethanol inhibited the radial growth of *F. poae* completely. Next to these extracts, *Proboscidea parviflora* (Woot.) Woot. & Standl. (double claw) also showed good fungal inhibition (86.6%), followed by *Baccharis glutinosa* Pers. (saltmarsh baccharis) (79.6%), compared to the alcoholic controls (0% inhibition).

In legumes, soybean and lentil, carvone (monoterpene compound), among other tested compounds, manifested the highest antimicrobial influence with complete inhibition to *F. oxysporum*. It showed broad-spectra activity against all the tested isolates of fungal strains at low concentrations. Peppermint, *T. capitatus* and caraway oils also demonstrated a high control effect against *Fusarium* sp. (Abdelmonem *et al.*, 2001). Also, the antifungal activity of the essential oils of thyme, clove, peppermint, soybean and peanut (*Arachis hypogaea* L.) were tested against *F. oxysporum* and *F. equiseti in vitro* on cowpea (*V. unguiculata* (L.) Walp.) (Kritzinger *et al.*, 2002). Likewise, plant leaf extracts (crude and aqueous) of basil, bitter leaf (*Vernonia amygdalina* Del.), neem and pawpaw (*Carica papaya* L.) reduced the incidence of *F. moniliforme* significantly and increased seed germination and seedling emergence of African yam bean (*Sphenostylis stenocarpa* (Hochst ex. A. Rich) Harms) when compared with the untreated controls (Nwachukwu and Umechuruba, 2001).

Regarding cereals, several natural plant compounds have been identified as having antifungal activity against seedborne fungi. The essential oils of *C. citratus*, *O. gratissimum* and *T. vulgaris* have proved valuable in protection against the seedborne fungus, *F. moniliforme* in rice. This study evaluated the ability to control seedborne infection and seed–seedling transmission in naturally infected seeds (Nguefack *et al.*, 2004). The extracts applied controlled seed infection and seedling transmission of the pathogen and increased the germination capacity of the treated seeds. In the field, as a result of extracts seed treatment as compared to the non-treated control, reduction of disease incidence and important increases in yield were recorded. After rice seeds inoculated with *F. moniliforme* were soaked in seven plant essential oils at ten different concentrations, anise, ginger, clove and cinnamon oils at 500 ppm provided the best results in controlling the fungus. The percentage of seed germination and the number of normal seedlings was significantly high when compared with the control. Anise and clove also showed the highest seedling dry weight (Palaoud, 2006).

In another study on wheat, ten plant extracts were tested for their efficacy *in vitro* against seedborne fungi; alcoholic extract of neem and garlic controlled the infection of *Fusarium* sp. complctely. Good results of these treatments contributed to increased seed germination (Hasan *et al.*, 2005). Furthermore, botanicals from male fern (*Dryopteris filix mas* (L.) Scott.) suppressed completely the population of *F. oxysporum* in the seed mycoflora of wheat (Rake *et al.*, 1989). Putative mycotoxicogenic fungi, such as *F. moniliforme*, were partially or totally sensitive to different essential oils extracted from 12 medicinal plants (Soliman and Badeaa, 2002). Results indicated that oils of thyme, cinnamon and anise (< or = 500 ppm), marigold (*Calendula officinalis* L.) and caraway (< or = 2000 ppm), spearmint and basil (3000 ppm) inhibited this fungus completely.

Bioassays using a poisoning technique were carried out with *C. citratus*, *O. gratissimum* and *T. vulgaris* for the control of seedborne fungi infecting maize (*Zea mays* L.) seeds. The results disclosed the fungicidal properties of theses oils against *F. verticillioides*. These natural products control the seedborne inoculum of the pathogen by 90% to 100%. Field trials conducted in the humid forest and the warm savannah zones of Cameroon have shown that these products are potential seed treatments that could be used as substitutes for synthetic fungicides, which are usually unaffordable to resource-limited farmers (Tagne *et al.*, 2008). Seed treatment with cinnamon, palmarosa and lemongrass oils at 500 mg/kg showed antimycotoxigenic ability against fumonisin B_1 accumulation by isolates of *F. verticillioides* and *F. proliferatum* (Marín *et al.*, 2003). Furthermore, different effects of oregano (*Origanum vulgare* L.) and herb Louisa (*Aloysia triphylla* (L'Herit) Britton) essential oils were observed on *F. verticillioides* M 7075 fumonisin B_1 production in corn grain in Argentina (López *et al.*, 2004). As alternative preharvest natural fungicides, Velutti *et al.* (2004) showed the antimycotoxigenic activity of the essential oils against *F. graminearum* on corn infested seed. The effect of oregano, cinnamon, lemongrass, clove and palmarosa on growth rate, zearalenone (ZEA) and deoxynivalenol (DON) production was assessed at two concentrations (500 and 1000 mg/kg), at different water activity and temperature levels. DON production in general was inhibited by all essential oils at 30°C and, although palmarosa and clove were the only essential oils with statistically significant inhibitory effect on ZEA production, an inhibitory trend was observed when cinnamon and oregano oils were added to maize grain. Studies on the efficacy of indigenous plant extracts against seedborne infection of *F. moniliforme* on maize demonstrated that aqueous extracts of leaves of *O. gratissimum*, *Acalypha ciliata* Forssk., *V. amygdalina*, *M. indica* L. (mango tree) and *A. indica* had significant inhibitory growth effects on the fungal pathogen. *A. ciliata* extract was more effective than other plant extracts and compared favourably with benomyl in the control of the pathogen (Owolade *et al.*, 2000). Furthermore, to determine whether essential oils can be used as a contact fungicide seed treatment for organic corn, the essential oils of 18 plants were screened for their fungicidal properties. Five oils, cinnamon, clove, *O. minutiflorum* O. Schwarz and P.H. Davis, savoury (*Satureja montana* L.), and thyme, controlled *Fusarium* completely *in vitro*. The minimum inhibitory concentration (MIC) was 800 µl/l and seedlings presented no phytotoxicity symptoms in the germination test at rates up to 64 µl/kg active ingredient (MIC × 20). Field emergence of inbred and hybrid seeds treated with the essential oils were significantly lower than seeds treated with the commercial fungicides, Maxim XL {fludioxonil [4-(2,2-difluoro-1,3-benzodioxol-4-yl)-1*H*-pyrrole-3-carbonitrile] 21.4%, mefenoxam [(*R*)-2-[(2,6-dimethylphenyl) methoxyacetylamino] propionic acid methyl ester] 8.4%}, which is a conventional fungicide, and Natural 2 (proprietary ingredients), which is an organic fungicide, but were not different from the organic fungicide, Yield Shield (*Bacillus pumilus* GB34 0.28%) or an untreated control (Christian and Goggi, 2008).

Bioactivity of different plant extracts on *F. thapsinum* pathogen of sorghum was evaluated on seeds contaminated with the fungi. Cinnamon, clove, epazote (*Teloxys ambrosioides* (L.) Weber), oregano and thyme, singly and in combination, as well as the essential oils of *Mentha* sp. and rue (*Ruta chalepensis* L.) and the combination of clove with cinnamon, had a fungicidal effect. Nevertheless, only thyme did not affect either seed germination or sorghum seedling height. The rest of the oils were phytotoxic (Montes-Belmont and Flores Monctezuma, 2001). Additionally, plant extracts were also tested on naturally infected sorghum seeds for controlling *F. moniliforme*. More than 50% of the growth of this fungus was reduced by *C. citratus* essential oil on seeds, whereas *Eucalyptus camaldulensis* Dehnh. (*Eucalyptus*) essential oil was less efficient, even at high concentrations (Somda *et al.*, 2007). Elisabeth *et al.* (2008) investigated the efficacy of essential oils extracted from *C. schoenanthus*, *C. nardus* and *C. giganteus* in controlling *Fusarium* sp. on seeds of sorghum and pearl millet. The results indicated that all the essential oils reduced seed contamination of both cereals significantly. The lowest rates of infected seeds were recorded on seeds treated with 10 μl and/or 15 μl of essential oil/g seeds. Most of the time, these doses were as efficient as the chemical control and oil of *C. giganteus* used at 15 μl/g seeds eliminated pearl millet seed infection by *Fusarium* completely.

In another experiment, de Souza *et al.* (2003) analysed the mycoflora and physiological quality of cotton (*Gossypium hirsutum* L.) seeds treated with chemical fungicides and aroeira (*Astronium urundeuva* L.) extract. Pure extract did not control the fungal population but, when mixed with the fungicides, captan, benomyl and tolylfluanid, they showed reduction in the incidence of *Fusarium* sp.

From *Leguminosae* members, leaf extracts of *Delonix regia* (Bojer) Raf., flamboyant tree, *Pongamia glabra* Vent. (Karanja) and *Acacia nilotica* (L.) Willd. ex Delile (gum arabic tree) significantly inhibited spore germination, mycelial growth and spore production of *F. solani* from sunflower (*Helianthus annuus* L.) seeds (Thiribhuvanamala and Narasimhan, 1998). The same pathogen could be controlled using crude leaf extracts of *A. indica* and *O. gratissimum* to protect egusi melon (*Cucumeropsis mannii* Naudin) seed. After 6 months incubation, all the seeds treated with leaf extracts showed no *Fusarium* infection (Adekunle and Uma, 2005).

Efficacy of some plant extracts in controlling seedborne *Fusarium* infections of mustard (*B. nigra* (L.) W.D.J. Koch) was evaluated. It was found that garlic and neem extracts were the most effective in controlling the pathogen among the plant extracts studied. The effectiveness of garlic extract was comparable to the fungicide, Rovral (Latif *et al.*, 2006).

Macrophomina

Several natural plant compounds have been identified as having antifungal activity against *M. phaseolina*. The work of Ahmed and Shultana (1984) reported that garlic oil produced inhibition in both mycelial growth and spore germination of *M. phaseolina*, an important seedborne fungal pathogen of jute. In sunflower seeds, the leaf extracts of the flamboyant tree, karanja and gum arabic tree significantly inhibited the germination of fungal spores, mycelial growth and spore production as well (Thiribhuvanamala and Narasimhan, 1998).

In vitro experiments conducted by Dwivedi and Singh (1999) confirmed the fungitoxicity of some higher plant extracts against the mycelial growth of *M. phaseolina*. Among the plant products, the essential oils of *T. ammi* exhibited absolute fungicidal effect at an MIC of 200 ppm.

Studies of Abdelmonem *et al.* (2001) also showed the inhibitory effect of the essential oils of *M. piperita*, *T. capitatus* and *C. carvi* against *M. phaseolina* associated with the seeds of soybean and lentil. Furthermore, when tested in infected cowpea seeds, *A. indica* extract was found to inhibit the incidence of the pathogen. After naturally infected seeds were immersed in a suspension containing neem tree oil at a concentration of 0.5% for 16 h, the infection incidence decreased to 50% in relation to controls using only water (Mello *et al.*, 2005).

Aspergillus and *Penicillium*

Putative mycotoxicogenic fungi of wheat grains were partially or completely sensitive to different essential oils extracted from 12 medicinal plants (Soliman and Badeaa, 2002). They were tested for inhibitory activity against *A. flavus*, *A. parasiticus* and

A. ochraceus. Results indicated that oils of thyme, cinnamon (< or = 500 ppm), marigold (< or = 2000 ppm), spearmint and basil (3000 ppm) inhibited all the tested fungi completely. Caraway was inhibitory at 2000 ppm against *A. flavus* and *A. parasiticus* and at 3000 ppm against *A. ochraceaus*. Also, the three species were suppressed by anise at < or = 500 ppm. An *in vitro* initial screening of a range of several spice hydrosols on inhibition of mycelial growth of *A. parasiticus* revealed that hydrosols of anise, cumin (*Cuminum cyminum* L.), fennel (*Foeniculum vulgare* Mill.), *Mentha* sp., oregano, savoury and thyme caused a stronger inhibitory effect on mycelial growth (Özcan, 2005).

When essential oil from oregano was applied as a fumigant against the mycelia and spores of *A. flavus*, *A. niger* and *A. ochraceus* on wheat, the oil vapour exhibited a fungicidal effect and a significant reduction in the per cent of infested grain was observed (Paster *et al.*, 1995). Plant extracts of *Z. officinale*, bulbs of *A. sativum* and *A. cepa*, leaves of *A. vasica*, *L. alba*, *A. indica*, *A. aspera*, stem of *C. reflexa*, root of *V. rosea* and seeds of *N. sativa* were tested for their efficacy *in vitro* against *Aspergillus* sp. and *Penicillium* sp. in wheat. All the plant extracts reduced the incidence of seedborne fungi significantly and increased seed germination, the number of healthy seedlings and the vigour index. Neem and garlic extracts controlled the intensity of the fungi completely (Hasan *et al.*, 2005).

A natural fungicide against aflatoxigenic fungi to protect stored rice using the essential oil of *C. citratus* was developed by Paranagama *et al.* (2003). Lemongrass oil was tested against *A. flavus* and the test oil was fungistatic and fungicidal against the test pathogen at 0.6 and 1.0 mg/ml, respectively. Aflatoxin production was completely inhibited at 0.1 mg/ml. The results obtained from the thin layer chromatographic bioassay and gas chromatography indicated citral a and b as the fungicidal constituents in lemongrass oil. During the fumigant toxicity assay of lemongrass oil, the sporulation and the mycelial growth of the test pathogen were inhibited at concentrations of 2.80 and 3.46 mg/ml, respectively. Therefore, lemongrass oil could be used to manage aflatoxin formation and fungal growth of *A. flavus* in stored rice. Besides, the essential oil of lemongrass inhibited growth of moulds like *A. flavus*, *A. fumigatus* and *P. chrysogenum* of maize and cowpea grains. Within a storage period of 10 days, seeds of maize and cowpea treated with lemongrass powder and essential oil showed no physical deterioration. Off-colour, off-odour and mouldiness, however, characterized untreated control seeds (Adegoke and Odelusola, 1996). Another assay on *A. flavus* determined optimal levels of dosages of 11 plant essential oils for maize kernel protection, effects of combinations and residual effects (Montes-Belmont and Carvajal, 1998).

Bankole (1997) showed that essential oils from *A. indica* and *Morinda lucida* Benth. (brimstone tree) inhibited the growth of a toxigenic *A. flavus* and reduced aflatoxin B_1 synthesis significantly in inoculated maize grains. Studies in experimental grain bins have demonstrated that soybean oil alone also reduces infection by storage fungi (White and Toman, 1994). After 12 months, kernel infection by *Penicillium* spp. and *Aspergillus* spp. was 83% and 63.7%, respectively, in untreated corn, compared to 60% and 46.2%, in soybean oil-treated corn at 200 ppm (McGee, 1989). Essential oils from aromatic plants such as cinnamon, clove, oregano, savoury and thyme inhibited the growth of the corn pathogen *Penicillium* sp. completely *in vitro*. The MIC of the essential oils in the laboratory was 800 ppm. The growing seedlings were not affected and no phytotoxicity symptoms were seen at rates up to 16,000 ppm concentration of the oils (Goggi *et al.*, 2008). A previous work was undertaken by Chatterjee (1990) to screen some essential oils for their inhibitory activity against fungal infection and mycelial growth in postharvest maize grains during storage. It was observed that the oils of *Cassia* sp. clove (30 μl/g grain and above), star anise (*Illicium verum* Hooker fil.) (40 μl/g grain and above), *Geranium* sp. (30 μl/g grain and above) and basil (50 μl/g grain) inhibited the *in vivo* mycelial growth of established seedborne infections of *A. flavus*, as well as preventing infection following inoculation with *A. flavus*, *A. glaucus*, *A. niger* and *A. sydowi*. These oils also preserved the grain from natural *A. flavus* infection during the experimental period. Christian and Goggi (2008) studied whether essential oils could be used as a contact fungicide seed treatment for organic corn. *In vitro*, the essential oils of cinnamon, clove, oregano, savoury and thyme controlled *Penicillium* completely. Soybean oil, applied at a rate used to suppress grain dust, reduced storage fungi growth in maize and soybeans in field storage bins. After 12 months, soybean seed infection by *Penicillium* spp. and *Aspergillus* spp. was 45.7% and 39.2%, respectively, in untreated seeds, 17.7% and 8.2% in soybean oil-treated seeds and 1.7% and 2% in soybean oil + thiabendazole-treated seeds (McGee, 1989; White and Toman, 1994). Also, soybean oil demonstrated its effectiveness in decreasing by 50% the levels of seed infection and physiological ageing by the storage fungus, *A. ruber*, on garden pea seeds (*Pisum sativum* L.) (Hall and Harman, 1991). Peppermint, thyme and clove oils were tested *in vivo* against *A. flavus*, *A. niger* and *P. chrysogenum* on different seed cultivars of cowpea. Antifungal activity was observed for the three oils, depending on cultivar and concentrations (Kritzinger *et al.*, 2002). In blackgram, essential oils extracted from wood chips of cedar and that from the seeds of ajowan exhibited absolute toxicity, inhibiting the mycelial growth of *A. niger* on storage seeds (Singh and Tripathi, 1999).

Major seedborne fungi associated with African yam bean like *A. niger* and *A. flavus* could be controlled by using leaf extracts (crude and aqueous) of basil, bitter leaf, neem and pawpaw. All the plants' leaf extracts reduced significantly the incidence of fungi tested and increased seed germination and seedling emergence when compared with the untreated controls. The crude extracts were most effective, mainly neem, which gave complete control of *A. niger* and *A. flavus*. In addition, seed germination was enhanced by this extract and reached nearly 90% (Nwachukwu and Umechuruba, 2001). *A. flavus* is also found infesting seeds of guar, *Cyamopsis tetragonoloba* (L.) Taub., a native plant of India, whose main commercial value is due to its seed gum (galactomannan gum). In this case, *A. flavus* was reduced by cumin oil extracted from seeds (Dwivedi *et al.*, 1991). Studies carried out have shown that cumin has powerful antimicrobial properties against diverse species of bacteria and fungi. The chemical studies indicated that the greater part of this antimicrobial activity might be attributed to the cuminaldehyde [*p*-isopropil benzaldehyde] that is present in the dried fruit of this plant (De *et al.*, 2003).

Another study (de Souza *et al.*, 2003) investigated the mycoflora and physiological quality of cotton seeds treated with chemical fungicides and aroeira extract. Pure extract did not control the fungal population but, mixed with the fungicides, captan, benomyl and tolylfluanid, it showed reduction in the incidence of *Aspergillus* sp. Garlic extract was also found to be effective in removal of the seedborne pathogens of mustard, including species of *Aspergillus* and *Penicillium* (Latif *et al.*, 2006).

Fungi of the genera *Aspergillus* and *Penicillium* are widely distributed storage fungi of egusi melon seeds, causing seed discoloration, decreased nutritive value, increase in free fatty acid and peroxide values, decreased seed germination and producing a number of toxic metabolites, including aflatoxin. Four mould species, *A. flavus*, *A. niger*, *A. tamarii* and *P. citrinum*, were inoculated on to shelled melon seeds. The essential oil of *C. citratus* at 0.1 and 0.25 ml/100 g seeds reduced deterioration and aflatoxin production significantly in shelled seeds inoculated with *A. flavus*. At higher dosages (0.5 and 1.0 ml/100 g seeds), the essential oil prevented aflatoxin production completely. After 6 months in farmers' stores, unshelled melon seeds treated with 0.5 ml/100 g seeds of essential oil had a significantly lower proportion of visibly diseased seeds and *Aspergillus* spp. infestation levels and significantly higher seed germination compared to the untreated seeds. The efficacy of the essential oil in preserving the quality of melon seeds in stores was statistically on a par with that of fungicide (iprodione) treatment (Bankole *et al.*, 2005).

Conclusions

While modern agricultural practices have resulted in higher and more stable yields, they have also weakened the natural balance between pests and their antagonists and have reduced soil fertility and health. Harmful chemicals threaten both the environment and human health alike. The benefits of pesticides, in terms of reduced crop losses, are often overestimated because the viability of alternative pest management approaches is not fully understood. Conversely, the costs of relying predominantly on synthetic pesticides in pest control, in terms of health, environment, agroecology and trade, are also not known completely and consequently are often underestimated (SP-IPM, 2008). Integrated pest management (IPM) has emerged as a way towards maintaining or increasing agricultural productivity without over-reliance on synthetic chemical pesticides, emphasizes the growth of a healthy crop with the least possible disruption of agroecosystems and encourages natural pest control mechanisms (FAO, 2002). In this context, development of simple and eco-friendly seedborne disease management methods is necessary to improve the quality of seed in general and farmers' saved-seed in particular (Elisabeth *et al.*, 2008).

Plant-derived compounds as crop protectants represent a vast and rapidly progressing resource. Botanical fungicides are best suited for use in industrialized countries when strict enforcement of pesticide regulations is impractical, or in the case of organic production. However, they can play a much greater role in protecting crops in developing countries, where human pesticide poisonings are most prevalent. Among the plant products, essential oils especially are a very attractive method of controlling plant diseases. Essential oils and their components are gaining increasing interest because of their relatively safe status, their wide acceptance by consumers and their exploitation for potential multi-purpose use. Besides, the problem of developing resistant strains of fungi may be solved by the use of essential oils of higher plants as fumigants in the management of fungal pathogens because of synergism between different components of the oils (Varma and Dubey, 1999; Dubey *et al.*, 2008).

In recent years, tremendous strides have been made in advancing the study of the natural control of plant pathogens, particularly seedborne fungi. As is shown in this chapter, plant metabolites and plant-based fungicides appear to be one of the better alternatives, as they are known to have minimal environmental impact and danger to consumers in contrast to synthetic pesticides. Despite the potential of these naturally occurring biochemicals as biorational fungicides, their practical development and implementation will require more detailed studies. Efforts should be made to search for indigenous plants as a source of antifungal compounds and to bioprospect the antifungal properties of these plant products, especially essential oils, towards seed fungi. Field trials are required to assess the practical applicability of botanical pesticides, together with bulk production, extensive usage of active compounds and interaction with other IPM components. Biosafety studies should be conducted to ascertain their toxicity to humans, animals and crop plants. Additional screenings might be focused on the quality assurance of botanicals and its regulation.

While it is unlikely that biopesticides will replace chemical pesticides completely in the foreseeable future, we can expect that there will be some decline in the use of chemicals, particularly in developed countries. Exploitation of naturally available chemicals from plants, which retard the reproduction of undesirable microorganisms, would be a more realistic and ecologically sound method for plant protection and will have a prominent role in the development of future commercial pesticides for crop protection strategies, with special reference to the management of plant diseases (Varma and Dubey, 1999; Gottlieb *et al.*, 2002). The prospect of botanical products as fungicides includes plant compounds with broad-spectrum activity to provide protection against a range of pathogenic fungi that attack the plant at the same or subsequent growth stages following their application. Furthermore, essential oils are made up of many components that may have synergistic effects; it may therefore be expected that blends of essential oils or oil components will be produced to control a wide range of fungal species (Szczerbanik *et al.*, 2007). In the coming years, we envisage a broader appreciation of the attributes of alternative methods and expect to see synergistic combinations of semi-chemicals with other technologies that will enhance the effectiveness and sustainability of integrated control.

References

Abdelmonem, A.M., Rasmy, M.R. and Hanafy, S.M. (2001) Non-chemical trials to minimize seedborne infection of some seed crops. *Proceedings of the 26th ISTA Congress,* 14–22 June 2001, Angers, France. International Seed Testing Association, Zürich, Switzerland, p. 58.

Adegoke, G.O. and Odelusola, B.A. (1996) Storage of maize and cowpea and inhibition of microbial agents of biodeterioration using the powder and essential oil of *Cymbopogon citratus. International Biodeterioration and Biodegradation* 37, 81–84.

Adekunle, A.A. and Uma, N.U. (2005) Effect of benlate solution, crude leaf extracts of *Azadarichta indica* and *Ocimum gratissimum* on growth of fungi and preservation of melon seeds. *Plant Pathology Journal* 4, 29–34.

Ahmed, M.F., Khalequzzaman, K.M., Islam, M.N., Anam, M.K. and Islam, M.T. (2002) Effect of plant extracts against *Bipolaris oryzae* of rice under *in vitro* conditions. *Pakistan Journal of Biological Sciences* 5, 442–445.

Ahmed, N. and Shultana, K. (1984) Fungitoxic effect of garlic on treatment of jute seed. *Bangladesh Journal of Botany* 13, 130–136.

Alice, D. and Rao, A.V. (1986) Management of seedborne *Drechslera oryzae* of rice with plant extracts. *International Rice Research Newsletter* 11, 19.

Arras, G. (1988) Antimicrobial activity of various essential oils against some citrus fruit disease agents. In: Goren, R. and Mendel, K. (eds) *Citriculture. Volume 2. Cultural Practices, Diseases and Nematodes. Proceedings of the Sixth International Citrus Congress, Middle-East, Tel Aviv, Israel.* Balaban Publishers, Rehovot, Israel, pp. 787–793.

Arras, G., Piga, A. and D'hallewin, G. (1993) The use of *Thymus capitatus* essential oil under vacuum conditions to control *Penicillium digitatum* development on citrus fruit. *Acta Horticulturae* 344, 147–153.

Asplund, R.O. (1968) Monoterpenes: relationship between structure and inhibition of germination. *Phytochemistry* 7, 1995–1997.

Bankole, S.A. (1997) Effect of essential oil from two Nigerian medicinal plants (*Azadirachta indica* and *Morinda lucida*) on growth and aflatoxin B_1 production in maize grain by a toxigenic *Aspergillus flavus. Letters in Applied Microbiology* 24, 190–192.

Bankole, S.A., Joda, O. and Ashidi, J.S. (2005) The use of powder and essential oil of *Cymbopogon citratus* against mould deterioration and aflatoxin contamination of 'egusi' melon seeds. *Journal of Basic Microbiology* 45, 20–30.

Beg, A.Z. and Ahmad, I. (2002) *In vitro* fungitoxicity of the essential oil of *Syzygium aromaticum. World Journal of Microbiology and Biotechnology* 18, 317–319.

Bhaskara, R.M.V., Angers, P., Gosselin, A. and Arul, J. (1998) Characterization and use of essential oil from *Thymus vulgaris* against *Botrytis cinerea* and *Rhizopus stolonifer* in strawberry fruits. *Phytochemistry* 47, 1515–1520.

Billerbeck, V.G. de, Roques, C.G., Bessiere, J.M., Fonvieille, J.L. and Dargent, R. (2001) Effects of *Cymbopogon nardus* (L.) W. Watson essential oil on the growth and morphogenesis of *Aspergillus niger. Canadian Journal of Microbiology* 47, 9–17.

Cabo, J., Crespo,, M.E., Jiménez, J. and Navarro, C. (1986) A study of the essences from *Thymus hyemalis* collected in three different localities. *Fitoterapia* 57, 117–119.

Chatterjee, D. (1990) Inhibition of fungal growth and infection in maize grains by spice oils. *Letters in Applied Microbiology* 11, 148–151.

Christensen, C.M. and Kaufmann, H.H. (1965) Deterioration of stored grains by fungi. *Annual Review of Phytopathology* 3, 69–84.

Christian, E.J. and Goggi, A.S. (2008) Aromatic plant oils as fungicide for organic corn production. *Crop Science* 48, 1941–1951.

De, M., De, A.K., Mukhopadhyay, R., Banerjee, A.B. and Miró, Y.M. (2003) Actividad antimicrobiana de *Cuminum cyminum* L. *Ars Pharmaceutica* 44, 257–269.
Di Pasqua, R. (2006) Attività antimicrobica di oli essenziali e dei loro costituenti contro microrganismi patogeni e alterativi di origine alimentare. Dottorato thesis, Università degli Studi di Napoli Federico II, Italy, 121 pp.
Di Pasqua, R., De Feo, V., Villani, F. and Mauriello, G. (2005) *In vitro* antimicrobial activity of essential oils from Mediterranean *Apiaceae*, *Verbenaceae* and *Lamiaceae* against pathogens and spoilage bacteria. *Annals of Microbiology* 55, 139–143.
Dorman, H.J.D. and Deans, S.G. (2000) Antimicrobial agents from plants: antibacterial activity of plant volatile oils. *Journal of Applied Microbiology* 88, 308–316.
Dubey, N.K., Srivastava, B. and Kumar, A. (2008) Current status of plant products as botanical pesticides in storage pest management. *Journal of Biopesticides* 1, 182–186.
Dwivedi, S.K. and Singh, K.P. (1999) Fungitoxicity of some higher plant products against *Macrophomina phaseolina* (Tassi) Goid. *Flavour and Fragrance Journal* 13, 397–399.
Dwivedi, S.K., Pandey, V.N. and Dubey, N.K. (1991) Effect of essential oils of some higher plants on *Aspergillus flavus* Link. infesting stored seeds of guar (*Cyamopsis tetragonoloba* (L.) Taub. *Flavour and Fragrance Journal* 6, 295–297.
Elisabeth, Z.P., Paco, S., Vibeke, L., Philippe, S., Irenee, S. and Adama, N. (2008) Importance of seed-borne fungi of sorghum and pearl millet in Burkina Faso and their control using plant extracts. *Pakistan Journal of Biological Sciences* 11, 321–331.
Faleiro, M.L., Miguel, M.G., Ladeiro, F., Venâncio, F., Tavares, R., Brito, J.C., Figueiredo, A.C., Barroso, J.G. and Pedro, L.G. (2003) Antimicrobial activity of essential oils isolated from Portuguese endemic species of *Thymus*. *Letters in Applied Microbiology* 36, 35–40.
FAO (2002) *International Code of Conduct on the distribution and use of pesticides* (revised version). Rome, Italy.
Ghfir, B., Fonvieille, J.L. and Dargent, R. (1997) Influence of essential oil of *Hyssopus officinalis* on the chemical composition of the walls of *Aspergillus fumigatus* (Fresenius). *Mycopathologia* 138, 7–12.
Goggi, S., Delate, K., Pollak, L. and Lamkey, K. (2008) Integration of natural seed treatments in organic and open-pollinated corn systems. *Leopold Center Progress Report* 17, 18–20.
Gottlieb, O.R., Borin, M.R. and Brito, N.R. (2002) Integration of ethnobotany and phytochemistry: dream or reality? *Phytochemistry* 60, 145–152.
Hall, J.S. and Harman, G.E. (1991) Efficacy of oil treatments of legume seeds for control of *Aspergillus* and *Zabrotes*. *Crop Protection* 10, 315–319.
Hamburger, M. and Hostettmann, K. (1991) Bioactivity in plants: the link between phytochemistry and medicine. *Phytochemistry* 30, 3864–3874.
Hasan, M.M., Chowdhury, S.P., Shahidul, A., Hossain, B. and Alam, M.S. (2005) Antifungal effects of plant extracts on seed-borne fungi of wheat seed regarding seed germination, seedling health and vigor index. *Pakistan Journal of Biological Sciences* 8, 1284–1289.
Kishore, G.K., Pande, S. and Harish, S. (2007) Evaluation of essential oils and their components for broad-spectrum antifungal activity and control of late leaf spot and crown rot diseases in peanut. *Plant Disease* 91, 375–379.
Knobloch, K., Pauli, A., Iberl, B., Weigand, H. and Weis, N. (1989) Antibacterial and antifungal properties of essential oil components. *Journal of Essential Oil Research* 1, 119–128.
Kritzinger, Q., Aveling, T.A.S. and Marasas, W.F.O. (2002) Effect of essential plant oils on storage fungi, germination and emergence of cowpea seeds. *Seed Science and Technology* 30, 609–619.
Latif, M.A., Saleh, A.K.M., Khan, M.A.I., Rahman, H. and Hossain, M.A. (2006) Efficacy of some plant extracts in controlling seed-borne fungal infections of mustard. *Bangladesh Journal of Microbiology* 23, 168–170.
López, A.G., Theumer, M.G., Zygadlo, J.A. and Rubinstein, H.R. (2004) Aromatic plants essential oils activity on *Fusarium verticillioides* fumonisin B_1 production in corn grain. *Mycopathologia* 158, 343–349.
McGee, D.C. (1989) Suppression of storage fungi in grain with soybean oil. (Abstr.) *Phytopathology* 79, 1140.
McGee, D.C. (1995) Advanced in seed treatment technology. Asian Seed '95. New Delhi, India, 27–29 September 1995 (www.green-seeds.com/pdf/seed_treat.pdf, accessed 15 January 2009).
Marín, S., Velluti, A., Muñoz, A., Ramos, A.J. and Sanchis, V. (2003) Control of fumonisin B_1 accumulation in naturally contaminated maize inoculated with *Fusarium verticillioides* and *Fusarium proliferatum*, by

cinnamon, clove, lemongrass, oregano and palmarosa essential oils. *European Food Research and Technology* 217, 332–337.

Mello, A.F.S., Athaide Sobrino, C. and Moraes, M.H.D. (2005) Effect of neem oil on the incidence of *Macrophomina phaseolina* in seeds of *Vigna unguiculata*. *XIII Congreso Latinoamericano de Fitopatologia y III Taller de la Asociación Argentina de Fitopatólogos, Villa Carlos Paz, Córdoba, Argentina, 19–22 April 2005.* Asociación Latinoamericana de Fitopatologia, Córdoba, Argentina, p. 297.

Montes-Belmont, R. and Carvajal, M. (1998) Control of *Aspergillus flavus* in maize with plant essential oils and their components. *Journal of Food Protection* 61, 616–619.

Montes-Belmont, R. and Flores Monctezuma, H. (2001) Combate de *Fusarium thapsinum* y *Claviceps africana* mediante semillas de sorgo tratadas con productos naturales (http://web.catie.ac.cr/informacion/RMIP/rev61/resinf2.htm, accessed 15 October 2007).

Neergaard, P. (1979) *Seed Pathology*, Volumes I and II. Revised edition. MacMillan Press, London, 1191 pp.

Nguefack, J., Leth, V., Amvam Zollo, P.H. and Mathur, S.B. (2004) Use of three essential oils as seed treatments against seed-borne fungi of rice (*Oryza sativa* L.) under laboratory and field conditions. (Abstract) *Phytopathology* 94, S75.

Nwachukwu, E.O. and Umechuruba, C.I. (2001) Antifungal activities of some leaf extracts on seed-borne fungi of African yam bean seeds, seed germination and seedling emergence. *Journal of Applied Sciences and Environmental Management* 5, 29–32.

Özcan, M. (2005) Effect of spice hydrosols on the growth of *Aspergillus parasiticus* NRRL 2999 strain. *Journal of Medicinal Food* 8, 275–278.

Owolade, O.F., Amusa, A.N. and Osikanlu, Y.O.K. (2000) Efficacy of certain indigenous plant extracts against seedborne infection of *Fusarium moniliforme* on maize (*Zea mays* L.) in south western Nigeria. *Cereal Research Communications* 28, 323–327.

Palaoud, M. (2006) Effect of essential oils on seedborne fungi and quality of rice seed cv. Khao Dawk Mali 105. MSc (Postharvest Technology), Postharvest Technology Institute, Chiang Mai University, Thailand, 163 pp.

Paranagama, P.A. (1991) *Analysis of Sri Lankan Essential Oils by Gas Chromatography and Mass Spectroscopy* (Senanayake, U.M. [ed.]). Industrial Technology Institute, Colombo, Sri Lanka, pp. 1–40.

Paranagama, P.A., Abeysekera, K.H., Abeywickrama, K. and Nugaliyadde, L. (2003) Fungicidal and anti-aflatoxigenic effects of the essential oil of *Cymbopogon citratus* (DC.) Stapf. (lemongrass) against *Aspergillus flavus* Link. isolated from stored rice. *Letters in Applied Microbiology* 37, 86–90.

Parimelazhagan, T. and Francis, K. (1999) Antifungal activity of *Clerodendrum viscosum* against *Curvularia lunata* in rice seeds. *Journal of Mycology and Plant Pathology* 29, 139–141.

Paster, N., Menasherov, M., Ravid, U. and Juven, B. (1995) Antifungal activity of oregano and thyme essential oils applied as fumigants against fungi attacking stored grain. *Journal of Food Protection* 58, 81–85.

Rahman, M.S. and Junaid, M. (2008) Antimicrobial activity of leaf extracts of *Eupatorium triplinerve* Vehl. against some human pathogenic bacteria and phytopathogenic fungi. *Bangladesh Journal of Botany* 37, 89–92.

Rajappan, K., Ushamalini, C., Subramanian, N., Narasimhan, V. and Abdul Kareem, A. (2001) Management of grain discoloration of rice with solvent-free EC formulations of neem and pungam oils. *Phytoparasitica* 29, 171–174.

Rake, K., Khanna, K.K., Sudhir, C., Khanna, R. and Chandra, R. (1989) Effect of homeopathic drugs on seed mycoflora of wheat. *National Academy of Science Letters* 12, 39–41.

Ranasinghe, L., Jayawardena, B. and Abeywickrama, K. (2002) Fungicidal activity of essential oils of *Cinnamomum zeylanicum* (L.) and *Syzygium aromaticum* (L.) Merr et L.M. Perry against crown rot and anthracnose pathogens isolated from banana. *Letters in Applied Microbiology* 35, 208–211.

Rennie, W.J. and Cockerell, V. (2006) Seedborne diseases. In: Cooke, B.M., Jones, D.G. and Kaye, B. (eds) *The Epidemiology of Plant Diseases*, 2nd edition. Springer, Netherlands, pp. 357–372.

Satish, S., Mohana, D.C., Ranhavendra, M.P. and Raveesha, K.A. (2007) Antifungal activity of some plant extracts against important seed borne pathogens of *Aspergillus* sp. *Journal of Agricultural Technology* 3, 109–119.

Schwinn, F.J. (1994) Seed treatment – a panacea for plant protection? In: Martin, T.J. (ed.) *Seed Treatment: Progress and Prospects*. Monograph No. 57. BCPC Publications, Thornton Heath, UK, pp. 3–14.

Shetty, S.A., Prakash, H.S. and Shetty, H.S. (1989) Efficacy of certain plant extracts against seedborne infection of *Trichoconiella padwickii* in paddy (*Oryza sativa*). *Canadian Journal of Botany* 57, 1956–1958.

Singh, J. and Tripathi, N.N. (1999) Inhibition of storage fungi of blackgram (*Vigna mungo* L.) by some essential oils. *Flavour and Fragrance Journal* 14, 1–4.

Sisterna, M. and Dal Bello, G. (2007) Natural plant extracts: an alternative control of seedborne fungi. In: Arya, A. and Mónaco, C. (eds) *Seedborne Diseases: Ecofriendly Management.* Scientific Publishers, India, pp. 15–36.

Sisterna, M. and Lori, G. (2005) Hongos parásitos débiles asociados al manchado del grano de trigo. *XIII Congreso Latinoamericano de Fitopatología y III Taller de la Asociación Argentina de Fitopatólogos, Villa Carlos Paz, Córdoba, Argentina, 19–22 April 2005.* Asociación Latinoamericana de Fitopatologia, Córdoba, Argentina, p. 464.

Sitara, U., Niaz, I., Naseem, J. and Sultana, N. (2008) Antifungal effect of essential oils on *in vitro* growth of pathogenic fungi. *Pakistan Journal of Botany* 40, 409–414.

Soliman, K.M. and Badeaa, R.I. (2002) Effect of oil extracted from some medicinal plants on different mycotoxigenic fungi. *Food Chemistry and Toxicology* 40, 1669–1675.

Somda, I., Leth, V. and Sérémé, P. (2007) Antifungal effect of *Cymbopogon citratus, Eucalyptus camaldulensis* and *Azadirachta indica* oil extracts on sorghum seed-borne fungi. *Asian Journal of Plant Sciences* 6, 1182–1189.

Souza, A.A. de, Lucena Alcântara Bruno R. de, Araújo, E. and Bandeira Bruno, G. (2003) Micoflora e qualidade fisiológica de sementes do algodoeiro tratadas com fungicidas químicos e extrato de aroeira. *Revista Brasileira de Sementes* 25, 56–64.

SP-IPM (2008) Incorporating integrated pest management into national policies. *IPM Research Brief No. 6.* SP-IPM Secretariat, International Institute of Tropical Agriculture (IITA), Ibadan, Nigeria.

Szczerbanik, M., Jobling, J., Morris, S. and Holford, P. (2007) Essential oil vapours control some common postharvest fungal pathogens. *Australian Journal of Experimental Agriculture* 47, 103–109.

Tagne, A., Feujio, T.P. and Sonna, F.P. (2008) Essential oil and plant extracts as substitutes to synthetic fungicides in the control of fungi. ENDURE International Conference 2008 Diversifying Crop Protection. La Grande Motte, France, 12–15 October 2008, C.O. 42 (http://www.endure-network.eu/international_conference_2008/proceedings/wednesday_october_15, accessed 2 February 2009).

Tequida-Meneses, M., Cortez-Rocha, M., Rosas-Burgos, E.C., López-Sandoval, S. and Corrales-Maldonado, C. (2002) Efecto de extractos alcohólicos de plantas silvestres sobre la inhibición de crecimiento de *Aspergillus flavus, Aspergillus niger, Penicillium chrysogenum, Penicillium expansum, Fusarium moniliforme* y *Fusarium poae. Revista Iberoamericana de Micología* 19, 84–88.

Thiribhuvanamala, G. and Narasimhan, V. (1998) Efficacy of plant extracts on seedborne pathogens of sunflower. *Madras Agricultural Journal* 85, 227–230.

Vanneste, J.L., Hill, R.A., Kay, J.S., Farrel, R.L. and Holland, P.T. (2002) Biological control of sapstain fungi with natural products and biological control agents: a review of the work carried out in New Zealand. *Mycological Research* 106, 228–232.

Varma, J. and Dubey, N.K. (1999) Prospective of botanical and microbial products as pesticides of tomorrow. *Current Science* 76, 172–179.

Velluti, A., Sanchis, V., Ramos, A.J., Turon, C. and Marín, S. (2004) Impact of essential oils on growth rate, zearalenone and deoxynivalenol production by *Fusarium graminearum* under different temperature and water activity conditions in maize grain. *Journal of Applied Microbiology* 96, 716–724.

White, D.G. and Toman, J. Jr (1994) Effects of postharvest oil and fungicide application on storage fungi in corn following high-temperature grain drying. *Plant Disease* 78, 38–43.

Zambonelli, A., D'Aulerio, A.Z., Severi, A., Benvenuti, S., Maggi, L. and Bianchi, A. (2004) Chemical composition and fungicidal activity of commercial essential oils of *Thymus vulgaris* L. *Journal of Essential Oil Research* 16, 69–74.

Part II

Disease Control Through Resistance

6 Resistance to *Septoria* Leaf Blotch in Wheat

María R. Simón
Cerealicultura, Facultad de Ciencias Agrarias y Forestales, Universidad Nacional de La Plata, La Plata, Argentina

Abstract
Mycosphaerella graminicola (Fuckel) Schroeter, in Cohn, is the causal agent of *Septoria* leaf blotch, an important disease in many wheat-producing areas of the world which causes significant yield losses. Breeding for resistance is the most economical approach to control the disease. Advances in the genetics of resistance and genetic variation of the pathogen population, as well as the new tools for a more efficient incorporation of resistance in breeding programmes, are discussed.

Introduction

Bread wheat (*Triticum aestivum* L.) is the most widely grown and consumed food crop in the world. It is the staple food of nearly 35% of the world population and the demand for wheat will grow faster than for any other major crop (Rajaram, 1999). The forecast global demand for wheat in the year 2020 varies between 840 (Rosegrant *et al.*, 1995) to 1050 Mt (Kronstad, 1998). To meet this demand, global production will need to increase by 1.6–2.6% annually from the present production level of 620 Mt.

Wheat breeding is focused on developing widely adapted, disease-resistant genotypes with high yields that are stable across a wide range of environments. Incorporating durable resistance is a priority since breeding for stable yields without adequate resistance against the major diseases would be impossible (Rajaram, 1999).

Diseases of wheat, mostly caused by fungal pathogens and a few by viruses and bacteria, are important production constraints in almost all wheat-growing environments (Rajaram and van Ginkel, 1996; McIntosh, 1998). Globally important fungal diseases of wheat caused by obligate parasites include the three rusts (leaf rust, caused by *Puccinia triticina* Eriks., yellow rust caused by *P. striiformis* West f. sp. *tritici* Eriks. and stem rust caused by *P. graminis* Pers. f. sp. *tritici* Eriks & Henn); powdery mildew caused by *Blumeria graminis tritici* (DC) Speer; asexual form *Oidium monilioides* (Nees) Link; stinking smut (*Tilletia caries* (DC) Tul. and C. Tul. and *T. foetida* (Wallr. Liro.); loose smut (*Ustilago tritici* (Pers.) Rostr.); *U. nuda* (J.L. Jensen) Kellerm. and Swingle. Those caused by facultative parasites include leaf blotch, *M. graminicola* (Fuckel) J. Schröt., in Cohn, asexual form *S. tritici* Rob ex Desm.; glume blotch (*Phaeosphaeria nodorum*, asexual form *Stagonospora nodorum* blotch); spot blotch (*Cochliobolus sativus* (Ito and Kuribashani) Drechs. ex Dastus, asexual form *Bipolaris sorokiniana* (Sacc.) Shoem.); tan spot,

 Management of Fungal Plant Pathogens (eds A. Arya and A.E. Perelló)

Pyrenophora tritici-repentis (Died.) Drechs., asexual form *Drechslera tritici-repentis* (Died.) Shoemaker; *Alternaria* spp. (belonging to the *A. infectoria* species groups); scab, *Fusarium graminearum* Schwabe, take all (*Gaeumannomyces graminis* (Sacc.) von Arx and Olivier var. *tritici* Walker).

Leaf blotch causes important yield losses in many countries. Yield reductions range from 31 to 54% (Eyal *et al.*, 1987), from 10 to 45% (Caldwell and Narvaez, 1960) and even yield losses higher than 60% have been reported (Shipton *et al.*, 1971). Sanderson (1972) proved the connection between the two stages and the sexual (teleomorph) form has been reported in several countries (Hunter *et al.*, 1999). The sexual stage in Argentina was reported by Cordo *et al.* (1990).

Mycosphaerella graminicola is a hemibiotrophic pathogen; early infection is biotrophic, followed by a switch to necrophic growth just prior to symptom expression. The sexual stage is also known to play a role in the disease cycle. It causes most of the initial infection of winter wheat crops during autumn in the UK (Shaw and Royle, 1989) and the USA (Schuh, 1990). In Argentina, an increase in ascospores at harvest time has been reported, suggesting that the sexual stage may be important to initiate the infection in the next growing season. Following stem elongation, infection of the upper leaves of a crop has been thought to be entirely due to the asexual stage of the fungus, in which pycnidia give rise to splash-dispersed pycnidiospores, which are splash-dispersed from infected basal tissue to the upper leaves by raindrops. However, more recent work has shown that upward movement of inoculum can occur in the absence of splashy rainfall, being influenced by the position of developing leaves in relation to infected leaf layers (Lovell *et al.*, 1997). Another possible means of spread within a crop during summer is by airborne ascospores, which may play a role more important than previously recognized (Hunter *et al.*, 1999).

Types of Resistance

Although several control methods, including cultural practices and the use of fungicides, may reduce the effect of *S. tritici* blotch, genetic resistance is the most cost-effective and environmentally safe technique to manage the disease.

Monogenic or oligogenic and polygenic resistance coexist in the pathosystem *T. aestivum*/*M. graminicola*. Monogenic or oligogenic resistance is generally near complete, isolate specific, follows the 'gene-for-gene' mode of inheritance and has been found in several genotypes (Rillo and Caldwell, 1966; Rosielle and Brown, 1979; Lee and Gough, 1984; Somasco *et al.*, 1996; Arraiano *et al.*, 2001; Brading *et al.*, 2002; McCartney *et al.*, 2002). Polygenic resistance is generally partial and isolate non-specific and is also present in several genotypes (Jlibene *et al.*, 1994; Simón and Cordo, 1997, 1998; Brown *et al.*, 2001; Zhang *et al.*, 2001; Chartrain *et al.*, 2004b).

Partial resistance is expressed as a reduced epidemic development and is supposed to be durable. Several components contribute to the epidemic-retarding effect. Parlevliet (1979) mentioned four partial resistance components: infection frequency, latent period, spore production and infection period. The earliest studies on this type of resistance, previous to the mapping of genes and QTLs, investigated the gene effects conditioning these components.

Several of the components of partial resistance to *M. graminicola* may be controlled by just a few genes (Jlibene *et al.*, 1994). Danon and Eyal (1990) determined that additive effects for pycnidial coverage were the major variance component, although dominance effects were also significant. Jlibene *et al.* (1994) found that general combining ability (GCA) effects accounted for most of the variation of percentage pycnidial coverage, although specific combining ability (SCA) effects were detected in some crosses. Simón and Cordo (1997, 1998) determined that GCA was preponderant for incubation period, latent period, pycnidial coverage and spore production, although SCA was also significant. Incubation period was inherited independently of maturation period and pycnidial coverage. Those components that are genetically different and independent could be combined into the same genetic background by crossing (van Ginkel and Rajaram,

1999), increasing the level of durable resistance. Significant correlations were found between pycnidia/cm^2 and spore/ml, indicating the feasibility of selecting for a lower pycnidial density in order to obtain a reduction in spore production (Simón and Cordo, 1998). Heritability tends to be only moderate (Simón *et al.*, 1998), but progress in breeding for resistance may still be possible. Major genes are interesting because of the high level of resistance and thus an almost complete absence of symptoms in the host; partial resistance, however, is very important due to its putative durability and its expression under a broad spectrum of isolates of the pathogen. A few genes may be enough to confer resistance that will hold up in farmers' fields (Dubin and Rajaram, 1996).

Resistance conditioned by a single dominant gene was assigned to some cultivars as Bulgaria 88 (Rillo and Caldwell, 1966), Oasis (Shaner and Buechley, 1989), Veranopolis (Wilson, 1979) and others. Later, genes were located and it was found for example that *Stb1* conditioned resistance in Bulgaria 88 and Oasis, *Stb2* in Veranopolis, etc. Some other cultivars showed resistance conditioned by several major genes as Kavkaz 4500 L.6.A.4. (Jlibene *et al.*, 1992) and the genes were identified (*Stb6*, *Stb7*, *Stb10* and *Stb12*; Chartrain *et al.*, 2005a). Also, three major genes were identified in the Portuguese line TE 9111 (*Stb6*, *Stb7* and *Stb11*; Chartrain *et al.*, 2005b). Furthermore, commercially grown cultivars range from moderately resistant to susceptible, indicating the presence of partial resistance. Chartrain *et al.* (2004b) found high partial resistance levels in several wheat cultivars from Europe and Mexico (Arina, Milan, Senat). Simón *et al.* (2005a) also found high levels of partial resistance in some Argentinian cultivars effective to several isolates (Klein Volcán, Klein Dragón) in adult stage. Some germplasm as the Portuguese line TE 9111 (Chartrain *et al.*, 2005b) also has been proved to carry several major genes together with partial resistance.

Location of the Resistance

Studies on the location of resistance began during the past decade. Some of them investigated the chromosomal location of resistance using substitution lines. Resistance was found to be located on chromosome 7D from a synthetic hexaploid wheat (*T. dicoccoides* × *T. tauschii*) in seedling and adult stage to some specific isolates (Simón *et al.*, 2001, 2005b). Also, resistance was found in chromosomes 1B at the seedling stage and on 5D at the seedling and adult stage of the *T. aestivum* cv. Cheyenne (Simón *et al.*, 2001, 2005b); on the 2B, 3A and 3B of the *T. aestivum* cv. Cappelle-Desprez at the seedling stage and on 6D and 7D of *T. spelta* with some specific isolates (Simón *et al.*, 2001, 2005b).

During the past decade, several genes (Table 6.1) and QTLs (Table 6.2) have been located. Some of them have proved to be effective to isolates from several regions in the world. Simón *et al.* (2007) tagged, using isolates from Argentina, a gene in the 7D chromosome of *Aegilops tauschii*, which is likely *Stb5*. This would indicate that the presence of *Stb5* ensures resistance against some isolates from both Europe (Portugal, The Netherlands) (Arraiano *et al.*, 2001) and South America (Argentina).

Breeding for Resistance

The incorporation of resistance to the pathogen has been slow for several reasons, among them:

1. The high variability of the pathogen population.
2. The lack of knowledge of the virulence spectrum.
3. The lack of relationship in the expression of resistance in seedling and adult stage.
4. The influence of heading date and plant height on resistance and the difficulty in assessing real values in breeding programmes.

Variability of the pathogen population

The population of the pathogen has been studied and a high variability has been found. Variation in virulence patterns within

Table 6.1. Major genes conditioning resistance to *Mycosphaerella graminicola* identified in hexaploid wheat.

Locus	Chromosomal location	Linked markers	Reference
Stb1	5BL	*Xbarc74*, *Xgwm335*	Adhikari *et al.*, 2004c
Stb2	3BS	*Xgwn389*, *Xgwm533.1*, *Xbarc133*, *Xbarc75*, *Xgwm493*	Adhikari *et al.*, 2004b
Stb3	6DS	*Xgdm132*	Adhikari *et al.*, 2004b
Stb4	7DS	*Xgwm44*, *RC3*, *Xgwm11*, *Xgwm437*, *Xgwm121*	Adhikari *et al.*, 2004a
Stb5	7DS	*Xgwm44*, *RC3*, *Xgwm111*, *Xgwm437*, *Xgwm121*	Arraiano *et al.*, 2001
Stb6	3AS	*Xgwm369*, *Xwmc11*	Brading *et al.*, 2002
Stb7	4AL	*Xgwm160*, *Xwmc219*, *Xwmc313*	McCartney *et al.*, 2002
Stb8	7BL	*Xgwm146*, *Xgwm577*, *Xgwm611*	Adhikari *et al.*, 2003
Stb9	2B		Chartrain, 2004
Stb10	1D	*Xgwm848*	Chartrain *et al.*, 2005a
Stb11	1BS	*Xbarc008*, *Xbarc137*	Chartrain *et al.*, 2005b
Stb12	4AL	*Xwmc219*, *Xwmc313*	Chartrain *et al.*, 2005a
Stb13	7BL	*Xwmc396-7B*	Cowling *et al.*, 2007
Stb14	3BS	*Xwm632-3B*	Brule Babel, 2007
Stb15	6AS	*Xpsr563a*, *Xpsr904*	Arraiano *et al.*, 2007

Table 6.2. Quantitative trait loci (QTLs) conditioning resistance to *Mycosphaerella graminicola* in hexaploid wheat.

Locus	Chromosomal location	Linked marker	Reference
QStb.risø-2B	2BL	*Xwmc1575a-Xwmc175a*	Eriksen *et al.*, 2003
QStb.risø-3A.1	3AS	*Xgwm369*	Eriksen *et al.*, 2003
QStb.risø-3A.2	3BL	*Xwmc489-Xwmc505*	Eriksen *et al.*, 2003
QStb.risø-3B	3B	*M62/P38-373*	Eriksen *et al.*, 2003
QStb.risø-6B.2	6B	*Xwmc397-Xwmc341*	Eriksen *et al.*, 2003
QtStb.risø-7B	7B	*M49/P38-229-M49/P11-229*	Eriksen *et al.*, 2003
QStb.ipk-1D	1D (seedlings)	*Xmwg938a*	Simón *et al.*, 2004a
QStb.ipk-2D	2D (seedlings)	*Xcdo405a*	Simón *et al.*, 2004a
QStb.ipk-6B	6B (seedlings)	*Xksuh4b*	Simón *et al.*, 2004a
QStb.ipk-3D	3D (adult)	*Xbcd515*	Simón *et al.*, 2004a
QStb.ipk-7B	7B (adult)	*Xksud2a*	Simón *et al.*, 2004a

and between populations was shown by assessing host response on a selected set of cultivars, with little similarity between the results obtained with various sets of differentials (Eyal *et al.*, 1995). Evidence for specificity was also confirmed by several researchers (Danon and Eyal, 1990; Kema and van Silfhout, 1997; Simón *et al.*, 2005a). Non-specific resistance to a wide set of isolates was also found (Simón *et al.*, 2005a). During the past decades, the population has been studied using molecular markers and a high variability within populations has been confirmed (Chen and McDonald, 1996; Zhan *et al.*, 2001, 2003; Cordo *et al.*, 2007). The sexual state might have an impact on the virulence spectrum in regions where pseudothecia were found and ascospore dispersal coincided with the wheat growing cycle (Shaw and Royle, 1989; Lovell *et al.*, 1997). No attempts to determine races have been carried out.

Recently, the genome of the pathogen was sequenced completely (Goodwin *et al.*,

2007). The essentially finished sequence contains 18 chromosomes from telomere to telomere, plus five fragments, which presumably make up two additional chromosomes. A comparative bioinformatics analysis of *M. graminicola* with seven other sequenced fungal genomes revealed that it possessed fewer enzymes than expected for degrading plant cell walls. The frequency of transposable elements in the genome of the pathogen was intermediate between those of other sequenced fungi. Availability of the finished genome for *M. graminicola* should aid research on this organism greatly and will help in the understanding of its interaction with wheat.

Expression of resistance in seedlings and adults

Resistance is sometimes expressed in seedlings, sometimes at adult stage and sometimes at both stages (Kema and van Silfhout, 1997). Some germplasm with resistance at both stages have been found (Arama, 1996; Somasco *et al.*, 1996; Simón *et al.*, 2005a).

Influence of heading date and plant height on resistance

One complicating factor for the assessment of resistance level has been the influence of heading date and plant height on the expression of resistance. Several scientists have reported an increased disease level in earlier heading or shorter cultivars (Eyal *et al.*, 1987; van Beuningen and Kohli, 1990; Camacho Casas *et al.*, 1995; Chartrain *et al.*, 2004a). Baltazar *et al.* (1990) suggested a genetic association between shortness and susceptibility, while Eyal (1981) and Rosielle and Boyd (1985) assumed a genetic association between earliness and susceptibility. Arama *et al.* (1999), Simón *et al.* (2005a) and Arraiano *et al.* (2006) reported no genetic association among those traits. Simón *et al.* (2004b, 2005a) determined that there was no influence of heading date when cultivars were evaluated at the same development stage under similar weather conditions and found that the relationship between those traits was caused mainly by environmental and epidemiological factors. Associations between pycnidial coverage percentage and days to heading were positive or negative, depending on whether weather conditions before the evaluations were more conducive to the development of the disease in late or early heading cultivars, respectively. Negative associations with plant height were only present in the years where weather conditions were less conducive to the development of the disease. Inconducive conditions and longer distances between leaves in tall cultivars could have reduced the rain-splash dispersal of pycnidiospores, thus causing this negative association, mainly when the sexual form is not present.

In most cases, previous reported associations between heading date and resistance could be attributed to the fact that the disease was scored at the same time but not at the same growth stage, causing early maturing lines to be exposed to inoculum for a longer period than later maturing leaves. Simón *et al.* (2009 unpublished) mapped a population derived from *T. spelta* 7D/Chinese Spring where QTLs conditioning resistance were found, but no genes for heading date were present. Also, some QTLs for resistance were mapped in a Synthetic 6 × (*T. tauschii* × Altar 84) × Opata 85 (Simón *et al.*, 2004a), which did not coincide with the regions where QTLs for flowering time were previously mapped. Eriksen *et al.* (2003) located in a double haploid population originated from the cross of Savanah and Senat, a QTL for increasing plant height linked to a QTL for resistance. Although associations could exist in some germplasm, pleiotropic effects have not been detected and breeders can select for *S. tritici* blotch resistance within a range of heading dates and plant heights.

Resistance and Integrated Management

It is necessary to consider that integrated management can contribute to the durability of resistance. Epidemiological advantages can be derived by combining management practices and through disease management on a regional scale. Diversifying sources of partial resistance, on a field or regional basis, might slow pathogen adaptation. Populations of *M. graminicola* sampled from mixtures of a susceptible and a partially resistant wheat cultivar were all less fit than populations derived from the same cultivars grown in pure stand (Mundt *et al.*, 2002).

Cultural practices such as adequate tillage method, planting density and *N*-fertilization conditions, together with fungicide applications, are important to the appropriated expression of resistance. The planting of no-till wheat may increase the level of *Septoria* leaf blotch because increasing levels of crop residue on the soil surface potentially increase primary inoculum of plant pathogens, mainly under continuous wheat production or wheat/soybean sequences in the same year. Since the pathogen can survive in infested wheat residues for about 2 years, a rotation where wheat is planted in only 1 of 3 years is recommended. Although there are contrasting results, several reports indicate that, under conducive conditions for the development of the disease, an increase in *N*-fertilization causes a slight increase in severity (Hayden *et al.*, 1994; Howard *et al.*, 1994; Leitch and Jenkins, 1995; Simón *et al.*, 2002, 2003).

Conclusions

Research on *Septoria* leaf blotch has expanded greatly in the past decades. New molecular tools enable the exploration of biological issues associated with the pathogen, the host and the host–pathogen interaction. Several genes and QTLs have been identified and mapped. The major challenge to wheat breeders and plant pathologists is the selection and development of cultivars with durable resistance. To achieve this goal, the incorporation of marker-assisted selection into breeding programmes will speed pyramiding several genes or QTLs effective at different stages of plant development into single wheat cultivars to develop broad-spectrum and durable resistance.

Management of cultivars should be optimized to minimize the associations between heading date, height and resistance, but selection of early and short lines with high levels of quantitative resistance is possible. Progress in the analysis of variability and virulence patterns of the pathogen population is also necessary to test the available germplasm with representative isolates. Durability of the resistance can be enhanced by appropriate cultural practices and diversifying sources of resistance.

References

Adhikari, T.B., Anderson, J.M. and Goodwin, S.B. (2003) Identification and molecular mapping of a gene in wheat conferring resistance to *Mycosphaerella graminicola*. *Phytopathology* 93, 1158–1164.

Adhikari, T.B., Cavaletto, J.R., Dubcovsky, J., Gieco, J., Schlatter, A.R. and Goodwin, S.B. (2004a) Molecular mapping of Stb4 gene for resistance to *Septoria tritici* blotch in wheat. *Phytopathology* 94, 1198–1206.

Adhikari, T.B., Wallwork, H. and Goodwin, S.B. (2004b) Microsatellite markers linked to the *Stb2* and *Stb3* genes for resistance to *Septoria tritici* blotch in wheat. *Crop Science* 44, 1403–1411.

Adhikari, T.B., Yang, X., Cavaletto, J.R., Hu, X., Buechley, G., Ohm, H.W., Shaner, G. and Goodwin, S.B. (2004c) Molecular mapping of *Stb1*, a potentially durable gene for resistance to *Septoria tritici* blotch in wheat. *Theoretical and Applied Genetics* 109, 944–953.

Arama, P.F. (1996) Effects of cultivar, isolate and environment on resistance of wheat to *Septoria tritici* blotch in Kenya. PhD thesis, Wageningen University, Wageningen, The Netherlands, 115 pp.

Arama, P.F., Parlevliet, J.E. and van Silfhout, C.H. (1999) Heading date and resistance to *Septoria tritici* blotch in wheat not genetically associated. *Euphytica* 106, 63–68.

Arraiano, L.S., Worland, A.J., Ellerbrook, C. and Brown, J.K.M. (2001) Chromosomal location of a gene for resistance to *Septoria tritici* blotch (*Mycosphaerella graminicola*) in the hexaploid wheat 'Synthetic 6x'. *Theoretical and Applied Genetics* 103, 758–764.

Arraiano, L.S., Brading, P.A., Dedryver, F. and Brown, J.K.M. (2006) Resistance of wheat to *Septoria tritici* blotch (*Mycosphaerella graminicola*) and associations with plant ideotype and the 1BL-1RS translocation. *Plant Pathology* 55, 54–61.

Arraiano, L.S., Chartrain. L., Bossolini. E., Slatter. H.N., Keller. B. and Brown, J.K.M. (2007) A gene in European wheat cultivars for resistance to an African isolate of *Mycosphaerella graminicola. Plant Pathology* 56, 73–78.

Baltazar, B., Scharen, A.L. and Kronstad, W.E. (1990) Associations between dwarfing genes *Rht1* and *Rht2* and resistance to *Septoria tritici* blotch in winter wheat (*Triticum aestivum* L. em Thell). *Theorethical and Applied Genetics* 79, 422–426.

Beuningen, L.T. van and Kohli, M.M. (1990) Deviation from the regression of infection on heading and height as a measure of resistance to *Septoria tritici* blotch in wheat. *Plant Disease* 74, 488–493.

Brading, P.A., Verstappen, E.C.P., Kema, G.H.J. and Brown, J.K.M. (2002) A gene for gene relationship between wheat and *Mycosphaerella graminicola*, the *Septoria tritici* blotch pathogen. *Phytopathology* 92, 439–445.

Brown, J.K.M., Kema, G.H.J., Forrer, H.R., Verstappen, E.C.P., Arraiano, L.S., Brading, P.A., Foster, E.M., Fried, P.M. and Jenny, E. (2001) Resistance of wheat cultivars and breeding lines to *Septoria tritici* blotch caused by isolates of *Mycosphaerella graminicola* in field trials. *Plant Pathology* 50, 325–338.

Brule-Babel, A.L. (2007) Catalogue of gene symbols for wheat. *Annual Wheat Newsletter* 53, 171.

Caldwell, R.M. and Narvaez, I. (1960) Losses to winter wheat from infection by *Septoria tritici. Phytopathology* 50, 630.

Camacho Casas, M.A., Kronstad, W.E. and Scharen, A.L. (1995) *Septoria tritici* resistance and associations with agronomic traits in a wheat cross. *Crop Science* 35, 971–976.

Chartrain, L. (2004) Genes for isolate-specific and partial resistance to *Septoria tritici* blotch in wheat. PhD thesis, John Innes Centre, University of East Anglia, Norwich, UK, 151 pp.

Chartrain, L., Brading, P.A., Makepeace, J.C. and Brown, J.K.M. (2004a) Sources of resistance to *Septoria tritici* blotch and implications for wheat breeding. *Plant Pathology* 53, 454–460.

Chartrain, L., Brading, P.A., Widdowson, J.P. and Brown, J.K.M. (2004b) Partial resistance to *Septoria tritici* blotch (*Mycosphaerella graminicola)* in winter cultivars Arina and Riband. *Phytopathology* 94, 497–504.

Chartrain, L., Berry, S.T. and Brown, J.K.M. (2005a) Resistance of wheat line Kavkaz-K4500 L.6.A.4. to *Septoria tritici* blotch controlled by isolate-specific resistance genes. *Phytopathology* 95, 664–671.

Chartrain, L., Joaquin, P., Berry, S.T., Arraiano, L.S., Azanza, F. and Brown, J.K.M. (2005b) Genetics of resistance to *Septoria tritici* blotch in the Portuguese wheat breeding line TE 9111. *Theoretical and Applied Genetics* 110, 1138–1144.

Chen, R.S. and McDonald, B.A. (1996) Sexual reproduction plays a major role in the genetic structure of populations of the fungus *Mycospharella graminicola. Genetics* 142, 1119–1127.

Cordo, C.A., Perelló, A.E., Alippi, H.E. and Arriaga, H.O. (1990) Presencia de *Mycosphaerella graminicola* (Fuckel) Schroeter, teleomorfo de *Septoria tritici* Rob ex Desm. en trigos maduros de la Argentina. *Revista de la Facultad de Agronomía de La Plata* 66/67, 49–55.

Cordo, C.A., Linde, C., Zhan, J. and McDonald, B. (2007) Diversidad genotípica del patógeno de la mancha de la hoja del trigo (*Septoria tritici*) en la provincia de Buenos Aires. *Boletín de la Sociedad Argentina de Botánica* 41, 293–305.

Cowling, S.G., Brule-Babel, A.L., Somers, D.J. and Lamari, A.L. (2007) Catalogue of gene symbols. *Annual Wheat Newsletter* 53, 171.

Danon, T. and Eyal, Z. (1990) Inheritance of resistance in spring wheat and winter wheats to two *Septoria tritici* isolates. *Euphytica* 47, 203–214.

Dubin, H.J. and Rajaram, S. (1996) Breeding disease-resistant wheat for tropical highland and lowlands. *Annual Review Phytopathology* 34, 503–506.

Eriksen, L., Borum, F. and Jahoor, A. (2003) Inheritance and localisation of resistance to *Mycospharella graminicola* and plant height in the wheat (*Triticum aestivum* L.) genome with DNA markers. *Theoretical and Applied Genetics* 107, 515–527.

Eyal, Z. (1981) Integrated control of Septoria diseases of wheat. *Plant Disease* 65, 763–768.

Eyal, Z., Scharen, A.L., Prescott, J.M. and Ginkel, M. van (1987) *The Septoria diseases of wheat: concepts and methods of disease management.* CIMMYT, Mexico DF, 46 pp.

Eyal, Z., Scharen, A.L., Huffman, M.D. and Prescott, J.M. (1995) Global insights into virulence frequencies of *Mycosphaerella graminicola. Phytopathology* 75, 1456–1462.

Ginkel, M. van and Rajaram, S. (1999) Breeding for resistance to the Septoria/Stagonospora blights of wheat. In: Ginkel, M. van, McNab, A. and Krupinsky, J. (eds) *Septoria and Stagonospora Diseases of Cereals. A Compilation of Global Research.* CIMMYT, Mexico, pp. 117–126.

Goodwin, S.B., Ponomarenko, A.L., Dhillon, B., Grigoriev, I. and Kema, G.H.J. (2007) The finished genomic sequence of the *Septoria tritici* blotch pathogen *Mycosphaerella graminicola.* National Wheat Conference Proceedings Abstract No. 26. *Annual Wheat Newsletter* 54, 27–28.

Hayden, N.J., Jones, D.G. and Gillison, L.J. (1994) The role of legume-fixed nitrogen and mixed cropping systems in the management of *Septoria tritici.* In: Arseniuk, E., Goral, T. and Czembor, P. (eds) *Proceedings of the 4th International Workshop on Septoria of cereals.* IHAR, Radzikow, Poland, pp. 243–245.

Howard, D.D., Chambers, A.Y. and Logan, J. (1994) Nitrogen and fungicide effects on yield components and disease severity in wheat. *Journal of Production Agriculture* 7, 448–454.

Hunter, T., Coker, R.R. and Royle, D.J. (1999) Studies on the sexual phase of leaf blotch in UK winter wheat. In: Lucas, A.J., Bowyer, P. and Anderson, H.M. (eds) *Septoria of Cereals: A Study of Pathosystems.* IACR, Long Aston, University of Bristol, UK, pp. 108–114.

Jlibene, M., Gustafson, J.P. and Rajaram, S. (1992) A field disease evaluation method for selecting wheat resistant to *Mycosphaerella graminicola. Plant Breeding* 108, 26–32.

Jlibene, M., Gustafson, J.P. and Rajaram, S. (1994) Inheritance of resistance to *Mycosphaerella graminicola* in hexaploid wheat. *Plant Breeding* 112, 301–310.

Kema, G.H.J. and van Silfhout, C.H. (1997) Genetic variation for virulence and resistance in the wheat-*Mycosphaerella graminicola* pathosystem III. Comparative seedling and adult plant experiments. *Phytopathology* 87, 266–272.

Kronstad, W.E. (1998) Agricultural development and wheat breeding in the 20th century. In: Braun, H.J., Altay, F., Kronstad, W.E., Beniwal, S.P.S. and McNab, A. (eds) *Wheat: Prospects for Global Improvement. Proceedings of the 5th International Wheat Conference, Ankara, Turkey.* Developments in Plant Breeding. Kluwer Academic Publishers, Dordrecht, Netherlands, pp. 1–10.

Lee, S. and Gough, F.J. (1984) Inheritance of Septoria leaf blotch (*S. tritici*) and Pyrenophora tan spot (*P. tritici repentis*) resistance in *Triticum aestivum* cv. Carifen 12. *Plant Disease* 68, 848–851.

Leitch, M.H. and Jenkins, P.D. (1995) Influence of nitrogen on the development of Septoria epidemics in winter wheat. *Journal of Agricultural Science* 124, 361–368.

Lovell, D.J., Parker, S.R., Hunter, T., Royle, D.J. and Coker, R.R. (1997) Influence of crop growth and structure on the risk of epidemics by *Mycosphaerella graminicola* (anamorph *Septoria tritici*) in winter wheat. *Plant Pathology* 46, 126–138.

McCartney, C.A., Brule-Babel, A.L. and Lamari, L. (2002) Inheritance of race-specific resistance to *Mycosphaerella graminicola* in wheat. *Phytopathology* 92, 138–144.

McIntosh, R.A. (1998) Breeding wheat for resistance to biotic stresses. In: Braun, H.J. *et al.* (eds) *Wheat Prospects for Global Improvement.* Kluwer Academic Press, Dordrecht, Netherlands, pp. 71–86.

Mundt, C.C., Cowger, C. and Garret, K. (2002) Relevance of integrated disease management to resistance durability. *Euphytica* 124, 245–252.

Parlevliet, J.E. (1979) Components of resistance that reduce the rate of epidemic development. *Annual Review of Phytopathology* 17, 203–222.

Rajaram, S. (1999) Historical aspects and future challenges of an international wheat program. In: Ginkel, M. van, McNab, A. and Krupinsky, J. (eds) *Septoria and Stagonospora Diseases of Cereals. A Compilation of Global Research.* CIMMYT, Mexico, pp. 1–17.

Rajaram, S. and Ginkel, M. van (1996) A guide to the CIMMYT bread wheat section. In: *Wheat Special Report No. 5.* CIMMYT, Mexico, DF.

Rillo, A.O. and Caldwell, R.M. (1966) Inheritance of resistance to *Septoria tritici* in *Triticum aestivum* subsp. *vulgare*, Bulgaria 88 (Abst.). *Phytopathology* 56, 897.

Rosegrant, M.W., Agcaoili-Sombilla, A. and Pérez, N. (1995) *Global food projections to 2020.* Discussion Paper 5. IFPRI, Washington, DC.

Rosielle, A.A. and Boyd, W.J.R. (1985) Genetics of host pathogen interactions to *Septoria* species of wheat. In: Scharen, A.L. (ed.) *Septoria of Cereals.* USDA-ARS Publication 12. USDA, Washington, DC, pp. 9–12.

Rosielle, A.A. and Brown, A.G.P. (1979) Inheritance, heritability and breeding behavior of three sources of resistance to *Septoria tritici* in wheat. *Euphytica* 28, 385–392.

Sanderson, F.R. (1972) A Mycospharella species as the ascogenous state of *Septoria tritici* Rob ex Desm. *New Zealand Journal of Agricultural Research* 21, 277–281.

Schuh, W. (1990) Influence of tillage systems on disease and spatial pattern of *Septoria* leaf blotch. *Phytopathology* 80, 1337–1340.

Shaner, G. and Buechley, G. (1989) Inheritance of resistance to *Mycosphaerella graminicola* in wheat. In: Fried, P.M. (ed.) *Proceedings 3rd International Workshop: Septoria Diseases of Cereals.* Swiss Federal Station for Agronomy, Zurich, Switzerland, pp. 133–155.

Shaw, M.W. and Royle, D.J. (1989) An epidemiology based forecasting scheme for *Septoria tritici.* In: Fried, P.M. (ed.) *Proceedings 3rd International Workshop: Septoria Diseases of Cereals.* Swiss Federal Station for Agronomy, Zurich, Switzerland, 189 pp.

Shipton, W.A., Boyd, W.J.R., Rosiello, B.I. and Shearer, B.I. (1971) The common *Septoria* diseases of wheat. *Botanical Review* 27, 231–262.

Simón, M.R. and Cordo, C.A. (1997) Inheritance of partial resistance to *Septoria tritici* in wheat (*Triticum aestivum* L.): limitation of pycnidia number and spore production. *Agronomie* 17, 343–347.

Simón, M.R. and Cordo, C.A. (1998) Diallel analysis of the resistance components to *Septoria tritici* in *Triticum aestivum. Plant Breeding* 117, 123–126.

Simón, M.R., Perelló, A.E. and Cordo, C.A. (1998) Response to selection in F2 populations of wheat crosses for resistance to *Septoria tritici. Cereal Research Communications* 26, 275–279.

Simón, M.R., Worland, A.J., Cordo, C.A. and Struik, P.C. (2001) Chromosomal location of resistance to *Septoria tritici* in seedlings of a synthetic hexaploid wheat, *Triticum spelta* and two cultivars of *Triticum aestivum. Euphytica* 119, 151–155.

Simón, M.R., Perelló, A.E., Cordo, C.A. and Struik, P.C. (2002) Influence of *Septoria tritici* on yield, yield components and test weight of wheat under two Nitrogen fertilization conditions. *Crop Science* 42, 1974–1981.

Simón, M.R., Cordo, C.A., Perelló, A.E. and Struik, P.C. (2003) Influence of nitrogen supply on the susceptibility of wheat to *Septoria tritici. Journal of Phytopathology* 151, 283–289.

Simón, M.R., Ayala, F.M., Cordo, C.A., Röder, M.S. and Börner, A. (2004a) Molecular mapping of quantitative trait loci determining resistance to *Septoria tritici* blotch caused by *Mycosphaerella graminicola* in wheat. *Euphytica* 138, 41–48.

Simón, M.R., Worland, A.J. and Struik, P.C. (2004b) Influence of plant height and heading date on the expression of the resistance to *Septoria tritici* blotch in near isogenic lines of wheat. *Crop Science* 44, 2078–2085.

Simón, M.R., Perelló, A.E., Cordo, C.A., Putten, P.E.L. van der and Struik, P.C. (2005a) Association between *Septoria tritici* blotch, plant height and heading date in wheat. *Agronomy Journal* 97, 1072–1081.

Simón, M.R., Worland, C.A. and Struik, P.C. (2005b) Chromosomal location of two components encoding for resistance to *Septoria tritici* blotch (*Mycosphaerella graminicola*) in seedlings and in the adult stage of wheat. *Netherlands Journal of Agricultural Science* 53, 113–140.

Simón, M.R., Ayala, F.M., Cordo, C.A., Röder, M.S. and Börner, A. (2007) The exploitation of wheat (*Triticum aestivum*)/*Aegilops tauschii* introgression lines for the detection of gene(s) determining resistance to *Septoria tritici* blotch (*Mycosphaerella graminicola*). *Euphytica* 154, 249–254.

Somasco, O.A., Qualset, C.O. and Gilchrist, D.G. (1996) Single-gene resistance to *Septoria tritici* blotch in the spring wheat cultivar 'Tadinia'. *Plant Breeding* 115, 261–267.

Wilson, R.E. (1979) Resistance to *Septoria tritici* in two wheat cultivars, determined by independent, single dominant gene. *Australasian Plant Pathology* 8, 16–18.

Zhan, J., Mundt, C.C. and McDonald, B.A. (2001) Using restriction fragment length polymorphisms to assess temporal variation and estimate the number of ascospores that initiate epidemics in field populations of *Mycosphaerella graminicola. Phytopathology* 91, 1011–1017.

Zhan, J., Pettway, R.E. and McDonald, B.A. (2003) The global genetic structure of the wheat pathogen *Mycosphaerella graminicola* is characterized by high nuclear diversity, low mitochondrial diversity, regular recombination, and gene flow. *Fungal Genetics Biology* 38, 286–297.

Zhang, X., Halye, S.D. and Jin, Y. (2001) Inheritance of *Septoria tritici* blotch resistance in winter wheat. *Crop Science* 41, 323–326.

7 Barley and Wheat Resistance Genes for *Fusarium* Head Blight

S.A. Stenglein and W.J. Rogers
Laboratorio de Biología Funcional y Biotecnología (BIOLAB)-CEBB, Facultad de Agronomía, Universidad Nacional del Centro de la Provincia de Buenos Aires (UNICEN), Buenos Aires, Argentina and *Consejo Nacional de Investigaciones Científicas y Técnicas (CONICET), Argentina*

Abstract

The genetic control of resistance to *Fusarium* head blight (FHB) in barley and wheat is reviewed. This disease, which can reach epidemic proportions under certain climatic conditions, is caused by various *Fusarium* species and affects grain yield and quality detrimentally, resulting in important economic losses in both crops. Furthermore, FHB infection poses a serious threat to human and animal health, due to the presence of toxic trichothecenes, of which deoxynivalenol and its derivatives appear to be the most important. Marker-based mapping studies have identified numerous quantitative trait loci (QTLs) for FHB resistance, located on all the chromosomes of both species. Only a relatively small number of these can be detected consistently over a wide range of different environments and genetic backgrounds. None the less, where genetic effects have been characterized, they have been shown to be mainly additive in nature, meaning that the accumulation of several QTL factors in a single line ought to be effective in achieving raised levels of resistance. Indeed, marker-assisted selection has been directly shown to be feasible for some QTL. A number of QTLs for FHB resistance are associated with other agronomic characters, such as heading date (HD), flowering time and plant height. In some cases, QTL alleles favourable for resistance are associated detrimentally with alleles for these characters, although there appear to be sufficiently large numbers of QTLs for resistance acting independently of these characters to imply that reasonable genetic gains for resistance ought to be achievable in the future. While most studies in barley have addressed Type I resistance (initial infection) and in wheat Type II (spread between spikelets), or a combination of both Type I and Type II, more recent studies have addressed other types of resistance, such as Type III (effects on kernel size and characteristics), Type IV (yield tolerance) and Type V (decomposition or non-accumulation of mycotoxins such as deoxynivalenol). Besides identifying additional QTLs, these latter studies offer insights into the mechanisms of the different types of resistance observed, in some cases blurring the distinctions between them. Other prospects for improvement in FHB resistance, additional to those offered by marker-assisted selection, are also discussed.

Introduction

Fusarium head blight (FHB) or scab is a destructive disease of wheat and barley in environments with prolonged wet climatic conditions from flowering through the soft-dough stage of kernel development (Parry *et al.*, 1995; McMullen *et al.*, 1997). The disease is of worldwide importance. FHB epidemics have been documented in 26 US states and five Canadian provinces. Economic losses in wheat since 1990 were estimated at US$2.5bn (Windels, 2000). Wheat yields in 1993 were reduced by about 50% in north-eastern North Dakota and 40% in north-western Minnesota, compared with 1992 (National Agricultural Statistics Service, 1993–1999). Barley losses have been equally devastating, with estimated losses from 1993 to 1999 totalling in excess of US$400m (Windels, 2000). In China, FHB has affected more than 7m ha wheat and has caused yield losses of more than 1 Mt in severe epidemics (Leonard and Bushnell, 2003). In Argentina, during the past 60 years, several FHB epidemics of varying severity have occurred in the central-north area, where yield losses were estimated to average between 20 and 50%.

FHB is a preharvest disease, but *Fusarium* species can grow in postharvest phase if wet grain is not dried efficiently and quickly. More than 17 *Fusarium* species have been isolated from naturally infected wheat or barley spikes and have been associated with FHB (Parry *et al.*, 1995; Leonard and Bushnell, 2003). *Fusarium graminearum* (teleomorph *Gibberella zeae*) is the most frequently encountered pathogen and the most virulent species, although *F. avenaceum* (teleomorph *G. avenacea*), *F. culmorum* and *F. poae* are reported to be prevalent in some European and North and South American countries (Leonard and Bushnell, 2003; Barreto *et al.*, 2004; Bourdages *et al.*, 2006). The distribution and predominance of a *Fusarium* species in a region is thought to be determined by climatic factors, competition among various *Fusarium* spp. sharing the same ecological niches, fertilizer use, cropping sequence and practices and vegetation type (Snyder and Nash, 1968; Nelson *et al.*, 1981; Doohan *et al.*, 2003).

FHB reduces kernel set and kernel weight. Invasion of the kernel by *Fusarium* destroys the starch granules and cell walls and affects endosperm storage proteins, resulting in a poor quality product (Fig. 7.1).

Fig. 7.1. Shrivelled lightweight seeds of wheat affected by FHB (left) and healthy wheat seeds (right).

Germination rate and seedling vigour are reduced when the seeds are infected.

In addition to causing significant yield losses, FHB is of greater significance under certain conditions because of the associated mycotoxin accumulation which can occur in infected grain. *Fusarium graminearum*, *F. avenaceum*, *F. culmorum* and *F. poae* can produce a range of mycotoxins and contaminated grain is unsuitable for animal and human consumption because of the adverse effects of such toxins on health (Placinta *et al.*, 1999; Gutleb *et al.*, 2002). Within *Fusarium* mycotoxins, some of the most important from the point of view of animal health and productivity, are the trichothecenes, zearalenone and the fumonisins (D'Mello *et al.*, 1999). Type A and B trichothecenes represent the most important members of these mycotoxins. Type A trichothecenes include T-2 toxin, HT-2 toxin, neosolaniol (NEO) and diacetoxyscirpenol (DAS), while type B trichothecenes include deoxynivalenol (DON, also known as vomitoxin) and its 3-acetyl and 15-acetyl derivates (3-DON and 15-DON, respectively), nivalenol (NIV) and fusarenon-X (FUS-X). A common feature of many *Fusarium* species is their ability to synthesize zearalenone (ZEN or F-2 toxin) and its co-occurrence with certain trichothecenes raises important issues regarding additivity and/or synergism in the aetiology of mycotoxicosis in animals (Placinta *et al.*, 1999). Fumonisins are an increasingly important group of toxins as they have been postulated as the causative agent for several endemic diseases, both in humans and animals (Sydenham *et al.*, 1990; Chu and Li, 1994).

Host resistance has long been considered the most practical and effective means of disease handling, but breeding for FHB resistance has been hindered by a lack of effective resistance genes and by the complexity of the resistance in identified sources (Mesterhazy, 1997). No source of complete resistance is known and current sources provide only partial resistance.

Resistance types are generally classified as either morphological or physiological. Head anatomy or positioning that contributes to higher humidity around the spikelets is often associated with more diseases. Generally, awned genotypes with short peduncule and a compact spike have faster disease spread than genotypes that are awnless, have a long peduncule and a lax spike (Rudd *et al.*, 2001). In addition, short saturated genotypes with a long grain-filling duration generally get more disease than tall genotypes that have rapid grain fill (Mesterhazy, 1995). These morphological characteristics contribute to resistance, but are often considered nuisance factors in screening nurseries, and it is generally agreed that they are of minor significance compared with physiological resistance (Rudd *et al.*, 2001). However, morphological traits have also been associated with FHB resistance in barley. Two-rowed barley is more resistant to FHB than six-rowed barley and, in crosses between six-rowed and two-rowed genotypes, two-rowed progenies are most resistant, followed by genotypes heterozygous for spike type. Six-rowed types are most susceptible (Takeda and Heta, 1989; Xihang *et al.*, 1991).

Mesterhazy (1995) described five types of physiological resistance, expanding the two types described by Schroeder and Christensen (1963). These include Type I resistance to initial infection. It may be passive, involving morphological characteristics of wheat head. Alternatively, Type I resistance may be active and include defence reactions such as the activation of enzymes degrading the fungal cell wall or pathogenesis-related (PR) proteins (Nicholson *et al.*, 2005). This type of resistance is estimated by spraying a spore suspension over flowering spikes and counting diseased spikelets. Type II refers to the resistance of movement of the pathogen from one infected spikelet to another via the rachis. The mechanisms involved in Type II resistance are thought to be active, but again may be due to morphological characteristics. This type of resistance is estimated by delivering conidia into a single floret of a spike and counting the blighted spikelets after a period of time. The other types of resistance include: kernel size and number retention (Type III), yield tolerance (Type IV) and decomposition or non-accumulation of mycotoxins (Type V). Type III resistance is measured by threshing infected spikes and observing the damage to the kernels. Kernel number reduction, kernel weight, test weight, or visual estimates of *Fusarium*-damaged kernels (tombstones) are common measurements used to assess this resistance. Type IV resistance, or yield tolerance, can be assessed by measuring grain yield of naturally or artificially inoculated spikes or plots and comparing the data with spikes or plots that do not show disease symptoms (Rudd *et al.*, 2001). Finally, Type V resistance is identified by measuring DON concentration at a given level of FHB (Rudd *et al.*, 2001). This resistance is important from a grain utilization perspective, for example for malting barley, because even trace levels of DON may reduce beer quality significantly.

Considerable progress in the search for host resistance has been made. Improvement of cultivar resistance has become a major breeding objective worldwide. Recent developments in genomic research and biotechnology hold promise for understanding the genetic mechanisms of FHB resistance and allow more effective utilization of FHB resistance genes to develop new resistant wheat and barley cultivars.

Genetics of FHB Resistance in Barley

Few sources of FHB resistance have been found in barley and the level of their resistance is modest. Although FHB in barley usually does not spread from spikelet to spikelet within a spike (up and down the spike), barley seems to be very susceptible to initial infection. Severe disease usually results from multiple initial infections in the spike.

Of primary importance to barley breeders are data on FHB severity and DON concentration, since these are traits that affect the marketing of grain in malting most severely. The first sources of resistance used were the breeding lines Gobernadora from ICARDA/CIMMYT in Mexico and Zhedar 1 and Zhedar 2 from China. All three lines had the two-rowed spike morphology. Other two-rowed barley with low DON content were CI 4196, Svanhals and Imperial. CI 4196 was identified as one of the most resistant two-rowed barley accessions and also accumulated low concentrations of DON (Urrea *et al.*, 2005).

Six-rowed types are preferred for malting, but they are generally more susceptible to FHB than two-rowed barley. Chevron, an old cultivar from Switzerland, is a six-rowed malting barley and a popular parent in barley breeding programmes. It has high resistance to kernel discoloration, which is a disease complex caused by several different fungal pathogens, including *Fusarium*.

In China and Japan, over 10,000 barley accessions from different countries have been screened for FHB resistance, but only several dozen accessions have a low level of FHB (Xihang *et al.*, 1991; Zhou *et al.*, 1991). To date, no wild species of *Hordeum* have shown greater resistance than that of two-rowed barley. DON content in even the best sources of resistance are still well above the specification for the brewing industry (< 0.5 mg/kg), but much lower than that of current commercial malting barley cultivars (Leonard and Bushnell, 2003).

Investigation of the genetics of resistance to FHB in barley has not been very extensive and published reports on the identification of loci controlling FHB resistance and DON accumulation are limited (Rudd *et al.*, 2001). Barley producers currently attempt to manage the disease through crop rotation and fungicide application. However, these measures alone are not sufficient to reduce the risk of the disease. Resistant barley cultivars are the most cost-effective measures for controlling the disease, but breeding for FHB resistance has been difficult for several reasons. One, genetic resistance is complex. There seem to be many QTLs that have relatively small effects and are subject to genotype × environment interactions. Two, FHB screening experiments are labour-intensive and expensive. Three, assessing FHB severity in both the field and the greenhouse is difficult. Disease severity is correlated strongly with HD and other agronomic and spike morphology traits. Since infection can occur only after the spike emerges from the boot, differences in HD make it difficult to distinguish 'true' disease resistance from 'apparent' resistance that is due to host escape from the pathogen. Both of these problems necessitate the identification of molecular markers linked to QTLs for FHB resistance that can be used in marker-assisted breeding. In addition, since disease expression is influenced strongly by the environment, comparisons among barley genotypes that differ in HD are themselves confounded by the effect of the environment on disease development. However, because of the complex nature of genetic resistance to FHB, QTL identification is not always very robust. Therefore, validation of these QTLs is important before implementing marker-assisted selection in a breeding programme.

To gain a genetic understanding of FHB resistance in barley, multiple sources of resistance including Chevron (de la Pena *et al.*, 1999; Ma *et al.*, 2000), Gobernadora (Zhu *et al.*, 1999), Fredrickson (Mesfin *et al.*, 2003; Smith *et al.*, 2004), Zhedar 2 (Dahleen *et al.*, 2003) and CI 4196 (Horsley *et al.*, 2006) have been used in QTL mapping studies.

QTLs providing resistance to FHB and DON accumulation in barley have been identified on all seven chromosomes. QTLs for FHB resistance were identified on chromosomes 1(7H), 2(2H), 3(3H), 4(4H), 5(1H) and 7(5H) in the Chevron (resistant)/M69 (susceptible) population (de la Pena *et al.*, 1999). A major QTL on chromosome 2(2H) explains 13.5% of the phenotypic variation for FHB resistance. However, this QTL is also associated with HD and the resistant allele is linked to late heading. Ma *et al.* (2000) used a population derived from the cross Chevron/Stander and reported nine QTLs for FHB resistance located on chromosomes 1(7H), 2(2H), 3(3H), 6(6H) and 7(5H). A QTL on chromosome 2(2H) was detected consistently in five environments and explained 11.8–20.7% of the phenotypic variation for FHB resistance. This QTL, in addition to the QTL on chromosome 2(2H) discovered by de la Pena *et al.* (1999), is also associated with days to heading. Using a population derived from the two-rowed parents, Gobernadora and CMB 643, Zhu *et al.* (1999) found QTLs for FHB resistance on all barley chromosomes except chromosome 7(5H). The largest QTL explained 33% of the phenotypic variation and was found on chromosome 2(2H). The QTL on chromosome 4(4H) explains 4–12% of the phenotypic variation for FHB resistance. This QTL was also associated significantly with morphological traits including plant height, seeds per inflorescence, inflorescence density and lateral floret size. In each of the previous mapping studies, QTLs for accumulation of DON in harvested grain were also detected. These QTLs were also distributed throughout the genome and were, in some cases, coincident with FHB QTL. Taken together, these studies indicate resistance is conditioned by many loci and that there is a strong association between certain morphological traits and FHB resistance.

Two major traits associated with FHB severity are spike type and HD. The *Vrs1* and *Int-c* loci control lateral floret fertility and hence determine whether a spike is two-rowed (*Vrs1*; *int-c/int-c*) (Lundqvist and Franckowiak, 1997) or six-rowed (*vrs1/vrs1*; *Int-c/Int-c*) (Hockett and Nilan, 1985). In several studies, the two-rowed spike type has been associated with FHB resistance (Chen *et al.*, 1991; Xihang *et al.*, 1991; Steffenson *et al.*, 1996; de la Pena *et al.*, 1999). In a genetic study, Takeda (1990) demonstrated an association between the *Vrs1* locus and FHB resistance. In two-rowed barley (*Vrs1*) with the *Int-c/Int-c* genotype, the laterals can be inflated and lateral floret size has been associated with FHB severity (Zhu *et al.*, 1999). The FHB mapping studies published to date have used populations derived from either six-rowed × six-rowed or two-rowed × two-rowed crosses (de la Pena *et al.*, 1999; Zhu *et al.*, 1999; Ma *et al.*, 2000). Therefore, the *Vrs1* locus was not segregating in these populations. HD may also strongly influence the severity of FHB on barley and QTLs for HD and FHB resistance are coincident (de la Pena *et al.*, 1999; Ma *et al.*, 2000). Generally, late heading plants tend to have lower severity, while early heading plants have higher severity, indicating that the late heading plants are exposed to the inoculum for a shorter period of time (Leonard and Bushnell, 2003).

In all of these studies except the one using Gobernadora, the bin 8 region of the long arm of chromosome 2H designated by

Horsley *et al.* (2006) as *Qrgz-2H-8* was associated consistently with FHB severity, HD and DON concentration. The approximate size for the overlapping QTL region ranged from 22cM in the Fredrickson/Stander population (Mesfin *et al.*, 2003) to 45cM in Chevron/M69 (de la Pena *et al.*, 1999) and CI 4196/Foster (Horsley *et al.*, 2006) populations. Depending on the population and the environment, *Qrgz-2H-8* explained 7–60% of the variation in FHB resistance, 12–30% of the variation in HD and 10–30% of the variation in DON concentration. In all of the studies, FHB severity and DON concentration were correlated negatively with HD. In a validation study of this QTL, the Chevron introgression at the *Qrgz-2H-8* region reduced FHB by 42% and increased HD by 3.8 days (Canci *et al.*, 2004).

The association between lower FHB severity and late heading may be due to shorter inoculum exposure (pleiotropy) or tight linkage of separate genes for flowering time and disease resistance (Leonard and Bushnell, 2003). To determine if the association between late HD and FHB resistance is due to linkage or pleiotropy, Nduulu *et al.* (2007) constructed a fine map for the chromosome 2(2H) QTL region using recombinant near isogenic lines (rNILs) derived from a cross between a BC_5 line carrying the Chevron alleles for markers at the *Qrgz-2H-8* region and the recurrent parent M69, and concluded that the relationship between FHB and HD at the *Qrgz-2H-8* region was likely due to tight linkage rather then pleiotropy.

Genetics of FHB Resistance in Wheat

Besides similar considerations as for barley regarding the detrimental effects of FHB on grain yield and quality in general, and the effects of mycotoxins on human and livestock health, the fact that the disease results in the degradation of the endosperm storage proteins means specifically that the quality of bread, biscuit, pasta and other industrial products can be seriously prejudiced. Worldwide, the species *F. graminearum* predominates, but *F. avenaceum*, *F. culmorum* and *F. poae* are also the cause of the disease in some environments. Epidemics may cause major losses when climatic conditions are favourable after flowering (Paillard *et al.*, 2004). As in barley, agricultural management and fungicide treatments, while reducing the damage (Gervais *et al.*, 2003), are not wholly effective (Stack, 1989; Bai and Shaner, 1994; Parry *et al.*, 1995). Unfortunately, complete FHB resistance is unknown, although long-term control of the disease is probably most likely to be achieved through genetic resistance research, involving QTL mapping and other procedures (see below), and its consequent application in the breeding of resistant cultivars. This appears to be the case, in spite of the complexity of the genetic control involved, the presence of confounding environmental effects, the influence of genotype × environment interaction and the fact that laborious inoculation and evaluation procedures in mature host plants are required in order to identify useful marker associations (Snidjers, 1990; van Ginkel *et al.*, 1996; del Blanco *et al.*, 2003). A further complication is that associations between FHB resistance with HD, flowering time (FT) and plant height (PH) have also been observed (Mesterhazy, 1997; Hilton *et al.*, 1999; Buerstmayr *et al.*, 2000).

For breeding purposes, three broad origins of resistant germplasm have been recognized (Gilbert and Tekauz, 2000; Paillard *et al.*, 2004): (i) spring wheat from Asia (e.g. cv. Ning 7840 [China], cv. Sumai 3 [China], cv. Nobeokabozu [Japan]); (ii) spring wheat from South America (e.g. cv. Frontana [Brazil]); and (iii) winter wheat from Europe (e.g. Arina, Praag-8, Novokrumka). Further examples of individual resistant cultivars are given in the studies described below, which are all concerned with bread wheat, unless specified otherwise.

In contrast to barley, FHB generally spreads between spikelets (although it is currently unclear whether this is so for *F. poae*) and most genetic research has therefore concentrated on Type II resistance (most frequently evaluated after single-spikelet inoculation with *F. graminearum*), although combined evaluation of Type I and Type II resistance through spray inoculation has also been widely carried out. However, there are an increasing number of studies that address other types of resistance, such as the ability to detoxify DON (Type V) and the ability to maintain grain yield in spite of disease symptoms (Type IV).

The first QTL mapping studies were carried out in the mid-1990s (Bai, 1995; Moreno-Sevilla *et al.*, 1997), involving the use of RFLP and RAPD markers to map Type II resistance. However, the marker associations identified individually accounted for only a small proportion of the variation, perhaps due to the relatively low level of polymorphism observed for the markers employed (Bai *et al.*, 1999). Subsequently (Bai *et al.*, 1999), AFLP markers were applied to a mapping population involving the relatively resistant cv. Ning 7840 (Type II resistant cultivar), where the main specific character measured was the area under disease progress curve (AUDPC) after *F. graminearum* single-spikelet inoculation. One major QTL was identified accounting for up to 60% of the observed variation, which, although originally thought to be located on 7B, was identified subsequently as being equivalent to the QTL identified on chromosome arm 3BS (designated *Qfhs.ndsu-3BS*) (Waldron *et al.*, 1999) and present in one of the ancestral cultivars of cv. Ning 7840, namely cv. Sumai 3. Two years later (Anderson *et al.*, 2001), the same group verified the presence of this QTL (up to 41.6% of the variation accounted for) in Sumai 3 and located two further QTLs from Sumai 3 on 6AS (up to 11.6%) and 6BS (up to 9.2%). The susceptible parent, cv. Stoa, was also shown to carry two QTLs for resistance, on 2AL (up to 14.3%) and 4BS (up to 7.2%). A further QTL from a third line, ND2603 (partially resistant), was located on 3AL (up to 9.1%), in this case in a cross with the susceptible cv. Butte 86. These studies referred to Type II resistance (0–100% FHB severity scale after *F. graminearum* single-spikelet inoculation).

During this period, in crosses between six resistant Chinese bread wheat cultivars with two susceptible cultivars (Bai *et al.*, 2001), where AUDPC was evaluated after *F. graminearum* single-spikelet inoculation, it was shown, from joint scaling tests aimed at explaining the differences in means between parental, F_1, F_2 and backcross generations (Mather and Jinks, 1982), that most of the observed genetic variation could be explained by additive effects, where dominant and epistatic effects accounted for only a small proportion of the genetic effects present in the crosses analysed. The authors pointed out that this implied that it should be possible to accumulate different genes to improve resistance to FHB. The mainly additive nature of genetic effects was also observed in the soft red winter wheat, Ernie (Liu *et al.*, 2005).

In a subsequent study involving Type II resistance after inoculation with *F. graminearum* and *F. culmorum* (applied separately) of a mapping population derived from the bread wheat cross cv. CM-82036 (resistant, a line derived from Sumai 3) × cv. Remus (susceptible) and using RFLP, AFLP, SSR and endosperm storage protein markers (Buertsmayr *et al.*, 2002), the large effect of *Qfhs.ndsu-3BS* (up to 60% of variation accounted for) was again confirmed and two further QTLs were located to 5A and 1B. The 3BS and 5A QTLs were flanked with SSR markers and the 1B QTL associated with the *Glu-B1* locus encoding high molecular weight glutenin subunits. In a second part of this study (Buertsmayr *et al.*, 2003), the authors extended the analysis to include combined Type I and Type II resistance; they found that, under spray inoculation, *Qfhs.ndsu-3BS* had a much larger effect than the 5A QTL, which they named *Qfhs.ifa-5A*, whereas after single-spikelet inoculation, the two loci showed effects of similar magnitude. *Qfhs.ndsu-3BS* appeared to be associated mainly with resistance to fungal spread (Type II), whereas *Qfhs.ifa-5A* appeared to be associated principally with fungal penetration, and might contribute primarily towards Type I resistance and, to a lesser extent, towards Type II. In both these studies, no isolate × wheat genotype interaction was observed, consistent with the previously observed non-specific or horizontal nature of resistance (Mesterhazy, 1995; van Eeuwijk *et al.*, 1995), which was particularly interesting in this case since the two isolates used belonged to different species

(*F. graminearum* and *F. culmorum*). The authors concluded that FHB resistance depended on a few (2–3) major QTLs, operating together with unknown numbers of minor genes. They pointed out that marker-assisted selection (MAS) for the major QTLs ought to be a feasible method of accelerating the development (through breeding that included use of backcrosses) of resistant cultivars that combined Type I and Type II resistance. They felt that marker-mediated transfer of the QTL to durum wheat also ought to be feasible, given that no D genome chromosomes were involved in the QTL identified.

The effect of *Qfhs.ndsu-3BS* was also observed in several other studies (Kolb *et al.*, 2001; Zhou *et al.*, 2002; Bourdoncle and Ohm, 2003; del Blanco *et al.*, 2003; Shen *et al.*, 2003a; Xie *et al.*, 2007). Effects on 2A and 2B have also been observed in analyses involving Sumai 3 (Zhou *et al.*, 2002). In one study (Yu *et al.*, 2006), it was suggested that the 3BS, 5AS and 6BS resistance QTLs of Sumai 3 were derived from the Chinese landrace, Taiwan Xiaomai.

QTLs on chromosomes 2A, 3A, 3B and 5A, which had been observed previously in Asian wheats, were also observed in RILs derived from a cross between the European winter wheat cultivars, Renan (resistant) and Récital (susceptible), using spray inoculation of *F. culmorum* (Gervais *et al.*, 2003). In the same study, new QTLs were identified on 2BS and 5AL. Although co-localization of QTLs for resistance with awnedness (5A), PH (5A) and FT (2B) was observed, the authors considered that it should be possible to produce resistant lines independent of these characters.

In RILs obtained from the Swiss winter wheat cross cv. Arina (resistant) × cv. Forno (susceptible) characterized with microsatellite and RFLP markers and subjected to spray inoculation with *F. graminearum* (combined Type I and II resistance), eight QTLs were identified that together explained 47% of the variation (Paillard *et al.*, 2004). Three of these were considered of major effect: 6DL (22%), 5BL (14%, contributed by the susceptible parent) and 4AL (10%). The authors considered that these were different from QTLs previously reported. The other QTLs detected were located on 2AL, 3AL, 3BL, 3DS and 5AL. The authors concluded that FHB resistance was polygenic, rather than the bimodal distribution observed in some previous studies (Bai *et al.*, 1999; Waldron *et al.*, 1999; Buertsmayr *et al.*, 2002). The 2AL QTL was located at the same map position as one originating from cv. Stoa (Waldron *et al.*, 1999; Anderson *et al.*, 2001) and the 5AL QTL in the same position as one identified previously (Gervais *et al.*, 2003). In contrast, the 3DS QTL was located differently compared to one identified previously on this arm (Shen *et al.*, 2003a). The major 6D and 5B QTL overlapped completely with a QTL for HD and the 6D QTL overlapped partially with a QTL for PH. However, QTLs for PH were identified that were not associated with FHB resistance. The data could not distinguish pleiotropic effects from linkage. A further study involving cv. Arina (crossed to cv. NK93604) failed to detect the same QTL (Semagn *et al.*, 2007); instead, QTLs on 1BL and 6BS from Arina and on 1AL and 7AL from NK93604 were detected. A study of Arina crossed to the susceptible UK cultivar, Riband, identified at least 10 QTLs, very few of which were coincident with the other Arina studies; the most consistent was a major QTL on 4DS (Draeger *et al.*, 2007), detected in four of the five environments evaluated.

In the winter wheat cross cv. Patterson × cv. Fundulea F201R (resistant cultivar from Rumania), QTLs for Type II resistance were found on 1B, 3A, 3D and 5A, with the 1B and 3A consistent over experiments (Shen *et al.*, 2003b).

It appears that, whereas Sumai 3 and its derivatives have major QTLs on 3B and 5A, the three winter wheat populations so far characterized seem to depend more on the accumulation of moderate and minor QTLs. The 3BL QTL located in the Renan/Récital population may be the same as that observed in the Arina/Forno population.

In a cross of the resistant Brazilian cv. Frontana with the susceptible cv. Remus (Steiner *et al.*, 2004) inoculated with *F. graminearum* and *F. culmorum*, a major QTL accounting for 16% of the variation in FHB severity and incidence was located on

3A and a QTL accounting for 9% of the variation in FHB severity was located on 5A. Smaller effects for severity were located on 1B, 2A, 2B, 4B, 5A and 6B. The resistance of Frontana was found to be due principally to the inhibition of fungal penetration (Type I), but with a minor effect on fungal spread (Type II). PH, FT and spike morphology influenced FHB reaction, but co-localization of QTLs was observed only for minor QTL, and sufficient QTLs for FHB resistance acting independently of these characters were observed in order to allow selection of resistant lines with any height, flowering date and spike morphology.

Seven QTLs for Type I and II resistance were found on 1BS, 1DS, 3B, 3DL, 5BL, 7BS and 7AL in a cross between cv. Cansas (moderately resistant) and cv. Ritmo (susceptible). The 1DS QTL seemed primarily to involve resistance to fungal penetration, while the other QTLs were concerned mainly with resistance to fungal spread (Klahr *et al.*, 2007). Significant correlations with PH and HD were observed.

The *Qfhs.ndsu-3BS* region of Sumai 3 has been fine mapped and named *Fhb1* (Cuthbert *et al.*, 2006), as well as being validated by near-isogenic line studies (Cuthbert *et al.*, 2007). A second region on 6BS has also been fine mapped and named *Fhb2* (Pumphrey *et al.*, 2007).

Over recent years, attention has turned towards other types of resistance. For example, in the partially resistant cultivars, Wuhan-1 and Maringa, QTLs for the accumulation of DON (Type V) were located on 2DS and 5AS (as well as QTLs for FHB resistance on 2DL, 3BS and 4B) (Somers *et al.*, 2003). QTLs were located on 5A (12.4%), 2A (8.5%) and 3B (6.2%) for low DON content in the Chinese landrace, Wangshuibai (as well as QTLs for Type II resistance on 3B and 2A (Ma *et al.*, 2006)). In the previously cited study on Arina × NK93604 (Semagn *et al.*, 2007), the QTLs located on 1AL and 2AS were associated with DON content, although only 1AL was associated with FHB resistance. In the additional Arina study cited, involving Arina × Riband (Draeger *et al.*, 2007), the major 4DS QTL identified was found to affect AUDPC, DON content, fungal DNA content (FDNA), relative spikelet weight (RSW) and per cent of *Fusarium*-damaged kernels (FDK); although this may be due to linked genes, the authors considered it more likely to represent one resistance gene (which appeared to be linked to the *Rht-D1* locus, an association that may prejudice attempts to improve resistance in germplasm containing the *Rht-D1b* (*Rht*) semi-dwarfing allele). In this study, further QTLs were observed as follows, whose detected presence varied over environments: AUDPC: 1BL, 2B, 6BL, 7AL, 7BL, 7DL; DON content: 6BL, 7DL; FDNA: 3DL, 6BL, 7BL; RSW: 1BL, 2AS, 6BL, 7DL; FDK (Type III): 5AS, 7AL; yield loss (Type IV): 7AL.

In a study involving lines derived from the cross CM-82036 × Remus (Lemmens *et al.*, 2005), the QTL on 3BS derived from Sumai 3, closely associated with resistance to spread of the disease (Type II), appears to convert DON to DON-3-*O*-glucoside. The authors hypothesized that the 3BS QTL encoded a DON-glucosyl-transferase or regulated the expression of this.

In a cross involving cv. CJ 9306 (Jiang *et al.*, 2007), two QTLs were found for resistance to DON accumulation, *QFhs.nau-2DL*, explaining up to 20% of the observed variation, and *QFhs.nau-1AS*, explaining 4–6%. The QTLs, *QFhs.ndsu-3BS* (up to 23% of the variation) and *QFhs.nau-5AS* (4–6%) were also validated. QTL × environment interaction was found for *QFhs.nau-2DL* only. The authors suggested that marker-assisted selection would be effective and made suggestions for the particular markers to be used, either singly or in combination. They also validated *QFhs.ndsu-3BS* for resistance to grain yield loss (Type IV). No QTL independent of Type II resistance was found.

In many of the above studies, markers closely linked to the FHB resistance QTL were identified, enabling MAS to be contemplated. For example, SSR markers for the 3A and 5A QTL in Frontana have been identified, allowing these to be combined through MAS with the QTL in Sumai 3 and its derivatives. The feasibility of MAS has been directly demonstrated (Wilde *et al.*, 2007), involving the 3B and 5A resistance of Sumai 3 and the 3A resistance of Frontana;

MAS for the two Sumai 3 QTLs gave significant reductions in FHB severity and DON content, although MAS for the Frontana QTL had no effect. Additional phenotypic selection acting on other unmarked QTLs should give additional gain. Some markers have been used extensively in breeding programmes (Guo *et al.*, 2006).

Some of the above reports are particularly illuminating, since they appear to be showing that the various types of resistance are not necessarily truly distinct categories. For example, the Sumai 3 3BS resistance generally has been regarded as being of Type II. However, this locus may in fact be involved in detoxifying DON (Type V resistance). That is, it may be that at least a part of the mechanistic basis of the Type II resistance associated with this locus is its Type V nature.

The map-based cloning of QTL ought to contribute to understanding resistance mechanisms further (Liu and Anderson, 2003; Shen *et al.*, 2006). An expressed sequences tag (EST) rich in leucine and with low similarity to a protein kinase domain of the *Rpg1* gene in barley was identified on 3BS and might represent a portion of a gene for FHB resistance (Shen *et al.*, 2006). This EST could be used in MAS and for map-based cloning. Resistance gene analogues (RGA) associated with 1AL have been identified (Guo *et al.*, 2006); all RGA markers studied contained a heat shock factor that initiated the production of heat shock proteins. Other promising areas for improvements in FHB are: (i) the introduction of genes from related species (QTLs for FHB resistance have been identified on 3A in *Triticum dicoccoides* (Otto *et al.*, 2002) and on 4A in *T. macha* (Steed *et al.*, 2005)); and (ii) the genetic engineering of FHB resistance by, for example, the expression in wheat of *Arabidopsis* NPR1 (Makandar *et al.*, 2006).

The above studies (and others not included here due to space confines, some of which are cited in the 'Catalogue of gene symbols for wheat' [McIntosh *et al.*, 2003] and subsequent annual supplements published in the Annual Wheat Newsletter) demonstrate that QTLs for FHB resistance have been identified on all the chromosomes of wheat, whose detected presence and magnitude of effects depend greatly on environmental factors and the particular genetic background in which they are evaluated. In this sense, the genetic control of resistance appears to be complex, even though genetic effects appear to be mainly additive in nature. The situation may be set to become more complicated still: although, as mentioned previously, FHB resistance is thought to be non-specific or horizontal, recent studies indicate that interactions may be more complex (Xihang *et al.*, 1991).

Conclusions

Although handling of FHB requires the application of several different disease management strategies, substantial progress has been made in understanding the genetic basis of resistance to FHB in wheat and barley. Quantitative resistance usually is caused by the simultaneous segregation of several to many genes and diverse non-genetic factors. Of the several types of resistance that have been hypothesized or reported, Type II resistance is the most stable and well studied. The Chinese wheat cultivar, Sumai 3, and its derivates are one of the best sources of resistance to FHB and may provide the maximum degree of Type II resistance. The major QTL on chromosome 3BS is found in most of the resistant cultivars from China. However, QTLs located on all the other chromosomes have also been reported but, for many of them, their expression is not stable over different environments or in all genetic backgrounds.

Only a few barley cultivars have a relatively higher level of FHB resistance. Most of these resistant cultivars are two-rowed barley. Within six-rowed barley, which is preferred for malting, the cultivar, Chevron, has the best degree of resistance, but its DON level is still too high and far from meeting the safety requirements of the brewing industry. In contrast to wheat, Type I resistance is the major resistance type in barley. Molecular mapping indicates that many QTLs, spread over many chromosomes and with minor effects, control this resistance. Correlation between FHB severity and other spike-related traits has presented a major barrier to breeding for FHB resistance in barley. Using MAS for the Chevron allele at the *Qrgz-2H-8* locus should help breeders surpass this barrier (Nduulu *et al.*, 2007).

Marked-assisted selection may provide such a technique for dissecting and stacking different resistant QTLs for FHB resistance and the application of high-throughput markers for FHB-resistant QTLs may improve selection efficiency significantly. Moreover, recent developments in genomics and biotechnology hold promise for understanding the genetic mechanism of FHB resistance and for more effective development of resistant wheat and barley cultivars. Functional genomics tools such as microarray analysis and ESTs open a new way for genome-wide gene expression profiling.

References

Anderson, J.A., Stack, R.W., Liu, S., Waldron, B.L., Fjeld, A.D., Coyne, C., Moreno-Sevilla, B., Fetch, J.M., Song, Q.J., Cregan, P.B. and Frohberg, R.C. (2001) DNA markers for *Fusarium* head blight resistance QTL in two wheat populations. *Theoretical and Applied Genetics* 102, 1164–1168.

Bai, G.H. (1995) Scab of wheat: epidemiology, inheritance of resistance, and molecular markers linked to cultivar resistance. PhD thesis, Purdue University, West Lafayette, Indiana.

Bai, G.H. and Shaner, G.E. (1994) Wheat scab: perspective and control. *Plant Disease* 78, 760–766.

Bai, G.H., Kolb, F.L., Shaner, G.E. and Domier, L.L. (1999) Amplified fragment length polymorphism markers linked to a major quantitative trait locus controlling scab resistance in wheat. *Phytopathology* 89, 343–348.

Bai, G.H., Plattner, R., Desjardins, A. and Kolb, F.L. (2001) Resistance to *Fusarium* head blight and deoxynivalenol accumulation in wheat. *Plant Breeding* 120, 1–6.

Barreto, D., Carmona, M., Ferrazini, M., Zanelli, M. and Perez, B.A. (2004) Occurrence and pathogenicity of *Fusarium poae* in barley in Argentina. *Cereal Research Communications* 32, 53–60.

Blanco, I.A. del, Frohberg, R.C., Stack, R.W., Berzonsky, W.A. and Kianian, S.F. (2003) Detection of QTL linked to *Fusarium* head blight resistance in Sumai 3-derived North Dakota bread wheat lines. *Theoretical and Applied Genetics* 106, 1027–1031.

Bourdages, J.V., Marchand, S., Rioux, S. and Belzile, F.J. (2006) Diversity and prevalence of *Fusarium* species from Quebec barley fields. *Canadian Journal of Plant Pathology* 28, 419–425.

Bourdoncle, W. and Ohm, H.W. (2003) Quantitative trait loci for resistance to *Fusarium* head blight in recombinant inbred wheat lines from the cross Huapei 57-2/Patterson. *Euphytica* 131, 131–136.

Buerstmayr, H., Steiner, B., Lemmens, M. and Ruckenbauer, P. (2000) Resistance to *Fusarium* head blight in two winter wheat crosses: heritability and trait associations. *Crop Science* 40, 1012–1018.

Buertsmayr, H., Lemmens, M., Hartl, L., Doldi, L., Steiner, B., Stierschneider, M. and Ruckenbauer, P. (2002) Molecular mapping of QTLs for *Fusarium* head blight resistance in spring wheat. I. Resistance to fungal spread (type II resistance). *Theoretical and Applied Genetics* 104, 84–91.

Buerstmayr, H., Steiner, B., Hartl, L., Griesser, M., Angerer, N., Lengauer, D., Miedaner, T., Schneider, B. and Lemmens, M. (2003) Molecular mapping of QTLs for *Fusarium* head blight resistance in spring wheat. II. Resistance to fungal penetration and soread. *Theoretical and Applied Genetics* 107, 503–508.

Canci, P.C., Nduulu, L.M., Muehlbauer, G.J., Dill-Macky, R., Rasmusson, D.C. and Smith, K.P. (2004) Validation of quantitative trait loci for *Fusarium* head blight and kernel discoloration in barley. *Molecular Breeding* 14, 91–104.

Chen, X.M., Yang, Y.H. and Gao, D.S. (1991) Primary identification of resistance to scab of Chinese barley germplasm sources. *Zhejiang Agricultural Science* 2, 91–97.

Chu, F.S. and Li, G.Y. (1994) Simultaneous occurrence of fumonisin B1 and other mycotoxins in moldy corn collected from the People's Republic of China in regions with high incidences of esophageal cancer. *Applied and Environmental Microbiology* 60, 847–852.

Cuthbert, P.A., Somers, D.J., Thomas, J., Cloutier, S. and Brulé-Babel, A. (2006) Fine mapping *Fhb1*, a major gene controlling *Fusarium* head blight resistance in bread wheat (*Triticum aestivum* L.) *Theoretical and Applied Genetics* 112, 1465–1472.

Cuthbert, P.A., Somers, D.J. and Brulé-Babel, A. (2007) Mapping of *Fhb2* on chromosome 6BS: a gene controlling *Fusarium* head blight field resistance in bread wheat (*Triticum aestivum* L.). *Theoretical and Applied Genetics* 114, 429–437.

D'Mello, J.P.F., Placinta, C.M. and McDonald, A.M.C. (1999) *Fusarium* mycotoxins: a review of global implications for animal health, welfare and productivity. *Animal Feed Science and Technology* 80, 183–205.

Dahleen, L.S., Agrama, H.A., Horsley, R.D., Steffenson, B.J., Schwarz, P.B., Mesfin, A. and Franckowiak, J.D. (2003) Identification of QTLs associated with *Fusarium* head blight resistance in Zhedar 2 barley. *Theoretical and Applied Genetics* 108, 95–104.

Doohan, F.M., Brennan, J. and Cooke, B.M. (2003) The use of species-specific PCR-based assays to analyse *Fusarium* ear blight of wheat. *European Journal of Plant Pathology* 109, 755–768.

Draeger, R., Gosman, N., Steed, A., Chandler, E., Thomsett, M., Srinivasachary Schondelmaier, J., Buerstmayr, H., Lemmens, M., Schmolke, M., Mesterhazy, A. and Nicholson, P. (2007) Identification of QTLs for resistance to *Fusarium* head blight, DON accumulation and associated traits in the winter wheat variety Arina. *Theoretical and Applied Genetics* 115, 617–625.

Eeuwijk, E.A. van, Mesterhazy, A., Kling, C.I., Ruckenbauer, P., Saur, L., Buerstmayr, H., Lemmens, M., Keiser, L.C.P., Maurin, N. and Snijders, C.H. (1995) Assessing non-specificity of resistance in wheat to head blight caused by inoculation with European strains of *Fusarium culmorum*, *F. graminearum* and *F. nivale* using a multiplicative model for interaction. *Theoretical and Applied Genetics* 90, 221–228.

Gervais, L., Dedryver, F., Morlais, J.Y., Bodusseau, V., Negre, S., Bilous, M., Groos, C. and Trottet, M. (2003) Mapping of quantitative trait loci for field resistance to *Fusarium* head blight in an European winter wheat. *Theoretical and Applied Genetics* 106, 961–970.

Gilbert, J. and Tekauz, A. (2000) Review: recent developments in research on *Fusarium* head blight of wheat in Canada. *Canadian Journal of Plant Pathology* 22, 1–8.

Ginkel, M. van, Van Der Schaar, W., Yang, Z.P., and Rajaram, S. (1996) Inheritance of resistance to scab in two wheat cultivars from Brazil and China. *Plant Disease* 80, 863–867.

Guo, P.G., Bai, G.H., Li, R.H., Shaner, G. and Baum, M. (2006) Resistance gene analogs associated with *Fusarium* head blight resistance in wheat. *Euphytica* 151, 251–261.

Gutleb, A.C., Morrison, E. and Murk, A.J. (2002) Cytotoxicity assays for mycotoxins produced by *Fusarium* strains: a review. *Environmental Toxicology and Pharmacology* 11, 309–320.

Hilton, A.J., Jenkinson, P., Hollins, T.W. and Parry, D.W. (1999) Relationship between cultivar height and severity of *Fusarium* head blight in wheat. *Plant Pathology* 48, 202–208.

Hockett, E.A. and Nilan, R.A. (1985) Genetics. In: Rasmusson, D.C. (ed.) *Barley*. American Society of Agronomy, Madison, Wisconsin, pp. 190–231.

Horsley, R.D., Schmierer, D., Maier, C., Kudrna, D., Urrea, C.A., Steffenson, B.J., Schwarz, P.B., Franckowiak, J.D., Green, M.J., Zhang, B. and Kleinhofs, A. (2006) Identification of QTLs associated with *Fusarium* head blight resistance in barley accession CIho 4196. *Crop Science* 46, 145–156.

Jiang, G.L., Dong, Y., Shi, J.R. and Ward, R.W. (2007) QTL analysis of resistance to *Fusarium* head blight in the novel wheat germplasm CJ 9306. II. Resistance to deoxynivalenol accumulation and grain yield loss. *Theoretical and Applied Genetics* 115, 1043–1052.

Klahr, A., Zimmermann, G., Wenzel, G. and Mohler, V. (2007) Effects of environment, disease progress, plant height and heading date on the detection of QTLs for resistance to *Fusarium* head blight in an European winter wheat cross. *Euphytica* 154, 17–28.

Kolb, F.L., Bai, G.H., Muehlbauer, G.J., Anderson, J.A., Smith, K.P. and Fedak, G. (2001) Host plant resistance genes for *Fusarium* head blight: mapping and manipulation with molecular markers. *Crop Science* 41, 611–619.

Lemmens, M., Scholz, U., Berthiller, F., Dall'Asta, C., Koutnik, A., Schuhmacher, R., Adam, G., Buerstmayr, H., Mesterhazy, A., Krska, R. and Ruckenbauer, P. (2005) The ability to detoxify the mycotoxin deoxynivalenol colocalizes with a major quantitative trait locus for *Fusarium* head blight resistance in wheat. *Molecular Plant–Microbe Interactions* 18, 1318–1324.

Leonard, K.J. and Bushnell, W.R. (2003) *Fusarium Head Blight of Wheat and Barley*. APS Press, St Paul, Minnesota.

Liu, S. and Anderson, J.A. (2003) Markers assisted evaluation of *Fusarium* head blight resistant wheat germplasm. *Crop Science* 43, 760–766.

Liu, S., Abate, Z.A. and McKendry, A.L. (2005) Inheritance of *Fusarium* head blight resistance in the soft red winter wheat Ernie. *Theoretical and Applied Genetics* 110, 454–461.

Lundqvist, U. and Franckowiak, J.D. (1997) Stock number: BGS 178. *Barley Genetics Newsletter* 26, 200–201.

Ma, H.X., Zhang, K.M., Gao, L., Bai, G.H., Chen, H.G., Cai, Z.X. and Lu, W.Z. (2006) Quantitative trait loci for resistance to *Fusarium* head blight and deoxynivalenol accumulation in Wangshuibai wheat under field conditions. *Plant Pathology* 55, 739–745.

Ma, Z.P., Steffenson, B.J., Prom, L.K. and Lapitan, N.L.V. (2000) Mapping of quantitative trait loci for *Fusarium* head blight resistance in barley. *Phytopathology* 90, 1079–1088.

McIntosh, R., Yamazaki, Y., Devos, K., Dubcovsky, J., Rogers, W. and Appels, R. (2003) *Proceedings of the 10th International Wheat Genetics Symposium.* Paestum, Italy.

McMullen, M., Jones, R. and Gallenberg, D. (1997) Scab of wheat and barley: a re-emerging disease of devastating impact. *Plant Disease* 81, 1340–1348.

Makandar, R., Essig, J.S., Schapaugh, M.A., Trick, H.N. and Shah, J. (2006) Genetically engineered resistance to *Fusarium* head blight in wheat by expression of *Arabidopsis* NPR1. *Molecular Plant–Microbe Interactions* 19, 123–129.

Mather, K. and Jinks, J.L. (1982) *Biometrical Genetics: The Study of Continuous Variation.* Chapman and Hall, New York.

Mesfin, A., Smith, K.P., Dill-Macky, R., Evans, K.C., Waugh, R., Gustus, C.D. and Muehlbauer, G.J. (2003) Quantitative trait loci for *Fusarium* head blight resistance in barley detected in two-rowed by six-rowed population. *Crop Science* 43, 307–318.

Mesterhazy, A. (1995) Types and components of resistance to *Fusarium* head blight of wheat. *Plant Breeding* 114, 377–386.

Mesterhazy, A. (1997) Breeding for resistance to *Fusarium* head blight of wheat. In: Dubin, H.J., Gilchrist, L., Reeves, J. and McNab, A. (eds) *Fusarium Head Scab: Global Status and Future Prospects.* CIMMYT, Mexico, pp. 79–85.

Moreno-Sevilla, B., Anderson, J.A., Waldron, B.L., Stack, R.W. and Frohberg, R.C. (1997) RFLP mapping of *Fusarium* head blight resistance genes in wheat. In: Dill-Macky, R. and Jones, R.K. (eds) *Proceedings of the National Fusarium Head Blight Forum.* University of Minnesota, St Paul, Minnesota, 18 pp.

Nduulu, L.M., Mesfin, A., Muehlbauer, G.J. and Smith, K.P. (2007) Analysis of the chromosome 2(2H) region of barley associated with the correlated traits *Fusarium* head blight resistance and heading date. *Theoretical and Applied Genetics* 115, 561–570.

Nelson, P.E., Toussoun, T.A. and Cook, R.J. (1981) *Fusarium: Diseases, Biology and Taxonomy.* University Press, University Park, Pennsylvania.

Nicholson, P., Gosman, N., Draeger, R., Thomsett, M., Chandler, E. and Steed, A. (2005) In: Buck, H.T., Nisi, J.E. and Salomon, N. (eds) *7th International Wheat Conference.* Mar del Plata, Argentina.

Otto, C.D., Kianian, S.F., Elias, E.M., Stack, R.W. and Joppa, L.R. (2002) Genetic dissection of a major *Fusarium* head blight QTL in tetraploid wheat. *Plant Molecular Biology* 48, 625–632.

Paillard, S., Schnurbusch, T., Tiwari, R., Messmer, M., Winzeler, M., Keller, B. and Schachermayr, G. (2004) QTL analysis of resistance to *Fusarium* head blight in Swiss winter wheat (*Triticum aestivum* L.). *Theoretical and Applied Genetics* 109, 323–332.

Parry, D.W., Jenkinson, P. and McLeod, L. (1995) *Fusarium* ear blight (scab) in small grains cereals: a review. *Plant Pathology* 44, 207–238.

Pena, R.C. de la, Smith, K.P., Capettini, F., Muehlbauer, G.J., Gallo-Meagher, M., Dill-Macky, R., Somers, D.A. and Rasmusson, D.C. (1999) Quantitative trait loci associated with resistance to *Fusarium* head blight and kernel discoloration in barley. *Theoretical and Applied Genetics* 99, 561–569.

Placinta, C.M., D'Mello, J.P.F. and McDonald, A.M.C. (1999) A review of worldwide contamination of cereal grains and animal feed with *Fusarium* mycotoxins. *Animal Feed Science and Technology* 78, 21–37.

Pumphrey, M.O., Bernardo, R. and Anderson, J.A. (2007) Validating the *Fhb1* QTL for *Fusarium* head blight resistance in near-isogenic wheat lines developed from breeding populations. *Crop Science* 47, 200–206.

Rudd, J.C., Horsley, R.D., McKendry, A.L. and Brown, P.O. (2001) Host plant resistance genes for *Fusarium* head blight: sources, mechanisms, and utility in conventional breeding systems. *Crop Science* 41, 620–627.

Schroeder, H.W. and Christensen, J.J. (1963) Factors affecting resistance of wheat to scab caused by *Gibberella zeae. Phytopathology* 53, 831–838.

Semagn, K., Skinnes, H., Bjornstad, A., Maroy, A.G. and Tarkegne, Y. (2007) Quantitative trait loci controlling *Fusarium* head blight resistance and low deoxynivalenol content in hexaploid wheat population from 'Arina' and NK93604. *Crop Science* 47, 294–303.

Shen, X., Francki, M.G. and Ohm, H.W. (2006) A resistance-like gene identified by EST mapping and its association with a QTL controlling *Fusarium* head blight infection on wheat chromosome 3BS. *Genome* 49, 631–635.

Shen, X.R., Ittu, M. and Ohm, H.W. (2003a) Quantitative trait loci conditioning resistance to *Fusarium* head blight in wheat line F201R. *Crop Science* 43, 850–857.
Shen, X.R., Zhou, M., Lu, W. and Ohm, H. (2003b) Detection of *Fusarium* head blight resistance QTL in a wheat population using bulked segregant analysis. *Theoretical and Applied Genetics* 106, 1041–1047.
Smith, K.P., Evans, C.K., Dill-Macky, R., Gustus, C., Xie, W. and Dong, Y. (2004) Host genetic effect on deoxynivalenol accumulation in *Fusarium* head blight of barley. *Phytopathology* 94, 766–771.
Snidjers, C.H. (1990) The inheritance of resistance to head blight caused by *Fusarium culmorum* in winter wheat. *Euphytica* 50, 11–18.
Snyder, W.C. and Nash, S.M. (1968) Relative incidence of *Fusarium* pathogens of cereals in rotation plots at Rothamsted. *Transactions of the British Mycological Society* 51, 417–425.
Somers, D.J., Fedak, G. and Savard, M. (2003) Molecular mapping of novel genes controlling *Fusarium* head blight resistance and deoxynivalenol accumulation in spring wheat. *Genome* 46, 555–564.
Stack, R.W. (1989) Comparison of inoculum potential of ascospores and conidia of *Gibberella zeae. Canadian Journal of Plant Pathology* 11, 137–142.
Steed, A., Chandler, E., Thomsett, M., Gosman, N., Faure, S. and Nicholson, P. (2005) Identification of type I resistance to *Fusarium* head blight controlled by a major gene located on chromosome 4A of *Triticum macha. Theoretical and Applied Genetics* 111, 521–529.
Steffenson, B.J., Hayes, P.M. and Kleinhofs, A. (1996) Genetics of seedling and adult plant resistance to net blotch (*Pyrenophora teres* f. *teres*) and spot blotch (*Cochliobolus sativus*) in barley. *Theoretical and Applied Genetics* 92, 552–558.
Steiner, B., Lemmens, M., Griesser, M., Scholz, U., Schondelmaier, J. and Buerstmayr, H. (2004) Molecular mapping of resistance to *Fusarium* head blight in the spring wheat cultivar Frontana. *Theoretical and Applied Genetics* 109, 215–224.
Sydenham, E.W., Thiel, P.G., Marassas, W.F.O., Shephard, G.S., Schalkwyk, D.J. van and Koch, K.R. (1990) Natural occurrence of some *Fusarium* mycotoxins in corn from low and high esophageal cancer prevalence areas of the Transkei, Southern Africa. *Journal of Agricultural and Food Chemistry* 38, 1900–1903.
Takeda, K. (1990) Selection response and parent offspring correlation of the resistance to *Fusarium* head blight in barley. *Japanese Journal of Breeding* 40, 91–101.
Takeda, K. and Heta, H. (1989) Establishing the testing method and a search for the resistance varieties to *Fusarium* head blight in barley. *Japanese Journal of Breeding* 39, 203–216.
Urrea, C.A., Horsley, R.D., Steffenson, B.J. and Schwarz, P.B. (2005) Agronomic characteristics, malt quality, and disease resistance of barley germplasm lines with partial *Fusarium* head blight resistance. *Crop Science* 45, 1235–1240.
Waldron, B.L., Moreno-Sevilla, B., Anderson, J.A., Stack, R.W. and Frohberg, R.C. (1999) RFLP mapping of QTL for *Fusarium* head blight in wheat. *Crop Science* 39, 805–811.
Wilde, F., Korzun, V., Ebmeyer, E., Geiger, H.H. and Miedaner, T. (2007) Comparison of phenotypic and marker-based selection for *Fusarium* head blight resistance and DON content in spring wheat. *Molecular Breeding* 19, 357–370.
Windels, C.E. (2000) Current status of *Fusarium* taxonomy. *Phytopathology* 90, 17–21.
Xie, G.Q., Zhang, M.C., Chakraborty, S. and Liu, C.J. (2007) The effect of 3BS locus of Sumai 3 on *Fusarium* head blight resistance in Australian wheats. *Australian Journal of Experimental Agriculture* 47, 603–607.
Xihang, Z., Mingsheng, C. and Xunyi, L. (1991) Screening and testing of barley varieties for scab resistance. *Acta Phytophylacica Sinica* 18, 265.
Yu, J.B., Bai, G.H., Cai, S.B. and Ban, T. (2006) Marker-assisted characterization of Asian wheat lines for resistance to *Fusarium* head blight. *Theoretical and Applied Genetics* 113, 308–320.
Zhou, W.C., Kolb, F.L., Bai, G.H., Domier, L.L. and Yao, J.J. (2002) Effect of individual Sumai 3 chromosomes on resistance to scab spread within spikes and deoxynivalenol accumulation within kernels in wheat. *Hereditas* 137, 81–89.
Zhou, X., Chao, M. and Liang, X. (1991) Screening and testing of barley varieties for scab resistance. *Acta Phytophylacica Sinica* 18, 261–265.
Zhu, H.L., Gilchrist, L., Hayes, P., Kleinhofs, A., Kudrna, D., Liu, Z., Prom, L., Steffenson, B., Toojinda, T. and Vivar, H. (1999) Does function follow form? *Theoretical and Applied Genetics* 99, 1221–1232.

8 Sustainable Management of Rice Blast (*Magnaporthe grisea* (Hebert) Barr): 50 Years of Research Progress in Molecular Biology

S. Nandy,[1] N. Mandal,[2] P.K. Bhowmik,[1] M.A. Khan[3] and S.K. Basu[4]

[1]*Bioproducts and Bioprocesses, Lethbridge Research Center, Agriculture and Agri-Food Canada, Lethbridge, Canada;* [2]*Bidhan Chandra Krishi Vishavidalay, Nadia, India;* [3]*Department of Weed Science, NWFP Agricultural University, Peshawar, Pakistan;* [4]*Department of Biological Sciences, University of Lethbridge, Lethbridge, Canada*

Abstract

Rice blast fungus (*Magnaporthe grisea* (Hebert) Barr) as a species has a very broad host range, infecting more than 40 *Graminaceous* hosts and some other non-grass hosts. The seedling stage, the rapid tillering stage after transplanting and the flower emergence stage have been identified as the most susceptible to rice blast. In developing countries, poor farmers cannot afford to control blast disease by the application of expensive fungicides. Therefore, sustainable rice blast disease management is more important for environmental concern, as well as for better financial returns to farmers in Third World countries. During the past few decades, a substantial amount of research has been conducted all over the globe to cope with blast fungus. In this chapter, we emphasize specifically the molecular biological aspect of the study on rice blast fungus over the past 50 years.

Abbreviations used: BRV: blast-resistant varieties; HR: hypersensitive response; RBD: rice blast disease; RBF: rice blast fungus; RGAs: resistance gene analogues; ROI: reactive oxygen intermediates; PCR: polymerase chain reaction; RAPD: random amplification of polymorphic DNA; RFLP: restriction fragment length polymorphism.

Introduction

Many rice researchers consider blast to be the most important disease of rice worldwide (Valent and Chumley, 1994). This is because the disease is widely distributed (85 countries) and can be very destructive when environmental conditions are favourable. Rice blast causes between 10–30% yield losses worldwide in rice, posing a constant threat to the supply of this staple food for nearly one-half of the world's population (Zhu *et al.*, 2000; Talbot, 2003). The rice blast fungus (RBF), scientifically known as *M. grisea* (Hebert) Barr (anamorph: *Pyricularia grisea* Sacc.), is a filamentous *Ascomycetous* fungus that parasitizes over 40 grasses, including economically important crops like wheat, rice, barley and millet (Ou, 1985), but the pathogen is best known as the casual agent of the rice blast disease (RBD). RBD is one of the most serious diseases in all rice-growing regions of the world. Under heavy dew, all aerial parts of the plant can be affected; leaf surfaces become speckled with oval to globular lesions and severely infected plants are liable to lodging if stems are infected. The infected panicle results in severe yield loss (Ou, 1985). The fungus has the capacity to overcome resistance in a short period of time, soon after the release of a resistant cultivar, and thus has made breeding for resistance a constant and difficult challenge to address for rice breeders and pathologists (Shao *et al.*, 2008). Analysis of the existing genetic variation in plant pathogen populations is an important prerequisite for understanding the mechanism of co-evolution in the plant pathological system (McDonald *et al.*, 1989). Several populations of rice blast pathogen all over the globe have been studied for their characteristic phenotypic and genotypic variations (Levy *et al.*, 1991, 1993; Shull and Hamer, 1994; Chen *et al.*, 1995; Kumar *et al.*, 1999). Blast disease was first reported in China (1637) and then in Japan (1704), Italy (1828) and in the USA (1996) (Asuyama, 1965; Ou, 1985; CRRI Annual Report, 2001–2002). In this chapter, we discuss the 50 years of research on *M. grisea* and the available sustainable disease resistance management in rice.

Epidemiology of Blast Disease

Seedling stage, rapid tillering stage after transplanting and flower emergence stage were identified as the most susceptible to rice blast. The fact that the age of the leaves influences the susceptibility to blast was also brought out. The older the leaves on the plant, the more they are resistant to blast (Ou, 1985; CRRI Annual Report, 2001–2002). Excessive exposure to nitrogen and cold night temperatures predisposed susceptible varieties, but did not show any effect on highly resistant varieties. The critical range of temperature for penetration and establishment of infection was around 25–26°C, whereas germination of spores and appressoria formation occurred within 6–10 h at 20–30°C in the presence of water on the surface of the leaf (Asuyama, 1965; Ou, 1985). The formation of dew or a little rainfall or the occurrence of fog provided the necessary water required for the germination of spores. Analysis of the intensity of infection recorded in different long-term experiments of several years revealed that blast infection had occurred under natural conditions when the minimum temperature during the night was 26°C and below, with the concomitant occurrence of relative humidity of 90% and higher (CRRI Annual Report, 2001–2002).

Grouping of Blast Fungal Isolates

M. grisea as a species has a very broad host range, infecting more than 40 *Graminaceous* hosts and some other non-grass hosts (Asuyama, 1965; Ou, 1985). Ou (1980) studied variability in the pathogen and the host resistance of *M. grisea*. Monoconidial cultures showed continued segregation for virulence pattern and generated diverse lesion types on individual leaves. Conidial and mycelial cells of *M. grisea* were reported to contain nuclei with a different number of chromosomes. These observations offered the best genetic explanation for the variation. Latterell and Rosi (1986) studied the longevity and pathogenic stability of *M. grisea* for 30 years. They suggested that the species comprised a wide range of pathotypes (races), each characterized by its capacity to attack certain cultivars of rice, and that these races were basically stable and mutations (or parasexual recombination) were the exception rather than the rule, resulting in broader host range or increased sporulating capacity. The detection of parasexual DNA exchanges in wild-type strains and the existence of merodiploids in nature suggest that parasexual recombination occurs in field populations of *M. grisea* (Zeigler *et al.*, 1997).

Three DNA probes were developed by Hamer *et al.* (1989), which reliably and specifically identified the genetic backgrounds of the full spectrum of the rice blast fungal pathotypes. One of these probes consists of cloned fragments of repeated DNA obtained from the RBF genome and which are called MGR586 (*M. grisea* repeat elements, previously referred to as PCB586). The probe hybridizes with approximately 50 *Eco*RI fragments, ranging in size from 1.5–20.0 kb in the genome of all *M. grisea* isolates pathogenic to rice. Worldwide conservation of MGR586 sequences in RBF suggests that they descend from a common ancestral source, genetically isolated from other host-limited forms of *M. grisea*. The use of MGR shows that sequences are dispersed randomly on all chromosomes of the pathogens and segregate as genetic loci (Zeigler *et al.*, 1997; Suzuki *et al.*, 2007).

Borromeo (1990) studied the Philippine isolates of RBF with MGR586 and MGR613. Valent and Chumley (1994) discussed the recent application of tools for molecular genetic analysis of *M. grisea* and past and current research in the problem areas. Iwano (1990) and Chen (1993) reported that the racial composition in a field in Yongnan province, China, and the Philippines showed wide yearly fluctuations. Iwano (1990) claimed that isolates from the same lesion changed their reaction on a set of several cultivars annually. Silue *et al.* (1992) studied the patterns of inheritance of avirulence in *M. grisea* in seven different rice cultivars. Avirulence to four cultivars has been reported as being controlled by one gene, whereas for the other three cultivars, it was controlled by two genes.

In another study using DNA polymorphism, common ancestral patterns were found among *Magnaporthe* infecting rice isolates and their associated weed hosts (Borromeo *et al.*, 1993) However, the pathogenic populations infecting the weed hosts do not supply pathogenic inoculums for the rice. Weeds can act as alternative hosts for the disease in greenhouse tests; but their role in the field is not yet quite clear (Kato, 2001). Rice, as a widely and intensively cultivated crop, could be a potential target for parasitic 'host shifts' and a potential agent for 'shifts' to accompanying weeds (Couch *et al.*, 2005). The authors also reported the single origin of rice-infecting *M. oryzae* after a 'host shift' from a *Setaria*-millet and that it was probably closely followed by additional 'shifts' to weeds of rice, cutgrass and torpedo grass. Levy *et al.* (1993) studied the genetic diversity of RBF in a disease nursery in Colombia. DNA fingerprints using MGR586, 115 haplotypes from 151 fungal isolates were identified and partitioned into six discretely distinct genetic lineages. Xia *et al.* (1993) conducted a DNA fingerprinting study to examine microgeographic variations in the *M. grisea* population in two different rice fields in Arakans in South-east Asia. The DNA fingerprints of 113 isolates were grouped based on restriction fragment length polymorphism (RFLP) similarity. Seven distinct fingerprint groups were identified and four fingerprint groups were common in both fields.

A study examining the relationship between phylogeny and pathotypes for isolates of the RBF in the Philippines revealed that the distribution of virulence was nonrandom with respect to lineage for the cultivars under study (Zeigler *et al.*, 1995). Sivaraj (1995) reported six different lineages (L, A, B, E, F and H) from Karnataka in southern India, using the MGR DNA fingerprinting approach. The repetitive DNA element, MGR586, has been widely used for fingerprinting and phylogenetic analyses of *M. grisea*. George *et al.* (1998) developed a polymerase chain reaction (PCR)-based marker to DNA fingerprint the *Magnaporthe* species coming from different biogeographic zones. Roumen *et al.* (1997) studied the genetic variability among 41 isolates of the blast pathogen from five rice-growing countries from the European Union, including Spain, France, Hungary, Italy and Portugal. DNA fingerprinting grouped the isolates into five discrete lineages, which typically showed less than 65% band similarity. Srinivasachary *et al.* (1998) classified 27 single spore isolates of *M. grisea* from Karnataka in southern India over three different locations using random amplified polymorphic DNA (RAPD) primers. They found three clear groups at 70% similarity level. But Srinivasachary *et al.* (2002a,b) used 27 isolates from Ponnampet, Mandya and Bangalore for genetic analysis using 30 RAPD primers. Three distinct lineages were reported by the authors. Chadha and Gopalakrishna (2005) also used 20 isolates from seven different locations in India using

123 RAPD primers for cluster analysis. Scientists have sequenced the *M. grisea* genome and it is now available online at http://www-genome.wi.mit.edu/annotation/fungi/magnaporthe/. It is, however, important to note that for the first time in the USA, the genomic structure of a significant plant pathogen has been made publicly available.

Physiology of Disease Resistance

Plants develop defence mechanisms to recognize pathogens and protect them from attack. These defence reactions are triggered by the recognition of pathogens by plant disease resistance (*R*) genes. After the recognition of pathogens, a signalling pathway is activated, resulting in resistance to pathogens (Hammond-Kosack and Jones, 1997). During the early steps in *R* gene-mediated disease resistance, reactive oxygen intermediates (ROI) such as O_2^- and H_2O_2 are generated rapidly after infection; and, subsequently, hypersensitive response (HR) leading to cell death has been observed. An understanding of how pathogens induce disease, how the plants become diseased and how they defend themselves against the pathogens would help us to understand the functions of the genes governing resistance, which remains unknown, and eventually to develop novel methods for controlling RBD. The nature of resistance to blast disease operating at both the pre- and post-penetrative stages of the disease was investigated using several models involving cultivars differing in their reaction to the disease, nitrogen fertilization and temperature-induced tissue susceptibility and resistance induced by certain chemicals (CRRI Annual Report, 2001–2002). Four different mechanisms govern blast resistance in rice: (i) the epicuticular wax present on the surface of the leaves influences the infection by suppressing the appressorium formation by the pathogen, thus offering a partial resistance resulting in a reduced number of lesions being formed; (ii) free phenolic compounds and their oxidases toxify the tissue in the infected region: the speed and magnitude at which the toxification takes place in response to infection determines the tissue resistance to the pathogen; (iii) the presence of two toxic cinnamate derivatives (ferulate and coumarate) in the cell walls forming toxic oxidized products/polymers like lignin and melanin-like compounds on oxidation forming a mechanical barrier for the fungus and thereby arresting the spread of the pathogen to adjacent cells, thus restricting disease lesions; and (iv) the synthesis and accumulation of antimicrobial compound(s) (diterpenoid in nature) known as 'phytoalexins' in response to infection toxic to the growth of the pathogen. However, none of these mechanisms seemed to be universal in nature and the defence mechanism was dependent on the varieties tested (CRRI Annual Report, 2001–2002).

Finding the Right Gene

The generation of cultivars that possess non-specific resistance to *M. grisea* would provide an economically effective and environmentally sound approach to rice blast control. One promising approach to the achievement of non-specific resistance to *M. grisea* is to incorporate genes that elicit general defence responses in rice (Dang and Jones, 2001; Stuiver and Custers, 2001). Much effort has been devoted to understanding the genetic and molecular basis of resistance in RBF and several genes have been cloned (Parson *et al.*, 1987; Leung *et al.*, 1990; Khang *et al.*, 2008; Shao *et al.*, 2008).

Although earlier studies focused on pathotypic variability (Ou, 1985), later studies focused extensively on molecular markers to characterize population diversity (Nandy *et al.*, 2004). Extensive use of the MGR586 heterodispersed element (Roumen *et al.*, 1997; Kumar *et al.*, 1999; Correll *et al.*, 2000; Viji *et al.*, 2000; Srinivasachary *et al.*, 2002a,b; Chadha and Gopalakrishna, 2005) to delineate DNA fingerprint lineages has helped to identify and classify the genetic structure of this important pathogen. PCR-based molecular markers are useful tools for detecting genetic variation within populations of important plant pathogens (Vakalounakis and

Fragkiadakis, 1999; Kolmer and Liu, 2000; Srinivasachary *et al.*, 2002a,b; Chadha and Gopalakrishna, 2005). RAPD (Welsh and McClelland, 1990; Williams *et al.*, 1990) and markers have been widely used for estimating genetic diversity in wild populations (Annamalai *et al.*, 1995), mainly because the technique does not need previous molecular genetic information and increases marker density for evaluating genetic kinship. The RAPD technique has also been used to study genetic diversity among RBF from different geographical locations in the world (Lima, 1999; Suzuki *et al.*, 2007).

The dynamic virulence of the rice blast pathogen could be the main cause for the breakdown of resistance in several rice varieties. The diversity and variability of the pathogen population may originate from the clonal mode of reproduction, coupled with mutation, migration, selection or random drift, heteroploidy and parasexuality of the fungus (Gesnovesi and Magill, 1976; Dayakar *et al.*, 2000; Noguchi *et al.*, 2007). A repeat sequence termed MGR586 was identified in the genome of rice-infecting strains of *M. grisea* (Shull and Hamer, 1994). This sequence has been widely used for DNA fingerprinting of *M. grisea* to investigate the epidemiology of the RBD (Roumen *et al.*, 1997; Kumar *et al.*, 1999; Correll *et al.*, 2000; Viji *et al.*, 2000; Chadha and Gopalakrishna, 2005). Molecular analysis of isolates of *M. grisea* from different regions within a state (West Bengal, India) revealed the occurrence of a high level of polymorphism, indicating a wide and diverse genetic base (Mandal *et al.*, 2004). Overall, a high genetic diversity was also obtained in Indian RBF (Roumen *et al.*, 1997; Kumar *et al.*, 1999; Correll *et al.*, 2000; Mandal *et al.*, 2004, Chadha and Gopalakrishna, 2005).

Genetic mechanisms, namely simple mutations, meiotic recombination and parasexual recombination, could explain such genetic diversity (Yamasaki and Niizeki, 1965; Zeigler, 1998; Zeigler *et al.*, 2000, Khang, 2001). Some indirect evidence suggests that *M. grisea* has the potential for sexual reproduction in specific geographic zones and localities (Viji *et al.*, 2000, Adreit *et al.*, 2007). There have been few investigations on the perfect state of *M. grisea* in India (Dayakar *et al.*, 2000; Mandal *et al.*, 2004). The sexual cycle does not seem to be a source of variation for the rice blast pathogen in India (Kumar *et al.*, 1999). Similar results have also been reported from other corners of the globe (Valent *et al.*, 1986). The wide range of diversity among collected isolates of *M. grisea* from different locations in West Bengal can be explained mainly by evolution resulting from natural and stress-induced transposition (Ikeda *et al.*, 2001). Other mechanisms like horizontal gene transfer between RBF and its host (Kim *et al.*, 2001) may also be of importance because varieties deployed within a region are based on crop seasons, along with several other biotic and geographic factors (Babujee and Gnanamanickam, 2000).

Using Genetic Diversity of Disease Resistance

Genetic studies of qualitative resistance were started when Goto (1970) established the differential system for races of *P. grisea* or *M. grisea* in Japan. Thirteen major genes for qualitative resistance have been reported by several researchers (Kiyosawa *et al.*, 1981). Several rice cultivars with durable blast resistance have been identified and 'Moroberekan' have been cultivated in the world for many years without high losses from blast (Notteghem, 1985). These plants have been used as resistance donors in breeding programmes. Major resistance genes have been used successfully for developing blast resistance cultivars (Khush, 2004) and several dominant resistance genes have been identified which confer complete blast resistance (Kiyosawa *et al.*, 1981). Atkins and Johnson (1965) identified two independent genes designated *Pi*-1 and *Pi*-6. Hsieh *et al.* (1967) in China found four dominant genes for pathogen resistance in japonica cultivars, named as *Pi*-4, *Pi*-13, *Pi*-22 and *Pi*-25 using RFLP techniques. Yu *et al.* (1991) mapped three major resistance genes, namely *Pi*-1, *Pi*-2 and *Pi*-4 in the Philippines. Several genes from tropical cultivars like 'Tetep', 'Pai-kan tao', '5173', 'LAC23', Moroberekan and 'Apura' were identified and mapped using RFLPs (Yu *et al.*, 1991; Miyamoto *et al.*, 1996; Rybka *et al.*, 1997) (Table 8.1). Recent reports identified at least four clusters, with five to eight loci each, located on chromosomes 4, 6, 11 and 12 (Roumen *et al.*, 1997; Rybka *et al.*, 1997, Tabien *et al.*, 2000; Gao *et al.*, 2002).

Many pathogenic races have been identified in *M. grisea* and pathogenic variability has been cited as the principal cause for the breakdown of resistance in rice varieties (Baker *et al.*, 1997). Therefore, an artificial inoculation study can be practised in place of natural screening, which is quite cumbersome, time-consuming and season specific. There has been considerable achievement in the development of blast-resistant varieties (BRV), particularly using vertical-resistant genes (Nandy *et al.*, 2004). Nevertheless, durable resistance alone can protect irrigated rice crops in the tropics adequately. Exploitation of durable resistance has been proposed for less blast-conducive environments (Buddenhagen, 1983; Notteghem, 1985; Parlevliet, 1988; Bonman *et al.*, 1992). Artificial inoculation in Karnataka, southern India, was also carried out by Srinivasachary *et al.* (2002a) to study involving the

Table 8.1. List of blast disease-resistance genes with chromosome numbers, donor varieties and linked markers of rice.

Gene symbol	Chromosome number	Donor variety	Linked marker	Reference(s)
Pi-1(t)	11	LAC23, C101LAC	Npb181, RZ536	Atkins and Johnson (1965); Yu *et al.* (1991); Leung *et al.* (1998)
Pi-2(t)	6	BL245, C101A51, 5173	RG64	Yu *et al.* (1991); Sridhar *et al.* (1999)
Pi-4(t)	12	Tetep, Pai-kan-tao, BL245, C101PKT	RG869, RZ397	Yu *et al.* (1991); Hittalmani *et al.* (1995); Tabien *et al.* (2000)
Pi-5(t)	4	RIL 45, RIL 249, Moroberekan	RG498	Wang *et al.* (1994); Sridhar *et al.* (1999)
Pi-6	12	–	RG869	Causse *et al.* (1994); Atkins and Johnson (1965)
Pi-7(t)	11	Moroberekan, RIL 29	RG103	Wang *et al.* (1994)
Pi-9	6	*O. minuta* derivative WHD-IS-75-1-127	RG16	Leung *et al.* (1998); Khush *et al.* (1999)
Pi-10	5	Moroberekan	RRF6, RRH18, OPF6(2700)	Naqvi *et al.* (1995), Tabien *et al.* (2000)
Pi-11	8	Oryzica Llanos 5	BP127, RZ617	Zhu *et al.* (1992); Roca *et al.* (1996); Khush *et al.* (1999)
Pi-12	12	Moroberekan, RIL 10	RG869	Khush *et al.* (1999)
Pi-b	2	F-145-2	RZ123	Miyamoto *et al.* (1996); Khush *et al.* (1999)
Pi-z[5]	6	C101A51	RG64, RG612	Fukuoka and Okuno (1997); Leung *et al.* (1998); Sridhar *et al.* (1999)
Pi-k	11	F-129-1	–	Chao *et al.* (1999); Bryan *et al.* (2000)
Pi-ta and *Pi*-ta[2]	12	Taducan, C101PKT, IR64, F-124-1, F128-1	RZ397, RG241	Shigemura and Kitamura (1954); Rybka *et al.* (1997); Leung *et al.* (1998); Bryan *et al.* (2000)

reaction of representative single-spore culture PPT-4 to rice varieties Moroberekan, isolines of Co39, namely *Pi*-1, *Pi*-2, *Pi*-4, *Pi*-2 + *Pi*-1, *Pi*-1 + *Pi*-4, along with IRAT177, Apura and Doddi showed resistant reaction. Of these, *Pi*-1, *Pi*-2, *Pi*-4, *Pi*-2 + *Pi*-1 and *Pi*-1 + *Pi*-4 are known to contain major genes conferring resistance to blast disease. Yamada *et al.* (1976) and Kiyosawa *et al.* (1981) selected 12 differential varieties for resistance genes *Pi*-k^{s}, *Pi*-a, *Pi*-k, *Pi*-k^{m}, *Pi*-z, *Pi*-ta (*Pi*-4), *Pi*-ta^{2}, *Pi*-zt, *Pi*-k^{p}, *Pi*-b and *Pi*-t. These differential varieties were used in Japan especially, but were not readily available in other countries. Monogenic lines including only a single gene in each genetic background and targeting for 24 different resistance genes – *Pi*-a, *Pi*-b, *Pi*-i, *Pi*-k^{s}, *Pi*-k, *Pi*-k-h, *Pi*-k^{m}, *Pi*-k^{p}, *Pi*-sh, *Pi*-t, *Pi*-ta (*Pi*-4), *Pi*-ta^{2}, *Pi*-z, *Pi*-z^{5} (*Pi*-2), *Pi*-z^{t}, *Pi*-1, *Pi*-3, *Pi*-5(t), *Pi*-7(t), *Pi*-9, *Pi*-11(t), *Pi*-12(t), *Pi*-19 and *Pi*-20 – were developed by Tsunematsu *et al.* (2000) as the first international standard differential variety set. The polymorphic RG-64 marker was used by Hittalmani *et al.* (2001) to identify rice plants carrying *Pi*-2(t) from an F2 population derived from the cross between Co39 and C101A51. More than 30 blast-resistant genes (Babujee and Gnanamanickam, 2000) and QTLs have been identified in rice by conventional genetic studies based on linkage analyses and recombination frequencies (Kinoshita, 1991; Mackill *et al.*, 1993). Some major genes for blast resistance have been identified in recombinant inbred lines (RILs) (Wang *et al.*, 1994).

Zeigler *et al.* (1995) proposed that organization of the blast fungus population into well-defined lineages and their distribution in specific geographic locations have led to the employment of resistance genes targeted against pathogen populations prevalent in that region. This has been known as the 'lineage exclusion' hypothesis. Sivaraj *et al.* (1996) proposed a model to support gene pyramiding based on lineage exclusion. They consider traditional plant breeding as a strategy of pathotype exclusion, which leads to frequent resistance breakdown when appropriate pathotypes appear within 1 or 2 years after such resistance is deployed in large areas. In lineage exclusion, the conventional strategy is modified as a phylogenetic pathotype exclusion. Lineage exclusion presumes that lineage-specific avirulences represent an evolutionary genetic barrier to pathotype diversification within the lineage. IRAT212/N22, RR18-3/Bala, Bala/Tetep, Azucena/Gaurav and several lines from the natural cross of CR314-5-10 were resistant to leaf blast disease (CRRI, Annual Report 2000–2001). A combination of genes is also considered useful to confer resistance to the pathogen lineages prevalent in China, the USA and Latin America (Babujee and Gnanamanickam, 2000).

Molecular Genetic Analysis of the Pathogen

Plant disease resistance (*R*) genes confer resistance to a wide range of pathogens (fungi, viruses, bacteria and nematodes); they share various conserved motifs, suggesting the existence of a common defence signal transduction pathway in different plant–microbe interaction systems (Dang and Jones, 2001; Martin *et al.*, 2003). In general, the *R* genes fall into six distinct classes, the most prevalent of which is the nucleotide-binding site plus leucine-rich repeat (NBS–LRR) genes (Martin *et al.*, 2003; Qu *et al.*, 2006). The LRR domains are generally thought to be involved in the interaction with avirulence (AVR) proteins and to be the major determinant of resistance specificity (Hulbert *et al.*, 2001). The *AVR-Pita* avirulence gene family has been cloned recently at Kansas State University, USA, by Khang *et al.* (2008). They have studied isolates of the *M. grisea* species complex from diverse hosts and have found that *AVR-Pita* is a member of a gene family, which led them to rename it *AVR-Pita1*. Using the dominant DNA markers derived from portions of the *Pi-ta* gene, 141 rice germplasm accessions were rapidly determined and the results were confirmed by inoculating rice germplasm with an *M. grisea* strain containing *AVR-Pita* (Wang *et al.*, 2007). The *Pi-ta* gene was found in accessions from major rice-producing countries, including China, Japan, Vietnam, the Philippines, Iran and the USA.

In another recent study, Shao *et al.* (2008) have reported that the expression of a hairpin-encoding gene (*hrf1*), derived from *Xanthomonas oryzae* pv. *oryzae*, confers non-specific resistance in rice to the blast fungus, *M. grisea*. Transgenic plants and their T1–T7 progenies were highly resistant to all major *M. grisea* races in rice-growing areas along the Yangtze River, China. The expression of defence-related genes was activated in resistant transgenic plants and the formation of melanized appressoria, which is essential for foliar infection, was inhibited on plant leaves. These results suggest that hairpins may offer new opportunities for generating broad-spectrum disease resistance in other crops. However, occurrence of clustered multigene families is a major obstacle in the cloning of *R* genes (Dixon *et al.*, 1996; Ori *et al.*, 1997), which makes it even more difficult to determine the functional copy of these genes. Therefore, fine mapping of *R*-gene analogues on different chromosomes would be helpful in the identification of multigene families in rice, which in turn will lead to the establishment of correlation between the chromosomal position of known *R* genes and their analogues. Recently, Kumar *et al.* (2007) cloned and also carried out *in silico* mapping of resistance gene analogues (RGAs) isolated from rice lines containing known genes for blast resistance. They have amplified RGAs from the genomic DNA of 10 rice lines having varying degrees of resistance to *M. grisea* by using degenerate primers. Twenty RGAs were mapped near to the chromosomal regions containing known genes for rice blast, bacterial leaf blight and sheath blight resistance. Thirty-nine RGA sequences also contained an open reading frame representing the signature of potential disease-resistance genes.

Control Measures

Kato (2001) suggests burning and composting of infected plant parts; use of non-infected or certified healthy seeds and disease-resistant cultivars; appropriate regulation of fertilizer application; proper cultural control and avoiding damp or most soil with high moisture content for seed sowing, etc. However, chemical control is the most commonly used approach in most parts of the globe for effective disease control. Several fungicides are used against blast disease, including benomyl, fthalide, edifenphos, iprobenfos, tricyclazole, isoprothiolane, probenazole, pyroquilon, felimzone (= meferimzone), diclocymet, carpropamid, fenoxanil and metominostrobin, and antibiotics such as blasticidin and kasugamycin (Kato, 2001). The composition, quantity, time and application method of fungicides applied in field trials are dependent on the disease forecast for a particular region or zone, or on the local disease prevalence rate (Kato, 2001).

Carbendazim, chlorobenthiozone, coratop, fungorene, hinosan and kitazin fungicides and antibiotic kasumin were effective against foliar and neck blast in India (CRRI Annual Report, 2001–2002). Rice seed treatment with Carbendazim + TMTD 25 was effective in controlling seedborne blast (CRRI Annual Report, 2001–2002). The control of rice blast relies on the use of resistant cultivars and the application of fungicides, but neither approach is particularly effective in different geographic locations (Shao *et al.*, 2008) because management of rice blast via breeding BRV has had only short-term success due to the frequent breakdown of resistance under field conditions (Valent and Chumley, 1994). The frequent appearance of new races (or pathotypes) of the fungus that are capable of infecting previously resistant varieties has been proposed as the principal cause for the loss of resistance (Ou, 1980).

Host resistance in rice to *M. grisea* functions via a classical gene-for-gene interaction in which a single dominant resistance gene corresponds with a dominant avirulence gene in the pathogen (Hammond-Kosack and Jones, 1997; Talbot, 2003). Because of the apparent instability in the genome of *M. grisea*, new pathogenic races evolve rapidly and thus host resistance typically lasts for a few years only (Zhu *et al.*, 2000; Talbot, 2003). Few fungicides are available for the effective control of rice blast, but rapid mutation in the pathogen leads to the emergence of fungicide-resistant variants (Takagaki *et al.*, 2004); thus, higher-dose applications of fungicides pose risks both to humans and the environment.

Sustainable Rice Blast Disease Management

In developing countries, poor farmers cannot afford to control blast disease by the application of fungicides. Chemical control of plant pathogens is most effective and yet the use of chemicals is not generally desired due to the serious environmental threat it poses. Environmental effects and resistance are not considered a major concern in developing countries. Farmers are more interested in short-term strategy for disease control. However, the continuous use of fungicides leads to the resurgence of resistant races of the pathogen under selection pressure. Therefore, sustainable rice blast disease management is more important for environmental concern. Figure 8.1 shows the basic components that have been used for rice blast management for the past 50 years.

Conclusions

We have reviewed here the past 50 years of research progress in the genetics and molecular biology of rice blast disease, but different approaches can be taken for sustainable disease control with recent advances in genomics, proteomics and diverse genetic resistance mechanisms. Liu *et al.* (2002) recently reported the application of candidate defence genes to develop blast-resistant breeding lines with resistance to diverse pathogen populations. Several biocontrol agents for blast have been deployed successfully to combat the disease in the laboratory, greenhouse and field tests (Chatterjee *et al.*, 1996; Krishnamurthy *et al.*, 1998; Gnanamanickam *et al.*, 1999). The feasibility of such strategies on a commercial scale still remains to be tested. Hence, use of

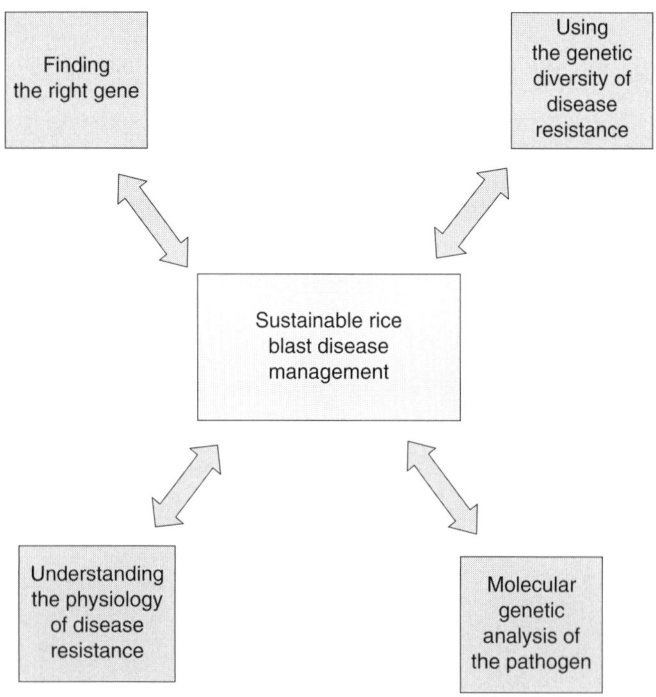

Fig. 8.1. The four basic components of sustainable rice blast management.

resistant cultivars is the best available alternative to overcome severe yield losses.

The objective of the green revolution has not changed; there is the added impetus that crop protection should be conducted in the context of improving the livelihood of rural people and preserving limited natural resources (Leung *et al.*, 2003). However, the gene revolution has opened up newer and better possible ways of preventing yield loss from pathogen attack, conservation and utilization of wild species for resistance genes. The variability of the pathogen and the history of resistance breakdown have led to the development of a number of different plant breeding and molecular approaches to achieve durable blast resistance. Combinations of resistance genes are thought to provide broader spectra of resistance through both ordinary gene action and quantitative complementation that result in durable resistance. Gene pyramiding is one of the strategies recommended to increase the durability of blast disease resistance (Robinson, 1973; Nelson, 1978; Buddenhagen, 1983; Pedersen and Leath, 1988).

Pyramided resistance will be durable in places where compatibility to the component resistance genes is distributed among the prevalent lineages. Agricultural practices such as soil preparation, low nitrogen fertilization, low sowing density, optimized use of water and seed selection contribute to reduce the virulence of *M. grisea* populations. Optimized integration of genetic resistance in agricultural management is the preferred strategy to protect cultivated rice from RBD in a way that is affordable, feasible, durable and, overall, compatible with environmental protection.

References

Adreit, H.S., Andriantsimialona, D., Utami, D.W., Notteghem, J.L., Lebrun, M.H. and Tharreau, D. (2007) Microsatellite markers for population studies of the rice blast fungus *Magnaporthe grisea*. *Molecular Ecology Notes* 7, 667–670.

Annamalai, P., Ishii, H., Lalithakumari, D. and Revathi, R. (1995) Polymerase chain reaction and its applications in fungal disease diagnosis. *Journal of Plant Diseases and Protection* 102, 91–104.

Asuyama, H. (1965) Morphology, taxonomy, host range and life cycle of *Pyricularia oryzae*. In: Chandler, R.F. Jr (ed.) *The Rice Blast Disease. Proceedings of Symposium, IRRI.* The Johns Hopkins University Press, Baltimore, Maryland, pp. 9–22.

Atkins, J.G. and Johnson, T.H. (1965) Inheritance in rice of reaction to races 1 and 6 of *Pyricularia oryzae*. *Phytopathology* 44, 993–995.

Babujee, L. and Gnanamanickam, S.S. (2000) Molecular tools for characterization of rice blast pathogen (*Magnaporthe grisea*) population and molecular marker assisted breeding for disease resistance. *Current Science* 78, 248–257.

Baker, B., Zambryski, P., Staskawicz, B. and Dinesh-Kumar, S.P. (1997) Signaling in plant–microbe interactions. *Science* 276, 726–733.

Bonman, J.M., Khush, G.S. and Nelson, R.J. (1992) Breeding rice for resistance to pests. *Annual Review of Phytopathology* 30, 507–528.

Borromeo, F.S. (1990) Molecular characterization of *Pyricularia oryzae* Cav. population from rice and other hosts. PhD thesis, University of Philippines, Los Banos, Laguna, The Philippines.

Borromeo, F.S., Nelson, R.J., Boonman, J.M. and Leung, H. (1993) Genetic differentiation among isolates of *Pyricularia* infecting rice and weed hosts. *Phytopathology* 83, 393–399.

Bryan, G.T., Wu, K.S., Farrall, L., Jia, Y., Hershey, H.P., McAdams, S.A., Faulk, K.N., Donaldson, G.K., Tarchini, R. and Valent, B. (2000) A single amino acid difference distinguishes resistant and susceptible alleles of the rice blast resistance gene *Pi*-ta. *The Plant Cell* 12, 2033–2045.

Buddenhagen, I.W. (1983) Breeding strategies for stress and disease resistance in developing countries. *Annual Review of Phytopathology* 21, 385–409.

Causse, M.A., Fulton, T.M., Cho, Y.G., Ahn, S.N., Chunwongse, J., Wu, K., Xiao, J., Yu, Z., Ronald, P.C., Harrington, S.E., Second, G., McCouch, S.R. and Tanksley, S.D. (1994) Saturated molecular map of the rice genome based on interspecific backcross population. *Genetics* 138, 1251–1274.

Chadha, S. and Gopalakrishna, T. (2005) Genetic diversity of Indian isolates of rice blast pathogen (*Magnaporthe grisea*) using molecular markers. *Current Science* 88(9), 1466–1469.

Chao, C.T., Moldenhauer, K.A.K. and Ellingboe, A.H. (1999) Genetic analysis of resistance/susceptibility in individual F3 families of rice against strains of *Magnaporthe grisea* containing different genes for avirulence. *Euphytica* 109, 183–190.
Chatterjee, A., Valasubramanian, R., Vachhani, A.K., Mau, W.L., Gnanamanickam, S.S. and Chatterjee, A.K. (1996) Isolation of ANT+ mutants of *Pseudomonas flourescens* strain Pf7-14 altered in antibiotic production, cloning of ANT+ DNA, and evolution of the role of antibiotic production in the control of blast and sheath blight of rice. *Biological Control* 7, 185–195.
Chen, D. (1993) Population of *Pyricularia grisea* (Cook) Sacc. in two screening sites in the Philippines and characterization of resistance genes. PhD thesis, University of Philippines, Los Banos, Laguna, The Philippines, 161 pp.
Chen, D., Zeigler, R.S., Leung, H. and Nelson, R.J. (1995) Population structure of *Pyricularia oryzae* at two screening sites in the Philippines. *Phytopathology* 85, 1011–1020.
Correll, J.C., Harp, T.L., Guerber, J.C., Zeigler, R.S., Liu, B., Cartwright, R.D. and Lee, F.N. (2000) Characterisation of *Pyricularia grisea* in the United States using independent and molecular markers. *Phytopathology* 90, 1396–1404.
Couch, B.C., Isabelle, F., Marc-Henri, L., Didier, T., Barbara, V., Pham, V.K., Jean-Loup, N. and Kohn, Linda M. (2005) Origins of host-specific populations of the blast pathogen *Magnaporthe oryzae* in crop domestication with subsequent expansion of pandemic clones on rice and weeds of rice. *Genetics* 170(2), 613–630.
CRRI Annual Report (2001–2002) Rainfed upland rice (www.crri.nic.in, accessed 10 June 2008).
Dang, J.L. and Jones, J.D.G. (2001) Plant pathogens and integrated defense responses to infection. *Nature* 411, 826–833.
Dayakar, B.V., Naraynan, N.N. and Gnanamanickam, S.S. (2000) Cross-compatability and distribution of mating type alleles of the rice blast fungus *Magnaporthe grisea* in India. *Plant Disease* 84(6), 700–704.
Dixon, M.S., Jones, D.A., Keddle, J.S., Thomas, C.M., Harrison, K. and Jones, J.D.G. (1996) The tomato cf-2 disease resistance locus comprises two functional genes encoding leucine-rich repeat proteins. *Cell* 84, 451–459.
Fukuoka, S. and Okuno, K. (1997) QTL analysis for field resistance to rice blast fungus using RFLP markers. *Rice Genetics Newsletter* 14, 98–99.
Gao, W., Khang, C.H., Park, S.Y., Lee, Y.H. and Kang, S. (2002) Evolution and organization of a highly dynamic, subtelomeric helicase gene family in the rice blast fungus *Magnaporthe grisea. Genetics* 162, 103–112.
George, M.L.C., Nelson, R.J., Zeigler, R.S. and Leung, H. (1998) Rapid population analysis of *Magnaporthe grisea* using rep-PCR and endogenous repetitive DNA sequence. *Phytopathology* 88, 223–229.
Gesnovesi, A.D. and Magill, C.W. (1976) Heterokaryosis and parasextuality in *Pyricularia grisea* Cavara. *Canadian Journal of Microbiology* 22, 531–536.
Gnanamanickam, S.S., Priyadarisini, V.B., Narayanan, N.N., Vasudevan, P. and Kavitha, S. (1999) An overview of bacterial blight disease of rice and strategies for its management. *Current Science* 77(11), 1435–1444.
Goto, I. (1970) Genetic studies on the resistance of rice plant to the blast fungus. I. Inheritance of resistance in crosses Sensho XH-79 and Imochi-shirazu XH-79. *Annals of the Phytopathological Society of Japan* 36(5), 304–312.
Hamer, J.E., Farral, L., Orbach, M.J., Valent, B. and Chumley, F.G. (1989) Host species-specific conservation of a family of repeated DNA sequences in the genome of a fungal plant pathogen. *Proceedings of the National Academy of Sciences of the United States of America* 86, 9981–9985.
Hammond-Kosack, K.E. and Jones, J.D.G. (1997) Plant disease resistance genes. *Annual Review of Plant Physiology and Plant Molecular Biology* 48, 575–607.
Hittalmani, S., Foolad, M.R., Mew, T.V., Rodriguez, R.L. and Huang, N. (1995) Development of PCR-based marker to identify rice blast resistance gene, *Pi*-2(t) in a segregating population. *Theoretical and Applied Genetics* 91, 9–14.
Hittalmani, S., Foolad, M., Mew, T., Rodriguez, R. and Huang, N. (2001) Identification of blast resistance gene, *Pi*-2(t) in rice plants by flanking DNA markers. *Rice Genetics Newsletter* 11, 25–30.
Hsieh, S.C., Lin, M.H. and Liang, H.L. (1967) Genetics analysis in rice. VIII. Inheritance of resistance to races 4, 22 and 25 of *Pyricularia oryzae. Botany Bulletin of the Academy of Taipei* 8, 225–260.
Hulbert, S.H., Webb, C.A., Smith, S.M. and Sun, Q. (2001) Resistance gene complexes: evolution and utilization. *Annual Review of Phytopathology* 39, 285–312.

Ikeda, K., Nakayasiiki, H., Takagi, M., Tosa, Y. and Mayama, S. (2001) Heat shock, copper sulphate and oxidative stress activate the retrotransposon MAGGY resident in the plant pathogenic fungus *M. grisea*. *Molecular Genetics and Genomics* 266, 318–325.

Iwano, M. (1990) Distribution of pathogenic races of rice blast fungus and varietal resistance to blast in Yunnan Province. *Plant Protection* 44, 57–61.

Kato, H. (2001) Rice blast control. *Pesticide Outlook* 23–25 doi. 10.1039/b 100803j (http://www.rsc.org/ej/PO/2001/b100803j.pdf, accessed 6 July 2008).

Khang, C.H., Park, S.Y., Lee, Y.H., Valent, B. and Kang, S. (2008) Genome organization and evolution of the AVR-Pita avirulence gene family in the *Magnaporthe grisea* species complex. *Molecular Plant–Microbe Interactions* 21(5), 658–670.

Khang, S. (2001) Organization and distribution pattern of MGLR-3 a novel retrotransposon in the rice blast fungus *Magnaporthe grisea. Fungal Genetics and Biology* 32, 11–19.

Khush, G.S. (2004) Harnessing science and technology for sustainable rice-based production systems. In: FAO Rice Conference 04/CRS.14, 12–13 February 2004, Rome, Italy, 13pp (http://www.fao.org/rice2004/en/pdf/khush.pdf, accessed 16 June 2008).

Khush, G.S., Bennet, J., Datta, S.K., Brar, D.S. and Li, Z. (1999) Advances in rice genetics and biotechnology. *International Rice Commission Newsletter* 48 (http://www.fao.org/docrep/, accessed March 2008).

Kim, N.S., Park, N.I., Kim, S.H., Han, S.S. and Kang, K. (2001) Isolation of TC/AG repeat microsatellite sequences for the fingerprinting rice blast fungus and their possible horizontal transfer to plant species. *Molecular Cell* 10, 127–134.

Kinoshita, T. (1991) Report of the Committee on Gene Symbolization, Nomenclature, and Linkage Groups. *Rice Genetics Newsletter* 8, 2–37.

Kiyosawa, S., Ikehashi, S., Kato, H. and Ling, Z.Z. (1981) Pathogenecity tests of Philippine isolates of blast fungus using two sets of rice varieties. *Japanese Journal of Breeding* 31, 367–376.

Kolmer, J.A. and Liu, J.Q. (2000) Virulence and molecular polymorphisms in international collections of the wheat leaf rust fungus *Puccinia triticina. Phytopathology* 90, 427–436.

Krishnamurthy, K. and Gnanamanickam, S.S. (1998) Biological control of rice blast by *Pseudomonas* fluorescens strain Pf7-14: evaluation of a marker gene and formulations. *Biological Control* 13(3), 158–165.

Kumar, J., Nelson, R.J. and Zeigler, R.S. (1999) Population structure and dynamics of *M. grisea* in the Indian Himalayas. *Genetics* 152, 971–984.

Kumar, S.P., Dalai, V., Singh, N.K. and Sharma, T.R. (2007) Cloning and *in silico* mapping resistance gene analogues isolated rice lines containing known genes for blast resistance. *Journal of Phytopathology* 155, 273–280.

Latterell, F.M. and Rosi, A.E. (1986) Longevity and pathogenic stability of *Pyricularia oryzae. Phytopathology* 76, 231–235.

Leung, H., Lehtinen, U., Karjalainen, R., Skinner, D. and Tooley, P. (1990) Transformation of the rice blast fungus *Magnaporthe grisea* to hygromycin B resistance. *Current Genetics* 17, 409–411.

Leung, H., Bernardo, M., Ebron, L., Tsunematsu, H., Kato, H., Imbe, T., Nghia, L.T. and Quang, V.D. (1998) Use of conserved motifs of disease resistance genes to characterize elite genetic stocks and germplasm; IRRI program report for 1998 (http://www.irri.org/publications/program/pdfs/98programreport/crosseco.pdf, accessed 12 March 2008).

Leung, H., Zhu, Y., Revilla-Mollina, I., Fan, J.X., Chen, H., Pangga, I., Cruz, C.V. and Mew, T.W. (2003) Using genetic diversity to achieve sustainable rice disease management. *Phytopathology* 87(10), 1156–1168.

Levy, M., Romao, J., Marcheitti, M.A. and Hammer, J.E. (1991) DNA fingerprinting with dispersed sequence resolve pathotype diversity in the rice blast fungus. *Plant Cell* 3, 95–102.

Levy, M., Correa-Victoria, F.J., Zeigler, R.S., Xu, S. and Hamer, J.E. (1993) Genetic diversity of the rice blast fungus in a disease nursery in Colombia. *Phytopathology* 83, 1427–1433.

Lima, A. (1999) Population genetic diversity of *P. grisea* on rice in Portugal. In: *Minutes of the 2nd Biennial Meeting of the Portuguese Phytopathology Society, Portugal*, pp 47–55.

Liu, B., Zhu, X., Zhang, S., Yang, Q., Wu, S. and Leung, H. (2002) Identification of candidate genes associated with durable resistance to blast in rice (*Oryza sativa* L.). (Abstract) *Proceedings of the International Rice Congress, 16–20 September 2002, Beijing, China*. International Rice Research Institute, Philippines, 316 pp.

McDonald, B.A., McDermott, J.M. and Goodwin, S.B. (1989) The population biology of host pathogen interactions. *Annual Review of Phytopathology* 27, 77–94.

Mackill, D.J., Salam, M.A., Wang, Z.Y. and Tanksley, S.D. (1993) A major photoperiod-sensitivity gene tagged with RFLP and isozyme markers in rice. *Theoretical and Applied Genetics* 85, 536–540.

Mandal, N., Nandy, S. and Mitra, S. (2004) DNA fingerprinting of Indian pathogen population of *Magnaporthe grisea* Barr. *Asian Journal of Microbiology, Biotechnology and Environmental Science* 6(2), 325–327.
Martin, G.B., Bogdanove, A.J. and Sessa, G. (2003) Understanding the functions of plant disease resistance proteins. *Annual Review of Plant Biology* 54, 23–61.
Miyamoto, M., Ando, I., Rybka, K., Kodama, O. and Kawasaki, S. (1996) High resolution mapping of the indica-derived blast resistance gene I. Pi-b. *Molecular Plant–Microbe Interactions* 9, 6–13.
Nandy, S., Mandal, N. and Mitra, S. (2004) Selection of rice DHs and RILs for resistance to leaf blast disease. *Asian Journal of Microbiology, Biotechnology and Environmental Science* 6(2), 329–331.
Naqvi, N.L., Bonman, J.M., Mackill, D.J., Nelson, R.J. and Chattoo, B.B. (1995) Identification of RAPD markers linked to a major blast resistance gene in rice. *Molecular Breeding* 1(4), 341–348.
Nelson, R.R. (1978) Genetics of horizontal resistance to plants. *Annual Review of Phytopathology* 16, 359–378.
Noguchi, M.T., Yasuda, N. and Fujita, Y. (2007) Fitness characters in parasexual recombinants of the rice blast fungus *Pyricularia oryzae*. *Japan Agricultural Research Quarterly* 41(2), 123–131.
Notteghem, J.L. (1985) Definition of a strategy for use of resistance through genetic analysis of the host pathogen relationship: the case of the rice *Pyricularia oryzae* relation. *Agronomy Tropical* 40, 129–147.
Ori, N., Eshed, Y., Paran, I., Presting, G., Aviv, D., Tanksley, S., Zamir, D. and Fluhr, R. (1997) The 12C family from the wilt disease resistance locus I2 belongs to nucleotide binding, leucin-rich repeat superfamily of plant resistance genes. *Plant Cell* 9, 521–532.
Ou, S.H. (1980) Pathogen variability and host resistance in rice blast disease. *Annual Review of Phytopathology* 18, 167–187.
Ou, S.H. (1985) *Rice Diseases*, 2nd Edn. Mycological Institute, Kew, Surrey, UK, 380 pp.
Qu, S., Liu, G., Zhou, B., Bellizzi, M. and Zeng, L. (2006) The broad spectrum blast resistance gene Pi9 encodes a nucleotide-binding site-leucine-rich repeat protein and is a member of a multigene family in rice. *Genetics* 172, 1901–1914.
Parlevliet, J.E. (1988) *Plant Disease, Epidemiology, Genetics, Resistance and Management*. McGraw-Hill Co., USA, 377 pp.
Parson, K.A., Chumley, F.G. and Valent, B. (1987) Genetic transformation of the fungal pathogen responsible for rice blast disease. *Proceedings of the National Academy of Sciences of the United States of America* 84, 4161–4165.
Pedersen, W.L. and Leath, S. (1988) Pyramiding major genes for resistance to maintain residual effects. *Annual Review of Phytopathology* 26, 369–378.
Robinson, R.A. (1973) Horizontal resistance. *Annual Review of Phytopathology* 52, 483–501.
Roca, W., Victoria, F.C., Martínez, C., Tohme, J., Lentini, Z. and Levy, M. (1996) Developing durable resistance to rice blast: productivity and environmental considerations. Biotechnology seminar paper, ISNAR Biotechnology Service, 74–80 pp (http://www.agbios.com/ docroot/articles/02-254-003.pdf, accessed 16 June 2008).
Roumen, E., Levym, M. and Notteghemm, J.L. (1997) Characterization of European pathogen population of *P. grisea* by DNA fingerprings and pathotype analysis. *European Journal of Plant Pathology* 103, 363–371.
Rybka, K., Miyamoto, M., Ando, I., Saito, A. and Kawasaki, S. (1997) I. High resolution mapping of the indica derived rice blast resistance gene. II. Pi-ta2 and Pi-ta and a consideration of their origin. *Molecular Plant–Microbe Interactions* 10(4), 517–524.
Shao, M., Wang, J., Dean, R.A., Lin, Y., Gao, X. and Hu, S. (2008) Expression of a hairpin-encoding gene in rice confers durable non-specific resistance to *Magnaporthe grisea*. *Plant Biotechnology Journal* 6, 73–81.
Shigemura, S. and Kitamura, E. (1954) Breeding of blast resistant cultivars with crossing of japonica and indica rices. (In Japanese.) *Journal of Agricultural Science, Tokyo Nogyo Daigaku* 9, 321–323.
Shull, V. and Hamer, J.E. (1994) Genomic structure and variability in *Pyricularia grisea*. In: Zeigler, R.S., Leong, S.A. and Teng, P.S. (eds) *Rice Blast Disease*. CAB International, Wallingford, UK, pp. 65–86.
Silue, D., Tharreau, D. and Notteghem, J.L. (1992) Identification of *Magnaporthe grisea* avirulence genes to seven rice cultivars. *Phytopathology* 82, 1462–1467.
Sivaraj, R. (1995) Genetic, structural and pathotype organization of rice blast fungus *P. grisea* in Southern India. PhD thesis, University of Madras, India.
Sivaraj, R., Gnanamanickam, S.S. and Levy, M. (1996) Studies on the genetic diversity of *Pyricularia grisea*: a molecular approach for the management of rice blast. In: Khush, G.S. (ed.) *Rice Genetics III*. IRRI Publication, Manila, pp. 958–962.

Sridhar, R., Singh, U.D., Agrawal, P.K. and Reddy, J.N. (1999) Usefulness of blast resistance genes and their combinations in different blast-endemic locations in India. *Pest Science and Management*, IRRI, pp. 22–23 (http://www.irri.org/publications/irrn/pdfs/vol24no2/IRRN24-2Pestscience.pdf, accessed 12 March 2008).
Srinivasachary, S., Kumar, K.G., Shashidhar, H.E. and Hittalmani, S. (1998) Molecular characterization of rice blast pathogen. *DAE Symposium on Induced Mutations and Molecular Techniques in Improving Crop Productivity and Quality, Chennai, 21–23 January 1998*, 35 pp.
Srinivasachary, S., Hittalmani, S., Kumar, K.G., Shashidhar, H.E. and Vaishali, M.G. (2002a) Identification of quantitative trait loci associated with sheath rot resistance (*Sarocladium oryzae*) and panicle exertion in rice (*Oryza sativa* L.). *Current Science* 82(2), 133–135.
Srinivasachary, S., Hittalmani, S., Shivayogi, S., Vaishali, M.G., Sshidhar, H.E. and Kumarm, G.K. (2002b) Genetic analysis of rice blast fungus of southern Karnataka using DNA markers and reaction of popular rice genotypes. *Current Science* 82(6), 732–735.
Stuiver, M.H. and Custers, J.H.H.V. (2001) Engineering disease resistance in plants. *Nature* 411, 865–868.
Suzuki, F., Sarai, M. and Yamaguchi, J. (2007) Genetic analysis of *Pyricularia grisea* population by rep-PCR during development of resistance to sactylone dehydratase inhibitors of melanin biosynthesis. *Plant Disease* 91(2), 176–184.
Tabien, R.E., Li, Z., Paterson, A.H., Marchetti, M.A., Stansel, J.W. and Pinson, S.R.M. (2000) Mapping of four major rice blast resistance genes from 'Lemont' and 'Teqing' and evaluation of their combinatorial effect for field resistance *Theoretical and Applied Genetics* 101, 1215–1225.
Takagaki, M., Kaku, K., Watanabe, S., Kawai, K., Shimizu, T., Sawada, H., Kumakura, K. and Nagayama, K. (2004) Mechanism of resistance to carpropamid in *Magnaporthe grisea*. *Pest Management Science* 60, 921–926.
Talbot, N.J. (2003) On the trail of a cereal killer: exploring the biology of *Magnaporthe grisea*. *Annual Review of Microbiology* 57, 177–202.
Tsunematsu, H., Yanoria, M.J.T., Ebron, L.A., Hayashi, N., Ando, I., Kto, H., Imbe, T. and Khush, G.S. (2000) Development of monogenic lines of rice for rice blast resistance. *Breeding Science* 50, 229–234.
Vakalounakis, D.J. and Fragkiadakis, G.A. (1999) Genetic diversity of *Fusarium oxysporum* isolates from cucumber: differentiation by vegetative compatibility, pathogenicity and RAPD fingerprinting. *Phytopathology* 89, 161–168.
Valent, B. and Chumley, F.G. (1994) Avirulence genes and mechanisms of genetic instability in the rice blast fungus. In: Zeigler, R.S., Teng, P.S. and. Leong, S.A. (eds) *Rice Blast Disease*. CAB International, Wallingford, UK, pp. 111–153.
Valent, B., Crawford, M.S., Weaver, C.G. and Chumley, F.G. (1986) Genetic studies of fertility and pathogenicity in *Magnaporthe grisea*. *Iowa State Journal of Research* 60, 569–594.
Viji, G., Gnanamanickam, S.S. and Levy, M. (2000) DNA polymorphism of isolates of *Magnaporthe grisea* from India those are pathogenic to finger millet and rice. *Mycological Research* 104, 161–167.
Wang, G.L., Mackill, D.J., Bonman, J., McCouch, S.R., Champoux, M.C. and Nelson, R.J. (1994) RFLP mapping of genes conferring complete and partial resistance to blast in a durably resistant rice cultivars. *Genetics* 136, 1421–434.
Wang, Z., Jia, Y., Rutger, J.N. and Xia, Y. (2007) Rapid survey for presence of a blast resistance gene *Pi-ta* in rice cultivars using the dominant DNA markers derived portions of the *Pi-ta* gene. *Plant Breeding* 126, 36–42.
Welsh, J. and McClelland, M. (1990) Fingerprinting genomes using PCR with arbitrary primers. *Nucleic Acids Research* 18, 7213–7218.
Williams, J.G.K., Kubelik, A.R., Livak, K.J., Rafalski, J.A. and Tingey, S.V. (1990) DNA amplified by arbitrary primers are useful genetic markers. *Nucleic Acids Research* 18, 6531–6535.
Xia, J.Q., Correll, J.C., Lee, F.N., Marchetti, M.A. and Rhoads, D.D. (1993) DNA fingerprinting to examine micro geographic variation in the *Magnaporthe grisea* (*Pyicularia grisea*) population in two rice fields in Arkansas. *Phytopathology* 83, 1029–1035.
Yamada, M., Kiyosawa, S., Yamaguchi, T., Hirano, T., Kobayashi, T., Kushibuchi, K. and Watanabe, S. (1976) Proposal of a new method of differentiating races of *Pyricularia oryzae* Cavara. in Japan. *Annals of the Phytopathological Society of Japan* 42, 216–219.
Yamasaki, Y. and Niizeki, H. (1965) Studies on variation of the rice blast fungus *Pyricularia oryzae* Cav. I. Karyological and genetical studies on variation. *Bulletin of the National Institute of Agricultural Sciences* 13, 231–274.
Yu, Z.H., Mackill, D.J., Bonman, J.M. and Tanksley, S.D. (1991) Tagging genes for blast resistance in rice via linkage to RFLP marker. *Theoretical and Applied Genetics* 85, 443–451.

Zeigler, R.S. (1998) Recombination in *Magnaporthe grisea*. *Annual Review of Phytopathology* 36, 249–276.

Zeigler, R.S., Cuo, L.X., Scott, R.P., Bernardo, M., Chen, D.H., Valent, B. and Nelson, R.J. (1995) The relationship between lineage and virulence in *Pyicularia grisea* in the Philippines. *Phytopathology* 85, 443–476.

Zeigler, R.S., Scott, R.P., Leung, H., Bordeos, A.A., Kumar, J. and Nelson, R.J. (1997) Evidence of parasexual exchange of DNA in the rice blast fungus challenges its exclusive clonality. *Phytopathology* 87, 284–294.

Zeigler, R.S., Kumar, J., Leung, H. and Nelson, R.J. (2000) Evidence for recombination in *Magnaporthe grisea*: a case study from the Indian Himalayas. In: Tharreau, D., Lebrun, M.H., Talbot, N.J. and Notteghem, J.L. (eds) *Advances in Rice Blast Research*. Kluwer Academic Press, Dordrecht, Netherlands, pp. 243–247.

Zhu, L., Chen, Y., Ling, Z., Xu, Z. and Xu, J. (1992) Identification of molecular markers linked to a blast resistance gene in rice. In: *Proceedings of the Asia-Pacific Conference on Agricultural Biotechnology, 20–24 August 1992, Beijing*. China Science and Technology Press, Beijing, 213 pp.

Zhu, Y.Y., Chen, H.R., Fan, J.H., Wang, Y.Y., Li, Y., Chen, J.B., Fan, J.X., Yang, S.S., Hu, L.P., Leung, H., Mew, T.W., Teng, P.S., Wang, Z.H. and Mundt, C.C. (2000) Genetic diversity and disease control in rice. *Nature* 406, 718–722.

Part III

Biological Control Mechanisms

9 Postharvest Technology – Yeast as Biocontrol Agents: Progress, Problems and Prospects

Neeta Sharma and Pallavi Awasthi
Mycology and Plant Pathology Division, Department of Botany, University of Lucknow, Lucknow, India

Abstract

Storage losses of fruits in India are high owing to temperature and humidity conditions. Losses in fruits are estimated to vary between 20 and 30%, valued at nearly 8000 crores annually, depending on the fruit variety and the postharvest handling system. The application of fungicides to fruits after harvest to reduce decay has been increasingly curtailed due to the development of resistance in pathogens to many key fungicides, lack of replacement with better fungicides, negative public perception regarding the safety of pesticides and consequent restrictions on fungicide use. Biological control of postharvest diseases has emerged as an effective alternative and several products are available in the market. One of the major limitations with biological disease control is inconsistency in the efficacy of the product. The limitations of biocontrol products can be addressed by enhancing biocontrol through genetic and environmental manipulations and integration with other alternative methods that, alone, do not provide adequate protection but, in combination with biocontrol, provide additive or synergistic effects.

Introduction

Approximately half of the population in the Third World does not have access to adequate food supplies. There are many reasons for this, one of which is food losses occurring in the postharvest and marketing system. A study on 'Postharvest Food Losses in Developing Countries' conducted by a committee of the US National Research Council concludes that, 'postharvest losses are "enormous"'. The committee extrapolated from apparent loss patterns and expected production trends and projected postharvest food losses to be, at a minimum, 47,000,000 Mt of durable crops and 60,000,000 Mt of perishable crops. 'The average minimum losses reported for roots and tubers and fruits and vegetables were 16 per cent and 21 per cent, respectively; many more "qualitative" references, not included here, indicate estimates of 40–50 per cent and above.'

The application of effective fungicides just prior to or shortly after harvest generally controls postharvest decay (Eckert and Ogawa, 1988). About 23m kg of fungicides is applied to fruits and vegetables annually and it is generally accepted that production and marketing would not be possible without their use (Ragsdale and Sisler, 1994). However, use of fungicides has been restricted due to their carcinogenicity, teratogenicity, residual toxicity and long degradation period causing environmental pollution (Unnikrishnan and Nath, 2002).

 Management of Fungal Plant Pathogens (eds A. Arya and A.E. Perelló)

The Food Quality Protection Act (FQPA) in the USA, the Food and Environment Protection Act (FEPA) 1985 and Control of Substances Hazardous to Health (COSHH) Regulations 1988, made under the Health and Safety at Work Act, 1974, in the UK are the guiding forces in the regulation of pesticide use in their respective countries. Several countries have implemented their own specific policies to reduce pesticide use (Matteson, 1995). Similarly, the fruit industry worldwide has accepted the concept of integrated fruit production (IFP). IFP aims to produce high-quality fruit in harmony with the consumer and the environment. This implies minimum usage of chemicals, especially after harvest. Globally, greater restrictions on pesticide use in the developed nations have resulted in increasing trends for natural, non-chemical or organic approaches to disease control. Understandably, alternatives to chemical pesticides or products that allow reduced usage in terms of fewer or reduced rates of application are beginning to appear on the market in the form of biological control agents (BCA). The present chapter reviews the status of yeast as a biocontrol agent and the problems associated with its commercialization and registration.

Cook and Baker (1983), in their book on biological control, cited only one example of the biocontrol of postharvest disease of strawberry fruit rot using *Trichoderma* sp. Subsequently, Wilson and Pusey (1985) presented their initial research on *Bacillus subtilis* to control brown rot on peaches, caused by *Monilinia fructicola*, and the organism was patented. A number of microorganisms (bacteria, yeasts and fungi), which effectively control postharvest pathogens, have been identified for the control of postharvest diseases and some of these have been patented and registered (El-Ghaouth and Wilson, 1997, 2002). In several studies, yeast strains (*Aureobasidium pullulans*, *Candida oleophila*, *C. guilliermondii*, *C. sake*, *Cryptococcus laurentii*, *Debaryomyces hansenii*, *Metschnikowia pulcherrima*, *Pichia gulliermondii*, *Sporobolomyces roseus*) are reported for biocontrol of postharvest fungal decays of fruits caused by *Alternaria alternata*, *Botrytis cinerea*, *Geotrichum candidum*, *M. fructicola*, *Penicillium digitatum*, *P. italicum*, *P. expansum* and *Rhizopus stolonifer* (Droby *et al.*, 1989; Wisniewski *et al.*, 1991; Sharma, 1992, 1993, 2000; Mehrotra *et al.*, 1996, 1998; Sharma *et al.*, 1997; Spadaro *et al.*, 2002) . In the past 25 years, research on biological control of postharvest diseases has moved from laboratory to practical applications (Wisniewski and Wilson, 1992; Wilson and Wisniewski, 1994; Mari and Guizzardi, 1998; Droby *et al.*, 2001; Janisiewiez and Korsten, 2002; Korsten, 2006).

By early 2000, there were three postharvest biological products available in the market: Aspire™, a product developed from *C. oleophila* (limited to the USA and Israel); BioSave™, developed from *P. syringae* to control decay caused by *P. italicum* and *P. digitatum* (limited to the USA); and YieldPlus™ (limited to South Africa). Avogreen™, a commercial product of *B. subtilis*, was developed to control diseases caused by *Cercospora* spot and anthracnose of avocado.

Isolation of Antagonist

Often, carposphere, phylloplane, flowers and, in a few cases, other matrixes have provided the major source for antagonists (Filonow *et al.*, 1996; Sharma, 2003; Belve *et al.*, 2006). Various strategies have been employed to isolate antagonists and these include isolation from natural cracks on the fruit surface; agar plates containing apple juice that were seeded with a rot pathogen (Wilson *et al.*, 1993); freshly made wounds on apples in the orchard that were exposed to colonization by fruit-associated microbiota from 1 to 4 weeks before harvest (Janisiewiez, 1996); and from an apple juice culture resulting from seeding diluted apple juice with the orchard-colonized wounds and repeated reinoculation to fresh apple juice. Isolation of the antagonists can be improved by using fruit from unmanaged orchards (Falconi and Mendgen, 1994) where natural populations have not been disturbed by chemical usage and the pool of potential antagonists is greater than in a chemically managed orchard (Smolka, 1992).

Natural microflora maintains a balance among the microbes normally present and inhibits the growth of newer arrivals. Sharma (2005) reported that undiluted fruit washings when plated on agar plates exhibited a dense population of yeast and bacteria and, on dilution, filamentous fungi of the pathogenic type were isolated. This suggests that bacteria and yeast, naturally present on the surface, may inhibit the growth of other microorganisms, including plant pathogenic fungi. Later, it was observed that the citrus fruits, when washed and stored, rotted faster than the unwashed fruits, suggesting that these bacteria and yeast provide protection to fruits against postharvest pathogens.

Rather than *in vitro* screening of organisms in Petri plates, which favoured the identification of antibiotic-producing organisms, a selection strategy was developed to identify suitable yeast antagonists (Wilson *et al.*, 1993). The method involved placing washing fluids obtained from the surface of the fruit into fruit wounds that subsequently were inoculated with a rot pathogen. Organisms were then isolated from the surface of wounds that did not develop infections. These were plated out and isolated.

Pure cultures of potential antagonists were produced and then each organism was screened individually to assess its potential as a biocontrol agent. This method identified a number of antagonists that were studied more intensely and measured against the criteria set for suitability for commercial production, as outlined by Wilson and Wisniewski (1989) and Hofstein *et al.* (1994):

- genetically stable
- effective at low concentrations
- not fastidious in its nutrient requirements
- ability to survive adverse environmental conditions (including low temperature and controlled atmosphere storage)
- effective against a wide range of pathogens on a variety of fruits and vegetables
- amenable to production on an inexpensive growth medium
- amenable to a formulation with a long shelf life
- easy to dispense
- does not produce metabolites that are deleterious to human health
- resistant to pesticides
- compatible with commercial processing procedures
- does not grow at 37°C and is not associated with infections in humans
- non-pathogenic to host commodity.

Biocontrol Activity

Most antagonistic yeasts are efficient colonizers, even under adverse environmental conditions, as they utilize nutrients rapidly, produce extracellular materials that enhance their survival on fruit surfaces and restrict both colonization sites and flow of germination caused to fungal propagules (Dugan and Roberts, 1995). In order to optimize disease control, it is important to understand the mode of action of the antagonists so that these attributes can be utilized to improve performance. The antagonist activity can be expressed in a number of ways. The most common is antibiosis (production of metabolites such as pyrrolnitrin or iturins), attributed mainly to bacterial antagonists (Smilanick and Dennis-Arrue, 1992). The antibiotic pyrrolnitrin, produced by *Pseudomonas cepacia* LT-4-12W (Janisiewiez and Roitman, 1988), reduced *in vitro* growth and conidia germination and controlled the pome fruit pathogens, *P. expansum* and *B. cinerea*, and citrus fruit pathogen, *P. italicum*. However, the significance of the antibiotics in these biocontrol situations was not clear, since strain LT-4-12W still provided substantial control of blue mould decay on oranges inoculated with laboratory-derived mutants of *P. italicum* resistant to pyrrolnitrin. Spadaro *et al.* (2002), in studies on *M. pulcherrima*, found that in the *in vitro* antagonism studies on different substrates, the yeast could produce some metabolites toxic to the pathogen, as distinct from the application of culture filtrates *in vivo*. In recent years, the use of antibiotic-producing bacteria has been abandoned in order to prevent the appearance of resistance in pathogen strains for humans or animals.

Competition for nutrients and/or space is the major mechanism involved for *P. guilliermondii*, *C. laurentii*, *C. utilis*, *C. oleophila*, *D. hansenii* and several other yeasts employed as bioagents (Chalutz and Wilson, 1990; Arras, 1996; Arras *et al.*, 1997; Spadaro *et al.*, 2002; He *et al.*, 2003; Chan and Tian, 2005; Zhang *et al.*, 2005). Janisiewiez *et al.* (2000) developed a non-destructive method using tissue culture plates having a defusing membrane at the lower end of cylindrical inserts for *in vitro* study of competition for nutrients separated from the competition for space. Living cells of the antagonist are necessary to guarantee fungal control. The ability to prevent infection by pathogen was lost when the antagonist cells were killed. It was also observed that competition for nutrients was not visible when a surplus of nutrients was available. Therefore, the nutritional environment available at the wound site may create a favourable microenvironment for antagonists to colonize, multiply and compete effectively (Zheng *et al.*, 2004). The activity of an antagonist is dependent on the concentration of the antagonist: the higher the concentration, the more effective the control. The antagonist cell concentration of $10^6 - 10^8$ CFU/ml or more of *Candida* spp., *D. hansenii* and *Pantoea agglomerans* provided satisfactory levels of control (Droby *et al.*, 1989; McLaughlin *et al.*, 1990). However, different isolates of *M. pulcherrima* at 10^6 CFU/ml were not found to provide satisfactory levels of control against *B. cinerea* and *P. expansum* (Spadaro *et al.*, 2002).

While early studies indicated that nutrient competition and the fast growth rate of antagonists played a major role in biocontrol activity, subsequent studies indicated a much more complex interaction, such as direct interaction with the pathogen (Wisniewski *et al.*, 1991; Spadaro *et al.*, 2002), induced resistance in host tissue (Wilson *et al.*, 1994; Droby *et al.*, 2002) or a gamut of interactions between the antagonist, pathogen and commodity. *Pichia guilliermondii* US-7 (Droby *et al.*, 1989) and *M. pulcherrima* (Spadaro *et al.*, 2002) exhibited nutrient competition along with direct parasitism against *B. cinerea* in apples. *Pichia membranefaciens* and *C. albidus* exhibited tenacious attachment with pathogen hyphae, along with secretion of extracellular lytic enzymes (Chan and Tian, 2005). Ultrastructural and cytochemical studies on yeast, *C. saitoana*, when co-cultivated with *B. cinerea*, showed cytological damage as papillae and protuberances in the cell wall and degeneration of the cytoplasm. It was also found to stimulate structural defence response in the host. Host cell walls were well preserved and displayed an intense and regular cellulose-labelling pattern, as seen in transmission electron microscopy (El Ghaouth *et al.*, 1998).

Yeast cells are able to produce hydrolytic enzymes capable of attacking the cell walls of pathogens and extracellular polymers that appear to have antifungal activity. Yeast, *P. anomala* strain K, effective in the control of grey mould of apple, increased production of exo-β-1,3-glucanase threefold in the presence of cell wall preparations of *B. cinerea* in apple wounds. Higher β-1,3-glucanase and chitinase activity was also detected in apple wounds treated with strains of another antagonist, *A. pullulans*, effective in controlling various decays on apple, table grape and other fruits (Ippolito *et al.*, 2000; Castoria *et al.*, 2001). Yeast, *P. membranefaciens* and *C. albidus*, show β-1,3-glucanase and exo-chitinase activity in the presence of cell wall preparations of *R. stolonifer*, *M. fructicola* and *P. expansum* (Chan and Tian, 2005).

Yeasts like *C. famata* are reported to control green mould due to induction of phytoalexins, scoparone and scopolectin (Arras, 1996). However, the role of enzymes and phytoalexins in biocontrol activity warrants further investigation. Fajardo *et al.* (1998) reported differential induction of proteins in orange flavedo by biologically based elicitors. More recently, molecular approaches to examine the mode of action have been studied on the biocontrol agent. A transformation system for *C. oleophila* yeast produced yeast lines with either higher or lower levels of a β-1,3-glucanase gene/enzyme expression compared to the wild type. Biocontrol activity did not differ between the different yeast lines, but the results did not rule out a role for this gene in biocontrol activity. It was also demonstrated that overexpression of a lytic peptide belonging to the defensin family of antimicrobial peptides in yeast could enhance biocontrol activity (Segal *et al.*, 2002; Yehuda *et al.*, 2003).

Constraints in Product Development and Registration

In the early years, several yeast antagonists that had commercial potential were misidentified, such as strain US-7 of *C. guilliermondi*, which was misidentified originally as *D. hansenii*. This caused some confusion in the patenting process and emphasized the need to have at least two confirming identifications by reputable yeast taxonomic services. It also emphasized the weakness of using physiological tests as the basis for making taxonomic determinations (McLaughlin *et al.*, 1990). Also, few isolates of *C. guillliermondii* were abandoned because they were found to be pathogenic to humans.

Potential biocontrol agents often have some significant limitations: sensitivity to adverse environmental conditions such as extreme dryness, heat and cold, limited shelf life, limited biocontrol efficacy in situations where several pathogens are involved in decay development and an inability to control latent infections. For commercialization, several semi-commercial and commercial trials have to be conducted, for which large volumes of antagonist are required. The mass production of the bioagent by rapid, efficient and inexpensive fermentation of the antagonist is a key issue. Therefore, it is fundamental to find carbon and nitrogen sources that provide maximum biomass production at minimum cost, while maintaining biocontrol efficacy. Cheap industrial waste materials such as cottonseed meal, corn steep liquor, partially digested peptone, yeast extract, dry brewer's yeast, sucrose and molasses have been used as growth media for the multiplication of cells (Hofstein *et al.*, 1994; Costa *et al.*, 2001).

Large-scale production of any yeast depends on the amount of technical information available on that specific strain, such as osmotolerance, temperature, oxygen requirements, optimum pH and optimum growth rate. Growth rate of yeast is very high, but lower than that of bacteria; longer fermentation durations pose the risk of yeast cultures becoming contaminated. Yeast is also sensitive to low pH (below three), which is used generally as a measure to check bacterial contamination because pH above five is favourable for bacteria that may contaminate yeast culture. Aeration of fermentors, to fulfil the oxygen requirement for maximum output, can also be a source of contamination during the early phases of production and, to prevent such contamination, other technologies must be used. The contaminants should be identified at each stage of production and quantified in the end product.

Yeast fermentation is an exothermic process; therefore, the fermentation temperature can never be below ambient and, since yeasts appear sensitive to high temperatures (above 28°C), a cooling system more efficient than the evaporative system routinely used has to be employed. This, however, adds to the cost of production.

A major obstacle to the commercialization of biocontrol products is the development of a shelf-stable product that retains bioactivity similar to that of fresh cells. Formulations can influence the survival and activity of biocontrol agents. An accurate formulation has a profound effect on the efficacy of a biocontrol agent, including its shelf life, ability to grow and survive after application, effectiveness in disease control, ease of operation and application and the cost (Fravel *et al.*, 1998). A biofungicide should be effective for at least 6 months, and preferably for 2 years (Pusey, 1994). This can be achieved by supplementing the yeast with protectants, carriers or additives. Alternatively, yeast can be conditioned during fermentation by using an emulsifier.

Drying the product and maintenance in a dry environment or suspension in oil are common approaches. Products are available as wettable powder, as frozen cell concentrated pellets or as liquid formulations. It was found that freeze-dried cells were significantly less effective than fresh cells.

Certain freeze-drying protective agents and rehydration media enhanced the viability of the antagonist, *P. agglomerans* strain CPA-2, effective against blue mould and grey mould of pome fruits (Costa *et al.*, 2000). Survival of cells of the antagonistic yeast, *C. sake*, was improved from 0.2% to 30–40%, by using freeze-drying protective media consisting of skim milk and other protectants, such as 10% lactose or glucose and 10% fructose or sucrose. The presence of trehalose in liquid formulations appeared to help preserve the viability of *C. sake* during storage. It is known that intracellular trehalose exerts a protective effect on yeast under extreme environmental conditions such as desiccation, freezing, osmotic stress and heat shock, and it also provides thermal stability to the cells (Abadias *et al.*, 2001).

The application of adjuvant can protect and stimulate the establishing of the antagonist on the host surface. The addition of xanthan gum to *A. pullulans* L47, applied to strawberries in the field from bloom to fruit at the green stage, improved survival of the antagonist and increased biocontrol of storage rot caused by *B. cinerea* (Ippolito *et al.*, 1998). Formulations may include wetters (humectants) to facilitate reabsorption of moisture from air. Wetters not only make water spray stay on plants but, like oil carriers, they also enable organisms to reach otherwise inaccessible places such as depressions, stomata and lenticels, thereby improving the chances of establishing antagonists for disease control. Oil carriers are expensive, but formulations containing oils can enhance the reliability of biological control agents (Jones and Burges, 1998). Research is needed to determine the value of each additive alone and also in the presence of other ingredients, as well as to ensure the requirements for ecological safety.

One of the major limitations with biological disease control is the inconsistency in efficacy that is often observed when useful antagonists reach the stage of large-scale testing, and which can arise from a variety of causes reflecting the biological nature of the control microorganism. Essentially, the organism must first survive application and then retain activity in the environment of use throughout the period when active control is required, which may be several months for some pathogens. During this time, it must survive fluctuations in the physical environment and the action of the indigenous and competitive microbiota. The use of appropriate inoculum production, formulation and application technologies, together with quality control checks, should also help in this process. Nevertheless, even if reliable BCAs can be produced, they must still be easy to use and cost-effective or they will either never reach the marketplace or not be used by growers.

By early 2000, there were two yeast-based postharvest biological products available on the market: Aspire™ (*C. oleophila* I-182) and YieldPlus™ (El-Ghaouth and Wilson, 1997, 2002; Wilson and El-Ghaouth, 2002). The commercial development of Aspire by Ecogen-Israel Partnership Ltd, focused on the biocontrol of postharvest decays of citrus, mainly blue mould and green mould caused by *P. italicum* and *P. digitatum*, respectively, which invade through wounds after harvest. Throughout the course of developing Aspire™, considerable research went into finding methods to enhance the reliability and efficacy of the product and other selected antagonists as well.

As a result, second generation biocontrol products were developed using a combination of natural products along with a yeast antagonist to address the poor ability. Research efforts led to the development of two new products whose main components consisted of the yeast antagonist, *C. saitoana*, and either a derivative of chitosan (Biocoat) or lysozyme (Biocure) (El Ghaouth *et al.*, 2000a). Both compounds have been tested worldwide and have shown strong eradicant activity. Both products contain additional additives, such as sodium bicarbonate, to enhance efficacy and perform as well as the postharvest fungicides currently available.

Another constraint concerns registration. Currently, there are no fungal biocontrol products registered and sold worldwide. Some products are available in several countries, while others are sold in their respective countries. This reflects the problems associated with registration requirements in different countries and includes concerns about releasing non-indigenous microorganisms. The legislation drafted essentially for chemical pesticides is not always applicable to biological pesticides and the requirements for the registration of biological pesticides are currently under discussion for appropriate review.

The position of the biocontrol product in the market governs its future. For example, if the product enters the agrochemical market, it competes against synthetic fungicides that can kill pathogenic organisms, while yeast only-based products cannot do so and neither do they have systemic action. They act mainly as protectants that may also induce resistance in the hosts. The other option is to position the product in the 'all green' category in markets such as those of perishables, where no other option is available, thus eliminating any competition and fulfilling the principal objective of consumer and environmental safety.

Integrated Control

Since, biological agents alone are not capable of providing commercially acceptable levels of control, their integration with other control measures is expected to provide greater stability and effectiveness. It is also desirable that the use of antagonists must be compatible with current handling and storage practices which could otherwise cause a reduction in the effectiveness of antagonist strains. For biological control to be effective, use of antagonists must be compatible with other control measures. An effective biocontrol based on a mixture of several complementary and non-competitive antagonists has several advantages: apart from a wider spectrum of activity, they increase efficacy, are more reliable and allow reduction in application times and treatment costs. They also permit the combination of different genetic characteristics, minimizing the need for genetic engineering. In a study on apples, a broader spectrum of pathogens was controlled and less total biomass of the antagonist was needed to control decay (Janisiewiez, 1996) when a mixture of antagonists was applied. The mixtures are either paired at random or after screening, for minimum mutual niche overlap. To determine further compatibility of the strains selected, it is important to conduct coexistence studies using De Wit displacement series in fruit wounds (Wilson and Lindow, 1994). The benefits of this approach are clear, but its implementation requires approval from the industry. It also entails doubling of the cost. However, this can be overcome by using in the mixture at least one antagonist which has been commercialized.

Some exogenous substances, such as chitosan, amino acids, antibiotics, calcium salts and carbohydrates, have been studied to enhance the biocontrol capability of antagonists against fungal pathogens. Calcium chloride improved biological control of the yeast, *P. guilliermondii* (Droby *et al.*, 1997). Combining 0.2% glycolchitosan with the antagonist, *C. saitoana*, was more effective in controlling green mould of oranges and lemons, caused by *P. digitatum*, and grey and blue moulds of apples than either treatment alone (El-Ghaouth *et al.*, 2000a,b). In a recent study by the authors, a combination of chitosan and the yeast, *C. utilis*, was found effective in controlling postharvest pathogens on tomato (Sharma *et al.*, 2006). The studies also showed that several yeast genera were compatible with low concentrations of chitosan and the protection afforded by this combination was superior to the stand-alone treatments.

GRAS (generally recognized as safe) substances such as sodium carbonate, sodium bicarbonate and ethanol reduced conidial germination of *P. digitatum*, the causal agent of green mould of citrus. Ethanol at 10%, in combination with ethanol-resistant *S. cerevisiae* strains 1440 and 1749, reduced the incidence of grey mould decay on apples from more than 90% to close to 0%, respectively, whereas either treatment alone did not reduce decay. The same concentration of ethanol reduced green mould of lemons to less than 5% (Smilanick *et al.*, 1995, 1999). *A. pullulans*, in combination with calcium chloride or sodium bicarbonate, was found effective in controlling postharvest pathogens on sweet cherries (Ippolito *et al.*, 1998).

Pre-storage hot air treatment of apples reduced or eliminated blue mould decay caused by *P. expansum* and grey mould decay (Fallik *et al.*, 1995). Heat also improved biocontrol with heat-tolerant yeasts when applied to apples up to 24 h after inoculation with the pathogen. The heat treatment alone provided little residual protection, but the residual protection provided by Ca and the antagonist in combination enhanced the control by heat. When antagonists were applied to apple wounds before heat treatment, the heat reduced populations of *P. syringae* and increased populations of the two heat-tolerant yeasts more than tenfold.

Conclusions

Future lines of research should be directed to find methods of enhancing the reliability and efficacy of selected antagonists, and the field is gaining momentum. It should aim at finding additives or physical control methods that will act synergistically with the antagonist. This involves combining the product with a low-level of postharvest fungicide or GRAS substances. It has been reported that physical treatments such as hot air, curing, hot water brushing and combinations of the above with pressure infiltration of calcium could also increase the efficacy of antagonists. Using mixtures of antagonists, or combining antagonists with specific nutrients or sugar analogues, is also suggested as an approach to increase efficacy.

Genetic manipulation of antagonists is a field in its infancy. Current efforts are focused on developing efficient transformation procedures for yeast antagonists and inserting genes for tracking the antagonist in the environment rather than enhancing biocontrol (Yehuda *et al.*, 2001).

Other approaches could be: the insertion of the gene for amylase under the constitutive promoter in some BCAs to allow effective use of the fruit carposphere starch; biocontrol strains with a higher capability to exploit the nitrogen compounds present or with a higher transport or metabolism rate of the limiting factor can be developed, because nitrogen is often a limiting substance when the biocontrol mechanism of action is competition for nutrients; and use of mutants that use new substrates, not metabolized by the pathogen, to provide a nutritional advantage or attempt to obtain strains resistant to phenolic compounds (Bizeau *et al.*, 1989).

Early experiments in transformation for marker genes have been successful. *Metschnikowia pulcherrima* was transformed with the green fluorescent protein gene (Nigro *et al.*, 1999) and histidine auxotrophs of *C. oleophila* were transformed with *HIS3*, *HIS4* and *HIS5* genes (Chand-Goyal *et al.*, 1999). In all cases, the transformed antagonists maintained their biocontrol capability and there were no detectable differences between the wild type and the transformants. All these studies were accomplished only to obtain variants of the antagonistic strains with a genetically stable marker. Jones and Prusky (2002) investigated the possibility of expressing a DNA sequence in *S. cerevisiae* to allow the production of a cecropin A-based antifungal peptide. Yeast transformants inhibited the growth of germinated *Colletotrichum coccoides* spores and inhibited decay developments caused by the pathogen in tomato fruit. The lack of activity toward non-target organisms by the peptide and the use of *S. cerevisiae* as a delivery system suggest that this method could provide a safe alternative for postharvest disease control. However, attempts to overexpress genes involved in biocontrol, for example, lytic enzymes, or engineering strains with desired biocontrol traits may soon yield positive results. Biological control of plant diseases in general and on fruit after harvest in particular is a niche market, with a relatively small profit potential. However, it is clear that the stage is set for biological control agents to play a greater part in agriculture and horticulture. This approach undoubtedly would encourage environmentally desirable products that are desired by the public to reach the marketplace rapidly.

References

Abadias, M., Benabarre, A., Teixidó, N., Usall, J. and Viñas, I. (2001) Effect of freeze drying and protectants on viability of the biocontrol yeast *Candida sake*. *International Journal of Food Microbiology* 65, 173–182.

Arras, G. (1996) Mode of action of an isolate of *Candida famata* in biological control of *Penicillium digitatum* in orange fruits. *Postharvest Biology and Technology* 8, 191–98.

Arras, G., De Cicco, V., Arru, S. and Lima, G. (1997) Biocontrol of blue mould of citrus fruits and the mode of action of an isolate of *Pichia guilliermondiii*. *Journal of Horticultural Science and Biotechnology* 73, 413–418.

Belve, G., Grieco, F., Cozzi, G., Logrieco, A. and Visconti, A. (2006) Isolation of epiphytic yeasts with potential for biocontrol of *Aspergillus carbonarius* and *A. niger* on grape. *International Journal of Food Microbiology* 108, 204–209.

Bizeau, C., Drilleau, J.F. and Michel, J. (1989) *Study of Interactions between Yeast Involved in Cider-making*. Presented at the 7th International Symposium on Yeasts. Wiley, Chichester, UK, pp. 169–173.

Castoria, R., De Curtis, F., Lima, G., Caputo, L., Pacifico, S. and De Cicco, V. (2001) *Aureobasidium pullulans* (LS-30), an antagonist of postharvest pathogens of fruits: study on its modes of action. *Postharvest Biology and Technology* 22, 7–17.

Chalutz, E. and Wilson, C.L. (1990) Biocontrol of green and blue mold and sour rot of citrus fruit by *Debaryomyces hansenii*. *Plant Disease* 74, 134–137.

Chan, Z. and Tian, S. (2005) Interaction of antagonistic yeasts against postharvest pathogens of apple fruit and possible mode of action. *Postharvest Biology and Technology* 36, 215–223.

Chand-Goyal, T., Eckert, J.W., Droby, S., Glickmann, E. and Atkinson, K. (1999) Transformation of *Candida oleophila* and survival of a transformant on orange fruit under field conditions. *Current Genetics* 35, 51–57.

Cook, R.J. and Baker, K.F. (1983) *The Nature and Practice of Biological Control of Plant Pathogens*. American Phytopathological Society, St Paul, Minnesota.

Costa, E., Usall, J., Teixidó, N., Garcia, N. and Viñas, I. (2000) Effect of protective agents, rehydration media and initial cell concentration on viability of *Pantoea agglomerans* strain CPA-2 subjected to freeze drying. *Journal of Applied Microbiology* 89, 793–800.

Costa, E., Teixidó, N., Usall, J., Atares, E. and Viñas, I. (2001) Production of the biocontrol agent *Pantoea agglomerans* strain CPA-2 using commercial products and by-products. *Applied Microbiology and Biotechnology* 56, 367–371.

Droby, S., Chalutz, E., Wilson, C. and Wisniewski, M. (1989) Characterization of the biocontrol activity of *Debaryomyces hansenii* in the control of *Penicillium digitatum* on grapefruit. *Canadian Journal of Microbiology* 35, 794–800.

Droby, S., Wisniewski, M.E., Cohen, L., Weiss, B., Touitou, D., Eilam, Y. and Chalutz, E. (1997) Influence of $CaCl_2$ on *Penicillium digitatum*, grapefruit peel tissue, and biocontrol activity of *Pichia guilliermondii*. *Phytopathology* 87, 310–315.

Droby, S., Cohen, L., Wiess, B., Daus, A. and Wisniewski, M. (2001) Microbial control of postharvest diseases of fruits and vegetables – current status and future outlook. *Acta Horticulturae* 553, 371–376.

Droby, S., Vinokur, V., Weiss, B., Cohen, L., Daus, A., Goldschmidt, E.E. and Porat, R. (2002) Induction of resistance to *Penicillium digitatum* in grapefruit by the yeast biocontrol agent *Candida oleophila*. *Phytopathology* 92, 393–399.

Dugan, F.M. and Roberts, R.G. (1995) Etiology of preharvest colonization of Bing cherry fruit by fungi. *Phytopathology* 84, 1031–1036.

Eckert, J.W. and Ogawa, J.M. (1988) The chemical control of postharvest diseases: deciduous fruits, berries, vegetables and root/tuber crops. *Annual Review of Phytopathology* 26, 433–469.

El-Ghaouth, A. and Wilson, C. (1997) Biologically-based technologies for the control of postharvest diseases. *Journal of Industrial Microbiology and Biotechnology* 19, 160–162.

El-Ghaouth, A. and Wilson, C.L. (2002) *Candida saitoana* Compositions for Biocontrol of Plant Postharvest Decay. US Patent No. 6,419,922.

El-Ghaouth, A. Wilson, C.L. and Wisniewski, M. (1998) Ultrastructural and cytochemical aspects of the biological control of *Botrytis cinerea* by *Candida saitoana* in apple fruit. *Phytopathology* 88, 282–291.

El-Ghaouth, A., Smilanick, J.L. and Wilson, C.L. (2000a) Enhancement of the performance of *Candida saitoana* by the addition of glycolchitosan for the control of postharvest decay of apple and citrus fruit. *Postharvest Biology and Technology* 19, 103–110.

El-Ghaouth, A., Smilanick, J.L., Wisniewski, M. and Wilson, C.L. (2000b) Improved control of apple and citrus fruit decay with a combination of *Candida saitoana* and 2-deoxy-D-glucose. *Plant Disease* 84, 249–253.

Fajardo, J.E., McCollum, T.G., McDonald, R.E. and Mayer, R.T. (1998) Differential induction of proteins in orange flavedo by biologically based elicitors and challenged by *Penicillium digitatum* Sacc. *Biological Control* 13, 143–151.

Falconi, C.J. and Mendgen, K. (1994) Epiphytic fungi on apple leaves and their value for control of the postharvest pathogens *Botrytis cinerea, Monilinia fructigena* and *Penicillium expansum. Journal of Plant Pathology* 101, 38–47.

Fallik, E.S., Grinberg, S., Gambourg, M. and Lurie, S. (1995) Prestorage heat treatment reduces pathogenicity of *Penicillium expansum* in apple fruit. *Plant Pathology* 45, 92–97.

Filonow, A.B., Vishniac, H.S., Anderson, J.A. and Janisiewiez, W.J. (1996) Biological control of *Botrytis cinerea* in apple by yeasts from various habitats and their putative mechanisms of antagonism. *Biological Control* 7, 212–220.

Fravel, D., Connick, W.J. Jr and Lewis, J.A. (1998) Formulation of microorganisms to control plant diseases. In: Burges, H.D. (ed.) *Formulation of Microbial Biopesticides: Beneficial Microorganisms, Nematodes and Seed Treatments.* Kluwer, Boston, 496 pp.

He, D., Zheng, X., Yin, Y., Sun, P. and Zhang, H. (2003) Yeast application for controlling apple postharvest diseases associated with *Penicillium expansum. Botanical Bulletin of Academia Sinica* 44, 211–216.

Hofstein, R., Fridlender, B., Chautz, E., Wisnewski, M. and Wilson, C.L. (1994) Large-scale production and pilot testing of biological control agents for postharvest diseases. In: Wilson, C. and Wisniewski, M. (eds) *Biological Control of Postharvest Diseases of Fruits and Vegetables –Theory and Practice.* CRC Press, Boca Raton, Florida, pp. 89–100.

Ippolito, A., Nigro, F., Romanazzi, G. and Campanella, V. (1998) Field application of *Aureobasidium pullulans* against *Botrytis* storage rot of strawberry. In: Bertolini, P., Sijmons, P.C., Guerzoni, M.E. and Serra, F. (eds) *COST 914 – COST 915 Joint Workshop – Non-conventional Methods for the Control of Postharvest Disease and Microbial Spoilage.* European Community, Luxembourg, pp. 127–133.

Ippolito, A., El-Ghaouth, A., Wilson, C.L. and Wisniewski, M. (2000) Control of postharvest decay of apple fruit by *Aureobasidium pullulans* and induction of defense responses. *Postharvest Biology and Technology* 19, 265–272.

Janisiewiez, W.J. (1996) Ecological diversity, niche overlap, and coexistence of antagonists used in developing mixtures for biocontrol of postharvest diseases of apples. *Phytopathology* 86, 473–479.

Janisiewiez, W.J. and Korsten, L. (2002) Biological control of postharvest diseases of fruits. *Annual Review of Phytopathology* 40, 411–441.

Janisiewiez, W.J. and Roitman, J. (1988) Biological control of blue mold and gray mold on apple and pear with *Pseudomonas cepacia. Phytopathology* 78, 1697–700.

Janisiewiez, W.J., Tworkoski, T.J. and Sharer, C. (2000) Characterizing the mechanism of biological control of postharvest diseases on fruits with a simple method to study competition for nutrients. *Phytopathology* 90, 1196–2000.

Jones, K.A. and Burges, H.D. (1998) Technology of formulation and application. In: Burges, H.D. (ed.) *Formulation of Microbial Biopesticides.* Kluwer, Boston, 496 pp.

Jones, R.W. and Prusky, D. (2002) Expression of an antifungal peptide in *Saccharomyces*: a new approach for biological control of the postharvest disease caused by *Colletotrichum coccodes. Phytopathology* 92, 33–37.

Korsten, L. (2006) Advances in control of postharvest diseases in tropical fresh produce. *International Journal of Postharvest Technology and Innovation* 1(1), 48–61.

McLaughlin, R.J., Wilson, C.L., Chalutz, E., Kurtzman, C.P., Fett, W.F. and Osmond, S.F. (1990) Characterization and reclassification of yeasts used for biological control of postharvest diseases of fruits and vegetables. *Applied Environmental Microbioogy* 56, 35–86.

Mari, M. and Guizzardi, M. (1998) The postharvest phase: emerging technologies for the control of fungal diseases. *Phytoparasitica* 26(1), 59–66.

Matteson, P.C. (1995) The '50% pesticide cuts' in Europe: a glimpse of our future? *American Entomologist Winter* 41(4), 210–220.

Mehrotra, N.K., Sharma, N., Ghosh, N.R. and Nigam, M. (1996) Biological control of green and blue mould diseases of citrus fruit by yeast. *Indian Phytopathology* 49(4), 350–354.

Mehrotra, N.K., Sharma, N., Nigam, M. and Ghosh, N.R. (1998) Biological control of sour-rot of citrus fruits by yeast. *Proceedings of the National Academy of Sciences of the United States of America* 68(B), 133–139.

Nigro, F., Finetti Sialer, M.M. and Gallitelli, D. (1999) Transformation of *Metschnikowia pulcherrima* 320, biocontrol agent of storage rot, with the green fluorescent protein gene. *Journal of Plant Pathology* 81(3), 205–208.

Pusey, P.L. (1994) Enhancement of biocontrol agents for postharvest diseases and their integration with other control strategies. In: Wilson, C.L. and Wisniewski, M.E. (eds) *Biological Control of Postharvest Diseases. Theory and Practice.* CRC Press, Boca Raton, Florida, pp. 77–88.

Ragsdale, N.N. and Sisler, H.D. (1994) Social and political implications of managing plant diseases with decreased availability of fungicides in the United States. *Annual Review of Phytopathology* 32, 545–557.

Segal, E., Yehuda, H., Droby, S., Wisniewski, M. and Goldway, M. (2002) Cloning and analysis of CoEXG1, a secreted 1,3-ß-glucanase of the yeast biocontrol agent, *Candida oleophila*. *Yeast* 19, 1171–1182.

Sharma, N. (1992) Biological control of *Ulocladium chartarum* (Preuss) Simon, the fruit rot pathogen of apple. *Indian Journal of Plant Pathology* 10(1&2), 65–68.

Sharma, N. (1993) Postharvest biological control of citrus fruit rot. *Journal of Biological Control* 7(2), 84–86.

Sharma, N. (2000) Biological control of grey mold of tomato with antagonist yeast. *International Journal of Environmental Biology* 5, 47–51.

Sharma, N. (2003) Biocontrol of *Aspergillus* rot of mango by using potential antagonists. *Indian Journal of Plant Pathology* 21(1&2), 114–115.

Sharma, N. (2005) Management of postharvest diseases through eco-friendly yeast. In: *Third International Conference on Plants and Environmental Pollution. International Society of Environmental Botanists and NBRI, Lucknow, India, 28 November–2 December, 2005*, 55 pp.

Sharma, N., Ghosh, N.R. and Nigam, M. (1997) Postharvest biocontrol of *Penicillium* rot to table grapes by using antagonist *Debaryomyces hansenii* Zopf. *Journal of Biological Control* 11, 53–58.

Sharma, N., Verma, U.K. and Awasthi, P. (2006) A combination of the yeast *Candida utilis* and chitosan controls fruit rot in tomato caused by *Alternaria alternata* (Fr.) Keissler and *Geotrichum candidum* Link ex Pers. *Journal of Horticultural Science and Biotechnology* 81(6), 1052–1056.

Smilanick, J.L. and Denis-Arrue, R. (1992) Control of green mold of lemons with *Pseudomonas* species. *Plant Disease* 76, 481–485.

Smilanick, J.L., Margosan, D.A. and Henson, D. J. (1995) Evaluation of heated solutions of sulfur dioxide, ethanol, and hydrogen peroxide to control postharvest green mold of lemons. *Plant Disease* 79, 742–747.

Smilanick, J.L., Margosan, D.A., Milkota, F., Usall, J. and Michael, I. (1999) Control of citrus green mold by carbonate and bicarbonate salts and influence of commercial postharvest practices on their efficacy. *Plant Disease* 83, 139–145.

Smolka, S. (1992) Methods for *in vitro* assessment of pesticide effects on microorganisms of the phyllosphere. *Nachrichtenblatt Duetsch Pflalzenschutzdienst* 44, 252–264.

Spadaro, D., Vola, R., Piano, S. and Gullino, M.L. (2002) Mechanisms of action and efficacy of four isolates of the yeast *Metschnikowia pulcherrima* active against postharvest pathogens on apples. *Postharvest Biology and Technology* 24, 123–134.

Unnikrishnan, V. and Nath, B.S. (2002) Hazardous chemicals in foods. *Indian Journal of Dairy Bioscience* 11, 155–158.

Wilson, C.L. and El Ghaouth, A. (2002) Biological Coating With a Protective and Curative Effect for the Control of Postharvest Decay. US Patent No. 6,423,310.

Wilson, C.L. and Pusey, P.L. (1985) Potential for biological control of postharvest plant diseases. *Plant Disease* 69, 375–378.

Wilson, C.L. and Wisniewski, M.E. (1989) Biocontrol of postharvest diseases of fruits and vegetables: an emerging technology. *Annual Review of Phytopathology* 27, 425–441.

Wilson, C.L. and Wisniewski, M. (1994) *Biological Control of Postharvest Diseases: Theory and Practice.* CRC Press, Boca Raton, Florida, 182 pp.

Wilson, C.L., Wisniewski, M.E., Droby, E. and Chalutz, E. (1993) A selection strategy for microbial antagonists to control postharvest diseases of fruits and vegetables. *Scientia Horticulturae* 53, 183–189.

Wilson, C.L., El Ghaouth, A., Chalutz, E., Droby, S., Stevens, C., Lu, J.Y., Khan, V. and Arul, J. (1994) Potential of induced resistance to control postharvest diseases of fruits and vegetables. *Plant Disease* 78, 837–843.

Wilson, M. and Lindow, S. (1994) Ecological similarity and coexistence of epiphytic ice-nucleating (IceC) *Pseudomonas syringae* strains and a non-ice-nucleating biological control agent. *Applied Environmental Microbiology* 60, 3128–3137.

Wisniewski, M.E. and Wilson, C.L. (1992) Biological control of postharvest diseases of fruits and vegetables: recent advances. *HortScience* 27, 49–58.

Wisniewski, M.E., Biles, C., Droby, S., McLaughin, R., Wilson, C. and Chalutz, E. (1991) Mode of action of the postharvest biocontrol yeast, *Pichia guilliermondii*. I. Characterization of the attachment to *Botrytis cinerea*. *Physiology and Molecular Plant Pathology* 39, 245–258.

Yehuda, H., Droby, S., Wisniewski, M. and Goldway, M. (2001) A transformation system for the biocontrol yeast, *Candida oleophila*, based on hygromycin B resistance. *Current Genetics* 40, 276–281.

Yehuda, H., Droby, S., Bar-shimon, M., Wisniewski, M. and Goldway, M. (2003) The effect of under- and over-expressed CoEXG1-encoded-exo-glucanase secreted by *Candida oleophila* on the biocontrol of *Penicillium digitatium*. *Yeast* 20, 771–780.

Zhang, H., Zheng, X. and Xi, Y. (2005) Biological control of postharvest blue mold of oranges by *Cryptococcus laurentii* (Kufferath) Skinner. *BioControl* 50, 331–342.

Zheng, X., Zhang, H. and Xi, Y. (2004) Effects of *Cryptococcus laurentii* (Kufferath) Skinner on biocontrol of postharvest decay of arbutus berries. *Botanical Bulletin of Academia Sinica* 45, 55–60.

10 Biological Control of Plant Diseases: An Overview and the *Trichoderma* System as Biocontrol Agents

Abhishek Tripathi,[1] Neeta Sharma[2] and Nidhi Tripathi[1]
[1]*Department of Bioscience and Biotechnology, Banasthali University, Banasthali Vidyapith, India;* [2]*Mycology and Plant Pathology Division, Department of Botany, University of Lucknow, Lucknow, India*

Abstract

Biocontrol is the reduction of inoculum density or disease-producing activities of a pathogen in its active or dormant state by one or more microorganisms, accomplished naturally. Research on biological control agents has utilized naturally occurring saprophytic soil fungi to compete with and/or destroy soilborne pathogens. Biological control has attracted attention from researchers for over 30 years, primarily because of the interest in developing more 'environmentally friendly' means of disease management in the absence of agricultural pesticides. Despite considerable effort in the area of biological control, few practical applications have become established in agriculture for the control of plant diseases. Common biocontrol agents include *Trichoderma*, *Gliocladium*, *Aspergillus*, *Penicillium*, *Chaetomium*, *Dactylella*, *Glomus*, etc. Biological control is achieved by competition, hyperparasitism, induced resisitance, hypovirulence, etc. Mycoparasitism and production of volatile and non-volatile antibiotics are important mechanisms operating in the case of *Trichoderma*, besides commercial uses and mass multiplication of the novel biocontrol agent. The future of biocontrol lies perhaps with the development of better application methods and the use of genetic engineering to increase the efficacy of various wild strains.

Introduction

Empirical approaches to chemical disease control have been practised since ancient times, when concoctions consisting of salt brine, sulphur, lime, ashes and salts of copper, mercury and arsenic were used to combat disease. Reports of pesticide residues in food, soil, river and groundwater undermine consumers' trust. Thus, the increasing concern, particularly in developed nations, is that modern methods of crop protection have an overall negative impact on the environment and on society. Pathogen resistance against certain classes of fungicides has further reduced the number of disease control measures available. In recent years, it has become evident, as a result of public opinion and environmental laws, that new and safer alternatives to traditional synthetic pesticides are both desirable and mandated. Research emphasis has therefore been on the development of alternative approaches to control the pathogens and pests of ornamental crops using biocontrol agents.

There are considerably more success stories involving the control of insect pests. Garrett's (1965) definition of biological control

 Management of Fungal Plant Pathogens (eds A. Arya and A.E. Perelló)

of plant disease was, 'any condition under which or practice whereby survival or activity of a pathogen is reduced through the agency of another living organism (except man himself), with the result that there is a reduction in the incidence of the disease caused by the pathogen'. Although biological control consists of diverse methods and approaches to suppress plant disease, in most cases antagonists to pathogens are added to the agroecosystem. Various approaches of biocontrol are directed at suppressing initial disease induced by a soilborne pathogen or the application of an avirulent isolate of the pathogen that 'competes' with the virulent pathogen on or in the host. Biological control employs living agents (usually antagonists or competitors of the causal agent) to control plant diseases. Effective biological controls take advantage of the natural competition of living organisms for limited resources or ecological niches. Thus, two organisms cannot occupy the same space at the same time, they cannot consume the same resource (e.g. food source) at the same time and, in some cases, one organism produces compounds that are inhibitory to the growth and development of the other organism. Certain microorganisms that normally compete for and live off debris and dead animal and plant cells in the soil environment have developed, through mutation, the ability to invade a host plant and escape the effects of antagonists. These invading organisms are referred to as pathogens. The lack of survival of the pathogen and the superior competitiveness of the antagonists relative to pathogens brings promise to the theory of using antagonists to control pathogens.

Tubeuf (1914) coined the term 'biological control' in relation to plant pathogens, while Hartley (1921) first attempted to control the root diseases of plants with introduced microorganisms. Cook and Baker (1983) defined biological control as, 'the reductions of the amount of inoculum or disease-producing activity of a pathogen accomplished by one or more organisms other than man'. Microorganisms, which are used in the management of plant diseases, are referred to as 'biocontrol agents'. The important genera of fungi used as biocontrol agents are *Trichoderma*, *Gliocladium*, *Aspergillus*, *Penicillium*, *Neurospora*, *Chaetomium*, *Dactylella*, *Arthrobotrys* and *Glomus*, etc.

According to Baker (1987), biological control is the decrease of pathogen activity accomplished by one or more organisms including the host plant but excluding humans. Harman (2000) defined biological control as a critically needed component of plant disease management. Biocontrol agents are known as antagonists. The most important, well-studied antagonists against several plant pathogens are fungi like *Ampelomyces* sp., *Aspergillus* spp. (particularly *A. niger* and *A. terreus*), *Chaetomium globosum*, *Coniothyrium minitans*, *Fusarium* sp., *Gliocladium virens*, *Penicillium citrinum*, *Peniophora gigantea*, *Trichoderma* spp. (particularly *T. harzianum* and *T. viride*) and *Sporodesmium* sp.; and bacteria like *Agrobacterium radiobacter* strain K84, species of *Bacillus*, *Enterobacter*, *Micromonospora*, *Pseudomonas* and *Streptomyces*.

Mechanisms of Biological Control of Plant Diseases

Competition

Competition occurs between microorganisms when space or nutrients (i.e. carbon, nitrogen and iron) are limiting and its role in the biocontrol of plant pathogens has been studied for many years, with special emphasis on bacterial biocontrol agents. An important attribute of a successful rhizosphere biocontrol agent would be the ability to remain at a high population density on the root surface, providing protection of the whole root for the duration of its life. Mycorrhizal fungi can also be considered to act as a sophisticated form of competition or cross-protection, decreasing the incidence of root disease. *Fomes (Heterobasidion) annosum* colonizes stumps of freshly cut pine and other conifers and spreads via root grafts to other healthy trees, where it causes root rot (refer to Chapter 26). Spraying freshly cut stumps with spore suspensions of *Phlebia (Peniophora) gigantea* will prevent

H. annosum from getting a foothold, and this is standard practice in the UK.

Antibiosis

Antibiosis is the inhibition of an organism by a metabolic product (such as an antibiotic) from another organism. Many organisms, especially soil fungi and *Actinomycetes*, produce antibiotic substances. The production of antibiotics by *Actinomycetes*, bacteria and fungi is demonstrated very simply *in vivo*. Numerous agar plate tests have been developed to detect volatile and non-volatile antibiotic products by putative biocontrol agents and to quantify their effects on pathogens. In general, however, the role of antibiotic production in biological control *in vitro* remains unproved. Three diseases can be controlled by antibiosis: *Armillaria* root rot by *T. viride*, *Pythium* and *Rhizoctonia* damping off and stem and root rot diseases by *P. fluorescens* and crown gall by *A. radiobacter*. The most widely accepted commercial example is the control of crown gall using strain 84.

Hyperparasitism and mycoparasitism

Biological control can occur through direct parasitism. Parasitism involves the direct utilization of food of one organism by another organism. Hyperparasites are organisms parasitic on other parasites. Some have referred to this as 'natural biocontrol'. A few examples of hyperparasitism include: *Darluca* (*Sphaerellopsis*) *filum* parasitizes rust fungi and species of *Ampelomyces* parasitize powdery mildews; *Tuberculina maxima* parasitizes the aecial stage of *Cronartium ribicola*, cause of white pine blister rust; *T. viride*, and a number of other species, are known to parasitize hyphae of *R. solani*. The most common example of mycoparasitism is that of *Trichoderma* sp., which attack a great variety of phytopathogenic fungi responsible for the most important diseases suffered by crops of major economic importance worldwide.

Hypovirulence

Hypovirulence is a term used to describe reduced virulence found in some strains of pathogens. This phenomenon was first observed in *Cryphonectria* (*Endothia*) *parasitica* (chestnut blight fungus) on European *Castanea sativa* in Italy, where naturally occurring hypovirulent strains were able to reduce the effect of virulent ones. These slower-growing hypovirulent strains contain a single cytoplasmic element of double-stranded RNA (dsRNA) similar to that found in mycoviruses, which is transmitted by anastomosis in compatible strains through natural virulent populations of *C. parasitica*. Hypovirulence has also been reported in many other pathogens, including *R. solani*, *Gaeumannomyces graminis* var. *tritici* and *Ophiostoma ulmi*, but the transmissible elements responsible for hypovirulence or reduced vigour of the fungi are subject to debate and may be due to dsRNAs, plasmids or viruses.

Induced Resistance and Cross-Protection

Induced resistance is a plant response to challenge by microorganisms or abiotic agents such that, following the inducing challenge, *de novo* resistance to pathogens is shown in normally susceptible plants. Both localized and systemic-induced resistance are non-specific and can act against a whole range of pathogens, but whereas localized resistance occurs in many plant species, systemic resistance is limited to some plants. Cross-protection differs from induced resistance in that, following inoculation with avirulent strains of pathogens or other microorganisms, both inducing microorganisms and challenge pathogens occur on or within the protected tissue. The most commonly reported examples of cross-protection involving fungi are probably those used against vascular wilts. Inoculation with non-pathogenic *formae speciales* of *Fusarium* and *Verticillium* species, or with other fungi or bacteria, has shown different levels of cross-protection.

Predation

Predation has also been examined as a potential form of biocontrol. Nematode-trapping fungi and predaceous nematodes have been studied in detail as potential biological control agents, but ultimately have had little effect on the numbers of plant parasitic nematodes in the soil.

Mycorrhizae

Mycorrhizae are symbiotic (mutualistic) associations between fungi and plant roots. The increased surface area provided by mycorrhizal fungi allows for increased nutrient uptake, which indirectly benefits disease management derived by healthier, more vigorous roots. Because of the generally beneficial effect of mycorrhizae on plant growth and their common occurrence, many investigations have looked into the potential of root–fungus associations as potential biological control agents.

Vesicular arbuscular mycorrhizal (VAM) fungi were recognized and described in the last few decades of the 19th century. The term 'VAM' was changed to 'AM' by Draft and Nicolson (1974) because some species did not form vesicles. AM fungi occur throughout the terrestrial ecosystem in almost all the herbaceous and woody plants, forming a symbiotic relationship with the roots (Gerdemann, 1968; Trappe and Fogel, 1977). This symbiotic association has been reported to play an important role in plant mineral nutrition (Gianinazzi and Gianinazzi-Pearson, 1986). It has been observed by several workers that these fungi facilitate the uptake of many nutrients (phosphorus, zinc, copper, sulphur, potassium, iron, calcium, etc.), resulting in increased biomass (Wani and Lee, 1992). Nutrient content of N, P and K, and also Fe, Mn and Cu, increased due to AM inoculation in papaya. Among all the AM species, *G. mosseae* was recorded as the most efficient for nutrient uptake. Rajeshwari *et al.* (2001) reported that *G. fasciculatum* at low phosphorus level increased the root and shoot biomass. They also recorded that AM colonization was reduced at higher phosphorus level.

Biocontrol of Airborne Diseases

Many naturally occurring microorganisms have been used to control diseases on the aerial surfaces of plants. The most common bacterial species that have been used for the control of diseases in the phylloshpere include *P. syringae*, *P. fluorescens*, *P. cepacia*, *Erwinia herbicola* and *B. subtilis*. Fungal genera that have been used for the control of airborne diseases include *T. ampelomyces* and the yeasts, *Tilletiopsis* and *Sporobolomyces*.

Biocontrol agents normally must achieve a high population in the phyllosphere to control other strains, but colonization by the agent may be reduced by competition with the indigenous microflora. Integration of chemical pesticides and biocontrol agents has been reported with *Trichoderma* spp. and *P. syringae*. Biocontrol agents tolerant to specific pesticides could be constructed using molecular techniques. Resistance to the fungicide benomyl is conferred by a single amino acid substitution in one of the b-tubulins of *T. viride*, the corresponding gene thereby producing a biological control agent that could be applied simultaneously or in alternation with the fungicide.

Biocontrol of Soilborne Diseases

Chemical control of soilborne plant diseases is frequently ineffective because of the physical and chemical heterogeneity of the soil, which may prevent effective concentration of the chemicals from reaching the pathogen. Biological control agents colonize the rhizosphere, the site requiring protection and leave no toxic residues, as opposed to chemicals. Microorganisms have been used extensively for the biological control of soilborne plant diseases, as well as for promoting plant growth. Fluorescent Pseudomonads are the most frequently used bacteria for biological control and plant growth promotion, but *Bacillus* and *Streptomyces* species have also been commonly used. *Trichoderma*, *Gliocadium* and *Coniothyrium* are the most commonly used fungal biocontrol agents. Perhaps the most successful biocontrol agent of a soilborne pathogen is *A. radiobactor* strain K84, used against crown gall disease caused by *A. tumefaciens*.

Molecular techniques have also facilitated the introduction of beneficial traits into rhizosphere competent organisms to produce potential biocontrol agents. Chitin and β-(1,3)-glucan are the two major structural components of many plant pathogenic fungi, except *Oomycetes*, which contain cellulose in their cell wall and no appreciable levels of chitin. Biological control of some soilborne fungal diseases has been correlated with chitinase production. Bacteria producing chitinases or glucanases exhibit antagonism *in vitro* against fungi. A recombinant *Escherichia coli* expressing the *chiA* gene from *S. marcescens* was effective in reducing disease incidence caused by *Screrotium rolfsii* and *R. solani*. In other studies, chitinase genes from *S. marcescens* have been expressed in *Pseudomonas* spp. and the plant symbiont, *Rhizobium meliloti*. The modified *Pseudomonas* strain controlled the pathogens, *F. oxysporum* f. sp. *rodelens* and *G. graminis* var. *tritici*.

Commercial Biocontrol Agents

The following is a list of commercially available products formulated for the biocontrol of plant pathogens and/or plant growth promotion involving the induction of plant host defence. The list originated in 2000 through the efforts of Dr Deborah Fravel, USDA-ARS, and is now being updated by the APS Biological Control Committee (Table 10.1).

The *Trichoderma* System as Biocontrol Agents

Trichoderma spp. are free-living, saprophytic fungi that exhibit a high rate of interactions

Table 10.1. Fungi, bacteria, activators and their available commercial products.

	Commercial products
Fungi	
Ampelomyces quisqualis	AQ10
Candida oleophila	Aspire
Coniothyrium minitans	Contans, Intercept WG, KONI
Fusarium oxysporum	Biofox C, Fusaclean
Gliocladium sp.	Primastorp, SoilGard
Myrothecium verrucaria	DiTera
Paecilomyces lilacinus	Paecil
Phlebia gigantea	Rotstop
Pythium oligandrum	Polyversum
Trichoderma sp.	Bio Fungus, Binab T, Root Pro, RootShield/PlantShield, T-22G, T-22 Planter Box, Trichodex, Trichopel, Trieco
Bacteria	
Agrobacterium radiobacter	Galltrol, Nogall
Bacillus sp.	BioYield, Companion, EcoGuard, HiStick N/T, Kodiak, Rhizo Plus, Serenade, Subtilex, YieldShield
Burkholderia cepacia	Deny, Intercept
Pseudomonas sp.	BioJect Spot-Less, Bio-save, BlightBan, Cedomon
Streptomyces sp.	Actinovate, Mycostop
Activators of host defence	
Bacteria	Actinovate, BioYield, YieldShield
Bacterial protein	Messenger
Synthetic chemical	Actigard

with root, soil and foliar environments. The antagonistic nature of fungi from the genus *Trichoderma* was demonstrated more than 70 years ago. Furthermore, excellent progress has been made towards the improvement of *Trichoderma* sp. as a biological control agent in the past few years. Many *Trichoderma* isolates have been used as biocontrol agents against soilborne pathogens (Weindling, 1934). *Trichoderma* is a ubiquitous genus present in almost all types of habitat fungal antagonists. It comprises 3% of the total fungal population in forests and 1.5% of the total fungal population in other soils. It also exhibits the property of competition with fellow plant pathogenic fungi for key exudates from seeds that stimulate the germination of propagules of plant pathogenic fungi in soil, and also with soil microorganisms for nutrients and space. *Trichoderma* spp. act against a range of economically important aerial and soilborne plant pathogens. They have been used in the field and greenhouse against silver leaf on plum, peach and nectarine; Dutch elm disease on elms, honey fungus (*A. mellea*) on a range of tree species and against rots on a wide range of crops, caused by *Fusarium*, *Rhizoctonia*, *Pythium* and *Sclerotium* (Table 10.2). Lacicowa and Pieta (1994) reported that *Trichoderma* spp. and *Gliocladium* sp. gave significant control against soilborne pathogenic fungi of pea, which was better than that obtained with the use of chemicals. Spiers *et al.* (2004) described the mode of action of *Trichoderma* sp. against plant pathogens. Recently, Herrera-Estrella and Chet (2004) discussed the role of *Trichoderma* spp. as a biological control agent; the expression of mycoparasitism related genes (MRGs); antibiosis; the role of MRGs in biological control and strain improvement; competition; induced resistance; plant growth promotion; and *Trichoderma* spp. as a source of genes for crop improvement.

The biocontrol action is due largely to the inherent nature of inhibition or degradation of pectinases and other enzymes, which are deemed essential for phytopathogenic fungi in order to cause pathogenesis in plants. These direct effects on other fungi are remarkable yet complex and, until now, were attributed to being the basis for the action exerted by *Trichoderma* sp. on plant growth and development.

Mechanism of Action of *Trichoderma*

Several modes of action have been proposed to explain the suppression of plant pathogens by *Trichoderma* spp. These include mycoparasitism, antibiosis, competition, siderophore production, induction of systemic resistance, growth promotion, etc. (Dennis and Webster, 1971; Upadhyay and Mukhopadhyay, 1986; Chet, 1987).

Table 10.2. *Trichoderma* as biocontrol agents and their target pathogens which cause diseases in various host plants.

Biocontrol agent	Pathogens	Host crop
Trichoderma spp.	*Pythium* sp.	Bean, pea, cucumber
T. harzianum	*Fusarium oxysporum*	Cucumber, cotton, wheat, muskmelon, tomato, ginger
	Fusarium sp.	Lentil, cotton
	Pythium sp.	Pea, radish, cucumber, tomato
	Rhizoctonia solani	Pea, radish, snapbean
	Sclerotinia sclerotiorum	Cucumber, *Mentha* sp.
	Sclerotium rolfsii	Sugarbeet, groundnut, chickpea, *Mentha* sp.
	Gaeumannomyces sp.	Wheat
T. viride	*Pythium* sp.	White mustard
	R. solani	Potato

Direct action of biocontrol agent *Trichoderma*

Mycoparasitism

Mycoparasitism is the phenomenon in which fungal parasites attack other fungi. It is divided into necrotrophic (destructive) and biotropic (balanced) parasitism (Barnett and Binder, 1973). *Trichoderma* spp. are grouped in necrotrophic mycoparsites. Velikanov *et al.* (1994) noticed hyperparasitism with different strains of *T. viride*, *T. harzianum* and *G. virens*, which were tested against five phytopathogenic fungi, namely *F. oxysporum*, *F. solani*, *Pythium* sp., *R. solani* and *S. sclerotiorum* causing root rot of pea.

Trichoderma recognizes signals from the host fungus, triggering coiling and host penetration. Remote sensing is due at least partially to the sequential expression of cell wall degrading enzymes. Different strains can follow different patterns of induction, but the fungi apparently always produce low levels of an extracellular exochitinase. The possible role of agglutinins in the recognition process determining fungal specificity has been examined recently. Barak *et al.* (1985) proposed that lectins of plant pathogenic fungi might play a role in recognition. Inbar and Chet (1992) proved the role of lectins in recognition during mycoparasitism using a biometric system. Secretion of lytic enzymes, including β-1,3-glucanase(s), proteinase(s), chitinases and lipases, enables *Trichoderma* spp. to degrade the host cell wall, thereby reducing the incidence of disease (Harman, 2001). Ordentlinch *et al.* (1990) reported that there was no correlation between *in vivo* and separated *in vitro* dual culture or enzyme assays. Involvement of chitinase and β-1,3-glucanase in *Trichoderma*-mediated biological control was also reported by Harman (2001). Involvement of β-1,6-glucanases and β-1,4-glucanases may also play an important role in mycoparasitism (Thrane *et al.*, 1997). *T. harzianum*-mediated mycoparasitism may involve 20 separate genes and gene products; most of these gene products are synergistic with one another (Lorito, 1998).

It is considered that mycoparasitism is one of the main mechanisms involved in the antagonism of *Trichoderma* as a biocontrol agent. The process apparently includes:

1. Chemotropic growth of *Trichoderma*;
2. Recognition of the host by the mycoparasite;
3. Secretion of extracellular enzymes;
4. Penetration of the hyphae; and
5. Lysis of the host.

Antibiosis

The high percentage of effectiveness of the biocontrol ability of *Trichoderma* is conferred most likely by more than one exclusive mechanism. Another known mechanism of biocontrol is antibiosis, which is the release of antibiotics and other metabolites that are harmful to the pathogen and inhibit their growth. Many such substances have been isolated from *Trichoderma* sp., namely gliotoxin and glyoviridin from *T. viride* (Sharma and Dohroo, 1991), viridin, alkyl pyrones, isonitriles, polyketides, diketopiperazines and some steroids (Upadhyay and Mukhopadhyay, 1986). Many *Trichoderma* spp. are reported to produce volatile and non-volatile antibiotics, chloroform soluble antibiotics, including trichodermin, and peptide antibiotics active against a range of plant pathogenic fungi (Dennis and Webster, 1971). Indeed some isolates of *Trichoderma* excrete growth-inhibitory substances. In fact, it seems advantageous for a biocontrol agent to suppress a plant pathogen using multiple mechanisms.

Competition

This mode of action implies the competition among microorganisms for space and nutrients when these factors are limiting in nature. It is considered a 'classical' mechanism of biocontrol. The mechanism is considered involved when no evidence of either mycoparasitism or antibiosis is found in a particular interaction. Since *Trichoderma* is an omnipresent fungus and is found in agricultural and natural soils throughout the world, it is enough proof of it being an excellent competitor for space and nutritional resources. Excellent competitiveness for space and nutrition is supposed to be useful for biological control in the absence of mycoparasitism or antibiosis (Cook and Baker, 1983). Elad (2000) reported that when conidia of T-39 were sprayed on leaves, germination of conidia of *B. cinerea* was slowed down, because the pathogenic conidia required external nutrients for germination and infection.

Indirect action of biocontrol agents

In addition to the ability of *Trichoderma* spp. to attack or inhibit the growth of plant pathogens directly, recent discoveries indicate that they can also induce systemic and localized resistance to a variety of plant pathogens.

Biochemical elicitors of disease resistance and induced systemic resistance

Induced systemic resistance (ISR) is another phenomenon of biocontrol exhibited by the plant to combat the harmful effects of the pathogen. It implies the elicitation of resistance or plant response against the microorganism or abiotic agent, such that following the inducing challenge posed to the plant, *de novo* resistance to pathogens is shown in normally susceptible plants. Localized and systemic induced resistance occurs in all or most plants in response to attack by pathogenic microorganisms, physical damage due to insects or other factors, treatment with various chemical inducers and the presence of non-pathogenic rhizobacteria. Specific strains of fungi in the genus *Trichoderma* colonize and penetrate plant root tissues and initiate a series of morphological and biochemical changes in the plant, which are considered to be part of the plant defence response. Finally, it leads to ISR in the entire plant. The capability of *T. harzianum* to promote increased growth response was verified both in greenhouse experiments and in the hydroponic system. A 30% increase in seedling emergence was observed and these plants exhibited a 95% increase in root area. Similarly, an increase in P and Fe concentration was observed in *Trichoderma* inoculated plants.

In recent times, there has been tremendous progress related to pathways of resistance and much has been done to elucidate them. In many instances, salicylic acid or jasmonic acid, together with ethylene or nitrous oxide, induce a cascade of events that lead to the production of a variety of metabolites and proteins with diverse functions. Different pathways are induced by different challenges, although there seems to be crosstalk or competition between pathways.

There has been a great leap in explaining the ISR pathway activated by rhizobacteria; the best part is that it is the closest analogue of induced resistance activated by *Trichoderma*. The rhizobacteria-induced systemic resistance (RISR) pathway phenotypically resembles systemic acquired resistance (SAR) systems in plants. Heil (2001) defined ISR as the set of changes by which plants respond to an initial infection or elicitor treatment in becoming systemically resistant against pathogen attack. Several workers demonstrated that *Trichoderma* spp. could also affect the host plant, which shows an induced resistance-type response.

Chang *et al.* (1986) reported hastened flowering, increased number of blooms in *Chrysanthemum* and an increase in the height and weights of other plants as a result of *T. harzianum* inoculation in steamed soil. *Trichoderma viride*-coated seeds of broad bean resulted in increased fresh and dry weight of shoots, roots and nodules (Yehia *et al.*, 1985). Pea seeds treated with apple pomace-based *Trichoderma* inoculant extracts resulted in increased emergence, rapid plant growth, increased seedling vigour and phenolics content. The increase in overall phenolic content may contribute to improved lignification and antioxidant response (Zheng and Shetty, 2000). Altomore *et al.* (1999) reported for the first time the ability of a *Trichoderma* strain (T-22) to solubilize insoluble or sparingly soluble minerals by three possible mechanisms, namely acidification, production of chelating agents and redox activity. Further, they reported the solubilization of Fe_2O_3, MnO_2, Zn and rock phosphate by the cell-free culture filtrate of T-22. *Trichoderma* strains are also supposed to induce the production of hormone-like metabolites on release of nutrients from soil or organic matter (Kleifeld and Chet, 1992).

Chemicals Produced by *Trichoderma*

What has been stated above is the induced resistance exhibited by some plants that is a result of some microorganism, in this case, *Trichoderma*. In this context, it has been found that *Trichoderma* produces three classes of compounds to exert its effect and induce resistance in plants. These include:

Proteins with enzymatic or other functions

With regards to the first class of biochemical elicitors of *Trichoderma*, it is stated that much before the discovery of the induction of resistance by *Trichoderma*, a small 22-kDa xylanase protein was shown to induce ethylene production and plant defence. Working in the direction of *Trichoderma*, it has been found very recently that a series of proteins and peptides that are active in inducing terpenoid phytoalexin biosynthesis and peroxidase activity in cotton are produced by strains of *T. virens*.

Avr homologues

Another class is the protein product of *Avr* genes, which have been identified in a variety of fungal and bacterial plant pathogens. These are usually seen functioning as race- or pathovar-specific elicitors, possessing the capability of inducing hypersensitive responses and other defence-related reactions in plant cultivars that contain the corresponding resistance gene. Proteome analysis of T-22 identified proteins that are homologues of Avr4 and Avr9 from *Cladosporium fulvum*; *T. atroviride* strain P1 also produces similar proteins.

Oligosaccharides and low molecular weight compounds

Another finding in this sphere has been the transformation of *Trichoderma* mutants with reporter based on green fluorescent protein or specific enzymatic activities (glucose oxidase) under the control of biocontrol-related promoters. One of the advantages of this discovery has been the possibility that biomolecules released by the action of *Trichoderma* secreted cell wall degrading enzymes on the cell walls of fungal pathogens and plants can be isolated. These molecules function as inducers of the antagonistic gene-expression cascade in *Trichoderma* and some also function as elicitors of plant defence mechanisms.

Plant Growth Promotion

Fungal as well as bacterial biocontrol agents are reported and known to induce growth of various crops and also increase crop yield. *Trichoderma* spp., and other beneficial root-colonizing microorganisms, also enhance plant growth and productivity. Mukhopadhyay (1996) has reported increased growth of several crop plants following seed treatment with *T. harzianum* and *T. virens*. The reason attributed to this effect of *Trichoderma* and other microbes on plants has been explained based on the following arguments.

1. Suppression of harmful root microflora, including those not a direct causal organism of disease.
2. Production or activation of growth-stimulating factors.
3. Increased nutrient uptake through solubilization and sequestering of nutrients.

It is a well-established fact that microorganisms closely associated with the roots of a plant can influence plant growth and development directly. Although the ability of species of *Trichoderma* spp. to promote or inhibit plant growth directly has been noted for many years (Ozbay and Newman, 2004), efforts to define and exploit these influences have met with limited success. Many workers have reported plant growth promotion by different strains of *Trichoderma* spp. Chang *et al.* (1986) observed plant growth promotion resulting in enhanced germination, more rapid flowering, increased flowering and increased height and fresh weight in pepper, periwinkle, *Chrysanthemum* and several others after treatment of the soil with peat/bran inoculum or conidial suspension of *T. harzianum*.

Solubilization and Sequestration of Inorganic Plant Nutrients

It is a common natural occurrence that plant nutrients undergo a complex, intricately woven conversion from soluble to insoluble forms when in the soil; this is a precursor to the ease of access and absorption by roots. It is here that microorganisms may influence these transitions (Altomare *et al.*, 1999). The most commonly and extensively studied nutrients are iron and manganese. *Trichoderma* sp. has been reported to produce some compounds called siderophores (Sen, 2000). Iron chelated with these siderophores is in the unavailable and bound form for plant pathogens and so they do not have access to iron. On the contrary, plant roots are capable of absorbing iron in this form, so these are accessible to the plant. This is one of the mechanisms that operate for the growth of plants and the supply of nutrients to them. *Trichoderma* sp. increases the uptake and concentration of a variety of nutrients (copper, phosphorus, iron, manganese and sodium) in roots of hydroponic culture, even under axenic conditions. This increased uptake indicates an improvement in plant active uptake mechanisms.

Pesticide Susceptibility

Another aspect and quality of *Trichoderma* sp. lies in the fact that it possesses innate and natural resistance against most agricultural chemicals, including fungicides. The capability differs with strain. Some lines have been selected or modified to be resistant to specific agricultural chemicals.

Mass Multiplication of *Trichoderma*

The most critical obstacles to the application of biological control fungi as an effective means of disease management are the lack of knowledge of methods for mass culturing and a proper delivery system, which is needed to augment the soil directly with fungal antagonists (Papavizas, 1985; Singh *et al.*, 2002, 2004; Dissevelt and Ravensberg, 2004). Solid media for the experimental production of *Trichoderma* sp. and *Gliocladium* sp., two of the most common fungal antagonists, have been used frequently in laboratory and greenhouse studies (Bateman, 2004).

Some workers have tried composted hardwood bark as a substrate for the large-scale production of biocontrol fungi (Nelson and Hoitink, 1983). Sundheim (1977) used bark pellets as a medium for mass production of *Trichoderma* and *Gliocladium* sp. to control *Phomopsis sclerotioides* in cucumber. A variety of media have been used by various researchers for the production of *Trichoderma* sp. in stationary flasks, shakers (Jin *et al.*, 1991) and liquid fermenters (Jin *et al.*, 1996).

Backman and Rodriguez-Kabana (1975) used diatomaceous earth granules along with molasses for developing a formulation of biocontrol agents for application in soil. Hadar *et al.* (1979) used wheat bran formulations for mass-multiplying biocontrol agents for field application. Papavizas *et al.* (1984) developed a liquid fermentation technology for mass production of fungal antagonists by employing a combination of molasses and brewer's yeast. Sivan *et al.* (1984) developed a formulation of *T. harzianum* on wheat bran and peat. Mukhopadhyay *et al.* (1986) used sorghum grains to prepare the powdered formulations of fungal antagonists.

Tapioca rind, cow dung, biogas slurry, farmyard manure, paddy chaff, rice bran, groundnut shell, sugarcane bagasse, sheep manure, chickpea husk, maize cob, etc., are some of the substrates used for mass multiplication of *T. harzianum* and *T. viride* (Kousalya and Jeyarajan, 1990). Conway *et al.* (1996) used oat seeds for mass culturing of *T. harzianum* isolate OK-86. Alginate pellets were used for formulating a biomass of *G. virens* and *T. hamatum* and various food bases like wheat bran, maize cobs, groundnut hulls, soy fibre, castor pomace, cocoa hulls and chitin were used. They found that the pellets with *G. virens* and all the food bases with bran, soy fibres, castor, pomace or chitin resulted in stands similar to those of the control, except cocoa hull meal significantly reduced damping-off of *Zinnia* caused by *R. solani* and *P. ultimum*.

Kumar and Marimuthu (1997) tested the effect of decomposed coconut coir pith (DCCP) added to normal nursery media on the survival of *T. viride*. The pure DCCP gave efficient sporulation of *T. viride* population. Lewis *et al.* (1998) used commercially manufactured cellulose granules (Biodac) in a mixture with a sticker and fermenter-produced biomass of *Trichoderma* sp. and *G. virens* to produce a formulation in which chlamydospores in the biomass were activated with dilute acid. Tiwari *et al.* (2004) suggested that among the eight substrates, namely grains of *Sorghum vulgare* [*S. bicolor*], wheat, *Pennisetum typhoides* [*P. glaucum*], *S. vulgare* cv. M.P. Chari and *Sorghum* sp., a locally available millet; wheat bran; rice bran; and sugarcane bagasse were evaluated for the mass propagation of *T. viride*. *Sorghum* sp., a locally available millet, resulted in the greatest spore concentration, spore viability and total biomass of the fungal antagonist. The greatest spore concentration (8×10^9) was observed after 15 days of incubation at 27 ± 1°C. The spores of *T. viride* remained viable for 6 months at 5°C.

Commercial Use of *Trichoderma*

Commercialized systems for the biological control of plant diseases are few. It has been stressed that microbes cannot be used in isolation and exceptional results expected. On the contrary, a biocontrol system or consortia needs to be developed, which would be beneficial to a larger degree than individual components.

A primary obstacle in the commercial use of *Trichoderma* spp. for both disease control and growth enhancement is the mass production and delivery methods of its formation to the plants (Papavizas, 1985; Mukhopadhyay, 1996). The problem lies in the fact that biocontrol products represent living systems. A large number of growth media are reported to be suitable for the genus *Trichoderma*, but most of these are either food grains or are expensive. For solid-state fermentation substrates like sorghum grain, wheat grains, wheat bran, tea leaf waste, coffee husk, sawdust, etc., have been used (Gogoi and Roy, 1996; Mishra, 1998). A liquid fermentation method consisting of molasses, wheat bran and yeast is proposed for large-scale production of *Trichoderma* (Montealegre *et al.*, 1993). Bioefficacy of *T. harzianum* produced by solid fermentation, which contains only conidia, was found more effective than when produced by liquid fermentation, where a mixture of chlamydospores, hyphal fragments and conidia were present.

Conidia of *Trichoderma* in pyrophyllite survived better than alone at between –5 and 30°C. A temperature range from –5 to 5°C was found most suitable for an improved shelf life (Mukherjee, 1991). Mukherjee reported that shelf life of *T. virens* was almost constant on coated chickpea seeds at 5°C and, at room temperature, it was decreased by 12%. Chlamydospore-based formulations exhibited longer shelf life than conidia-based formulations (Mishra *et al.*, 2001).

Basic Components of Biocontrol Systems

There are three basic components of biocontrol systems. These are as follows:

Biocontrol strain

The first step towards successful biocontrol is to obtain or produce a highly effective biocontrol strain or other material (Table 10.3). For instance, the development of the T-22

Table 10.3. Inexpensive production and formulation of the biocontrol agent using various base materials.

Base material	Biocontrol agent	Formulation	References
Blackgram shell, shelled maize cob, coir pith, peat, gypsum, barley grains	*Trichoderma viride*, *T. harzianum*	Powder	Kumar and Marimuthu, 1997
Coffee fruit skin + biogas slurry	*T. harzianum*	Pellets	Sawant and Sawant, 1996
Coffee husk	*T. harzianum*, *T. viride*, *T. virens*	Pellets	Bhai *et al.*, 1994
Coffee berry husk	*T. harzianum*, *T. viride*, *T. virens*	Pellets	Sawant and Sawant, 1989
Fruit skin and berry mucilage	*T. harzianum*, *T. viride*, *T. virens*	Pellets	Sawant and Sawant, 1989
Groundnut shell	*T. viride*	Powder	
Mustard oil cake	*T. viride*	Pellets	
Soil	*T. harzianum*, *T. viride*	Powder	Singh, 2002
Sorghum grain	*T. harzianum*, *T. virens*	Powder	Upadhyay and Mukhopadhyay, 1986; Mishra, 1998
Sugarcane straw	*T. harzianum*, *T. viride*, *T. reesei*, *T. koningii*	Pellets	Singh *et al.*, 2004
Wheat bran	*T. virens*	Powder	Singh *et al.*, 2002
Rice husk, maize cob powder, spent tea leaves, wheat bran, citrus fruit pulp	*T. harzianum* (MTCC 3843)	Powder	Tripathi, 1998

strain of *T. harzianum* by Harman and fellow researchers was the result of a decade and more of hard work. Still, its commercial product, Root Shield, picked up pace in the late 20th century (Harman, 2000). Besides the usual properties of a biocontrol agent, the strain must also possess the following: (i) to be able to compete and persist in the environment in which it must operate and (ii) ideally, to be able to colonize and proliferate on existing and newly formed plant parts well after application. Sundaram (1996) developed fusants of two isolates of *T. harzianum* (Th-1 and Th-2), among them some showed morphological characters immediately between Th-1 and Th-3. When *T. harzianum* (Th-3) was fused with *T. virens*, many fusants were developed and few exhibited improved biocontrol activity (Ghosh, 1996) (Table 10.4).

Ease of delivery and application

Some delivery methods for *Trichoderma* are listed in Table 10.5.

Compatibility Testing of *Trichoderma*

The success of a biocontrol agent depends on its compatibility with other disease management systems. This requires holistic testing of biocontrol agents (BCA) in combination with other disease management practices in a system approach. Once the BCA is found to be compatible, it can be integrated successfully with the disease management modules for each cropping system. Csinos *et al.* (1983) evaluated the compatibility of *Trichoderma* spp. with fungicides for the management of *S. rolfsii* in groundnut. *T. harzianum*, *Rhizobium* and carbendazim were integrated successfully for the management of stem rot of groundnuts caused by *S. rolfsii*. A combination of either *Trichoderma* or *Gliocladium* with fungicides like carboxin or metalaxyl protected crop plants against soilborne pathogens and was emphasized by several workers (Sawant and Mukhopadhyay, 1990; Mukhopadhyay *et al.*, 1992). The alternation of BCA with fungicides was found to be more effective than mixtures.

Table 10.4. Commercial products of *Trichoderma* currently in the open market or under registration.

Product	Biocontrol agent	Effective against	Manufacturer/distributor
Antifungus	*Trichoderma* sp.	Various fungi	Grondortsmettigen De Cuester n.v., Belgium
Bas-derma	*T. viride* Basarass	Various fungi	Biocontrol Res. Lab., India
Binab T	*T. harzianum* (ATCC 20476) and *T. polysporum* (ATCC 20475)	Control of wound decay and wood rot	Bio-innovation AB, UK
Bioderma	*T. harzianum/T. viride*	Various fungi	Biotech International Ltd., India
Biofungus	*Trichoderma* sp.	*Sclerotinia*, *Phytophthora*, *Rhizoctonia solani*, *Pythium* spp., *Fusarium*, *Verticillium*	Grondortsmettigen De Cuester n.v., Belgium
Blo-trek 22G	*T. harzianum*	Various fungi	Bioworks, Inc. of Geneva, NY
Ecofit	*T. viride*	Various fungi	Hoechst Schering Agro Evo Ltd., India
Root pro, Root Protato	*T. harzianum*	*R. solani*, *Pythium* spp., *Fusarium* spp. and *Sclerotium rolfsii*	Efal Agr, Israel
Root shield, Plant shield, T-22 Planter Box	*T. harzianum Rifai* strain KRL-AG(T-22)	*Pythium* spp., *R. solani*, *Fusarium* spp.	Bioworks Inc., USA
RUTOPIA	*Trichoderma* sp.	Organic Soil Amendment Turfgrass Biostimulant	NaEx Corp/Poulenger USA, Inc
SoilGard (formerly GlioGard)	*Trichoderma* sp.	Damping-off diseases caused by *Pythium* and *Rhizoctonia* spp.	USA
Supresivit	*T. harzianum*	Various fungi	Borregaard and Reitzel, Czech Republic
T-22 G, T-22 HB	*T. harzianum* strain KRL-AG2	Various fungi	THT Inc., USA
Trichoderma 2000	*Trichoderma* spp.	*R. solani*, *S. rolfsii*, *Pythium* spp., *Fusarium* spp.	Myocontrol Ltd., Israel
Trichodex, Trichophel	*T. harzianum*	*Botrytis* of vegetables and grapevines	Makhteshim Chemical Works Ltd., USA
Trichophel, Trichoject, Trichodowels, Trichoseal	*T. harzianum* and *T. viride*	*Armillaria*, *Botryosphaeria*, *Chondrosternum*, *Fusarium*, *Nectria*, *Phytophthora*, *Pythium*, *Rhizoctonia*	Agrimm Technologies Ltd., New Zealand
Tri-control	*Trichoderma* sp.	Various fungi	Jeypee Biotechs, India
Trieco	*T. viride*	*Rhizoctonia* spp., *Pythium* spp., *Fusarium* spp., root rot, seedling rot, collar rot, red rot, damping-off *Fusarium* wilt	Ecosense Labs Pvt. Ltd., Mumbai, India
TY	*Trichoderma* spp.	Various fungi	Myocontrol, Israel
Tusal	*Trichoderma* spp.	Damping-off diseases caused by *Pythium*, *Phoma* and *Rhizoctonia* species, rhizomania disease of sugarbeet and drop of lettuce	Spain

Table 10.5. Mass production and delivery methods of *Trichoderma*.

Biocontrol agent	Mass production	Delivery method
Trichoderma viride	Commercially produced pellets (BINAB T SEPPIC). Also produced on wheat bran: sawdust and tap water (3:14). Have been produced on a variety of growth media (autoclaved rye, barley and sunflower seeds)	Applied directly to the soil along with food base
T. harzianum	As in *T. viride*; also produced on molasses and enriched clay granules as food base	Backman and Rodriguez-Kabana applied it @ 140 kg/ha after 70 days of planting

Integration of *T. harzianum* with a sublethal dose of methyl bromide (300 kg/ha) and soil solarization yielded maximum control of *Fusarium* crown and root rot of tomato caused by *F. oxysporum* f. sp. *radicis-lycopersici* (Sivan and Chet, 1993).

In order to get the maximum efficiency from *Trichoderma*, it is important that it should be applied properly. It is effective as a seed treatment with or without fungicides. The basic reason why this is used is its multifaceted nature and broad range. It colonizes roots, increases root mass and improves plant health, and consequently provides yield increases, which chemical fungicides applied at reasonable rates cannot do. It can also be used in conjugation with other microbes, which thereby increases its efficiency. The two-pronged advantage would be a reduction in the use of pesticides and limiting root-attacking diseases, plus protection of transplants in the field by virtue of its ability to colonize roots. Besides this, powdered formulations can be made and applied to the seed directly, and then the seeds are sown. This would reduce the amount of biocontrol agent used, as well as protect the plants from pathogen attack. Further, plant growth would also improve.

References

Altomare, C., Norvell, W.A., Björkman, T. and Harman, G.E. (1999) Solubilization of phosphates and micronutrients by the plant-growth promoting and biocontrol fungus *Trichoderma harzianum* Rifai 1295-22. *Applied and Environmental Microbiology* 65, 2926–2933.

Backman, P.A. and Rodriguez-Kabana, R. (1975) A system for growth and delivery of biological control agents to the soil. *Phytopathology* 65, 819–821.

Baker, K.F. (1987) Evolving concepts of biological control of plant pathogens. *Annual Review of Phytopathology* 26, 67–85.

Barak, R., Edal, Y., Mirelman, D. and Chet, I. (1985) Lectins: a possible basis for specific recognition in the interaction of *Trichoderma* and *Sclerotium rolfsii*. *Phytopathology* 75, 458–462.

Barnett, H.L. and Binder, F.L. (1973) The fungal host–parasite relationship. *Annual Review of Phytopathology* 11, 273–292.

Bateman, R. (2004) Constraints and enabling technologies for mycopesticide development. *Pest Management* 15(2), 64–69.

Bhai, S.R., Thomas, J. and Naidu, R. (1994) Evaluation of carrier media for field application of *Trichoderma* spp. in cardamom growing soils. *Journal of Plantation Crops* 22(1), 50–52.

Chang, Y.-C., Chang, Y.-C. and Baker, R. (1986) Increase growth of plants in the presence of the biological control agent *Trichoderma harzianum*. *Plant Disease* 70, 145–148.

Chet, I. (1987) *Trichoderma* – application, mode of action, and potential as a biocontrol agent of soil-borne plant pathogenic fungi. In: Chet, I. (ed.) *Innovative Approaches to Plant Disease Control*. Wiley Interscience, New York, pp. 137–160.

Conway, K.E., Tomasino, S. and Claypool, P.L. (1996) Evaluations of biological and chemical controls for southern blight of apple rootstock in Oklahoma nurseries. *Proceedings of the Oklahoma Academy of Science* 76, 9–15.

Cook, R.J. and Baker, K.F. (1983) *The Nature and Practices of Biological Control of Plant Pathogens.* APS Books, St Paul, Minnesota, 539 pp.

Csinos, A.S., Bell, D.K., Minton, N.A. and Wells, H.D. (1983) Evaluation of *Trichoderma* spp., fungicides, and chemical combinations for control of southern stem rot on peanuts. *Peanut Science* 10, 75–79.

Dennis, C. and Webster, J. (1971) Antagonistic properties of species group of *Trichoderma* I. Production of non-volatile antibiotics. *Transactions of the British Mycological Society* 57, 25–29.

Dissevelt, M. and Ravensberg, W. (2004) Challenges in the commercialisation of *Trichoderma harzianum* strain T-22, a new biocontrol agent for Europe. *Bulletin OILB/SROP* 27(1), 47–51.

Draft, M.J. and Nicolson, T.H. (1974) Arbuscular mycorrhizas in plant colonizing coal wastes in Scotland. *New Phytologist* 73, 1129–1138.

Elad, Y. (2000) Biological control of foliar pathogens by means of *Trichoderma harzianum* and potential modes of action. *Crop Protection* 19, 709–714.

Elad, Y., Shtienberg, D. and Niv, A. (1994) *Trichoderma harzianum* T39 integrated with fungicides: improved biocontrol of grey mould. *Proceedings of the Brighton Crop Protection Conference – Pests and Diseases*, pp. 1109–1114.

Garrett, S.D. (1965) Prelude to biological control. In: Baker, K.F. and Snyder, W.C. (eds) *Ecology of Soil-borne Plant Pathogens.* University of California Press, Berkeley, California, pp. 4–17.

Gerdemann, J.W. (1968) Vesicular arbuscular mycorrhiza and plant growth. *Annual Review of Phytopathology* 6, 397–418.

Ghosh, S. (1996) Biocontrol characterization of *Trichoderma harzianum*, Rifai, isolate-3 and its protoplast fusion with *Gliocladium virens*, Miller *et al.* MSc thesis, Govind Ballabh Pant University of Agriculture and Technology, Pantnagar, India, 75 pp.

Gianinazzi, S. and Gianinazzi-Pearson, V. (1986) Progress and headaches in endomycorrhiza biotechnology. *Symbiosis* 2, 139–149.

Gogoi, R. and Roy, A.K. (1996) Effect of soil pH and media on the antagonism of *Aspergillus terreus* to the rice sheath blight fungus. *Indian Phytopathology* 49(1), 32–36.

Hadar, Y., Chet, I. and Henis, Y. (1979) Biological control of *Rhizoctonia solani* damping-off with wheat bran culture of *Trichoderma harzianum. Phytopathology* 69, 64–68.

Harman, G.E. (2000) Myths and dogmas of biocontrol. Changes in perceptions derived from research on *Trichoderma harzianum* T22. *Plant Disease* 84, 377–393.

Harman, G.E. (2001) Microbial tools to improve crop performance and profitability and to control plant diseases. In: Tzeng, D.D.S. and Huang, J.W. (eds) *Proceedings of International Symposium on Biological Control of Plant Diseases for the New Century – Mode of Action and Application Technology.* National Chung Hsing University, Taichung City, Taiwan, pp. 71–84.

Hartley, C. (1921) Damping-off in forest nurseries. *US Department of Agriculture Bulletin* 934, 1–99.

Heil, M. (2001) The ecological concept of cost of induced systemic resistance (ISR). *European Journal of Plant Pathology* 107, 137–146.

Herrera-Estrella, A. and Chet, I. (2004) The biological control agent *Trichoderma* from fundamentals to applications. In: Arora, D.K. (ed.) *Fungal Biotechnology in Agricultural, Food, and Environmental Applications.* Marcel Dekker, Inc, New York, pp. 147–156.

Inbar, J. and Chet, I. (1992) Biomimics of fungal cell–cell recognition by use of lectin-coated nylon fibers. *Journal of Bacteriology* 174, 1055–1059.

Jin, X., Harman, G.E. and Taylor, A.G. (1991) Liquid coating formulation for the application of biological seed treatments of *Trichoderma harzianum. Biological Control* 1, 237–243.

Jin, X., Taylor, A.G. and Harman, G.E. (1996) Development of media and automated liquid fermentation methods to produce desiccation-tolerant propagules of *Trichoderma harzianum. Biological Control* 7, 267–274.

Kleifeld, O. and Chet, I. (1992) *Trichoderma harzianum* – interactions with plants and effect on growth response. *Plant and Soil* 144, 267–272.

Kousalya, G. and Jeyarajan, R. (1990) Mass multiplication of *Trichoderma* spp. *Journal of Biological Control* 4, 70–71.

Kumar, A. and Marimuthu, T. (1997) Decomposed coconut coir pith – a conducive medium for colonization of *Trichoderma viride. Acta Phytopathologica-et-Entomologica-Hungarica* 32(1–2), 51–58.

Lacicowa, B. and Pieta, D. (1994) Protective effect of microbiological dressing of pea seeds (*Pisum sativum* L.) against pathogenic fungi living in soil. *Annals Universitatis Mariae Curie-Sklodowska, Sectio EEE, Horticultura* 2, 165–171.

Lewis, J.A. and Papavizas, G.C. (1985) Effect of mycelial preparation of *Trichoderma* and *Gliocladium* on populations of *Rhizoctonia solani* and the incidence of damping-off. *Phytopathology* 75, 812–817.

Lewis, J.A., Larkin, R.P. and Rogers, D.L. (1998) A formulation of *Trichoderma* and *Glioclaldium* to reduce damping-off caused by *Rhizoctonia solani* and saprophytic growth of the pathogen in soilless mix. *Plant Disease* 82(5), 501–506.

Lorito, M. (1998) Chitinolytic enzymes and their genes. In: Harman, G.E. and Kubicek, C.P. (eds) *Trichoderma* and *Gliocladium*, Volume 2. Taylor and Francis, London, pp. 73–99.

Mishra, D.S. (1998) Comparative efficacy of some biocontrol agents against *Rhizoctonia solani* Kuhn, the cause of sheath blight of rice. MSc (Ag.) thesis, Govind Ballabh Pant University of Agriculture and Technology, Pantnagar, India, 242 pp.

Mishra, D.S., Singh, U.S. and Dwivedi, T.S. (2001) Comparative efficacy of normal seed treatment and seed bio-priming with commercial formulations of *Trichoderma* spp. In: *53rd Annual Meeting of Indian Phytopathological Society and National symposium on 'Ecofriendly Approaches for Plant Disease Management' held at Chennai, India, 21–23 January 2001.*

Montealegre, J., Varnero, M.T. and Sepulveda, C. (1993) A method for biomass production of *Trichoderma harzianum* strain V: growth evaluation. *Fitopathology* 28, 99–106.

Mukherjee, P.K. (1991) Biological control of chick-pea wilt complex. PhD thesis, Govind Ballabh Pant University of Agriculture and Technology, Pantnagar, India, 188 pp.

Mukhopadhyay, A.N. (1994) Biocontrol of soil-borne fungal plant pathogens – current status, future prospect and potential limitations. *Indian Phytopathology* 47(1), 1–8.

Mukhopadhyay, A.N. (1996) Recent innovations in plant disease control by ecofriendly biopesticides. In: *83rd Annual Meeting of Indian Science Congress, Patiala, India, 1–8 January 1996.*

Mukhopadhyay, A.N., Patel, G.J. and Brahbhatt, A. (1986) *Trichoderma harzianum*: a potential biocontrol agent for tobacoo damping-off. *Tobacco Research* 12, 26–35.

Mukhopadhyay, A.N., Shrestha, S.M. and Mukherjee, P.K. (1992) Biological seed treatment for control of soil borne plant pathogens. *FAO Plant Protection Bulletin* 40, 21–30.

Nelson, E.B. and Hoitink, H.A.J. (1983) The role of microorganisms in the suppression of *Rhizoctonia solani* in container media amended with composted hardwood bark. *Phytopathology* 73, 274–278.

Ordentlinch, A., Nachmias, A. and Chet, I. (1990) Integrated control of *Verticillium dahliae* in potato by *Trichoderma harzianum* and Captan. *Crop Protection* 9, 363–366.

Ozbay, N. and Newman, S.E. (2004) Effect of *Trichoderma harzianum* strains to colonize tomato roots and improve transplant growth. *Pakistan Journal of Biological Sciences* 7(2), 253–257.

Papavizas, G.C. (1985) *Trichoderma* and *Gliocladium*: biology, ecology and potential for biocontrol. *Annual Review of Phytopathology* 23, 13–54.

Papavizas, G.C., Dunn, M.T., Lewis, J.A. and Beagle, R.J. (1984) Liquid fermentation technology for experimental production of biocontrol fungi. *Phytopathology* 74, 1171–1173.

Rajeshwari, Z., Laha, J.K.J., Vanamgaundi, K.A. and Marayanam, R. (2001) Effect of arbuscular mycorrhizal and phosphorus on feeding growth of *Casuarina equisetifolia*. *Indian Phytopathology* 54(1), 85–87.

Sawant, I.S. and Mukopadhyay, A.N. (1990) Integration of metalaxyl with *Trichoderma harzianum* for the control of *Pythium* damping-off in sugarbeet. *Indian Phytopathology* 43, 535–541.

Sawant, I.S. and Sawant, S.D. (1989) Coffee fruit skin and cherry husk as substrates for mass multiplication of *Trichoderma harzianum* as antagonist to citrus *Phytophthora*. *Indian Phytopathology* 42, 336.

Sen, R. (2000) Budgeting for the wood-wide web. *New Phytologist* 145, 161–165.

Sharma, S.K. and Dohroo, N.P. (1991) Post harvest management of rhizome rot (*Fusarium oxysporum* f. sp. *zingiberi* Trujillo) of ginger through chemical and antagonists. *Indian Cocoa, Arecanut and Spices Journal* 14(4), 150–152.

Singh, H.B., Singh, A. and Nautiyal, C.S. (2002) Commercialization of biocontrol agents: problem and prospects. In: Rao, G.P., Manohachari, C., Bhat, D.J., Rajak, R.C. and Lakhanpal, T.N. (eds) *Frontiers of Fungal Diversity in India*. International Book Distributing Company, India, pp. 847–861.

Singh, H.B., Singh, S., Singh, A. and Nautiyal, C.S. (2004) Mass production formulation and delivery systems of fungal and bacterial organisms in India. In: Singh, S.P. and Singh, H.B. (eds) *Eco Agriculture with Bioaugmentation: An Emerging Concept.* DASP, Lucknow, India, pp. 53–69.

Singh, R., Palat, R., Singh, N. and Narayan, P. (2002) Effects of different solid media on the growth and sclerotial production of *Sclerotinia sclerotiorum*. *Progressive Agriculture* 2(2), 180–182.

Sivan, A. and Chet, I. (1993) Integrated control of *Fusarium* crown root rot of tomato with *Trichoderma harzianum* in combination with methyl bromide or soil solarisation. *Crop Protection* 12, 380–386.

Sivan, A., Elad, Y. and Chet, I. (1984) Biological control effects of a new isolate of *Trichoderma harzianum* on *Pythium aphanidermatum. Phytopathology* 74, 498–501.

Spiers, M., Hill, R. and Fullerton, B. (2004) Trials using *Trichoderma* in greenhouse vegetable crops. *Grower* 37–39.

Sundaram, R.M. (1996) Biocontrol characterization of *Trichoderma harzianum*, Rifai, isolate-1 and its protoplast fusion with *Trichoderma harzianum*, Rifai, isolate-3. MSc thesis, Govind Ballabh Pant University of Agriculture and Technology, Pantnagar, India, 91 pp.

Sundheim, L. (1977) Attempts at biological control of *Phomopsis sclerotioides* in cucumber. *Netherlands Journal of Plant Pathology* 83, 439–442.

Thrane, C., Tronsmo, A. and Jenson, D.F. (1997) Endo β-1,3 glucanase and cellulase from *Trichoderma harzianum*: biological activity against plant pathogenic *Pythium* spp. *European Journal of Plant Pathology* 103, 331–344.

Tiwari, A.K., Kumar, K., Razdan, V.K. and Rather, T.R. (2004) Mass production of *Trichoderma viride* on indigenous substrates. *Annals of Plant Protection Sciences* 12(1), 71–74.

Trappe, J.M. and Fogel, R.D. (1977) Ecosystematic functions of mycorrhizae. In: Marshall, J.K. (ed.) *The Belowground of Ecosystem: A Synthesis of Plant Associated Processes*. Range Science Department Science Series 26. Colorado State University, Fort Collins, Colorado, pp. 205–214.

Tripathi, A. (1998) Mycorrhizal diversity influencing growth and yield response of bitter gourd (*Momordica charantia* Linn.). MSc thesis (Botany), Govind Ballabh Pant University of Agriculture and Technology, Pantnagar, India.

Tubeuf, C.F. von (1914) Biologische Bekämpfung von Pilzkrankheiten der Pflanzen. *Naturwissenschaftliche Zeitschrift ffir Forst- und Landwirtschaft* 12, 11–19.

Upadhyay, J.P. and Mukhopadhyay, A.N. (1986) Biological control of *Sclerotium rolfsii* by *Trichoderma harzianum* in sugarbeet. *Tropical Pest Management* 32, 215–220.

Velikanov, L.L., Cukhonosenko, E.Yu, Nikolaeva, S.I. and Zavelishko, I.A. (1994) Comparison of hyperparasitic and antibiotic activity of the genus *Trichoderma* Pers. Fr. and *Gliocladium virens* Miller, Giddens et Foster isolates towards the pathogens causing root rot of pea. *Mikologiya i Fitopatologiya* 28(6), 52–56.

Wani, S.P. and Lee, K.K. (1992) Role of biofertilizers in upland crop production. In: Tandon, L.S. (ed.) *Fertilizers, Organic Manures, Recyclable Wastes and Biofertilizers.* Fertilizer Development and Consultation Organization, India, pp. 91–212.

Weindling, R. (1932) *Trichoderma lignorum* as a parasite of other soil fungi. *Phytopathology* 22, 834–845.

Weindling, R. (1934) Studies on a lethal principle effective in the parasitic action of *Trichoderma lignorum* on *Rhizoctonia solani* and other soil fungi. *Phytopathology* 24, 1153–1179.

Yehia, A.H., El-Hassan, S.A. and El Bahadli, A.H. (1985) Biological seed treatment to control *Fusarium* root rot of broad bean. *Egyptian Journal of Phytopathology* 14, 59–66.

Zheng, Z. and Shetty, K. (2000) Enhancement of pea (*Pisum sativum*) seedling vigor and associated phenolic content by extracts of apple pomace fermented with *Trichoderma* spp. *Crop Protection* 36(1–2), 79–84.

11 Physiological Specialization of *Ustilaginales* (Smut) of Genera *Bromus*, *Zea* and *Triticum* in Argentina

Marta M. Astiz Gassó[1] and María del C. Molina[1,2]
[1]*Instituto Fitotécnico de Santa Catalina (IFSC) and* [2]*Consejo Nacional de Investigaciones Científicas y Tecnológicas (CONICET), Facultad de Ciencias Agrarias y Forestales, Universidad Nacional de La Plata, Llavallol, Buenos Aires, Argentina*

Abstract
The objective of this project was to determine the existence of physiological forms of *Ustilaginales* in *Bromus*, *Zea* and *Triticum* types in Argentina. Studies were carried out on the physiological specialization of *Ustilago bullata* Berk on *Bromus* spp., *Zea* seedlings' reaction to inoculation with *U. maydis* (D.C.) Corda and physiological specialization of *Tilletia laevis* Wallr. (common bunt) on *Triticum* spp. The smut was collected in different agricultural and cattle-raising regions in the country, using *Ustilaginales* taxonomic keys for smut identification and classification. The experiments were carried out in greenhouses and in fields at the Instituto Fitotecnico de Santa Catalina (FCAyF-UNLP). For *U. bullata* and *T. laevis*, the techniques used were as follows: inoculation by sprinkling of teliospores on host seeds and inoculation by hypodermic syringe with suspension of *U. maydis* sporidia on plantlets of *Z. mays* and related wild species. As a result of said studies, it was determined that: (i) different physiological forms exist in each of the kinds of smut analysed; (ii) genetic variability exists in the hosts which have genes that express different degrees of resistance to the disease; and (iii) genetic improvement is the most efficient and least environmentally harmful method.

Introduction

Smuts are pathogens of plants that belong to Phylum *Basidiomycota*, Class Ustilaginomycetes, Order *Ustilaginales*. Smut has the characteristic of forming greyish-black powdery masses of teliospores (basidiospores) on different organs such as the seeds, stems, leaves, flowers and fruit of the hosts. Approximately 1400 species of smut are known, which attack around 75 families of *Angiospermae*; the most familiar diseases are those affecting Monocotyledoneae, especially cereals, where the pathogens produce important economic losses (Fischer and Holton, 1957; Hirschhorn, 1986; Snetselaar and Mims, 1992). Until the 20th century, they were considered, worldwide, one of the most serious causes of loss of grain and/or seeds, similar to the effects produced by rust.

In Argentina, between 1934 and 1995, Hirschhorn and collaborators carried out several studies on *Ustilaginales* covering the taxonomic classification of the species, geographical distribution, germination types and histopathology and cytology of the different

 Management of Fungal Plant Pathogens (eds A. Arya and A.E. Perelló)

species of smut (Hirschhorn, 1986). Currently, control of these diseases is by means of agrochemicals and, on a smaller scale, by obtaining species with resistant genes through improvement programmes and studies on variability of these pathogens.

The objective of this project was to determine the existence of physiological forms of *Ustilaginales* in *Bromus*, *Zea* and *Triticum* in Argentina:

1. Physiological specialization of *U. bullata* Berk on *Bromus* spp.
2. *Zea* seedlings' reaction to inoculation with *U. maydis* (D.C.) Corda.
3. Physiological specialization of *T. laevis* Wall. (common bunt) on *Triticum* spp.

Physiological Specialization of *Ustilago bullata* Berk on *Bromus* spp.

Head smut *(U. bullata* Berk) is a pathogen which affects the growth of various grass species, especially within the genus *Bromus*. The disease is initiated when fungal hyphae penetrate seedlings; the attack develops from the inflorescences at the expense of the ovaries, forming a typical sorus. Severe infection affects limbs and glumes, reducing seed and forage production. In the USA, Fischer and Holton (1957) and Hirschhorn (1977, 1986) verified experimentally the existence of genes for resistance, physiological forms and the ability to cross-breed *U. bullata* and *U. striiformis*. Kreizinger *et al.* (1947) recorded the different reactions of *U. bullata* on *Bromus* which grew in the mountains and *Bromus* which grew on the plains; these experiences indicated that resistant *Bromus* varieties and lines could be obtained by artificial infection under controlled conditions and in the field. Also, 13 physiological forms of the pathogen could be studied (Meinrs and Fischer, 1953). In New Zealand, Falloon (1976, 1979a,b) carried out studies on the effect of *U. bullata* infection on *B. catharticus*. Also, Falloon and Hume (1988) reported the effects of the pathogen on *B. willdenowii* productivity and endurance in the field. In Argentina, Hirschhorn (1977) studied teliospore morphological variations in samples of *B. catharticus*, *B. mollis*, *Hordeum jubatum* and *H. compresum*. The author also determined that *Bromus* head smut in Argentina was represented by *U. bullata*, *U. bullata* cv. *macrospora* (Hirschhorn, 1977, 1986). Astiz Gassó (1983, 1985, 1994) reported the presence of genes for resistance to the pathogen and *in vitro U. bullata* teliospore formation, a phenomenon unrecorded for this pathogen and uncommon in other smuts. The objective of this work was to determine the existence of physiological forms in *U. bullata* populations on several *Bromus* species.

In this experiment, we used seeds of *B. catharticus*. *B. parodii*, *B. brevis*, *B. auleticus* and *B. inermis* cv. *gombaszpuzta* were provided by the Instituto Fitotécnico de Santa Catalina, FCAyF and Department of Genetics, and the Experimental Estación of Pergamino (INTA). The seeds were de-infested with a 2% formaldehyde solution for 20 min and then washed in sterile water three times. For identification of the pathogen, spores from each isolate harvested from plants naturally infected in the field were examined microscopically (Table 11.1). Viability of teliospores was tested by plating them in PDA medium 2% (Fischer and Holton, 1957). The seeds were infested with teliospores ($1.8 \times 10\ ^3$g teliospore/g seed), placed in sulphite paper envelopes and shaken well, so that spores would stick to the seed. Precautions were taken to avoid contamination with the different isolates of the pathogen. Thirty live seeds per isolate in three replications were inoculated during 4 consecutive years. An uninoculated sample was also included during the study. Inoculated samples were sown in experimental plots 1.5 m × 0.40 m in three rows

Table 11.1. *Ustilago bullata* isolates collected in different localities in Argentina.

Locality	Province
Pergamino	Buenos Aires
Tres Arroyos	Buenos Aires
Llavallol	Buenos Aires
Gowland	Buenos Aires
General Roca	Río Negro
Check	Mixture

with a distance of 0.20 m between them. The experimental design used was a randomized complete block. Evaluations in the field were conducted by head countings, recording the percentage of infection based on the number of infected and healthy heads. Then, the average infection for the 4 years was calculated. The level of resistance/susceptibility was determined using a disease rating scale (Table 11.2).

Isolates showed an 80–90% teliospore germination, approximately 20–25 h after they were cultivated on PDA. The teliospore germination rate increased with temperature from 20 to 25°C, with significant among-population differences. Boguena *et al.* (2007) also obtained similar results when they examined the effect of temperature from teliospore germination. Table 11.3 shows the reaction of the *Bromus* species tested with the different *U. bullata* isolates. *Bromus catharticus* was susceptible to all isolates including the mixture and similar results were reported for Astiz Gassó and Aulicino (1999); *B. parodii* showed similar reactions, but the levels of infection were lower than *B. catharticus*; *B. brevis* gave a resistant reaction to isolate Gowland, a moderately resistant reaction to Pergamino, Llavallol and the mixture, a moderately susceptibie reaction to isolate Tres Arroyos and a susceptible reaction to General Roca. Similar results were reported previously by Astiz Gassó (1983). *B. auleticus* and *B. inermis* cv. *gombaszpuzta* were resistant to all isolates and the uninoculated check did not show any infection.

Four physiological forms in the populations of *U. bullata* are shown in Fig. 11.1: (i) Tres Arroyos; (ii) Pergamino and Llavallol; (iii) Gowland; and (iv) General Roca. The species *B. brevis* would be the differential host.

Table 11.2. Disease rating scale for *Ustilago bullata*.

Reaction	Infection (%)
Resistant (R)	0–5
Moderately resistant (MR)	6–10
Moderately susceptible (MS)	11–30
Susceptible (S)	31–100

Reaction to Inoculation with *Ustilago maydis* (D.C.) Corda on *Zea* seedlings

Ustilago maydis is a smut that promotes the development of galls in *Zea*, the relation with the host being necessary to fulfil its life cycle. Damage produced in plants by the presence of corn stunt is: chlorosis, seedling death and tumours in leaves, stems, ears and tassels. At first, it was considered that *U. maydis* attacked *Z. mays* and *Z. mexicana*, but it was later verified that it also attacked *Z. perennis*, *Z. diploperennis*, *Z. parviglumis*, *Z. luxurians* and their hybrids with the grown species (Hirschhorn, 1986; Duran, 1987).

Table 11.3. Reaction of *Bromus* species to different *Ustilago bullata* collected in different localities in Argentina.

	Ustilago bullata isolates					
HOSTS	Pergamino	Tres Arroyos	Gowland	Llavallol	General Roca	Mixture
Bromus catharticus	S	S	S	S	S	S
B. parodii	S	S	S	S	S	S
B. brevis	MR	MS	R	MR	S	MR
B. auleticus	R	R	R	R	R	R
B. inermis cv. *gombaszpuzta*	R	R	R	R	R	R
Nor-inoculated check	0	0	0	0	0	0

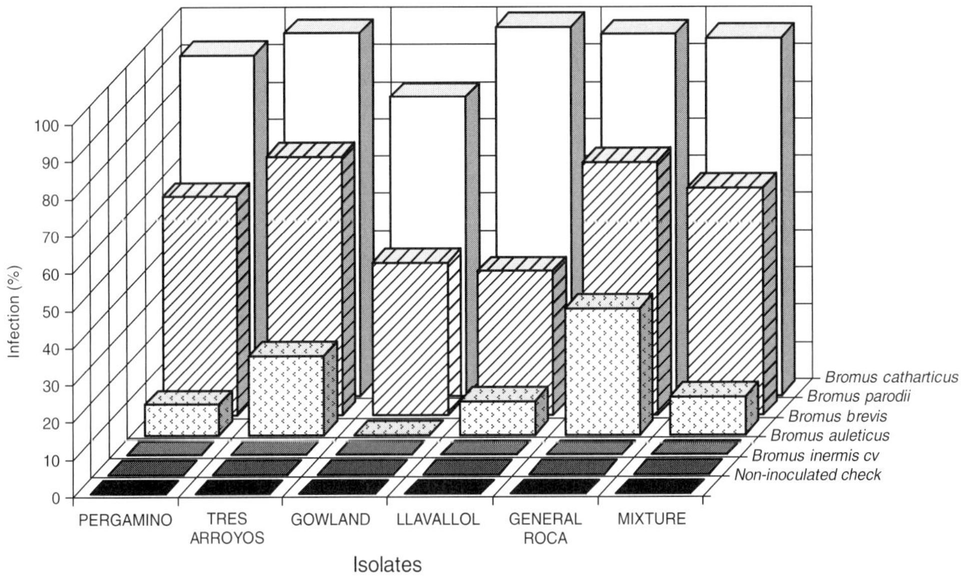

Fig. 11.1. Reaction of *Bromus* spp. to *U. bullata* isolates.

Until 1964, corn stunt did not any have incidence at the Instituto Fitotécnico de Santa Catalina, but in that year, a *Z. perennis* from Jalisco (México) was introduced and later on *Z. mexicana*, *Z. parviglumis*, *Z. luxurians* and *Z. diploperennis* were also grown and hybridized to *Z. mays*. As the hybrids are grown in the field as well as in the greenhouse, vegetative plants are available throughout the year (Astiz Gassó and Molina, 1996).

The pathogen multiplies on these plants with the corresponding increase in the number of spores disseminated by air and in the soil. Losses from corn smut range from 1% to up to 10% of all *Zea* species and hybrids are also attacked, depending on the environmental conditions favouring pathogen development; sweet corn may show losses approaching 100% from corn smut in localized areas (Callow and Ling, 1973; Hirschhorn, 1986; Banuett, 1995; Astiz Gassó and Molina, 1999).

In this chapter, the results from analysing the response of *Z. mays*, *Z. perennis* and *Z. diploperennis* seedlings when they are inoculated with six populations of *U. maydis* are presented. This was done with the purpose of determining resistance of the species and/or inbreds to *U. maydis*. The host materials used were the population 'Colorado Klein', the inbreds SC66, B73, E624A688 of *Z. mays*, as well as clones of *Z. perennis* and *Z. diploperennis*. Over a time period of 2 years, 1296 plants were inoculated with different strains of *U. maydis* isolated from the province of Buenos Aires (Santa Catalina, Balcarce, Necochea and 25 de Mayo), the province of Entre Ríos (Paraná) and the province of Córdoba (Río Cuarto). These strains were cultivated in a liquid medium of PDB 2% on a shaker for 18–24 h running at 25°C ± 2. The pathogen was inoculated by puncturing the base of the seedling with a hypodermic syringe and the sporidial suspension with concentrations 10^5–10^6 sporidia/ml was then forced up into the leaf whorl (Callow and Ling, 1973; Snetselaar and Mims, 1992, 1993; Banuett, 1995; Edmunds, 1998; du Toit and Pataky, 1999). In many previous works, this method was very successful in producing disease galls in seedlings (Astiz Gassó and Molina, 1999).

The trial involved three replications and a tester (non-treated plants). The plants were evaluated using a reaction scale to determine the mean percentage of infection with *U. maydis* (Table 11.4). The first symptoms in seedlings were observed 4–6 days after inoculation and gall development occurred 7–8 days after the treatment (Fig. 11.2).

The behaviour of the host when inoculated with six populations of *U. maydis* was analysed in Fig. 11.3. The hosts that reacted forming galls (grade 4) were cv. Colorado Klein: Necochea (8.34%) and Balcarce (2.78%); B73: Río Cuarto (14.15%), 25 de Mayo (11.11%), Santa Catalina (5.84%) and Balcarce (1.04%); E642A688: 25 de Mayo (8.33%) and Santa Catalina (3.34%); SC66: Río Cuarto (4.55%); *Z. perennis*: Santa Catalina (1.67%) and *Z. diploperennis*: 25 de Mayo (13.89%), Paraná (2.78) and Santa Catalina (1.67%).

Table 11.4. Reaction scale in hosts.

Behaviour	Host reaction
0 = Immune	No reaction
1 = Resistant	Partial chlorosis
2 = Medium resistant	Accent chlorosis and/or presence of stripe or anthocyanin stain
3 = Medium susceptibility	Necrosis and reduction of growth in plant
4 = Susceptibility	Formation of tumours

Fig. 11.2. Reaction of hosts after inoculations with *U. maydis*. (a) No reaction, immune; (b) partial chlorosis; (c–d) accent chlorosis and/or presence of stripe or anthocyanin stain; (e) necrosis and reduction of growth in plant; (f) formation of tumours (galls).

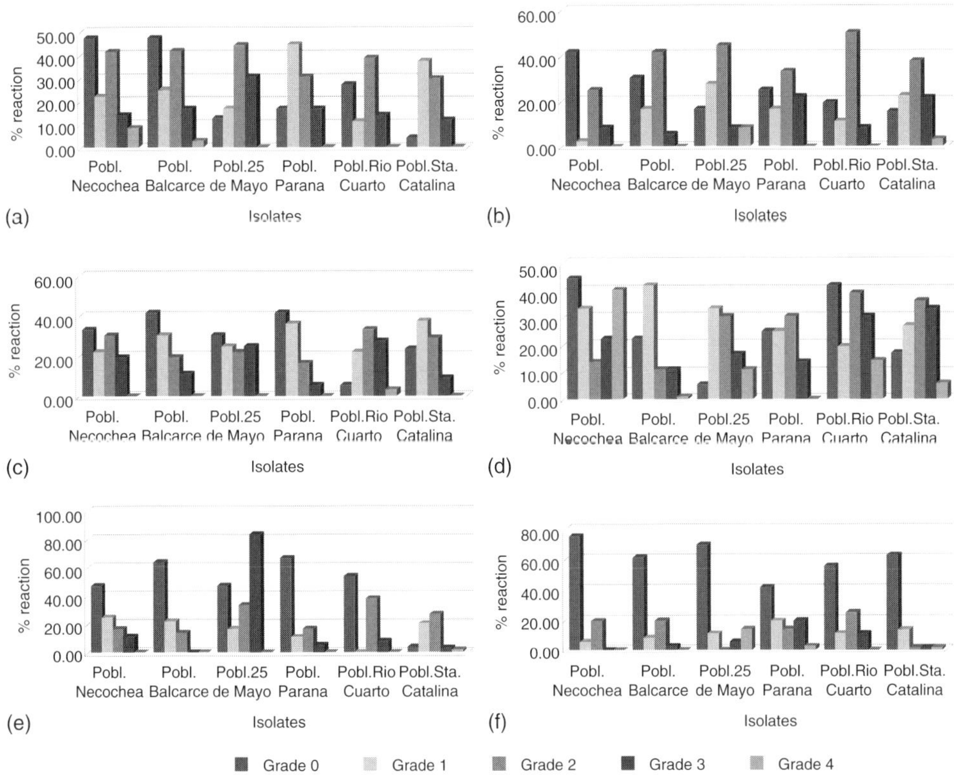

Fig. 11.3. Reaction of *Zea mays* (lines and populations), *Zea perennis* and *Zea diploperennis* to six strains of *U. maydis* isolates: (a) Colorado Klein (*Z. mays*); (b) Lines E642A688 (*Z. mays*); (c) Line SC66 (*Z. mays*); (d) Line B73; (e) *Z. perennis*; and (f) *Z. diploperennis*.

Physiological Specialization of *Tilletia laevis* Wallr. (Common Bunt) on *Triticum* spp. in Argentina

Common bunt of wheat is caused by *T. tritici* and *T. laevis*; infection takes place in the coleoptile when teliospores are found on the coleoptile surface and/or the ground (Fischer and Holton, 1957; Hirschhorn, 1986; Wilcoxson and Saari, 1996). Chemical control is achieved through seed treatments; however, the disease is aggravated due to inefficiency in the method of fungicide application and the widespread use of susceptible wheat cultivars. In common bunt, the spores survive in the soil for long periods and can cause infection of seedlings. The most effective control method is by genetic resistance to the pathogen and by establishing the variability or physiological specialization of *Tilletia* species. Historically, pathogenic races that are virulent to resistant cultivars have appeared, so new germplasm is screened continually for resistance. Investigations to determine disease resistance were incorporated into breeding programmes (Meinrs and Fischer, 1953; Kendirck, 1961; Metzger and Hoffmann, 1978; Gaudet, 1990; Johnsson, 1991; Gaudet *et al.*, 1994; Wilcoxson and Saari, 1996).

In Argentina, Hirschhorn and collaborators studied the morphology, taxonomy, symptomatology, spore germination, basidial cytology and geographical distribution of pathogens, *T. tritici* and *T. laevis*, to common bunt of wheat (Hirschhorn, 1986; Astiz Gassó, 1992; Astiz Gassó and Hirschhorn, 1994). The presence of 12 *T. foetida* (= *T. laevis*) physiological forms and cultivar wheat differentials for identification of *T. foetida* were reported by Astiz Gassó (1992, 1997a,b) and Astiz Gassó and Hirschhorn (1994). The objective of this work was to establish the physiological forms of *T. laevis* and to study the reaction of commercial wheat cultivars to the pathogen in Argentina.

In this experiment, we used ten hexaploid bread wheat cultivars with different levels of resistance and two tetraploid cultivars considered resistant. Seeds were deinfested with a formaldehyde solution (3:1) and washed in sterile water. Pathogens from 25 localities in the Argentine wheat belt were tried. Wheat cultivars were inoculated with 0.5 g of teliospore/100 g of seed. Experimental field plots consisted of three rows 2 m long per cultivar/pathogen isolate. Field evaluations were carried out by head countings and the percentage of infection was based on the number of infected and healthy heads. Results were transformed through the Arcosen and the average for 6 years of testing was calculated. Data were subjected to ANOVA (Statistix, 2008). Where significant differences were detected, treatment means were separated using HSD Tukey test ($P < 0.05$). Our field research to date indicates that *T. laevis* shows several physiological forms: Tandil, Rio Cuarto, Villa María, Cabildo, Castelar and Casilda. The rest of the 19 populations of common bunt showed homogeneous behaviour, so it could be considered as one physiological form (Table 11.5).

Tetraploid cultivar, Buck Cristal, proved the presence of resistant genes. The hexaploid wheat cultivars, Buck Ñapuca and Buck Yapeyú, were moderately resistant to pathogen incompatibility to different isolates (Table 11.6). The rest of the hexaploid

Table 11.5. Means of infections of 25 *T. laevis* populations.

Tilletia laevis populations	Province	Mean
1. Tandil	Buenos Aires	21.50 a
2. Río Cuarto	Córdoba	19.83 ab
3. Bordenave Col.1	Buenos Aires	17.12 abc
4. Sta Rosa	La Pampa	16.60 abcd
5. Venado Tuerto	Santa Fé	16.45 abcd
6. Tres Arroyos Col.1	Buenos Aires	15.70 abcde
7. Lincoln	Buenos Aires	15.58 abcde
8. Tres Arroyos Col.2	Buenos Aires	15.38 abcde
9. Laboulaye	Córdoba	14.48 abcde
10. Rafaela	Santa Fé	14.10 abcde
11. Pergamino	Buenos Aires	14.01 abcde
12. San Francisco	Córdoba	13.70 abcde
13. Villa María	Córdoba	12.92 bcde
14. Salliquelo	Buenos Aires	12.61 bcde
15. Marcos Juarez	Córdoba	11.89 bcde
16. Necochea	Buenos Aires	11.87 bcde
17. Cabildo	Buenos Aires	10.91 cde
18. Bordenave Col.2	Buenos Aires	10.73 cde
19. Cañada de Gomez	Santa Fé	10.30 cde
20. Bragado	Buenos Aires	9.49 cde
21. Río Tercero	Córdoba	9.07 cde
22. Paraná	Entre Ríos	8.32 cde
23. Castelar	Buenos Aires	8.28 de
24. Casilda	Santa Fé	7.29 e
25. Bolivar	Buenos Aires	7.17 e

Note: Means followed by different letters within column indicate significant differences according to Tukey's test ($P < 0.05$).

Table 11.6. Means of infection of common bunt in wheat cultivars.

Hosts	Mean
Buck Charrua	19.97 a
Buck Ombú	18.78 a
Buck Catriel	17.07 ab
Buck Bagual	16.58 ab
Buck Fogón	13.01 bcd
Buck Guaraní	11.21 bcd
Buck Ñapuca	8.88 cd
Buck Yapeyu	8.72 e
Buck Cristal	2.85 f

Note: Means followed by the same letter with a column indicate cultivars that are homogenous according to Tukey's test ($P < 0.05$).

wheat was moderately susceptible. Also, the interaction among wheat cultivar populations of *T. laevis* was significantly high and the interaction among pathogen population replications was significantly high according to Tukey's test ($P < 0.05$).

Conclusions

From this analysis, four physiological forms of *U. bullata* were found in the isolates studied: Tres Arroyos, Pergamino and Llavallol, Gowland and General Roca. *Bromus brevis* is the differential host for the fungus populations and shows genetic resistance to the Gowland isolate. *Bromus auleticus* and *B. inermis* cv. *gombaszpuzta* were resistant to all the fungus isolates. This was the first report in Argentina determining the physiological forms of smut *U. bullata* of *Bromus* spp. It can be concluded that the wild species and the grown species of the genus *Zea* reacted in different ways (tolerant and/or resistant to moderately susceptible), depending on the geographic origin of *U. maydis* populations. These results might be considered when selecting germplasm to obtain new forage plants from interespecific hybrids of the genus *Zea*. The wheat cultivars evaluated would also be used as differentials for identification of *T. laevis* races. Six physiological forms were detected among the used populations of *T. laevis*. This is the first report in Argentina determining the physiological forms of smut *T. laevis* of *Triticum* spp. The most effective methods to control the disease are genetic resistance and establishing the variability of the smut populations. Determination of the physiological forms of *U. bullata*, *U. maydis* and *T. laevis* and genetic improvement is the most efficient and least environmentally harmful method.

References

Astiz Gassó, M.M. (1983) Búsqueda de fuentes de resistencia en *Bromus* spp. a *Ustilago bullata* Berk. *V Jornadas Fitosanitarias Argentinas.* Resúmenes, 21 pp.

Astiz Gassó, M.M. (1985) Formación de clamidosporas 'in vitro' de *Ustilago bullata* Berk. *XII Jornadas Argentinas de Micología.* Resúmenes, 40 pp.

Astiz Gassó, M.M. (1992) Estudios sobre especialización fisiológica de las caries del trigo. *VIII Jornadas Fitosanitarias Argentinas.* Paraná provincia de Entre Ríos, Argentina. Resúmenes, 8 pp.

Astiz Gassó, M.M. (1994) Specialization physiological forms in *Ustilago bullata* Berk. of *Bromus* spp. In: Fuentes-Dávilas, G. (ed.) *Proceedings del IXth Biennial Workshop on the Smut Fungi.* CIMMYT, El Batán D.F., México, pp. 74–80.

Astiz Gassó, M.M. (1997a) Comportamiento de cultivares y líneas de trigo a las caries (*Tilletia foetida*). *Revista de Fitopatología* ALF 33(3), 16–17.

Astiz Gassó, M.M. (1997b) Variabilidad patógena de poblaciones de *Tilletia foetida* en *Triticum* spp. en Argentina. *Revista de Fitopatología* ALF 33(3), 18. Abstract.

Astiz Gassó, M.M. and Aulicino, M.B. (1999) Selección para resistencia al carbón de la panoja (*Ustillago bullata* Berk) líneas y poblaciones de *Bromus catharticus* Vhal de la provincia de Buenos Aires. *Actas 29º Congreso Argentino de Genética. III Jornadas Chileno-Argentino de Genética.* Resúmenes, 362 pp.

Astiz Gassó, M.M. and Hirschhorn, E. (1994) Physiologic specialization of *Tilletia foetida* Wallr (common bunt) in *Triticum* spp. in Argentina. In: Fuentes-Dávilas, G. (ed.) *Proceedings del IXth Biennial Workshop on the Smut Fungi.* CIMMYT, El Batán D.F., México, pp. 90–97.

Astiz Gassó, M.M. and Molina, M.C. (1996) Estudios preliminares para determinar el grado de resistencia a *Ustilago maydis* DC. Corda en especies cultivadas y silvestres del Gro *Zea. Proceedings Xth Biennial Workshop on the Smut Fungi.* University of Calgary, Calgary, Alberta, Canada, pp. 57–62.

Astiz Gassó, M.M. and Molina, M.C. (1999) *Zea* seedling reaction to inoculation with *Ustilago maydis* (DC) Corda. *Maize Genetics Cooperation Newsleter* 73, 58–60.

Banuett, F. (1995) Genetics of *Ustilago maydis*, a fungal pathogen that induces tumors in maize. *Annual Review of Genetics* 29, 179–208.

Boguena, T., Meyer, S.E. and Nelson, D. (2007) Low temperature during infection limits *Ustilago bullata* (*Ustilaginaceae, Ustilaginales*) disease incidence on *Bromus tectorum* (*Poaceae, Cyperales*). *Biocontrol Science and Technology* 17, 33–52.

Callow, J.A. and Ling, I.T. (1973) Histology of neoplasms and chlorotic lesions in maize seedlings following the infection of sporidia of *Ustilago maydis* (DC) Corda. *Physiological Plant Pathology* 3, 489–494.

Duran, R. (1987) *Ustilaginales of México. Taxonomy, Symptomatology, Spore Germination and Basidial Cytology.* Washington State University, Pullman, Washington, 331 pp.

Edmunds, L.K. (1998) Use of sporidial hypodermic infection to test sorghum for head smut resistance. *Plant Disease Report* 47, 903–913.

Falloon, R.E. (1976) Effect of infection by *Ustilago bullata* on vegetative growth of *Bromus catharticus. New Zealand Journal of Agricultural Research* 19, 249–254.

Falloon, R.E. (1979a) Description and illustration of *Ustilago bullata* growing in culture. *Transactions of the British Mycological Society* 73, 223–227.

Falloon, R.E. (1979b) Further studies on the effects of infection by *Ustilago bullata* on vegetative growth of *Bromus catharticus. New Zealand Journal of Agricultural Research* 22, 621–626.

Falloon, R.E. and Hume, D.E. (1988) Productivity and persistance of prairie grass *(Bromus willdenowii* Kunth) 1. Effects of the head smut fungus *Ustilago bullata* Berk. *Grass and Forage Science* 43, 179–184.

Fischer, G.W. and Holton, C.S. (1957) *Biology and Control of the Smut Fungi.* Ronald Press, New York, 622 pp.

Gaudet, D.A. (1990) Culm height and susceptibility of winter and spring wheat cultivars to common bunt (*Tilletia tritici* and *T. laevis*). *Proceedings of the Seventh Biennial Workshop on the Smut Fungi.* University of Maryland, Frederick, Maryland.

Gaudet, D.A., Puchalski, B.L. and Kozub, G.C. (1994) Reaction of CIMMYT and Candian red spring wheat cultivars to common bunt (*Tilletia tritici* and *T. laevis*). In: Fuentes-Dávilas, G. (ed.) *Proceedings del IXth Biennial Workshop on the Smut Fungi.* CIMMYT, El Batán, México, pp. 59–60.

Hirschhorn. E. (1977) Novedades sobre el carbón que ataca *Bromus* spp. en Argentina. *Boletín de la Sociedad Argentina de Botánica* 18, 56–64.

Hirschhorn, E. (1986) *Las Ustilaginales de la Flora Argentina.* Edit Comisión de Investigaciones Científica de la provincia de Buenos Aires. Publicación Especial. CIC, 530 pp.

Johnsson, L. (1991) Climate factors influencing attack of common bunt (*Tilletia caries* (D.C.) Tul) in winter wheat in 1940–1988 in Sweden. *Journal of Plant Diseases and Protection* 99(1), 21–28.

Kendirck, E.L. (1961) Race groups of *Tilletia caries* and *Tilletia foetida* for varietal resistance testing. *Phytopathology* 51, 537–540.

Kreizinger, E.J., Fischer, G.W. and Law, A.G. (1947) Reactions of mountain brome and Canada wild-rye strains to head smut (*Ustilago bullata*). *Journal of Agricultural Research* 75, 105–111.

Meinrs, J.P. and Fischer, G.W. (1953) Further studies of host specialization in the head smut of grasses, *Ustilago bullata. Phytopathogy* 43, 200–203.

Metzger, R.J. and Hoffmann, J.A. (1978) New races of common bunt useful to determine resistance of wheat to dwarf bunt. *Crop Science* 18, 49–51.

Snetselaar, K.M. and Mims, C.W. (1992) Sporidial fusion and infection of maize seedlings by the smut fungus *Ustilago maydis. Mycologia* 84, 193–203.

Snetselaar, K.M. and Mims, C.W. (1993) Infection of maize stigmas by *Ustilago maydis*: light and electron microscopy. *Phytopathology* 83, 843–850.

Statistix for Windows (2008) Analytical Software,Tallahassee, Florida.

Toit, L.J. du and Pataky, J.K. (1999) Variation associated with channel inoculation for common smut of sweet corn. *Plant Disease* 83, 727–732.

Wilcoxson, R.D. and Saari, E.E. (eds) (1996) *Bunt and Smut Diseases of Wheat. Concepts and Methods of Disease Management.* CIMMYT, México, 66 pp.

Part IV

Endophytes in Plant Disease Control

12 Status and Progress of Research in Endophytes from Agricultural Crops in Argentina

Silvina Larrán and Cecilia Mónaco
Centro de Investigaciones de Fitopatología (CIDEFI), Facultad de Ciencias Agrarias y Forestales, Universidad Nacional de La Plata, La Plata, Argentina

Abstract

Plants harbour a heterogeneous population of endogenous microorganisms, comprising both pathogens and non-pathogens, including fungi, bacteria, actinomycetes, etc. Their association has substantial impact on plant health and fitness. Endophytes reside inside healthy plant tissues without producing any disease symptoms. They are helpful in modifying biochemicals produced by plants and may add to their protection from insect herbivores, fungal pathogens and even grazing by animals. However, the ecological role of these endophytes is not yet fully understood. This chapter reports on endophytic fungi present in beet and tomato leaves. Isolation and analysis of endophytic microorganisms of soybean and wheat are also described. It is advocated that endophytes may have a definite role in the biological control of *Drechslera tritici-repentis*, responsible for tan spot disease in wheat.

Introduction

Before beginning, the term 'endophyte' must be defined. Literally, an endophyte is an organism which lives inside a plant, 'endo' meaning within and 'phyte' is derived from the Greek word 'phyton', meaning plant. There are several definitions of endophytes, such as 'endophyte' is an all-encompassing topographical term that includes all organisms that are living in plant tissues during a more or less long period of their life, colonizing symptomlessly the living internal tissues of their hosts (Petrini, 1991). Such infections are termed 'endophytic', particularly when the association is believed to be mutualistic or at least non-pathogenic, or 'latent infections', where a latent pathogen is involved (Cabral *et al.*, 1993). Therefore, the term endophyte has been used lately in a broad sense to include any fungi isolated from symptomless plant tissues, but the concepts of endophytic colonization and latent infection by fungi are clearly different. Endophytic colonization or infection cannot be considered as causing disease, since a plant disease is an interaction between the host, parasite, vector and the environment over time, which results in the production of disease signs and/or symptoms. Endophytic fungi may be described as mutualistic (Clay, 1991). Latent infecting fungi are parasitic but cannot be considered mutualistic. Latent infection is the state in which a host is infected with a pathogen, but does not show symptoms and persists until signs or symptoms are prompted to appear by environmental or nutritional conditions or by the state of

 Management of Fungal Plant Pathogens (eds A. Arya and A.E. Perelló)

maturity of the host or pathogen (Sinclair and Cerkauskas, 1996).

Petrini's definition of endophytes (1991) encompasses not only mutualistic and neutral symbionts, but also those pathogens known to live latently within their hosts. Therefore, Wilson (1995) has expanded the 'endophyte' definition to include internal bacteria that live inside plant tissues without causing disease. A wide range of bacterial genera has been isolated from healthy plant species of agricultural and horticultural crops (Chanway, 1996, 1998; Sturz *et al.*, 1998). Endophyte associations may range from intimate contact where the fungus inhabits the intercellular spaces and xylem vessels in the plant, to more or less superficial colonization of peripheral, often dying or dead tissues (Petrini, 1996). They may colonize single cells (Stone *et al.*, 1994) or tissues (Schulz *et al.*, 1999).

The endophytes of aerial parts of plants could be assembled in two different groups: the fungal endophytes of grasses and non-grass endophytes (fungi and bacteria). Grass endophytes are a particular type of systemic symbiosis, these are fungi of the family *Clavicipitaceae*, which grow between host cells in vegetative tissues, ovules and seeds and are seed transmitted vertically (Stone and Petrini, 1997). Most studies of endophytes have dealt with grasses due to their economic importance to livestock (Clay, 1988, 1991). The close association of an endophyte (*Neotyphodium coenophialum* (Morgan-Jones & Gams) Glenn, Bacon & Hanlin = *Acremonium coenophialum*) and tall fescue (*Festuca arundinacea* L.) has been widely studied. As a result of the association between host and fungus, alkaloids are produced. These are responsible for fescue toxicosis in livestock (Bacon *et al.*, 1977).

Since the initial work of Bacon *et al.* (1977), numerous researchers have come to understand further the relationship between fungal endophytes of grasses and animal toxicosis. On the other hand, it has been well documented that grass endophytes provide their host with a number of benefits that increase host fitness. The most intensely studied symbioses are tall fescue with *N. coenophialum* and perennial ryegrass with *N. lolii* (Latch, Samuels & Christensen) Glenn, Bacon & Hanlin. In tall fescue, *N. cocnophialum* causes enhanced tillering and root growth, increases drought tolerance (Arechavaleta *et al.*, 1989) and protects against certain nematodes (Kimmons *et al.*, 1990), fungal pathogens (Gwinn and Gavin, 1992) and insect herbivores (Rowan and Latch, 1994). The protective nature of endophytes is due to the presence of alkaloids, whereas these alkaloids are responsible for poisoning domestic animals. Ergovaline is associated with various maladies often observed among cattle that graze *N. coenophialum*-infected tall fescue and collectively called 'tall fescue toxicosis'. Likewise, lolitrem B is associated with the malady 'ryegrass staggers', most commonly observed in sheep grazing perennial rye grass in New Zealand (Schardl and Phillips, 1997).

On the other hand, symptomless endophytes of plants other than grasses have been known for more than 80 years (Lewis, 1924; Carroll and Carroll, 1978; Fisher *et al.*, 1992; Menendez *et al.*, 1995; Faeth and Hammon, 1997; Gasoni and Stegman, 1997; Fröhlich *et al.*, 2000; Larran *et al.*, 2007). Endophytes are found in all plants and are extremely abundant and very diverse. Endophytes of a non-grass host represent a broad range of genera. Taxonomically, the endophytic fungi recovered from plants belong mainly to the phylum *Ascomycota* and *Basidiomycota* (fungi) and some *Oomycetes* (phylum *Oomycota*, *Chromista*) have been isolated as endophytes (Sinclair and Cerkauskas, 1996), along with members of phylum *Ascomycota* and their conidial form or anamorphic form lacking a sexual state.

The strategy of endophytes is commonly characterized by early occupation of living host tissue, ensuring possession of the nutritional resource (Dingle and McGee, 2003). Host colonization by these fungi is frequently localized in foliage, roots, stems and bark and they are transmitted horizontally via spores. Frequently, colonization is more often non-systemic. These endophytic infections are often presumed to form mutualistic association with their hosts in a manner similar to the endophytes in grasses (Stone and Petrini, 1997). The plant tissues act as host for complex fungal communities. In the past few years, several works have provided evidence for the development of a highly specific endophytic assemblage for a given host (Bertoni and Cabral, 1988; Petrini and Fisher, 1988; Sieber *et al.*, 1988, 1991; McInroy and Kloepper, 1991; Pereira *et al.*, 1999; Larran *et al.*, 2000, 2001, 2002a,b). Organ specificity, probably the result of adaptation by some endophytes to the particular microecological and physiological conditions present in a given organ, has been demonstrated in several studies (Fisher *et al.*, 1991; Petrini *et al.*, 1992). Whereas a large number of species can be isolated from a given host, in general, only a few species are present in significant amounts (Petrini *et al.*, 1992). The ecological roles of endophytes are not yet clarified in all associations. Only the interaction of *Neotyphodium*/grass has been studied in depth, but less is known about other endophytic associations (Clay, 1990).

The endophytes may provide a rapidly evolving defence mechanism against herbivory (Carroll, 1988, 1991; Findlay *et al.*, 1995) and many are potential producers of secondary metabolites and enzymes that will probably find diverse applications in the most diverse fields of biology (Petrini *et al.*, 1992; Schulz *et al.*, 1995; Istifadah and McGee, 2006; Istifadah *et al.*, 2006). Several studies have demonstrated auxin and cytokinin production (Pugh, 1972; Bacon and De Battista, 1991) and antibiotic compounds (Clark *et al.*, 1989; Brunner and Petrini, 1992). Competition for infection site, their capacity to produce secondary metabolites and their potential to stimulate defence reactions may contribute to antagonism by the endophytes against pathogens living in the same tissues (Dingle and McGee, 2003; Istifadah and McGee, 2006).

Also, several authors have proposed that endophytes could be used as vectors of genes to be introduced artificially in the population of the host, due to natural genomes showing useful characteristics and attributes that could be selected. For example, endophytes used as vectors of genetic information could also be of particular interest for the development of mycoherbicides (Petrini *et al.*, 1992). The knowledge of endophyte distribution, biodiversity and biochemical characteristics could be important in improving plant fitness. Moreover, they could play an important role in the interactions present in an ecological agriculture.

In the past few years, research on endophytes has been carried out at the CIDEFI Research Centre in the city of La Plata, Buenos Aires, Argentina. It is thought that endophytes could be used as biocontrol agents. In Argentina nowadays, biological control is an attractive option for the management of some plant diseases. A considerable amount of knowledge on endophytes has been accumulated. Preliminary studies have focused mainly on determining the biodiversity of endophytes on economically important plants. Likewise, species composition from different organs has been investigated. Finally, research will be undertaken to test the antagonistic interactions between endophytes and plant pathogens. Significant research is summarized in this chapter.

Endophytic Fungi in Beet (*Beta vulgaris* var. *esculenta* L.) Leaves

The aim of this investigation was in order to document the species composition of endophytic fungi of healthy cultivated beet leaves; to determine their infection frequencies and to verify possible qualitative and quantitative changes of species isolated during the growing season (Larran *et al.*, 2000). Samples were collected from healthy beet leaves of plants cultivated in the experimental field of the Facultad de Ciencias Agrarias y Forestales, Universidad Nacional de La Plata (UNLP), Buenos Aires, Argentina. The plants were sampled three times during the growing season. Leaves were cut, surface-sterilized and then leaf disks were incubated on 2% potato dextrose agar (PDA) for 8 days. Nested ANOVA and Tukey tests were applied to evaluate the differences in infection frequencies for different fungi. Data were transformed according to $y = \arcsin R^2\ (P/100)$. Microscopic examinations were made from leaf disks previously surface-sterilized and then incubated in a humid chamber for 48 h. The disks were cleared and stained. Hyphae were the principal fungal structures observed (Fig. 12.1). They could be observed emerging through the stomata or growing intercellularly under the cuticle and could be followed between the layers of cells. No visible disruption or impairment of the plant cells by the fungi was noted. The endophytes isolated from healthy beet leaves are shown in Table 12.1.

Fungi colonized 100% of the leaves sampled. Twelve taxa of endophytic fungi were isolated and identified. Yeast, *Alternaria alternata*, *Pleospora herbarum*, *Stemphylium* sp. and *Epicoccum nigrum* were the most frequently isolated fungi. The frequency of *A. alternata* and *P. herbarum* increased significantly in time, whereas yeast decreased along the growth stages. There were no relevant quantitative changes in the frequency of colonization by other species. The diversity of isolated fungi species decreased from the first to the last sampling.

Endophytic Fungi in the Leaves of *Lycopersicon esculentum* Mill.

We have selected tomato plants for this investigation because both greenhouse and field production in La Plata horticultural area are economically important (Larran *et al.*, 2001).

Tomato production is used mainly for fresh consumption, as well as being a source of many value-added products. The investigation reports the endophyte frequencies from healthy tomato leaves (cultivar Tommy) cultivated in the field of the Facultad de Ciencias Agrarias y Forestales, UNLP, Buenos Aires, Argentina. Samples were collected for 2 years to determine possible qualitative and quantitative changes of species. Data were analysed by ANOVA for factorial experiments. Differences between means were separated with Tukey's test ($P \leq 0.05$). Likewise, different surface-sterilized techniques were evaluated previously and the technique selected was used. The diversity of isolated endophytes is shown in Table 12.2.

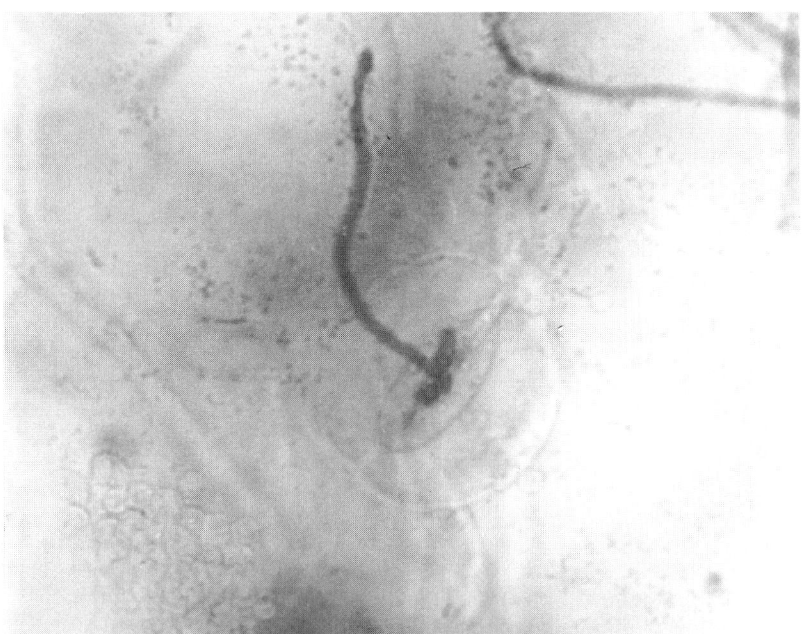

Fig. 12.1. Hyphae emerging from stomata.

Table 12.1. Mean density of colonization (%) of endophytic fungi from beet leaves at three different time intervals during the growing season.

	Sampling dates		
Endophytes	1	2	3
Alternaria alternata (Fr.) Keissler	12.5[a]**	23.0**	31.0**
Chaetomium sp.	1.0	0	0
Cladosporium spp.	1.0	1.0	1.0
Colletotrichum dematium (Pers.) Grove	1.0	0	0
Epicoccum nigrum Link.	5.3	3.0	3.0
Glomerella cingulata (Stonem.) Spaulding & Schrenk	2.0	1.0	0
Penicillium spp.	1.0	4.0	1.0
Phoma betae Fr.	0	1.0	0
Phomopsis sp.	1.0	0	0
Pleospora herbarum (Pers. ex Fr.) Rabenh.	7.3	9.0	11.0
Stemphylium sp.	6.1	9.0	8.0
Yeast	18.0	10.0	7.0
Sterile mycelia	0	1	0
Total number of endophytes	11	10	7
Total segments sampled: 300			

Note: [a]Mean of ten replications. Numbers followed by ** differ statistically according Tukey's test ($P \leq 0.05$).

Table 12.2. Mean frequencies (%) of endophytic fungi isolated from tomato leaves in 1998 and 1999.

	Mean frequencies (%)	
Endophytes	1998	1999
Alternaria alternata (Fr.) Keissler	8.75	25.8*
Arthrinium sp.	3.78	0
Bipolaris cynodontis (Marig.) Shoem.	0	1.44*
Chaetomiun globosum Kunze ex Fries	2.50	0
Cladosporium sp.	3.75	5.48
Colletotrichum coccodes (Wallr.) Hughes	2.50	0
C. gloeosporioides (Penz.) Sacc.	13.75*	0
Epicoccum nigrum Link.	0	1.59*
Cryptococcus sp.	0	1.87*
Nigrospora sphaerica (Sacc.) Mason	2.50	0
Penicillium spp.	2.50	2.55
Phomopsis sp.	3.75	0
Ulocladium alternariae (Cooke) Simmons	2.50	0
Stemphylium botryosum (Pers.ex Fr.) Rabenh.	1.25	0
Rhodotorula sp.	0	2.25*

Note: Means followed by * differ significantly according to Turkey's test ($P \leq 0.05$).
Total segments sampled at each growth stage: 75.

Different endophytic species were isolated in 1998 and 1999, although some of them were isolated in both years. This could be due to the different climatic conditions registered, as several authors observed that various climatic conditions – site moisture, rainfall and wind exposure – yielded distinct endophyte assemblages (Chapela,

1989; Petrini *et al.*, 1992). *Alternaria alternata* was the fungus isolated most frequently from tomato leaves in 1999, but it was the second most common species in 1998. In contrast, *C. gloeosporioides* was the fungus isolated most frequently in 1998, but it was not found in 1999. Species of other genera, such as *Cladosporium* and *Penicillium*, were isolated in both years. These two genera have been described as endophytes from other plants as well (Fisher *et al.*, 1992; Cabral *et al.*, 1993).

Endophytic Fungi in Healthy Soybean Leaves

Soybean (*Glycine max* (L.) Merr.) in Argentina is one of the most important crops, not only by its production but also because of the volume exported, and it is planted on about 16.5 m ha. A study (Larran *et al.*, 2002b) was undertaken to document the diversity of endophytic fungi of healthy cultivated soybean leaves and their infection frequency and to verify possible qualitative and quantitative changes of species isolated at two growth stages: R_2–R_3 and R_4–R_5 (according to Fehr *et al.*, 1971). Fifty asymptomatic plants were randomly sampled at each growth stage from a segregating population (F_3 generation) cultivated at the experimental field of the Facultad de Ciencias Agrarias y Forestales, UNLP, Buenos Aires, Argentina. Samples were surface-sterilized and incubated over 9 days. The student *t*-test and percentage differences test were used to evaluate differences in infection frequencies for various fungi. The results are shown in Table 12.3. Twelve genera of endophytic fungi were isolated and identified from healthy soybean leaves. In general, in both growth stages, the same species were isolated and most of them did not show significant differences in their infection frequencies, except for *Phomopsis* sp., *P. longicolla* and *Cladosporium* sp. The endophytic fungi isolated more frequently from healthy leaves of soybean were

Table 12.3. Mean percentage frequencies of endophytic fungi and their variations from soybean leaves at R_2–R_3 and R_4–R_5 stages (total segments sampled: 591).

	Frequencies (%)			
Endophytes	R_2–R_3 stage[a]	R_4–R_5 stage	Variation (%)	
Alternaria alternata (Fr.) Keissler	78.48[b]	68.79	−12.34	NS
A. tenuissima (Kunze ex Pers.) Wiltshire	0	1.60	–	NS
Bipolaris sorokiniana (Sacc.) Shoem.	0.94	0	−100.00	NS
Cladosporium sp.	0	2.06	–	*
Colletotrichum sp.	1.28	0	−100.00	NS
Curvularia lunata (Wakker) Boedijni	0	0.40	–	NS
Epicoccum nigrum Link.	1.23	1.93	+56.90	NS
Glomerella cingulata (Stoneman) Spauld. & Schrenk	17.20	14.04	−18.40	NS
G. glycines Lehm. & Wolf	0.51	0.40	−21.60	NS
Nigrospora sphaerica (Sacc.) Mason	1.33	2.00	+50.40	NS
Penicillium sp.	0	1.86	–	NS
Phomopsis longicolla Hobbs	1.99	0	−100.00	*
P. sojae Lehman	3.48	3.86	+10.90	NS
Phomopsis sp.	2.89	5.50	+90.30	**
Pleospora herbarum (Pers. ex Fr.) Rabenh.	0.66	0.99	+50.00	NS
Stemphylium sp.	3.32	2.99	−9.90	NS

Note: [a]Based on the plant stages designated by Fehr *et al.* (1971). [b]The infection frequency was calculated as the number of subsamples infected by a given fungus divided by the total number of subsamples incubated. *Significant difference ($P < 0.05$); **highly significant difference ($P < 0.01$); NS, no significant difference.

A. alternata and *G. cingulata*. Most of the fungi isolated in this work are cited as soybean pathogens in different parts of the world (Farr *et al.*, 1989). Because it is known that most fungal pathogens of soybean have an asymptomatic or latent period after infection or colonization, these fungi could be either avirulent or hypovirulent, or virulent but in a latent phase. Pathogenicity tests would be needed to investigate this hypothesis. Soybean leaves are hosts to an abundance of endophytic fungi, but only *A. alternata* is the dominant species. Further studies will be carried out to evaluate the potential use of endophytes from soybean leaves in biological control.

Isolation and Analysis of Endophytic Microorganisms in Wheat Leaves

The presence of endophytic fungi in healthy wheat crops has been demonstrated previously in other countries of the world. The present investigation was undertaken in order to document the spectrum of endophytes of healthy leaves from three wheat cultivars and to determine their infection frequencies at three growth stages in Argentina (Larran *et al.*, 2002a). Wheat cultivars, Buck Ombú, Klein Centauro and Klein Dragón, were grown in the experimental field of the Facultad de Ciencias Agrarias y Forestales, UNLP, Buenos Aires, Argentina. Ten asymptomatic plants of each cultivar were randomly sampled at three defined growth stages: second node detectable, medium milk and soft dough stages (32, 75 and 85, according to Zadoks *et al.*, 1974). Samples were surface-sterilized and incubated on 2% PDA and, after 9 days, identifications were made. Data were analysed by ANOVA for factorial experiments. Differences between means were separated by LSD ($P \leq 0.05$).

From the 450 wheat leaf segments incubated, 3 bacterial isolates and 130 fungal isolates were obtained (Table 12.4). From all the isolates, 19 fungal species were identified. There were significant differences between microorganisms, stages of growth and stages × microorganism interactions. Differences between cultivars, stages × cultivars, microorganisms × cultivars and the triple interaction were not significant. The frequency of the microorganisms isolated increased with crop age, but it was statistically similar for the three wheat cultivars tested. *Rhodotorula rubra*, *A. alternata*, *C. herbarum* and *E. nigrum* were isolated in the highest frequency. The other microorganisms were present at intermediate or low values. Most fungal endophyte isolates from wheat leaves have been described as endophytes of wheat and others plants (Sieber *et al.*, 1988; Petrini *et al.*, 1992; Gindrat and Pezet, 1994).

A variation in the number of taxa isolated was recorded along the growing season of wheat. A change in species composition from the three growth stages was observed; however, no differences were noted between cultivars. Further studies were needed to analyse endophyte composition and variation from other organs and cultivars. Therefore, the following study was undertaken.

Endophytic Fungi from Wheat (*Triticum aestivum* L.)

In this work, five wheat cultivars (Buck Poncho, B. pronto, Klein Cobre, K. Dragón and Pro INTA Federal) were grown in the experimental field of the Facultad de Ciencias Agrarias y Forestales, UNLP, Buenos Aires, Argentina. The purpose of this investigation was to document the diversity of endophytes from different cultivars and to determine their infection frequencies from different plant organ (leaves, stems, glumes and grains) (Larran *et al.*, 2007). Samples were collected at five growth stages from crop emergence to harvest (GS 2, GS 8, GS 10.5, GS 11.1 and GS 11.4) (Large, 1954), with the aim of verifying possible qualitative and quantitative changes of the species isolated. Pieces of tissues were surface-sterilized and incubated on 2% PDA over 9 days. An ANOVA including organs, microorganisms, cultivars and growth stages as a source of variation was carried out but, due to differences between organs, an ANOVA was performed considering each organ separately. Differences between means were separated by LSD

Table 12.4. Frequencies of endophytes isolated from wheat leaves of three cultivars at three growth stages.

Endophytes	Samplings			
	Gs. 35*	Gs. 75	Gs. 85	Average
Alternaria alternata (Fr.) Keissler	0 a	0.67 ab	14.0 d	4.89 de
Alternaria sp. I	0 a	0 a	0.67 ab	0.22 a
Alternaria sp. II	0 a	0 a	2.0 abc	0.67 a
Arthrinium sp.	1.33 a	0 a	0 a	0.44 a
Aspergillus sp.	0 a	0 a	0.67 ab	0.22 a
Bipolaris sp.	0 a	2.67 abc	2.0 abcd	1.56 ab
B. cynodontis (Marig.) Shoem.	0 a	3.33 bc	2.0 abcd	1.78 abc
B. sorokiniana (Sacc.) Shoem.	0 a	0.67 ab	0.67 ab	0.44 a
Chaetomium globosum Kunze ex Fries	0 a	0 a	1.33 abc	0.44 a
Cladosporium herbarum (Pers.: Fr.) Link.	0 a	3.33 bc	7.33 f	3.56 cd
Cryptococcus sp.	0 a	4.0 c	1.33 abc	1.78 abc
Epicoccum nigrum Link.	0 a	5.33 c	4.67 de	3.33 bcd
Fusarium sp.	0 a	2 abc	0 a	0.67 a
Penicillium sp.	0 a	0 a	0.67 ab	0.22 a
Phoma sp.	0.67 a	0 a	0 a	1.33 a
Phomopsis sp.	0.67 a	0 a	0 a	0.22 a
Pleospora herbarum (Fr.) Raben.	0 a	0 a	4 cde	0.22 a
Rhodotorula rubra Harrison	0 a	9.33 d	6.67 ef	5.33 e
Stemphylium sp.	0 a	0.67 ab	3.33 bcd	1.33 a
SM I	0 a	0.67 ab	0 a	0.44 a
SM II	0 a	0 a	1.33 abc	0.22 a
Bacillus sp.	0 a	0.67 ab	1.33 abc	0.67 a
Average of growth stages	0.12 a[a]	1.52 b	2.45 c	

Gs. 85: soft dough stage. *Growth stages according to Zadoks *et al.* (1974). Data are the mean of 150 leaf pieces (5 pieces × 10 replications × 3 cultivars)/growth stage. Means followed by same letter in the same column are not statistically different according to LSD ($P \leq 0.05$). [a]For the average of growth stages means followed by the same letter in the same row are not statistically different ($P \leq 0.05$). Gs.35: second node detectable. Gs.75: medium milk.

($P \leq 0.05$). A total of 1750 plant segments were processed from wheat tissues and 33 microbes were recovered. Three bacteria, 27 fungal taxa and 3 non-sporulating mycelia, assigned as 'sterile mycelia', were registered (Tables 12.5 and 12.6). *A. alternata*, *C. herbarum*, *E. nigrum*, *Cryptococcus* sp., *R. rubra*, *Penicillium* sp. and *Fusarium graminearum* were the fungi that showed the highest colonization frequency in all the tissues and organs analysed. As is shown, the bacterial isolates (*Serratia* sp., *Bacillus* sp. and unidentified yellow bacteria) were registered with high frequencies. The results of this statistical analysis showed that organs, microorganisms and interaction of organs × microorganisms were significant.

On the other hand, as results of ANOVA from each organ, we obtained that the number of taxa isolated was greater in the leaves than in the other organs analysed. Respectively, 25, 17, 12 and 15 were the number of taxa recovered from leaves, stems, glumes and grains. Few species were dominant from grains, whereas they had the highest percentages of isolates from the total samples analysed.

Likewise, a variation occurs in the species composition of endophytes isolated from different organs and growth stages. No significant differences between cultivars were obtained, except when the glumes were analysed. Whereas *Bacillus* sp. was isolated from stems and grains, *Serratia* sp. and yellow bacteria were recovered from all organs analysed.

Although most of the microorganisms followed a similar pattern in the four organs,

Table 12.5. Frequency (means) of microorganisms isolated from leaves, stems, glumes and grains on five wheat cultivars.

Endophytes	Means (all organs) and growth stages
Alternaria alternata (Fr.) Keissler	8.48 e*
A. infectoria species group	0.56 a
Arthrinium sp.	0.58 ab
Bacillus sp.	1.26 ab
Bipolaris sorokiniana (Sacc.) Shoem.	0.73 ab
B. spicifera (Bainier) Subramanian	0.00 a
Bipolaris sp.	0.26 a
Candida albicans (C.P. Robin) Berkhout	0.04 a
Cephalosporium sp.	0.06 a
Chaetomium globosum Kunze ex Fries	0.19 a
Cladosporium herbarum (Pers.:Fr.) Link.	6.55 d
Cryptococcus sp.	2.14 b
Cochliobolus spicifer Nelson	0.14 a
Curvularia lunata (Wakker) Boedijni	0.01 a
Epicoccum nigrum Link.	4.38 c
Fusarium oxysporum Schlechtend.: Fr.	0.53 a
F. graminearum Schwabe	1.01 ab
Helicocephalum sp.	0.00 a
Nigrospora sp.	0.04 a
Penicillium sp.	1.16 ab
Phoma sp.	0.00 a
Pleospora herbarum (Fr.) Raben.	0.00 a
Rhodotorula rubra Harrison	1.27 ab
Septoria tritici Roberge in Desmaz.	0.00 a
Serratia sp.	8.95 e
Stachybotrys sp.	0.00 a
Stemphylium botryosum Wallr.	0.09 a
Trichoderma hamatum (Bonord.) Bainier	0.17 a
Ulocladium sp.	0.04 a
SM 1	0.00 a
SM 2	0.00 a
SM 3	0.38 a
Yellow bacteria	4.33 c
Organs	
Leaves	0.94 a
Stems	1.27 a
Glumes	0.98 a
Grains	2.03 b
Cultivars	
Klein Dragon	1.54 a
Buck Pronto	1.37 a
Klein Cobre	1.33 a
Buck Poncho	1.39 a
Pro INTA Federal	0.91 a

Note: *Means followed by the same letter in the same column within the same treatment are not statistically different according LSD ($P \leq 0.05$). SM, sterile mycelia.

Table 12.6. Means of the frequencies of microorganisms, cultivars and growth stages for each organ (leaves, stems, glumes and grains) of five wheat cultivars.

Endophytes	Leaves	Stems	Glumes	Grains
Alternaria alternata (Fr.) Keissler	4.8 e*	2.4 efg	9.33 e	17.6 d
A. infectoria species-group	1.4 abc	0.0 a	0.26 ab	0.8 a
Arthrinium sp.	1.2 abc	0.16 ab	0.00 a	1.2 a
Bacillus sp.	0.0 a	3.68 gh	0.00 a	1.6 a
Bipolaris sorokiniana (Sacc.) Shoem.	1.0 abc	0.0 a	0.53 abc	1.6 a
B. spicifera (Bainier) Subramanian	0.2 a	0.0 a	0.00 a	0.0 a
Bipolaris sp.	0.4 a	0.48 abc	0.00 a	0.4 a
Candida albicans (C.P. Robin) Berkhout	0.0 a	0.0 a	0.00 a	0.4 a
Cephalosporium sp.	0.0 a	0.48 abc	0.00 a	0.0 a
Chaetomium globosum Kunze ex Fries	0.2 a	0.0 a	0.00 a	0.8 a
Cladosporium herbarum (Pers.: Fr.) Link.	1.4 abc	1.28 abcde	2.13 cd	21.6 e
Cryptococcus sp.	2.4 cd	4.8 h	1.60 abc	0.0 a
Cochliobolus spicifer Nelson	0.0 a	0.0 a	0.00 a	0.4 a
Curvularia lunata (Wakker) Boed. Boedijni	0.4 a	0.0 a	0.26 ab	0.0 a
Epicoccum nigrum Link.	3.0 d	2.88 fg	1.86 bc	10.8 c
Fusarium oxysporum Schlechtend.: Fr.	2.2 cd	0.16 ab	0.00 a	0.0 a
F. graminearum Schwabe	1.0 abc	2.88 fg	0.00 a	0.8 a
Helicocephalum sp.	0.0 a	0.16 ab	0.00 a	0.0 a
Nigrospora sp.	0.4 a	0.0 a	0.00 a	0.0 a
Penicillium sp.	2.0 bcd	2.08 def	0.80 abc	0.0 a
Phoma sp.	0.2 a	0.0 a	0.00 a	0.0 a
Pleospora herbarum (Fr.) Raben.	0.2 a	0.0 a	0.00 a	0.0 a
Rhodotorula rubra Harrison	2.4 cd	3.04 fg	0.26 ab	0.0 a
Septoria tritici Roberge in Desmaz.	0.2 a	0.0 a	0.00 a	0.0 a
Serratia sp.	3.3 d	13.6 i	12.53 f	6.8 b
Stachybotrys sp.	0.2 a	0.0 a	0.00 a	0.0 a
Stemphylium botryosum Wallr.	0.2 a	0.0 a	0.00 a	0.4 a
Trichoderma hamatum (Bonord.) Bainier	0.6 ab	0.64 abcd	0.26 ab	0.0 a
Ulocladium sp.	0.0 a	0.0 a	0.00 a	0.4 a
SM 1	0.2 a	0.0 a	0.00 a	0.0 a
SM 2	0.0 a	0.0 a	0.00 a	0.0 a
SM 3	0.0 a	1.76 cdef	0.00 a	0.0 a
Yellow bacteria	3.0 d	1.6 bcdef	3.73 d	9.6 bc
Cultivars				
Klein Dragon	0.90 a	1.21 a	1.53 c	2.73 a
Buck Pronto	1.09 a	1.01 a	0.89 ab	2.73 a
Klein Cobre	0.83 a	1.40 a	1.31 bc	2.18 a
Buck Poncho	1.12 a	1.26 a	1.05 abc	2.36 a
Pro INTA Federal	0.97 a	1.48 a	0.48 a	1.39 a
Growth stages				
2	0.73 a	0.85 a		
8	0.65 a	2.16 b		
10.5	0.80 a	1.02 a	0.24 a	
11.1	1.76 b	1.26 a	1.87 c	1.72 a
11.4		1.09 a	0.94 b	2.83 b

Note: *Means followed by the same letter in the same column within the same treatment are not statistically different according to LSD ($P \leq 0.05$). SM, sterile mycelia.

there were some, *A. alternata* for example, with higher values in grains and glumes than in leaves and stems. The spectrum of species isolated ranges from potential saprobes over taxa that probably are present as natural symbionts to known pathogens (Fisher *et al.*, 1992). Whereas *A. alternata*, *C. herbarum* and *E. nigrum* are species commonly abundant in the phylloplane and are considered primary saprobes and minor pathogens, others like *B. sorokiniana*, *C. lunata* and *F. graminearum* are economically important pathogens of wheat (Zillinsky, 1984).

Due to the fact that some of these endophytes adapted to a given organ may benefit the host against pathogens, further studies were undertaken.

A Biological Control Approach to Infection of *Drechslera tritici-repentis* in Wheat

The investigation was carried out to study the interactions between some endophytes isolated from healthy wheat plants and *Drechslera tritici-repentis* and to determine its possible significance in the biological control of tan spot (Larran *et al.*, unpublished). Endophytes isolated previously from wheat cultivars in Buenos Aires Province, Argentina, were selected for the assay. They were: *A. alternata*, *Bacillus* sp., *C. globosum*, *C. herbarum*, *E. nigrum*, *Penicillium* sp., *R. rubra*, *Trichoderma hamatum* and *P. lilacinus*. Mycelial and conidial morphological alterations and inhibition of colony growth of *D. tritici-repentis* were registered under *in vitro* conditions. Likewise, greenhouse experiments were also carried out. The results obtained from all tests have demonstrated that endophytes may have a role as biocontrol agents against *D. tritici-repentis*.

Conclusions

The study of endophytes began with the aim of studying their biodiversity and distribution from different hosts. We confirmed that endophytes were present in all the hosts evaluated. Then, we found that endophytes colonized distinct ecological niches and could suggest their organ specificity according to several authors (Sieber, 1988; Fisher *et al.*, 1991). On the other hand, in our studies, we have isolated a large number of species from healthy tissues of beet, tomato, soybean and wheat but only few species were dominant, in agreement with Petrini *et al.* (1992). Distinct endophyte assemblages were obtained from healthy tomato leaves in 1998 and 1999, which could be explained because of the different climatic conditions prevailing in both years.

Endophytes could be adapted to their hosts and be antagonists for their pathogens and, depending on their antagonistic capacity, they would be able to displace, reduce, suppress or induce resistance against them.

Nowadays, in accordance with the status of our investigation, we consider that further studies are needed to evaluate the possible use of endophytes as biocontrol agents against pathogens of agricultural crops. Intensive work is needed to understand the role of endophytes and, mainly, their possible use as agents of biocontrol. Likewise, it is very important to study the nature of plant–endophyte–pathogen interactions and the mechanism of antagonism (antibiosis, hyperparasitism, competition) with the aim of improving the efficiency of the biological control of pathogens.

References

Arechavaleta, M., Bacon, C.W., Hoveland, C.S. and Radeliffe, D.E. (1989) Effect of the tall fescue endophyte on plant response to environmental stress. *Agronomy Journal* 81, 83–90.

Bacon, C.W. and De Battista, J. (1991) Endophytic fungi of grasses. In: Arora, D.K., Rai, B., Mukerji, K.G. and Knudsen, G.R. (eds) *Handbook of Applied Mycology*. Volume I. Marcel Dekker, New York, pp. 231–256.

Bacon, C.W., Porter, J.K., Robbins, J.D. and Luttrell, E.S. (1977) *Epichloe typhina* from toxic tall fescue grasses. *Applied and Environmental Microbiology* 34, 576–581.
Bertoni, M.D. and Cabral, D. (1988) Phyllosphere of *Eucalyptus viminalis* II: distribution of endophytes. *Nova Hedwigia* 46, 491–502.
Brunner, F. and Petrini, O. (1992) Taxonomy of some *Xylaria* spp. and xylariaceous endophytes by isozyme electrophoresis. *Mycological Research* 96, 723–733.
Cabral, D., Stone, J.K. and Carroll, G.C. (1993) The internal mycobiota of *Juncus* spp.: microscopic and cultural observations of infection patterns. *Mycological Research* 97, 367–376.
Carroll, G.C. (1988) Fungal endophytes in stems and leaves: from latent pathogen to mutualistic symbiont. *Ecology* 69, 2–9.
Carroll, G.C. (1991) Fungal associates of woody plants as insect antagonists in leaves and stems. In: Barbosa, P., Krischick, V.A. and Jones, C.G. (eds) *Microbial Mediation of Plant–Herbivore Interactions.* Wiley, New York, pp. 253–272.
Carroll, G.C. and Carroll, F.E. (1978) Studies on the incidence of coniferous needle endophytes in the Pacific Northwest. *Canadian Journal of Botany* 56, 3034–3043.
Chanway, C.P. (1996) Endophytes: they are not just fungi. *Canadian Journal of Botany* 74, 321–322.
Chanway, C.P. (1998) Bacterial endophytes: ecological and practical implications. *Sydowia* 50, 149–170.
Chapela, I.H. (1989) Fungi in healthy stems and branches of American beech and aspen: a comparative study. *New Phytologist* 113, 65–75.
Clark, C.L., Miller, J.D. and Whitney, N.J. (1989) Toxicity of conifer needle endophytes to spruce budworm. *Mycological Research* 93, 508–512.
Clay, K. (1988) Fungal endophytes of grasses: a defensive mutualism between plants and fungi. *Ecology* 69, 10–16.
Clay, K. (1990) Fungal endophytes of grasses. *Annual Review of Ecology and Systematics* 21, 275–297.
Clay, K. (1991) Fungal endophytes, grasses and herbivores. In: Barbosa, P., Krischik, V.A. and Jones, C.G. (eds) *Microbial Mediation of Plant–Herbivore Interactions.* John Wiley & Sons, Inc., New York, pp. 199–252.
Dingle, J. and McGee, P.A. (2003) Some endophytic fungi reduce the density of pustules of *Puccinia recondita* f. sp. *tritici* in wheat. *Mycological Research* 107, 310–316.
Faeth, S.H. and Hammon, K.E. (1997) Fungal endophytes in oak trees: long-term patterns of abundance and association with leafminers. *Ecology* 78, 810–819.
Farr, D.F., Bills, G.F., Chamuris, G.P. and Rossman, A.Y. (1989) *Fungi on Plants and Plant Products in the United States.* APS Press, St Paul, Minnesota, pp. 209–216.
Fehr, W.R., Caviness, C.E., Burmood, D. and Penington, J.S. (1971) Stage of development descriptions for soybeans, *Glycine max* (L.) Merr. *Crop Science* 11, 929–931.
Findlay, J.A., Buthelezi, S., Lavoie, R. and Peña-Rodriguez, L. (1995) Bioactive isocoumarins and related metabolites from conifer endophytes. *Journal of Natural Products* 58, 1759–1766.
Fisher, P.J., Petrini, O. and Webster, J. (1991) Aquatic hyphomycetes and other fungi in living aquatic and terrestrial roots of *Alnus glutinosa. Mycological Research* 95, 543–547.
Fisher, P.J., Petrini, O. and Lappin Scott, H.M. (1992) The distribution of some fungal and bacterial endophytes in mays (*Zea mays* L.). *New Phytologist* 122, 299–305.
Fröhlich, J., Hyde, K.D. and Petrini, O. (2000) Endophytic fungi in palms. *Mycological Research* 104, 1202–1212.
Gasoni, L. and Stegman de Gurfinkel, B. (1997) The endophyte *Cladorrhinum foecundissimum* in cotton roots: phosphorus uptake and host growth. *Mycological Research* 101, 867–870.
Gindrat, D. and Pezet, R. (1994) Le paraquat, un outil pour la révélation rapide d´infections fongiques latentes et de champignons endophytes. *Journal of Phytopathology* 141, 86–98.
Gwinn, K.D. and Gavin, A.M. (1992) Relationship between endophyte infestation level of tall fescue seed lots and *Rhizoctonia zeae* seedling disease. *Plant Disease* 76, 911–914.
Istifadah, N. and McGee, P.A. (2006) Endophytic *Chaetomium globosum* reduces development of tan spot in wheat caused by *Pyrenophora tritici-repentis. Australasian Plant Pathology* 35, 411–418.
Istifadah, N., Saleeba, J.A. and McGee, P.A. (2006) Isolation of endophytic *Chaetomium* spp. inhibit the fungal pathogen *Pyrenophora tritici-repentis in vitro. Canadian Journal of Botany* 84, 1148–1155.
Kimmons, C.A., Gwinn, K.D. and Bernard, E.C. (1990) Nematode reproduction on endophyte-infected and endophyte-free tall fescue. *Plant Disease* 74, 757–761.
Larran, S., Mónaco, C. and Alippi, H.E. (2000) Endophytic fungi in beet (*Beta vulgaris* var. *esculenta* L.) leaves. *Advances in Horticultural Science* 14, 193–196.
Larran, S., Mónaco, C. and Alippi, H.E. (2001) Endophytic fungi in leaves of *Lycopersicon esculentum* Mill. *World Journal of Microbiology and Biotechnology* 17, 181–184.

Larran, S., Perelló, A., Simón, M.R. and Moreno, V. (2002a) Isolation and analysis of endophytic microorganisms in wheat (*Triticum aestivum* L.) leaves. *World Journal of Microbiology and Biotechnology* 18, 683–686.
Larran, S., Rollán, C., Bruno Angeles, H.J., Alippi, H.E. and Urrutia, M.I. (2002b) Endophytic fungi in healthy soybean leaves. *Investigación Agraria Producción y Protección Vegetales* 17, 173–178.
Larran, S., Perelló, A.E., Simón, M.R. and Moreno, V. (2007) The endophytic fungi from wheat (*Triticum aestivum* L.). *World Journal of Microbiology and Biotechnology* 4, 565–572.
Lewis, F.J. (1924) An endophytic fungi in the coniferae. *Nature* 114, 860.
McInroy, J.A. and Kloepper, J.W. (1991) Endophytic bacteria from field-grown corn and cotton. *Phytopathology* 81, 812–813.
Menendez, A., Bertoni, M.D. and Cabral, D. (1995) Comparative occurrence of fungal endophytes in *Juncus* species of Argentina. *Nova Hedwigia* 60, 583–588.
Pereira, J.O., Carneiro Vieira, M.L. and Azevedo, J.L. (1999) Endophytic fungi from *Musa acuminata* and their reintroduction into axenic plants. *World Journal of Microbiology and Biotechnology* 15, 37–40.
Petrini, O. (1991) Fungal endophytes of tree leaves. In: Andrews, J.A. and Hirano, S.S. (eds) *Microbial Ecology of Leaves.* Springer Verlag, New York, pp. 179–197.
Petrini, O. (1996) Ecological and physiological aspects of host-specificity in endophytic fungi. In: Redlin, S.C. and Carris, L.M. (eds) *Endophytic Fungi in Grasses and Woody Plants. Systematics, Ecology, and Evolution.* APS Press, St Paul, Minnesota, pp. 87–99.
Petrini, O. and Fisher, P.J. (1988) A comparative study of fungal endophytes in xylem and whole stems of *Pinus sylvestris* and *Fagus sylvatica. Transactions of the British Mycological Society* 91, 233–238.
Petrini, O., Sieber, T.N., Toti, L. and Viret, O. (1992) Ecology, metabolite production, and substrate utilization in endophytic fungi. *Natural Toxins* 1, 185–196.
Pugh, G.J.F. (1972) Saprophytic fungi and seeds. In: Heydecker, W. (ed.) *Seed Ecology.* Butterworth, London, pp. 337–345.
Rowan, D.D. and Latch, G.C.M. (1994) Utilization of endophyte-infected perennial ryegrasses for increased insect resistance. In: Bacon, C.W. and White, J.F. Jr (eds) *Biotechnology of Endophytic Fungi of Grasses.* CRC Press, Boca Ratón, Florida, pp. 169–183.
Schardl, C.L. and Phillips, T.D. (1997) Protective grass endophytes. Where are they from and where are they going? *Plant Disease* 81, 430–438.
Schulz, B., Sucker, J., Aust, H.J., Krohn, K., Ludewig, K., Jones, P.G. and Döring, D. (1995) Biologically active secondary metabolites of endophytic *Pezicula* species. *Mycological Research* 99, 1007–1015.
Schulz, B., Rommert, A.K., Daumman, U., Aust, H.J. and Strack, D. (1999) The endophyte–host interaction: a balanced antagonism? *Mycological Research* 103, 1275–1283.
Sieber, T.N. (1988) Endophytische pilze in nadel von gesunden und geschädigten fichten (*Picea abies* (L.) Karsten). *European Journal for Pathology* 18, 321–342.
Sieber, T., Riesen, T.K., Müller, E. and Fried, P.M. (1988) Endophytic fungi in four winter wheat cultivars (*Triticum aestivum* L.) differing in resistance against *Stagonospora nodorum* (Berk.) Cast. and germ.= *Septoria nodorum* (Berk.) Berk. *Journal of Phytopathology* 122, 2289–2306.
Sieber, T.N., Sieber-Canavesi, F. and Dorworth, C.E. (1991) Endophytic fungi of red alder (*Alnus rubra*) leaves and twigs in British Columbia. *Canadian Journal of Botany* 69, 407–411.
Sinclair, J.B. and Cerkauskas, R.F. (1996) Latent infection vs. endophytic colonization by fungi. In: Redlin, S.C. and Carris, L.M. (eds) *Endophytic Fungi in Grasses and Woody Plants. Systematics, Ecology and Evolution.* APS Press, St Paul, Minnesota, pp. 3–29.
Stone, J. and Petrini, O. (1997) Endophytes of forest trees: a model for fungus–plant interactions. In: Carroll, G.C. and Tudzynski, P. (eds) *The Mycota V. Part. B.* Springer-Verlag, Berlin Heidelberg, pp. 129–140.
Stone, J.K., Viret, O., Petrini, O. and Chapela, I.H. (1994) Histological studies of host penetration and colonisation by endophytic fungi. In: Petrini, O. and Ouellette, G.B. (eds) *Host Wall Alterations by Parasitic Fungi.* American Phytopathological Society Press, St Paul, Minnesota, pp. 115–126.
Sturz, A.V., Christie, B.R. and Matheson, B.G. (1998) Association of bacterial endophyte populations from red clover and potato crops with potential for beneficial allelopathy. *Canadian Journal of Microbiology* 44, 162–167.
Wilson, D. (1995) Endophyte: the evolution of a term, and clarification of its use and definition. *Oikos* 73, 274–276.
Zadoks, J.C., Chang, T.T. and Konzak, K. (1974) A decimal code for the growth stages of cereals. *Weed Research* 14, 415–421.
Zillinsky, F.J. (1984) *Enfermedades comunes de los cereales pequeños. Una guía para su clasificación* CIMMYT, El Batán, México.

13 Effect of Tillage Systems on the Arbuscular Mycorrhizal Fungi Propagule Bank in Soils

Santiago Schalamuk[1,2] and Marta N. Cabello[1,3]
[1]*Instituto de Botánica 'Spegazzini'*; [2]*CONICET (Consejo Nacional de Investigaciones Cientificas y Technicas), Universidad Nacional de La Plata, La Plata, Argentina;* [3]*CICBA (Comision de Investigaciones Cientificas de la Provincia de Buenos Aires), Argentina*

Abstract

In this chapter we discuss the effects of tillage and no-tillage systems on the characteristics of the arbuscular mycorrhizal fungi (AMF) propagule bank in soils. These fungi, which belong to the phylum *Glomeromycota*, are of great interest in agriculture. AMF are often assumed to be solely beneficial; however, in certain environmental conditions, growth depressions related to AMF have been observed. In soils under no-tillage, an intact hyphal network is present, whereas under conventional tillage, this network can be damaged and AMF spores may remain as propagule sources. Some direct effects of tillage on AMF propagules are: (i) disruption of the hyphal network; (ii) dilution of the propagule-rich topsoil; and (iii) accelerated root decomposition. Spore counts in soils should be considered as useful indicators for AMF activity *in situ*; however, the presence of spores does not always imply recent activity of AMF and mechanical disturbance may change their spatial distribution in the soil profile. Therefore, the information about spore numbers in agricultural systems needs to be analysed cautiously. The different environmental conditions and direct effects related with tillage and no-tillage on AMF communities generate shifts not only in the composition of the AMF soil propagule bank, but also in its diversity. If the differential use of the various types of propagules by the *Glomeromycota* families, as many authors suggest, is confirmed, the lack of disruption of the hyphal network in no-tillage can help to explain the differences in *Glomeromycota* diversity that are found in field experiments.

Importance of AMF in Agriculture

Arbuscular mycorrhizae (AM) show symbioses between plant roots and fungi belonging to the phylum *Glomeromycota* (Schübler *et al.*, 2001). These fungi are obligate biotrophs and form associations with most plant species (Trappe, 1987). AM associations are the most frequent symbioses in nature because of their broad association with plants and their cosmopolitan distribution (Harley and Smith, 1983). They have been found from the Antartic Peninsula to the tropics (Huante *et al.*, 1993; Cabello *et al.*, 1994). The wide host range of these fungi and their ability to grow in different environments are the reason why arbuscular mycorrhizal fungi (AMF) are usually considered 'generalists' with low host specificity (Smith and Read, 1997). Studies have confirmed that mycorrhizal fungi colonize most agricultural plants and that they can have a substantial impact on crop productivity (Johnson, 1993).

The interaction between the fungus and its host plant consists mainly in nutrient transfer: the plant provides the fungus with carbon compounds, while the fungus delivers nutrients to the plant. The increased nutrient uptake from the soil, particularly of phosphorus and nitrogen, is the main benefit attributed to mycorrhizal symbioses (Smith and Read, 1997; Govindarajulu *et al.*, 2005). Other benefits may include enhancement of resistance to root parasites (Borowicz, 2001), improvement of drought tolerance (Augé, 2001) and reduction of the impact of environmental stresses such as salinity (Ruiz-Lozano *et al.*, 1996). AMF also have an important role in the improvement of soil stability, which can possibly diminish erosion (Rillig *et al.*, 2002).

AM fungi are often assumed to be solely beneficial, since they are widely thought to function as mutualists. However, their effects on host growth often depend on environmental conditions such as nutrient availability and soil moisture (Peng *et al.*, 1993; Al-Karaki *et al.*, 1998; Graham and Abbott, 2000; Valentine *et al.*, 2001). As AMF draws C from the host, the overall effect on host growth depends on the cost–benefit relationship of the symbiosis (Johnson *et al.*, 1997; Grimoldi *et al.*, 2005). Consequently, in fertile soils, growth patterns of mycorrhizal plants often do not differ significantly from those of non-mycorrhizal ones (Newsham *et al.*, 1995) and even growth depressions related to AMF have been observed in many plant species (Johnson *et al.*, 1997; Allen *et al.*, 2003). In such plant–AMF interactions, only the fungal symbiont has a net benefit, and this has sometimes been interpreted as parasitism (Johnson *et al.*, 1997).

AM fungi are grouped into genera that encompass more than 150 species described to date and the effects that they have on their host plants, or 'effectivity', differ greatly between fungal strains or species (Miller *et al.*, 1985; Modjo and Hendrix, 1986). Since a single root can be colonized simultaneously by various *Glomeromycota* species, AMF root colonization is mediated by interspecific fungal interactions, such as competition, antagonism and dominance (Allen *et al.*, 2003). Because of the importance of AMF in agrosystems, their study is relevant both for the manipulation of indigenous AMF in the field through appropriate agricultural practices and for the development of a successful inoculation.

Agricultural Practices and Mycorrhizae

Agricultural practices for annual crops, such as crop rotations, tillage, sowing, fertilization, pest, weed and disease control, and harvest, generate changes that affect the microbial communities in the rhizosphere. Conventional tillage is characterized by the use of disc or mouldboard ploughs, followed by harrowing for seedbed preparation. In no-tillage, seeds are drilled directly into the soil with an appropriate planting machine (Crovetto, 1992). No-tillage systems are characterized by the accumulation of crop residues on the soil surface, leading to greater carbon, nitrogen and surface water, compared to conventional tillage (Doran and Linn, 1994). Several changes in soil properties have been reported with no-tillage management systems: improved aggregate stability, moisture availability with residue retention, changes in the distribution of organic matter residues down the soil profile, for example, a more even distribution of organic matter in cultivated soil as compared to that in non-tilled soil, where residues are concentrated on the surface (Alvarez *et al.*, 1998). One of the problems that may occur in no-tillage is the nutritional deficiency because of the reduced mineralization of the soil organic matter (Fox and Bandel, 1986).

In the case of AMF, the lack of soil physical disturbance in no-tillage might wrongly suggest that soils with annual crops under this system may be similar to those of natural grasslands. However, agroecosystems have particular characteristics which influence AMF activity. Natural ecosystems present various plant species hosting AMF, at different phenological stages. Annual crops, however, inherently represent a change for AMF, because of the reduction in host biodiversity. In addition, cropped systems show two clearly different periods: a period with high density of host plants of the same species growing simultaneously and, after harvesting, the fallow period with no host or, in some cases, scarce presence of spontaneous vegetation (i.e. weeds). As obligate symbionts, *Glomeromycota* relies on the plant host for the supply of C assimilates required for its growth, maintenance and functioning. Therefore, dynamics and biodiversity are clearly affected by agricultural practices (Kurle and Pfleger, 1994).

Significance of the AMF Propagule Bank on Root Colonization

Effect of tillage

Colonization of roots by AM fungi can arise from three sources of inoculum: spores, colonized root fragments and hyphae. The propagules in soils therefore may be called a 'propagule bank' that is 'waiting' for suitable conditions to germinate, grow and eventually colonize new plant roots (Öpik, 2004; Schalamuk, 2005). Most of the host plant benefits obtained by AM symbiosis, mainly phosphorus acquisition, depend on the early colonization of roots. The rapid colonization is related to AMF propagule density and composition, i.e. the so-called propagule bank. A graph of the percentage of the root length colonized against time has a sigmoid form showing three phases: lag phase, linear phase and a plateau (Sieverding, 1991). A higher AMF propagule density often reduces the length of the lag phase and thereby accelerates the process of mycorrhizal colonization (Smith and Read, 1997).

Numerous studies have shown that mycorrhizal colonization is affected negatively by tillage (Douds *et al.*, 1995; McGonigle and Miller; 1996a; Kabir *et al.*, 1998; Mozafar *et al.*, 2000). Soil disturbance reduces AMF propagule density since tillage of soil breaks up the AM fungi hyphal network and consequently lowers mycorrhizal colonization (McGonigle and Miller, 1996a). At the final crop stages, the AMF colonization levels in no-tillage and conventional tillage often do not differ significantly; however, at the early stages, crop plants under no-tillage often show higher mycorrhizal colonization (Schalamuk *et al.*, 2004). As already mentioned, in no-tillage systems, the reduced mineralization of the soil organic matter often generates plant nutritional deficiencies. Nevertheless, a higher nutrient concentration related to a rapid AMF colonization has been observed under no-tillage systems (McGonigle and Miller, 1996a; Mozafar *et al.*, 2000; Schalamuk *et al.*, 2004). By using the method of Plenchette *et al.* (1989), we have previously found higher levels of mycorrhizal soil infectivity in no-tillage systems (Schalamuk *et al.*, 2004). As already pointed out, colonization of roots by AM fungi can arise from different sources of inoculum. Colonized root fragments (Rives *et al.*, 1980), spores (Gould and Liberta, 1981; Jasper *et al.*, 1987, 1988) and hyphae (Jasper *et al.*, 1989) lose their ability to initiate colonization with soil disturbance, which can be related to physical damage to the propagules by tillage and/or unfavourable conditions for germination or colonization after disturbance (Stahl *et al.*, 1988; Bellgard, 1993).

Mycorrhizal soil infectivity (MSI) (Plenchette *et al.*, 1989) compares the ability of different soils to induce colonization in plants and depends on the activity of all the propagule types in soil. It is difficult to distinguish the relative contributions of the different types of propagules to the colonization of root systems (Smith and Read, 1997), and mycorrhizal infectivity does not provide information about the relevance of each propagule type in any particular field situation. Although a number of different propagule types exist in the soil, they may not be equally effective at producing new infection units (Klironomos and Hart, 2002). In many habitats, the hyphal network in the soil, together with root fragments, is probably the main means by which plants become colonized, even when significant spore populations are also present (Hepper, 1981;

Tommerup and Abbott, 1981; Birch, 1986; Jasper *et al.*, 1992). Studies have shown that AMF extraradical hyphae are affected severely by soil disturbance at tillage (Fairchild and Miller, 1990; McGonigle and Miller, 1996b; Kabir *et al.*, 1997; Wright and Upadhyaya, 1998). Jasper *et al.* (1989) have stated that due to the importance of the AMF hyphal network as inoculum in undisturbed soil, a lower infectivity of soil propagules after the disturbance usually can be determined by the damage on the network, rather than on spores and colonized root fragments. Another effect of tillage on the AMF propagule bank, which occurs simultaneously with the disruption of the hyphal network, is the dilution of the topsoil rich in propagules, with the poorest part in the subsurface (Sieverding, 1991). Clearly, mechanical soil mixing affects all types of AMF propagules.

As a conclusion, it is suggested that tillage affects all types of AMF propagules directly, to a greater or lesser extent, through different mechanisms acting together: (i) disruption of the hyphal network; (ii) dilution of the propagule-rich topsoil; and (iii) accelerated root decomposition. Through all these direct effects, tillage may reduce soil mycorrhizal infectivity and thereby AM root colonization at the early stages of crop growth.

Effects of Tillage and Cropping on AMF Spore Densities in Soils

AMF spores are formed by differentiation of vegetative hyphae in soil or roots and appear to be long-term survival structures. In agricultural systems with annual crops, other propagule types (i.e. hyphae inside and outside the roots) seem to be more important to start colonization in particular conditions. Nevertheless, spore counts in soils should be considered as useful indicators for AMF activity *in situ*. Several studies have found higher spore numbers in no-tillage than in conventional tillage (Crovetto, 1985; Kabir *et al.*, 1998; Jansa *et al.*, 2002; Schalamuk *et al.*, 2003). In agroecosystems with annual crops, the number of spores generally increases during the growing cycle (Cabello, 1987) and sporulation is frequently linked to host phenology in the field (e.g. maximum spore production occurs near the middle or the end of a growing season) (Morton *et al.*, 2004). At the early stages of the crop, higher spore densities are usually found in untilled soils, in comparison with conventional systems, whereas at the more advanced phenological stages, differences between tillage systems are reduced (Schalamuk *et al.*, 2003).

It is well known that spores can survive in soils for several years (Sieverding, 1991). Thus, spore counts reflect both the sporulation and the action of many factors that affect their survival and accumulation in the soil. Consequently, spore density is a result of a complex balance and, while sporulation is probably related to the recent activity of the AMF, spore counts in the soil include structures formed at different times.

Spore production depends on carbon supply from the host to the fungus (Furlan and Fortin, 1977; Daft and El Giahmi, 1978). Douds *et al.* (1993) have indicated that the production of fungal AM spores can decrease when soils are tilled. Increases in spore numbers have been associated with root growth (Hayman, 1970) and/or with host maturity or senescence (Hayman, 1970; Koske and Halvorson, 1981; Giovannetti, 1985; Gemma *et al.*, 1989; Troeh and Loynachan, 2003). Agricultural practices generate disturbances that affect AMF colonization and, in turn, spore formation in soils (Kurle and Pfleger, 1994). Therefore, tillage, either through changes in mycorrhizal colonization or through indirect effects, such as changes in the soil environment and plant growth, largely affect AMF spore production in soils.

The survival of a spore depends on its morphological traits, determined mainly by the species of *Glomeromycota* to which it belongs, as well as on the characteristics of the soil environment. Spore survival is an important factor determining the variations in AMF spore counts in soils; however, information about spore survival is scarce as compared to that about sporulation (Lee and Koske, 1994a). In natural ecosystems, decreases in spore numbers have been attributed mainly to their germination, the activity of macro and micro fauna and their destruction by other soil fungi and parasites (Gerdemann and Trappe, 1974; McIlveen and Cole, 1976; Ross and Ruttencutter, 1977; Ross and Daniels, 1982; Rabatin and Stinner, 1985, 1988). AMF spores are commonly infected either by other fungi (Daniels and Menge, 1980; Lee and Koske, 1994a; Rousseau *et al.*, 1996) or by *actinomycetes* (Lee and Koske, 1994b), and environmental conditions have a strong influence on these processes (Janos, 1980; Koske, 1988). In agricultural systems, another effect that directly reduces spore counts is the dilution of the topsoil rich in spores with the part in the subsurface poorer in propagules (Crovetto, 1985; Sieverding, 1991). For all these reasons, spore survival and accumulation may have a great influence on spore counts, and the largest spore numbers in no-tillage at the early stages may be the result of either higher or faster sporulation and/or the presence of residual spores produced during the fallow or the previous crop. As the presence of spores does not always imply recent activity of AMF, and mechanical disturbance may change their spatial distribution in the soil profile, the information about spore numbers in agricultural systems is useful, but needs to be analysed cautiously.

AMF Propagule Bank and Biodiversity

As already pointed out, tillage may alter the AMF propagule bank in several ways and the lack of disturbance in continuous no-tillage systems can generate accumulative effects. Therefore, in soils under no-tillage, an intact hyphal network can be present, whereas under conventional tillage, this network can be damaged and AMF spores may remain as propagule sources. Little information exists on the effect of tillage systems on *Glomeromycota* diversity (Jansa *et al.*, 2002; Schalamuk *et al.*, 2006). Several studies have shown that *Glomeromycota* taxa may vary in their colonization strategies and that these variations can be associated with the utilization of different propagule types by AMF families (i.e. *Acaulosporaceae*, *Gigasporaceae* and *Glomeraceae*) (Tommerup and Abbott, 1981; Biermann and Linderman, 1983; INVAM, 1993; Braunberger *et al.*, 1996; Brundrett *et al.*, 1999; Klironomos and Hart, 2002; Hart and Reader, 2002, 2004). Jansa *et al.* (2002), in an intensively used agricultural soil under long-term reduced tillage management, found that the presence of certain AMF species, especially those that did not belong to *Glomus* spp., had a tendency to increase. However, we have found that the contribution of species belonging to the *Glomeraceae* family increases in no-tillage plots, to the detriment of *Acaulosporaceae* and *Gigasporaceae* (Schalamuk *et al.*, 2006). In that experiment, the greatest contribution of *Glomeraceae* species in no-tillage indicated a lower equitability in the distribution among the families of *Glomeromycota*, and thereby a lower diversity, in comparison with conventional tillage. These findings differ from those of Jansa *et al.* (2002). Nevertheless, it is important to point out that mycorrhizal communities are site-specific and that each AMF species can be affected in several ways by different agricultural management practices; therefore, generalization is difficult.

De Souza (2005), based on life history strategy studies, suggested that members of the *Gigasporaceae* family were '*K*' strategists in contrast to single spore-producing '*Glomus*' species. Hart and Reader (2004) found that the *Gigasporaceae* family was less sensitive to soil disturbance than the *Glomeraceae*. The basis for this difference between both families is due probably to differences in their colonization strategies. AM fungi in the *Gigasporaceae* colonize primarily from spores, whereas those belonging to the *Glomeraceae* can colonize from hyphae (Tommerup and Abbot, 1981; Biermann and Lindermann, 1983). Hyphae are more sensitive to soil disturbance than spores and thus subsequent colonization of additional roots is affected more.

Tillage or the lack of disturbance in continuous no-tillage determine different environmental conditions and direct effects on AMF communities, and thereby shifts in the composition of the AMF soil propagule bank. Consequently, if the differential use of the various types of propagules by the *Glomeromycota* families, as many authors suggest, is confirmed, the lack of disruption of the hyphal network in no-tillage for a period of several years can help to explain the higher proportions of *Glomeraceae* that have been found previously in the system (Schalamuk *et al.*, 2006).

Conclusions

Tillage and continuous no-tillage systems change the composition of the AMF propagule banks in the soil, whereas mechanical soil mixing affects all types of AMF propagules. Continuous no-tillage systems favour the presence of an intact hyphal network in soils. Possible differences in colonization strategies among *Glomeromycota* taxa might have a great influence on the impacts of tillage on AMF diversity.

References

Allen, E.B., Swenson, W., Querejeta, J.I., Egerton-Waburton, L.M. and Treseder, K.K. (2003) Ecology of mycorrhizae: a conceptual framework for complex interactions among plants and fungi. *Annual Review of Phytopathology* 41, 271–303.

Alvarez, R., Russo, M.E., Prystupa, P., Scheiner, J.D. and Blotta, L. (1998) Soil carbon pools under conventional and no-tillage systems in the Argentine rolling pampa. *Agronomy Journal* 90, 138–143.

Al-Karaki, G.N., Al-Radad, A. and Clark, R.B. (1998) Water stress and mycorrhyzal isolate effects on growth and nutrient acquisition of wheat. *Journal of Plant Nutrition* 21, 891–902

Augé, R.M. (2001) Water relations, drought and VA mycorrhizal symbiosis. *Mycorrhiza* 11, 3–42.

Bellgard, S.E. (1993) Soil disturbance and infection of *Trifolium repens* roots by vesicular-arbuscular mycorrhizal fungi. *Mycorrhiza* 3, 25–29.

Biermann, B. and Linderman, R.G. (1983) Mycorrhizal roots, intraradical vesicles and extraradical vesicles as inoculum. *New Phytologist* 95, 97–105.

Birch, C.P.D. (1986) Development of VA mycorrhizal infection in seedlings in semi-natural grassland turf. In: Gianinazzi-Pearson, V. and Gianinazzi, S. (eds) *Physiological and Genetical Aspects of Mycorrhizae*. INRA, Paris, pp. 233–237.

Borowicz, V.A. (2001) Do arbuscular mycorrhizal fungi alter plant–pathogen relations? *Ecology* 82, 3057–3068.

Braunberger, P.G., Abbot, L.K. and Robson, A.D. (1996) Infectivity or arbuscular mycorrhizal fungi after wetting and drying. *New Phytologist* 134, 673–684.

Brundrett, M.C., Abbot, L.K. and Jasper, D.A. (1999) Glomalean fungi from tropical Australia. I. Comparison of the effectiveness of isolation procedures. *Mycorrhiza* 8, 305–314.

Cabello, M.N. (1987) Mycorrizas vesiculo-arbusculares en un cultivo de girasol. *Revista Facultad de Agronomía, La Plata* 63, 46–52.

Cabello, M.N., Gaspar, M.L. and Pollero, R. (1994) *Glomus antarcticum sp*-nov, a vesicular-arbuscular mycorrhizal fungus from Antarctica. *Mycotaxon* 58, 123–128.

Crovetto, C. (1985) *Cero labranza, extraordinaria alternativa para el cultivo de cereales en suelos erosionados. VII International Conference of Soil Conservation, Maracaibo, Venezuela. IICA, Montevideo, Uruguay*, pp. 461–472.

Crovetto, C. (1992) *Rastrojos sobre el suelo. Una introducción a la cero labranza.* Editorial Universitaria, Santiago.

Daft, M.J. and El Giahmi, A.A. (1978) Effects of arbuscular mycorrhiza on plant growth. VIII. Effects of defoliation and light on selected hosts. *New Phytologist* 80, 365–372.

Daniels, B.A. and Menge, J.A. (1980) Hyperparasitism of vesicular-arbuscular mycorrhizal fungus. *Phytopathology* 70, 584–588.

De Souza, F.A. (2005) Biology, ecology and evolution of the family *Gigasporaceae*, arbuscular mycorrhizal fungi (*Glomeromycota*). PhD thesis, Leiden University, The Netherlands.

Doran, J.W. and Linn, D.M. (1994) Microbial ecology of conservation management systems. In: Hatfield, J.L. and Stewart, B.A. (eds) *Advances in Soil Science. Soil Biology: Effects on Soil Quality.* Lewis Publishers/CRC Press, Boca Raton, Florida, pp. 1–27.

Douds, D.D. Jr, Janke, R. and Peters, S.E. (1993) VAM fungus spore populations and colonization of roots of maize and soybean under conventional and low-input sustainable agriculture. *Agriculture, Ecosystems and Environment* 43, 325–335.
Douds, D.D. Jr, Galvez, L., Janke, R.R. and Wagoner, P. (1995) Effect of tillage and farming systems upon populations and distribution of vesicular–arbuscular mycorrhizal fungi. *Agriculture, Ecosystems and Environment* 52, 111–118.
Fairchild, G.L. and Miller, M.H. (1990) Vesicular-arbuscular mycorrhizas and the soil disturbance induced reduction of nutrient absorption in maize. III. Influence of P amendments to soil. *New Phytologist* 114, 641–650.
Fox, R.H. and Bandel, V.A. (1986) Nitrogen utilization with no-tillage. In: Sprague, M.A. and Triplett, G.B. (eds) *No Tillage and Surface Tillage Agriculture. The Tillage Revolution.* John Wiley and Sons, New York, pp. 116–148.
Furlan, V. and Fortin, J.A. (1977) Effects of light intensity on the formation of vesicular-arbuscular mycorrhizas on *Allium cepa* by *Gigaspora calospora. New Phytologist* 79, 335–340.
Gemma, J.N., Koske, R.E. and Carreiro, M. (1989) Seasonal dynamics of selected species of V-A mycorrhizal fungi in a sand dune. *Mycological Research* 92, 317–321.
Gerdemann, J.W. and Trappe, M. (1974) *The Endogonaceae in the Pacific Northwest.* Mycologia Memoir No. 5. The New York Botanical Garden, Bronx, New York, 76 pp.
Giovannetti, M (1985) Seasonal variations of vesicular-arbuscular mycorrhizas and endogonaceous spores in a maritime sand dune. *Transactions of the British Mycological Society* 84, 679–684.
Gould, A.B. and Liberta, A.E. (1981) Effects of topsoil storage during surface mining on the viability of vesicular-arbuscular mycorrhiza. *Mycologia* 73, 914–922.
Govindarajulu, M., Pfeffer, P.E., Jin, H., Abubaker, J., Douds, D.D., Allen, J.W., Bucking, H., Lammers, P.J. and Shachar-Hill, Y. (2005) Nitrogen transfer in arbuscular mycorrhizal symbiosis. *Nature* 435, 819–823.
Graham, J.H. and Abbott, L.K. (2000) Wheat responses to aggressive and non-aggressive arbuscular mycorrhizal fungi. *Plant and Soil* 220, 207–218.
Grimoldi, A.A., Kavanová, M., Lattanzi, F.A. and Schnyder, H. (2005) Phosphorus nutrition-mediated effects of arbuscular mycorrhiza on leaf morphology and carbon allocation in perennial ryegrass. *New Phytologist* 168, 435–444.
Harley, J.L. and Smith, S.E. (1983) *Mycorrhizal Symbiosis.* Academic Press, London,
Hart, M.M. and Reader, R.J. (2002) Taxonomic basis for variation in the colonization strategy of arbuscular mycorrhizal fungi. *New Phytologist* 153, 335–344.
Hart, M.M. and. Reader, R.J (2004) The role of the external mycelium in early colonization for three arbuscular mycorrhizal fungal species with different colonization strategies. *Pedobiologia* 49, 269–279.
Hayman, D.S (1970) *Endogone* spore numbers in soil and vesicular-arbuscular mycorrhiza in wheat as influenced by season and soil treatment. *Transactions of the British Mycological Society* 54, 53–63.
Hepper, C.M. (1981) Techniques for studying the infection of plants by vesicular-arbuscular mycorrhizal fungi under axenic conditions. *New Phytologist* 88, 641–647.
Huante, P., Rincón, E. and Allen, E.B. (1993) Effect of vesicular-arbuscular mycorrhizae on seedling growth of four tree species from the tropical deciduos forest in Mexico. *Mycorrhiza* 2, 141–145.
INVAM (1993) *Properties of infective propagules at the suborder level* (*Glomineae* versus *Gigasporineae*). INVAM Newsletter 3: September.
Janos, D.P. (1980) Vesicular-arbuscular mycorrhizae influence lowland tropical rain forest plant growth. *Ecology* 61, 151–162.
Jansa, J., Mozafar, A., Anken, T., Ruh, R., Sanders, I.R. and Frossard, E. (2002) Diversity and structure of AMF communities as affected by tillage in a temperate soil. *Mycorrhiza* 12, 225–234.
Jasper, D.A., Robson, A.D. and Abbott, L.K. (1987) VA mycorrhizal fungi in revegetation after soil disturbance by mining. In: Sylvia, D.M., Hung, L.L. and Graham, J.H. (eds) *Mycorrhizae in the Next Decade.* Institute of Food and Agricultural Sciences, Gainsville, Florida, pp. 148–152.
Jasper, D.A., Robson, A.D. and Abbott, L.K. (1988) Revegetation in an iron-ore mine – nutrient requirements for plant growth and the potential role of vesicular-arbuscular (VA) mycorrhizal fungi. *Australian Journal of Soil Research* 26, 497–507.
Jasper, D.A., Abbott, L.K. and Robson, A.D. (1989) Soil disturbance reduces the infectivity of external hyphae of vesicular-arbuscular mycorrhizal fungi. *New Phytologist* 112, 93–99.
Jasper, D.A., Abbott, L.K. and Robson, A.D. (1992) Soil disturbance in native ecosystems – the decline and recovery of infectivity of VA mycorrhizal fungi. In: Read, D.J., Lewis, D.H., Fitter, A.H. and Alexander, I.J. (eds) *Mycorrhizas in Ecosystems.* CAB International, Wallingford, UK, pp. 151–155.

Johnson, N.C. (1993) Can fertilization of soil select less mutualistic mycorrhizae? *Ecological Applications* 3, 749–757.
Johnson, N.C., Graham, J.H. and Smith, F.A. (1997) Functioning of mycorrhizal associations along the mutualism–parasitism continuum. *New Phytologist* 135, 575–586.
Kabir, Z., O'Halloran, I.P. and Hamel, C. (1997) Overwinter survival of arbuscular mycorrhizal hyphae is favored by attachment to root but diminished by disturbance. *Mycorrhiza* 7, 197–200.
Kabir, Z., O'Halloran, I.P., Widden, P. and Hamel, C. (1998) Vertical distribution of arbuscular mycorrhizal fungi under corn (*Zea mays* L.) in no-till and conventional tillage systems. *Mycorrhiza* 8, 53–55.
Klironomos, J.N. and Hart, M.M. (2002) Colonization of roots by arbuscular mycorrhizal fungi using different sources of inoculum. *Mycorrhiza* 12, 181–184.
Koske, R.E. (1988) VA mycorrhizae of some Hawaiian dune plants. *Pacific Science* 42, 217–229.
Koske, R.E. and Halvorson, W.L. (1981) Ecological studies of vesicular-arbuscular mycorrhizae in a barrier sand dune. *Canadian Journal of Botany* 59, 1413–1422.
Kurle, J.E. and Pfleger, F.L. (1994) The effects of cultural practices and pesticides on VAM fungi. In: Pfleger, F.L. and Linderman, R.G. (eds) *Mycorrhizae and Plant Health.* APS Press, St Paul, Minnesota, pp. 101–132.
Lee, P.J. and Koske, R.E. (1994a) *Gigaspora gigantea*: seasonal abundance and ageing of spores in a sand dune. *Mycological Research* 98, 453–457.
Lee, P.J. and Koske, R.E. (1994b) *Gigaspora gigantea*: parasitism of spores by fungi and actinomycetes. *Mycological Research* 98, 458–466.
McGonigle, T.P. and Miller, M.H. (1996a) Mycorrhizae, phosphorus absorption, and yield of maize in response to tillage. *Soil Science Society of America Journal* 60, 1856–1861.
McGonigle, T.P. and Miller, M.H. (1996b) Development of fungi below ground in association with plants growing in disturbed and undisturbed soils. *Soil Biology and Biochemistry* 28, 263–269.
McIlveen, W.D. and Cole, H. (1976) Spore dispersal of Endogonaceae by worms, ants, wasps and birds. *Canadian Journal of Botany* 54, 1486–1489.
Miller, D.D., Domoto, P.A. and Walker, C. (1985) Mycorrhizal fungi at eighteen apple rootstock plantings in the United States. *New Phytologist* 100, 379–391.
Modjo, H.S. and Hendrix, J.W (1986) The mycorrhizal fungus *Glomus macrocarpun* as a cause of tobacco stunt disease. *Phytopathology* 76, 688–691.
Morton, J.B., Koske, R.E., Sturmer, S.L. and Bentivenga, S.P. (2004) Mutualistic arbuscular endomycorrhizal fungi. In: Mueller, G.M., Bills, G.F. and Foster, M.S. (eds) *Biodiversity of Fungi: Inventory and Monitoring Methods.* Elsevier Academic Press, Washington, DC, pp. 313–332.
Mozafar, A., Anken, T., Ruh, R. and Frossard, E. (2000) Tillage intensity, mycorrhizal and non-mycorrhizal fungi, and nutrient concentrations in maize, wheat, and canola. *Agronomy Journal* 92, 1117–1124.
Newsham, K.K., Fitter, A.H. and Watkinson, A.R. (1995) Multi-functionality and biodiversity in arbuscular mycorrhizas. *Trends in Ecology and Evolution* 10, 407–441.
Öpik, M. (2004) Diversity of arbuscular mycorrhizal fungi in the roots of perennial plants and their effect on plant performance. PhD thesis, Faculty of Biology and Geography, University of Tartu, Estonia.
Peng, S., Eissenstat, D.M., Gram, J.H., Williams, K. and Hodge, N.C. (1993) Growth depression in mycorrhizal citrus at high-phosphorus supply: analysis of carbon costs. *Plant Physiology* 101, 1063–1071.
Plenchette, C., Perrin, R. and Duvert, P. (1989) The concept of soil infectivity and a method for its determination as applied to Endomycorrhizas. *Canadian Journal of Botany* 67, 112–115.
Rabatin, S.C. and Stinner, B.R. (1985) Arthropods as consumers of vesicular-arbuscular mycorrhizal fungi. *Mycologia* 77, 320–322.
Rabatin, S.C. and Stinner, B.R. (1988) Indirect effects of interactions between VAM fungi and soil-inhabiting invertebrates on plant processes. *Agriculture, Ecosystems and Environment* 24, 135–146.
Rillig, M.C., Wright, S.F. and Eviner, V.T. (2002) The role of arbuscular mycorrhizal fungi and glomalin in soil aggregation: comparing effects of five plant species. *Plant and Soil* 238, 325–333.
Rives, C.S., Bajwa, M.I., Liberta, A.E. and Miller, R.M. (1980) Effects of topsoil storage during surface mining on the viability of VA mycorrhiza. *Soil Science* 129, 253–257.
Ross, J.P. and Daniels, B.A. (1982) Hyperparasitism of endomycorrhizal fungi. In: Schenck, N.C. (ed.) *Methods and Principles of Mycorrhizal Research.* American Phytophatological Society, St Paul, Minnesota, pp. 55–58.
Ross, J.P. and Ruttencutter, R. (1977) Population dynamics of two vesicular-arbuscular mycorrhizal fungi and the role of hyperparasitic fungi. *Phytopathology* 67, 490–496.
Rousseau, A., Benhamou, N., Chet, I. and Piché, Y. (1996) Mycoparasitism of the extramatrical phase of *Glomus intraradices* by *Trichoderma harzianum. Phytopathology* 86, 434–434.

Ruiz-Lozano, J.M., Azcón, R. and Palma, J.M. (1996) Superoxide dismutase activity in arbuscular mycorrhizal *Lactuca sativa* plants subjected to drought stress. *New Phytologist* 134, 327–333.
Schalamuk, S. (2005) Dinámica y biodiversidad de hongos formadores de micorrizas arbusculares (HFMA): efecto de sistemas de labranza y fertilización en cultivos extensivos. (Dynamics and biodiversity of arbuscular mycorrhizal fungi (AMF): effect of tillage systems and fertilization in extensive crops). PhD thesis, Facultad de Ciencias Naturales y Museo, Universidad Nacional de La Plata, Argentina.
Schalamuk, S., Velázquez, S., Chidichimo, H. and Cabello, M. (2003) Efecto de diferentes sistemas de labranza y fertilización sobre la simbiosis micorrizica arbuscular en cultivo de trigo. *Boletín micológico. Valparaíso, Chile* 18, 15–19.
Schalamuk, S., Velázquez, S., Chidichimo, H. and Cabello, M. (2004) Effect of no-till and conventional tillage on mycorrhizal colonization in spring wheat. *Boletín de la Sociedad Argentina de Botánica* 39, 13–20.
Schalamuk, S., Velázquez, S., Chidichimo, H. and Cabello, M. (2006) Fungal spore diversity of arbuscular mycorrhizal fungi associated with spring wheat: effect of tillage. *Mycologia* 98, 22–28.
Schübler, A., Schwarzott, D. and Walker, C. (2001) A new fungal phylum, the *Glomeromycota*: phylogeny and evolution. *Mycological Research* 105, 1413–1421.
Sieverding, E. (1991) *Vesicular-Arbuscular Mycorrhiza Management in Tropical Agrosystems.* Deutche Gesellschaft für Technische Zusammenarbeit, GTZ No 224, Eschborn.
Smith, S.E. and Read, D.J. (1997) *Mycorrhizal Symbiosis.* Second Edition. Academic Press, London.
Stahl, P.D., Williams, S.E. and Christiensen, M. (1988) Efficacy of native vesicular-arbuscular mycorrhizal fungi after severe soil disturbance. *New Phytologist* 110, 347–354.
Tommerup, I.C. and Abbott, L.K. (1981) Prolonged survival and viability of VA mycorrhizal hyphae after root death. *Soil Biology and Biochemistry* 13, 431–433.
Trappe, J.M. (1987) Phylogenetic and ecologic aspects of mycotrophy in the angiosperms from an evolutionary standpoint. In: Safir, G.R. (ed.) *Ecophysiology of VA Mycorrhizal Plants.* CRC Press, Baton Rouge, Louisiana, pp. 5–25.
Troeh, Z.I. and Loynachan, T.E. (2003) Endomycorrhizal fungal survival in continuous corn, soybean, and fallow. *Agronomy Journal* 95, 224–230.
Valentine, A.J., Osborne, B.A. and Mitchell, D.T. (2001) Interactions between phosphorus supply and total nutrient availability on mycorrhizal colonization, growth and photosynthesis of cucumber. *Scientia Horticulturae* 88, 177–189.
Wright, S.F. and Upadhyaya, A. (1998) A survey of soils for aggregate stability and glomalin, a glycoprotein produced by hyphae of arbuscular mycorrhizal fungi. *Plant and Soil* 198, 97–107.

14 Mechanism of Action in Arbuscular Mycorrhizal Symbionts to Control Fungal Diseases

Arun Arya, Chitra Arya and Renu Misra
Department of Botany, Faculty of Science, The Maharaja Sayajirao University of Baroda, Vadodara, India

Abstract

Currently, the world over, especially in developing countries, maintenance of soil fertility and control of plant diseases have become crucial issues in meeting the biomass needs for food, fodder and fuel, as well as preserving a clean environment. An ideal fertile soil is characterized not only by optimum physical properties and chemical constituents conducive for plant growth, but also by microbiological processes that are maintained in equilibrium. More than 90% of land plants are estimated to form arbuscular mycorrhizal (AM) associations with soilborne fungi in the phylum *Glomeromycota*. They have a wide host range, yet certain host and fungal combinations are more effective from either the perspective of the fungus, i.e. greater spore/hyphae production, or from that of the host, i.e. enhanced growth, nutrient acquisition or pathogen resistance. Besides improving uptake of phosphorus, AM fungi improve plant health through improved resistance to various biotic and abiotic stresses. Of particular importance is the bioprotection conferred to plants against many soilborne pathogens, such as species of *Aphanomyces*, *Cylindrocladium*, *Fusarium*, *Macrophomina*, *Phytophthora*, *Pythium*, *Rhizoctonia*, *Sclerotium*, *Thielaviopsis* and *Verticillium*, as well as various nematodes by AM fungal colonization of the plant roots.

Achieving the effective and sustainable control of plant diseases remains a formidable challenge for all agricultural systems. Despite the continued release of resistant cultivars and pesticides, pathogens still cause crop damages and losses that exceed 12% worldwide. Studies have shown that root rot in wheat caused by *S. rolfsii* was prevented by the inoculation of *Glomus fasciculatum*. Reduced quantum of lesioned roots was found in take-all diseases caused by *Gaeumannomyces graminis tritici* due to *G. deserticola* in wheat. The association of *G. radiatum* with apple has been studied in the USA. It was found that soilborne fungi, *Cylindrocarpon*, *Pythium* and the parasitic nematode, *Pratylenchus* spp., were common with replant diseases of apple. In this disease, young trees are stunted and develop fewer branches than healthy trees.

The exact mechanisms by which AM fungal colonization confers the protective effect are not completely understood, but a greater understanding of these beneficial interactions is necessary for the exploitation of AM fungi in organic and/or sustainable farming systems. The mechanisms employed by AM fungi indirectly to suppress plant pathogens include enhanced nutrition to plants; morphological changes in the root; increased lignification; changes in the chemical composition of the plant tissues like antifungal chitinases, isoflavonoids, etc.; alleviation of abiotic stress and changes in the microbial composition in the mycorrhizosphere. Bioprotection within AM fungal-colonized plants is the outcome of complex interactions between plants, pathogens and AM fungi. In this chapter, the different diseases of cereals, pulses, fruits and vegetables and the potential mechanisms by which AM fungi contribute to bioprotection against plant soilborne pathogens are discussed.

 Management of Fungal Plant Pathogens
(eds A. Arya and A.E. Perelló)

Introduction

Arbuscular mycorrhizal (AM) symbiosis is the most commonly occurring underground symbiosis in plants. It can be found in a large majority of terrestrial plants (Newman and Reddell, 1987) and in almost a quarter of a million plant species. It is as normal for the roots of plants to be mycorrhizal as it is for the leaves to photosynthesize (Mosse, 1986). The AM fungi are included in the phylum *Zygomycota*, order *Glomales* (Redecker *et al.*, 2000), but recently they have been classified into the phylum *Glomeromycota* (Schussler *et al.*, 2001). The phylum is divided into 4 orders, 8 families, 10 genera and 150 species; the common genera are *Aculospora*, *Gigaspora*, *Glomus* and *Scutellospora* (Schussler, 2005). They are characterized by the presence of extra-radical mycelium, branched haustoria-like structures within the cortical cells, termed arbuscules. These are the main sites of nutrient transfer between the two symbiotic partners (Hock and Verma, 1995; Smith and Read, 1997).

AM fungi colonize plant roots and penetrate the surrounding soil, extending the root depletion zone and the root system. They supply water and mineral nutrients from the soil to the plant, while AM benefit from carbon compounds provided by the host plant (Smith and Read, 1997). AM fungi are associated with improved growth of host plant species due to increased nutrient uptake, production of growth-promoting substances, tolerance to drought, salinity and synergistic interactions with other beneficial microorganisms (Sreenivasa and Bagyaraj, 1989). The beneficial role of AM fungi in plant biomass production is associated with their capacity to reduce or prevent the development of plant disease (Manoharachary, 2004). The protective ability of mycorrhizae is generally observed against soilborne diseases and is often related to the nature of the host plant, mycorrhizal symbionts, plant pathogens and the condition of the soil (Tello *et al.*, 1987). AM fungi are helpful in controlling disease; however, Ross (1972) reported increased development of *Phytophthora* root rot in soybean. The known interaction may include a number of mechanisms such as exclusion of pathogens, lignification of plant cell wall and change in phosphorus nutrition, leading to exudation by roots and the formation of inhibitory low molecular weight compounds. The mycorrhizal fungi can produce certain compounds that inhibit or kill the pathogenic fungi.

Interaction of AM Fungi with Fungal Pathogens

Cereal crops

Achieving the effective and sustainable control of plant disease remains a formidable challenge for all agricultural systems. Despite the continued release of resistant cultivars and pesticides, pathogens still cause crop damages and losses that exceed 12% worldwide (Johar, 2005). Root rot in wheat caused by *S. rolfsii* was prevented by inoculation of *G. fasciculatum* (Harlapur *et al.*, 1990). Graham and Menge (1982) reported reduced quantum of lesioned roots in take-all disease caused by *G. graminis tritici* due to *G. deserticola* in wheat.

It was found that root dry weight of paddy was not affected by *R. solani* in mycorrhizal plants, but the pathogen caused 29% loss in root dry weight in non-mycorrhizal plants (Khadge *et al.*, 1990). Also, the pathogen multiplied less in mycorrhizal plants. *Cochliobolus sativus* negated the effect of VAM inoculation in locally adapted WI 2291 cultivar of barley, whereas in the absence of the pathogen, AM inoculation increased grain yield from 31.9 g to 46.6 g in phosphorus fertilized plants but did not have fertilized plants (Grey *et al.*, 1989). Contrary results were obtained by Schonbeck and Dehne (1979), who observed increase in disease due to *Erysiphe graminis* and *Helminthosporium sativum* in barley. The severity of common root caused by *Bipolaris sorokiniana* in barley was reduced by three species of *Glomus* (Boyethko and Tewari, 1990).

Pulses and oil crops

Gigaspora calospora exerted an inhibitory effect on the development of pigeon pea blight caused by *P. drechsleri* f. sp. *cajani* (Bisht *et al.*, 1985). Similarly, in Tamil Nadu Agricultural University, India, studies showed that another AM fungus, *G. etunicatum*, induced tolerance to cowpea (*Vigna unguiculata*) against *Macrophomina* root rot. Disease incidence was 16% in inoculated plants as against 33% in uninoculated plants (Ramraj *et al.*, 1988). Rosendahl (1985) observed a decrease in disease incidence in peas due to *Aphanomyces euteiches*. Similar results were observed for soybean (Zambolin and Schenck, 1983) and groundnut (Abdalla and Abdel-Fattah, 2000) due to *F. solani*. Krishna and Bagyaraj (1983) observed a reduction in disease due to *M. phaseolina* in soybean. Studies conducted at the University of Bayreuth, Germany, showed that in leachates of AM rhizospheric soil of *Zea mays* and *Trifolium subterraneum*, fewer sporangia and zoospores were produced by *P. cinnamomi* as compared to non-AM plants, suggesting that sporangium-induced microorganisms declined or sporangium inhibitors increased (Meyer and Linderman, 1983).

Pandey and Upadhyay (2000) studied the effect of microbial populations on the development of pigeon pea in Pusa, Bihar, India. Screening for resident antagonists was carried out and the mode of mycoparasitism was studied. Dual inoculation with AM endophyte (*G. mosseae*) and *M. phaseolina* restricted the progression of the pathogen significantly in the roots of mungbean (*V. radiata*). Disease incidence was reduced from 77.9% in pathogen inoculated to 13.3% in AM + pathogen inoculated plants (Jalali *et al.*, 1990). *G. fasciculatum* reduced the number of sclerotia produced by *S. rolfsii* in groundnuts (*Arachis hypogaea*) (Krishna and Bagyaraj, 1983).

Horticultural crops

The early wilt symptoms caused by *F. oxysporum* on tomato appeared 8–10 days earlier in mycorrhizal plants. However, 2 months later, disease severity was reduced significantly in these plants. Between the two species tested, *G. etunicatum* was more effective than *G. mosseae* (Sharma and Johri, 2002). *Brassica oleracea* infected with AM fungus had lower infection by *R. solani*; higher moisture content (25%) enhanced disease incidence (Iqbal *et al.*, 1988). Studies conducted at the University of Jordan, Jordan, showed that the mycorrhizal plants of tomato inoculated with *F. oxysporum* had significantly higher root and shoot weights and plant heights than plants inoculated with *F. oxysporum* only (Al-Momany and Al-Radded, 1988). Only the presence of *G. intraradices* resulted in a significant decrease in the population of *F. oxysporum* and root necrosis (Caron *et al.*, 1986). Early infestation of *G. fasciculatum* enhanced tomato plant growth and reduced *Fusarium* wilt (Manian *et al.*, 2006). They also observed that the percentage disease index was less in mycorrhizal than in non-mycorrhizal tomato plants when inoculated with *Alternaria solani*.

The presence of *G. mosseae* decreased both weight reduction and root necrosis in tomato caused by *P. nicotianae* var. *parasitica* (Trotta *et al.*, 1996). *In vitro* experiments in which Ri T-DNA transformed roots of alfalfa were inoculated with AM fungi showed normal mycorrhizal formation by *G. intraradices* and hypersensitivity-like response to *G. margarita*. Colonized cells became necrotic and HPLC studies indicated concentration of phenolics and isoflavonoids in these roots. The data strongly support the existence of a degree of specificity between AM fungi and the host (Douds *et al.*, 1998).

Onion pink rot caused by *Pyrenochaeta terrestris* and tomato root rot caused by *T. basicola* are controlled by mycorrhizal fungi (Vidhyasekaran, 2004). Inoculation of *G. mosseae* in tomato and eggplant seedlings controlled the incidence of *Verticillium* wilt caused by *V. dahliae* in Greece (Karagiannidis *et al.*, 2002). Trotta *et al.* (1996) studied the interaction between the soilborne root pathogen *P. nicotinae* var. *parasitica* and the arbuscular mycorrhizal fungus *G. mosseae* in tomato plants. Treatment with *Phytophthora* resulted in a visible reduction in plant weight and in a widespread root necrosis in plants without mycorrhiza. The presence of AM fungus decreased both weight reduction and root necrosis. The percentage reduction of root necrosis ranged between 63 and 89%.

Utkhede *et al.* (1992) studied the effect of *G. mosseae* on replant disease of apple. It was found by Graham and Egel (1988) in Florida, USA, that *G. intraradices* did not increase the resistance or tolerance of sweet orange seedlings to *Phytophthora* root rot unless mycorrhizae conferred a phosphorus nutritional advantage over the non-mycorrhizal plants. Citrus root rot caused by *P. parasitica* and *T. basicola* can be controlled by AM fungi (Vidhyasekaran, 2004). Prior root colonization by mycorrhizal fungi, *G. margarita* or *G. macrocaropum*, reduced the damage caused by *P. parasitica* in two citrus root stocks, Carrigo citrage and Sour orange (Schenck *et al.*, 1977). To ensure good mycorrhizal establishment in citrus roots, plants were exposed for 110 days to mycorrhizal fungi before challenging them with the pathogen. In phalsa (*Grewia subinaequalis*), better root growth and feeding sites of nematodes during the rainy season promoted better colonization of AM fungi (Hasan and Khan, 2006).

Cash crops

Studies conducted at the Rajasthan Agriculture University, India, showed that *Cuminum cyminum* in association with *G. calospora*, *G. fasciculatum*, *G. mosseae* and *Acaulospora laevis* enhanced nutrient uptake and reduced wilt severity due to *F. oxysporum* f. sp. *cumini* (Champawat, 1991). In Germany, *G. etunicatum* reduced leaf blight in rubber plants caused by *Microcycles ulei* (Feldmann *et al.*, 1990). *G. monosporum* inoculated tobacco plants showed better tolerance against *T. basicola* (Giovannetti *et al.*, 1991). Sivaprasad *et al.* (2006) controlled foot rot of black pepper by inoculation of *G. monosporum*. Two other species, *G. etunicatum* and *G. mosseae*, exhibited a medium level of resistance to the disesases. Rhizome rot of ginger caused by *P. aphanidermatum* was controlled by *G. mosseae* and *G. fasciculatum* (Sivaprasad *et al.*, 2006).

Field application of a commercially available formulation of AM marketed as Josh by Cadila Pharmaceuticals, Agro Division, was tried for the management of charcoal stump rot disease caused by *Ustulina zonata* (Chakraborty *et al.*, 2005). Commercial production of the medicinal plants in arid and semi-arid areas of the Thar Desert is affected mostly by the soilborne plant pathogens ready to attack any seedlings transplanted into the field. Mycorrhizal symbiosis resulted in significant disease severity in *Chlorophytum borivillianum*, *Convolvulus microphyllous* and *Withania somnifera* (Vyas, 2005).

Role of AM fungi in forestry

Studies conducted at the Northern Forest Research Centre, Canada, showed that *Fusarium* wilt disease severity in *Albizia procera* and *Dalbergia sissoo* was reduced significantly when inoculated with mycorrhizal fungi (Chakravarty and Mishra, 1986). The effect of AM fungi, *Pseudomonas* and *Rhizobium*, was observed on the rate of photosynthesis and colonization in *D. sissoo* (Bisht *et al.*, 2006). The rate of photosynthesis was significantly higher in plants inoculated with AM consortium. Arya and Chaterjee (1995–1996) found better plant biomass and good growth of neem seedlings after inoculation of *G. fasciculatum*. Arya (2006) recorded a change in soil mycoflora after inoculation of AM fungus in neem seedlings. Fungi like *Aspergillus fumigatus*, *A. nidulans*, *A. ochraecous* and *F. pallidoroseum* were not recorded after 3 months.

A significant increase in dry weight of *Santalum* (Krishnamurthy *et al.*, 1998) and *Tamarindus* (Bagyaraj and Reena, 1990) seedlings has been observed after inoculation of AM fungi. In ectomycorrhizae, the presence of a mantle around the root prevents the entry of pathogens, while in endomycorrhizae, the better nutrient uptake makes the plant more resistant to various pathogens.

Fungi are harmful agents to humans but mycorrhizal fungi are indispensable for luxuriant growth of forest trees. Contrary to popular belief, the luxuriance of rainforest is not because the rainforest soil is more fertile (as torrential rains over millennia leach out soluble minerals), but because the roots associate with fungi, whose spreading hyphae increase the area of absorption of scarce nutrients and transport this to the plant in return for photosynthetically fixed carbon (Maheshwari, 2005). In Ghana and the Mopri Forest Reserve of Cote d'Ivoire, *Terminalia ivorensis* plantations are susceptible to dieback, the cause of which is unknown; poor mycorrhizal infection may be a contributory factor (Wilson *et al.*, 1994).

Signalling Pathway in Mycorrhiza

The signalling pathway to activate the mycorrhiza-specific phosphate transporter has its origin in the PL (phospholipid) PC (phosphatidylcholine), imager component of membranes of plants and probably, also of the AM fungus. However, PC is not active in itself. It gains activity only after treatment with PLA_2 and PC from plants, fungus or both remains to be explored further. Several PLA_2s have been identified in plants and all are secretory proteins. Their regulation and substrate specificity are unknown. This might hint at extracellular production of the LPC (lyso-phosphatidylcholine) signal might be generated more specifically in the arbuscules containing cells. LPCs are highly mobile within the intact cells and LPC is therefore a good candidate for a cytoplasmic messenger that transduces signals to activate downstream processes and gene expression in the nucleus (Drissner *et al.*, 2007).

Bioprotectant Nature of AM Fungi

Plant diseases can be controlled by manipulation of indigenous microbes or by introducing antagonists to reduce the disease-producing propagules (Linderman, 1992). AM fungi and their associated interactions with plants reduce the damage caused by plant pathogens (Harrier and Watson, 2004). These interactions have been documented for many plant species. With the increasing cost of inorganic fertilizers and the environmental and public health hazards associated with pesticides and pathogens resistant to chemical pesticides, AM fungi may provide a more suitable and environmentally acceptable alternative for sustainable agriculture (Table 14.1).

Mechanism of Disease Control

Any one or more mechanisms may be operative in plants, imparting them with resistance against pathogens.

1. Physical alteration in plant body.
2. Physiological changes.
3. Biochemical mechanisms

Physical alteration in plant body

According to some scientists, AM affects soilborne plant pathogens on the basis of physical alterations. Lignification of cell wall and production of other polysaccharides has been reported, which prevents penetration of mycorrhizal plants by *F. oxysporum* (Dehne and Schonbeck, 1979) and *Phoma terrestris* (Becker, 1976). Mycorrhizal inoculation improves plant growth. Arya (2006) found better growth of neem seedlings after inoculation with three isolates of *G. fasciculatum*. It has also been suggested that a stronger vascular system of the mycorrhizal plants is likely to increase the flow of nutrients, impart greater mechanical strength and diminish the effect of vascular pathogens (Schonbeck, 1979). A few electron opaque structures resembling the deposits were found in some cells and intercellular spaces of non-infected mycorrhizal carrot roots, but were absent in infected, non-mycorrhizal carrot roots. Restriction of pathogen growth, together with an increase in hyphal alteration and accumulation of new plant products in mycorrhizal roots, but absent in

Table 14.1. Effects of AM fungi on fungal diseases of certain crops.

Crop	AM fungi	Pathogen	Reference
Tomato	*Glomus intraradices*	*Fusarium oxysporum* f. sp. *lycopersici*	Caron *et al.*, 1986; Akkopru and Demir, 2005
	G. mosseae	*F. oxysporum*	Al-Momany and Al-Raddad, 1988
	G. etunicatum	*F. oxysporum* f. sp. *lycopersici*	Bhagawati *et al.*, 2000
	G. mosseae	*Phytophthora parasitica*	Pozo *et al.*, 2002
	G. intraradices	*Rhizoctonia solani*	Berta *et al.*, 2005
Banana	*Glomus* sp.	*Cylindrocladium*	Declerck *et al.*, 2002
	G. proliferum	*Spathiphylli*	Declerck *et al.*, 2002
	G. etunicatum	*R. solani*	Yao *et al.*, 2002
Cucumber	*G. etunicatum* *Glomus* sp.	*Pythium ultimum*	Rosendahl and Rosendahl, 1990
	G. etunicatum	*F. oxysporum* f. sp. *cucumerinum*	Hao *et al.*, 2005
Pepper	*G. mosseae*	*F. oxysporum*	Al-Momany and Al-Raddad, 1988
		P. capsici	Ozgonen and Erkilic, 2007
Onion	*Glomus* sp.	*Sclerotium cepivorum*	Torres-Barragan *et al.*, 1996
Pea	*G. fasciculatum*	*Aphanomyces euteiches*	Rosendahl, 1985
Cowpea	*G. fasciculatum*	*Macrophomina phaseolina*	Devi and Goswami, 1992
	G. fasciculatum	*F. oxysporum*	Sundaresan *et al.*, 1993
Chickpea	*G. fasciculatum*	*F. oxysporum* f. sp. *ciceris*	Siddiqui and Singh, 2004
	G. intraradices	*M. phaseolina*	Akhtar and Siddiqui, 2006
	G. fasciculatum	*M. phaseolina*	Akhtar and Siddiqui, 2007
Cotton	*G. mossae* *G. vesiformae*	*V. dahliae*	Liu, 1995

non-mycorrhizal roots, shows that mycorrhizal infection is responsible at least in part for the plant defence system which provides protection against pathogen attack (Benhamon *et al.*, 1994).

Physiological changes

AM fungi can interact directly with the pathogens through phenomen like antagonism, antibiosis or predation. The studies conducted so far suggest that they affect the host–pathogen relationship indirectly through physiological alteration or by competing for space or host resources. Through increased P nutrition, AM fungi enhance root growth, expand the absorptive capacity of the root system for nutrients and water and affect cellular processes in roots (Hussey and Roncadori, 1982; Reid *et al.*, 1984; Smith and Gianinazzi, 1988). In addition to phosphorus, AM fungi are known to enhance uptake of Ca, Cu, S and Zn (Gerdemann, 1968; Sharma, 1990). *Glomus monosporum* was found effective against *P. capsici* in black pepper (Sivaprasad *et al.*, 2006). The authors found resistance due to improved nutrient uptake. Host susceptibility to infection by the pathogen and tolerance to disease is influenced by the nutritional status of the host and the fertility status of the soil (Wallace, 1973). For example, nematode-damaged plants frequently show deficiencies of B, N, Fe, Mg and Zn (Good, 1968). High levels of P fertilization in the absence of AM fungi can interact with minor elements, creating a deficiency situation which predisposes plants to root knot nematodes (Smith *et al.*, 1986). AM fungi may, therefore, also increase host tolerance to pathogens by increasing uptake of essential nutrients other than P which are otherwise deficient in non-mycorrhizal plants. Production of siderophore can suppress root pathogens (Sharma and Johri, 2002). Higher levels of amino acids, especially arginine, in combination with root exudates of the mycorrhizal plant have been reported to reduce chlamydospore production of *T. basicola* (Baltruschat and Schoenbeck, 1975). Increased levels of phenylalanine and serine have been observed in tomato roots inoculated with *G. fasciculatum*. High concentrations of orthodihydroxy (O-D) phenols in mycorrhizal roots suppressed the growth of *S. rolfsii* (Goodman *et al.*, 1967; Krishna and Bagyaraj, 1983). The presence of HCN precursors has been observed in rubber plant infected with *G. etunicatum* (Lieberei and Feldmann, 1990).

Biochemical mechanisms

The production of phytoalexins in AM-containing plants has been demonstrated conclusively. Enhanced accumulation of glyceollin I, a highly antifungal phytoalexin, has been reported in the roots of mycorrhizal soybeans (Morandi *et al.*, 1984). According to Sharma and Johri (2002), it is not clearly understood how AM fungi induce the production of phytoalexins and elicitors. It may be possible that mycorrhizal fungi perturb root tissues so that the plant elicitors are liberated. Cell damage, which is closely associated with the production of isoflavanoids in legumes (Bailey, 1982), has been observed rarely in mycorrhizal soybean roots. The concentration of coumestrol increased in mycorrhizal roots (25 μg/g) and was much greater than that of glyceollin I (Morandi *et al.*, 1984); coumestrol inhibits the growth of bacteria and nematodes.

According to Chakraborty *et al.* (2005), induction of disease resistance in pea plants against charcoal stump rot was associated with the accumulation of defence enzymes, followed by stimulation of antifungal phenolics.

Roots colonized by an AM fungi exhibit high chitinolytic activities. These enzymes can be effective against other fungal pathogens under the direct influence of mycorrhizal fungi and root tissues become more resistant to pathogenic attack. Since the first report of mycorrhiza-related chitinase in tobacco (Dumas-Gaudot *et al.*, 1992), additional ones have been demonstrated in various plant species. Lambias and Mehdy (1996) evaluated the expression of mycorrhiza-specific chitinases and ß-1,3-glucanases in soybean root infected with *G. intraradices*.

The efficacy of six AM species, *A. morrawae*, *G. margarita*, *G. fasciculatum*, *G. macrocarpum*, *S. calospora* and *Sclerocystis rubiformis* obtained from rhizosphere of *C. microphyllus* was evaluated for enhancement of PRO (peroxidase), PPO (polyphenol oxidase) effect, with *S. calospora* being the most promising of all the fungi. Good results were observed with *G. fasciculatum* in *W. somnifera*.

AM fungi ensure protection against certain soilborne pathogens (Diop, 1996). An AM fungus influences microbial populations and improves soil texture by the secretion of mucilaginous compounds (Strullu *et al.*, 1991). Vesicles are lipid-filled and are initiated after the formation of arbuscules, but live longer after the senescence of arbuscules (Diouf *et al.*, 2003). In *Medicago truncatula*, at an early stage of arbuscule development by *G. versiforme*, bright diffuse florescence is seen around the arbuscular branches following antitubulin labelling. At later stages of development, short microtubules are closely associated with plasma membrane surrounding the labyrinthine surface of the arbuscule. γ Tubulin has been shown to be associated with the nuclear envelope and perifungal membrane in tobacco arbuscular mycorrhizas (Genre and Bonfante, 1999). Mycorrhizosphere changes in populations of antagonists to specific pathogens depend on having those antagonists present in background soil. If antagonists are absent and deleterious microbes are present in significant numbers and enhanced by AM, the incidence of disease can be increased (Sharma *et al.*, 2002).

Conclusions

The use of AM fungi as a biofertilizer is the only alternative for successful farming.

With this approach, we can obtain maximum return, not only for a season or a year, but also over centuries. Mycorrhizal–disease interaction has been studied in different plants by various workers. In many cases, a better growth of plant is reported after inoculations of an exotic AM fungi. A low level of fungal aggressiveness and a weak plant reaction are no doubt two key factors that help in the establishment of a successful symbiotic relationship between the two organisms.

Increase in peroxide activity which is localized in plant vacuoles and the cell wall (Schloss *et al.*, 1987) in AM-infected root is one example of a mechanism of plant resistance to microorganisms. Likewise, production of phytoalexins (glyceolin I, daidzein and coumestrol), chitinase, ß-1,3-glucanases and b_1 (PR) protein are important components of AM-induced changes leading to the resistance of the host to other pathogens. The systemic effect of AM fungi to *Phytophthora* infection in tomato has been demonstrated by Pozo *et al.* (2002). Spores of AM fungi are reported from different soils in the country. These symbionts improve plant growth. Their utilization as biocontrol agents is gaining importance after many successful trials. Multiplication and inoculum production of indigenous efficient AM fungi should be undertaken. Efforts are needed to commercialize these novel microbes to bring about a second green revolution in the country.

References

Abdalla, M.E. and Abdel-Fattah, G.H. (2000) Influence of the endomycorrhizal fungus *Glomus mosseae* on the development of peanut pod rot disease in Egypt. *Mycorrhiza* 10, 29–35.

Akhtar, M.S. and Siddiqui, Z.A. (2006) Effect of phosphate solubilizing microorganisms on the growth and root rot disease complex of chickpea. *Mikologiya i Fitopatologiya* 40, 246–254.

Akhtar, M.S. and Siddiqui, Z.A. (2007) Effects of *Glomus fasiculatum* and *Rhizobium* sp. on the growth and root rot disease complex on chickpea. *Archives of Phytopathology and Plant Protection* 40, 37–43.

Akkopru, A. and Demir, S. (2005) Biocontrol of *Fusarium* wilt in tomato caused by *Fusarium oxysporum* f.sp. *lycopersici* by AMF *Glomus intraradices* and some rhizobacteria. *Journal of Phytopathology* 153, 544–550.

Al-Momany, A. and Al-Radded, A. (1988) Effect of vesicular arbuscular mycorrhiza on *Fusarium* wilt of tomato and pepper. *Alexandria Journal of Agricultural Research* 33, 249–261.

Arya, A. (2006) Application of arbuscular mycorrhizal fungi for biomass production in neem. In: Prakash, A. and Mehrotra, V.S. (eds) *Mycorrhiza.* Scientific Publishers, Jodhpur, India, pp. 259–264.

Arya, A. and Chatterjee, T.J. (1995–1996) VAM and its effect on neem seedlings. *J.M S. University of Baroda,* 42–43 (Sc. Tech. & Medicine Number) Publ. 1998, pp. 95–101.

Bagyaraj, D.J. and Reena, J. (1990) Growth stimulation of *Tamarindus indica* by selected VA mycorrhizal fungi. *World Journal of Microbiology and Biotechnology* 6(1), 59–63.

Bailey, J.A. (1982) Mechanism of phytoalexins accumulation. In: Bailay, J.A. and Manfield, J.W. (eds) *Phytoalexins.* Blackie and Son, Glasgow, UK, pp. 289–318.

Baltruschsat, H. and Shoenbeck, F. (1975) The influence of endotrophic mycorrhiza on the infestation of *Thielaviopsis basicola* in tobacco root. *Phytopathologische zeitschriff* 84, 172–188.

Becker, W.N. (1976) Quatification of onion vesicular arbuscular mycorrhiza and their resistance to *Pyrenochaeta terrestris.* PhD thesis, University of Illinos, Urbana.

Benhamon, N., Fortin, J.A., Hamel, C., St-Arnaud, M. and Shatilla, A. (1994) Resistance responses of mycorrhizal Ri T-DNA transformed carrot roots to infection by *Fusarium oxysporum* f. sp. *Chrysanthem. Phytopathlogy* 84, 958–968.

Berta, G., Sampo, S., Gamalero, E., Musasa, N. and Lemanceau, P. (2005) Separation of *Rhizoctonia* root rot of tomato by *Glomus mosseae* BEG 12 and *Pseudomonas flourescens* A6RI is associated with their effect on the pathogen growth and on the root morphogenesis. *Eur. Journal of Plant Pathology* 111, 279–288.

Bhagawati, B., Goswami, B.K. and Singh, S. (2000) Management of disease complex of tomato caused by *Meloidogyne incognita* and *Fusarium oxysporum* f. sp. *lycopersici* through bioagents. *Indian Journal of Nematology* 30, 16–22.

Bisht, V.S., Krishna, K.R. and Nene, Y.L. (1985) Interaction between vesicular-arbuscular mycorrhiza and *Phytophthora drechsleri* f. sp. *cajani. International Pigeonpea Newsletter* 4, 63–64.

Bisht, R., Johri, B.N. and Srivastava, P.C. (2006) Arbuscular mycorrhizal, *Pseudomonas* and *Rhizobium* tripartite interaction on the growth of *Dalbergia sissoo.* In: Prakash, A. and Mehrotra, V.S. (eds) *Mycorrhiza.* Scientific Publishers, Jodhpur, India, pp. 133–138.

Boyethko, S.H. and Tewari, J.P. (1990) Effect of phosphorus and VA mycorrhizal fungi on common root rot of barley. *8th NACOM Innovation and Hierarchical Integration, 5–8 September, 1990. Jackson, Wyoming,* 33 pp.

Caron, M., Fartian, J.A. and Richard, C. (1986) Effect of inoculation sequence on the interaction between *Glomus intraradices* and *Fusarium oxysporum* f.sp. *radicis lycopersici* in tomatoes. *Canadian Journal of Plant Pathology* 8, 12–16.

Chakaravarty, P. and Mishra, R.R. (1986) Influence of endotrophic mycorrhiza on the fusarial wilt of *Cassia tora* L. *Phytopathology* 115(2), 130–133.

Chakraborty, B.N., Chakraborty, U., Barman, B.C. and Rai, K. (2005) Field application of VAM for management of charcoal stump rot disease of tea. In: Khetarpal, R.K., Gupta, K., Chalam, V.C., Gour, H.N., Thakore, B.B.L. and Bhargava, S. (eds) *Second Global Conference on Plant Health – Global Wealth, 25–29 November 2005. Indian Society of Mycology and Plant Pathology, Udaipur, India,* pp. 204–205.

Champawat, R.S. (1991) Interaction between vesicular arbuscular mycorrhizal fungi and *Fusarium oxysporum* f. sp. cumin. *Proceedings of the Indian National Science Academy* 57b(1), 59–62.

Declerck, S., Risede, J.M., Rufyikiri, G. and Delvaux, B. (2002) Effects of arbuscular mycorrhizal fungi on the severity of root of bananas caused by *Cylindrocladium spathiphylli. Plant Pathology* 51, 109–115.

Dehne, H.W. and Schonbeck, F. (1979) Investigation of the influence of endotrophic mycorrhizas on plant disease, phenol metabolism and lignification. *Phytopathologishe Zeitschrift* 95, 210–216.

Devi, T.P. and Goswami, B.K. (1992) Effect of VA mycorrhiza on the disease incidence due to *Macrophomina phaseolina* and *Meledogyne incognita* on cowpea. *Annals of Agricultural Research* 13, 253–256.

Diop, T.A. (1996) Les mycoorhizes a vesicules at arbuscules. *Journal Faculte des Sciences et Techniques* (Universite Cheikh Anta Diop, Dakar, Senegal) 2, 49–64.

Diouf, D., Diop, T.A. and Ndoye, I. (2003) Actinorhizal, mycorhizal and rhizobial symbiosis: how much do we know? *African Journal of Biotechnology* 2, 1–7.

Douds, D.D. Jr, Galvez, L., Becard, G. and Kapulnik, Y. (1998) Regulation of arbuscular mycorrhizal development by plant host and fungus species in alfalfa. *New Phytologist* 138, 27–35.

Drissner, D., Kunze, G., Callewaert, N., Gehrig, P., Tamasloukht, M.D., Boller, T., Felix, G., Amrhein, N. and Bucheir, M. (2007) Lysophophatidycholine is a signal in the arbuscular mycorrhizal symbiosis. *Science* 318, 265–268.

Dumas-Gaudot, E., Furlan, V., Grenier, U. and Asselin, A. (1992) New acidic chitinase isoforms induced in tobacco roots by vesicular arbuscular mycorrhiza fungi. *Mycorrhiza* 1, 133–136.

Feldmann, F., Janqueira, N.T.V., Grenier, U. and Asselin, A. (1990) Utilization of VA – mycorrhiza as a factor in integrated plant protection. *Agriculture, Ecosystems and Environment* 29(1–4), 131–135.

Genre, A. and Bonfante, P. (1999) Cytoskeleton related proteins in tobacco mycorrhizal cells: γ tubulin and clathrin localization. *European Journal of Histochemistry* 43, 105–111.

Gerdemann, J.W. (1968) Vesicular arbuscular mycorrhiza and plant growth. *Annual Review of Phytopathology* 6, 397–418.

Giovannetti, M., Tosiu, L., Torre, G.D. and Zaggerini, A. (1991) Histological, physiological and biochemical interaction between vesicular arbuscular an *Thielaviopsis basicola* on tobacco plants. *Journal of Phytopathology* 131(4), 265–274.

Good, J.M. (1968) Relation of plant parasitic nematodes to soil management practices. In: Smart, G.C. and Parry, V.G. (eds) *Tropical Nematology.* University of Florida, Gainesville, Florida, 152 pp.

Goodman, R.N., Kiralay, Z. and Zaitlin, M. (1967) *The Biochemistry and Physiology of Infectious Plant Disease.* Van Nostrand Company, Inc, Princeton, New Jersey, pp. 187–231.

Graham, J.H. and Egel, D.S. (1988) *Phytophthora* root rot development on mycorrhizal and phosphorus-fertilized non-mycorrhizal sweet orange seedlings. *Plant Disease* 72(7), 611–614.

Graham, J.H. and Menge, J.A. (1982) Influence of vesicular arbuscular mycorrhizae and soil phosphorus on take all diseases of wheat. *Phytopathology* 72, 95–98.

Grey, W.E., Leur, J.A.G., Kashour, G. van and El-Naime, M. (1989) The interaction of vesicular arbuscular mycorrhizae and common root rot (*Cochliobolus sativus*) in barley. RACHIS 8, 18–20.

Hao, Z., Christie, P., Qin, L., Wang, C. and Li, X. (2005) Control of Fusarium wilt of cucumber seedlings by inoculation with arbuscular mycorrhizal fungus. *Journal of Plant Nutrition* 28, 1961–1974.

Harlapur, S.I., Kulkarni, S., Hedge, R.K., Sreenivasa, M.N. and Rajeshkhara, E. (1990) Interaction between vesicular arbuscular (VA) mycorrhiza and *Sclerotium rolfii* sacc. in wheat. *Karnataka Journal of Agricultural Sciences* 3, 203–207.
Harrier, L.A. and Watson, C.A. (2004) The potential role of arbuscular mycorrrhizal fungi (AM) in the bioprotection of plants against soil borne pathogens in organic and/or other sustainable farming systems. *Pest Management Science* 60, 149–157.
Hasan, A. and Khan, M.N. (2006) Seasonal dynamics of vesicular arbuscular mycorrhiza and parasitic nematode *Helicotylenchus indicus* on phalsa. *Mycorrhiza News* 18, 9–12.
Hock, B. and Verma, A. (1995) *Mycorrhiza Structure Function, Molecular Biology and Biotechnology.* Springer, Berlin.
Hussey, R.S. and Roncadori, R.W. (1982) Vesicular arbuscular endomycorrhizae may limit nematode activity and improve plant growth. *Plant Disease* 66, 9–14.
Iqbal, S.H., Ghazala, N. and Mubashir, N. (1988) Role of vesicular arbuscular mycorrhiza as a deterrent to damping off caused by *Rhizoctonia solani* in *Brassica oleracea. Biologia Pakistan* 34, 79–84.
Jalali, B.L., Chhabra, M.L. and Singe, R.P. (1990) Interaction between vesicular-arbuscular mycorrhizal endophytes and *Macrophomina phaseolina* in mungbean. *Indian Phytopathology* 43, 527–530.
Johar, G.S. (2005) Thinking of unthinkable: building plants resistant to all diseases. In: Khetarpal, R.K., Gupta, K., Chalam, V.C., Gour, H.N., Thakore, B.B.L. and Bhargava, S. (eds) *Second Global Conference on Plant Health – Global Wealth, 25–29 November 2005.* Indian Society of Mycology and Plant Pathology, Udaipur, India, 276 pp.
Karagiannidis, N., Blestor, F. and Stavropoulos, N. (2002) Effect of verticillium wilt (*Verticillium dohliae kleb.*) and mycrorrhiza (*Glomus mosseae*) on root colonization, growth and nutrient uptake in tomato and egg plant seedling. *Scientia Horticulturae* 94(1–2), 145–156.
Khadge, B.R., Ilag, L.L. and Mew, T.W. (1990) Interaction study of *Glomus mosseae* and *Rhizoctonia solani.* In: Jalali, B.L. and Chand, H. (eds) *Proceedings of the National Conference on Mycorrhiza. Current Trends in Mycorrhizal Research, 14–15 February 1990, Haryana Agriculture University, Hissar, India,* pp. 94–95.
Krishna, K.R. and Bagyaraj, D.J. (1983) Interaction between *Glomus fasciculatum* and *Sclerotium rolfsii* in peanut. *Canadian Journal of Botany* 61, 2349–2351.
Krishnamurthy, K.V., Nelson, R. and Kumar, S. (1998) VAM association in sandal hemi-root parasite. *Mycorrhiza News* 9(4), 11–12.
Lambias, M.R. and Mehdy, M.C. (1996) Soyabean roots infected by *Glomus intraradices* strains differing in infectivity exhibit differential chitinase and β-1,3-glucanase expression. *New Phytologist* 134, 531–538.
Lieberei, R. and Feldmann, F. (1990) Physiological change in roots colonized by vesicular arbuscular mycorrhizal fungi reactions in mutualistic and parasitic interactions. *Agriculture, Ecosystems and Environment* 29, 251–255.
Linderman, R.G. (1992) VA mycorrhizae and soil microbial interactions. In: Bethelenfalvay, G.J. and Linderman, R.G. (eds) *Mycorrhiza in Sustainable Agriculture.* ASA Special Publication No. 54, Madison, Wisconsin, pp. 45–70.
Liu, R.J. (1995) Effect of vesicular-mycorrhizal fungi on Verticillium wilt of cotton. *Mycorrhiza* 5, 293–297.
Maheshwari, R. (2005) Fungal biology in 21st century. *Current Science* 88, 1406–1418.
Manian, S., Thompson, T.E. and Arunkumar, A. (2006) Management of pests and diseases of tomato (*Lycopersicon esculentum* Mill.) seedling vesicular arbuscular mycorrihza. In: Prakash, A. and Mehrothra, V.S. (eds) *Mycorrhiza.* Scientific Publishers, Jodhpur, India, pp. 127–131.
Manoharachary, C. (2004) Biodiversity, taxonomy, ecology, conservation and biotechnology of arbuscular mycorrhizal fungi. *Indian Phytopathology* 57(1), 1–6.
Meyer, J.R. and Linderman, R.G. (1983) Response of subterranean clover to dual inoculation with vesicular arbuscular mycorrhizal fungi and a plant growth promoting bacterium *Pseudomonas putida. Soil Biology and Biochemistry* 22, 1085–1088.
Morandi, D., Bailey, J.A. and Gianinazzi-Pearson, V. (1984) Isoflavanoid accumulation in soyabean roots infected with vesicular arbuscular mycorrhizal fungi. *Physiological Plant Pathology* 24, 257–364.
Mosse, B. (1986) Mycorrhiza in a suitable agriculture. *Biology, Agriculture and Horticulture* 3, 191–209.
Newman, E.I. and Reddell, P. (1987) The distribution of mycorrhizae among the families of vascular plants. *New Phytologist* 106, 745–751.
Ozgonen, H. and Erkilic, A. (2007) Growth enhancement and *Phytophthora* blight (*Phytophthora capsici* L.) control by arbuscular mycorrhizal fungal inoculation in pepper. *Crop Protection* 26, 1682–1688.

Pandey, K.K. and Upadhayay, J.P. (2000) Microbial population from rhizosphere and non-rhizosphere soil of pigeon pea: screening for resident antagonist and mode of mycoparasitism. *Journal of Mycology and Plant Pathology* 33, 7–10.

Pozo, M.J., Cordier, C., Dumas-Gaudot, E., Gianinazzi, S., Baria, J.M. and Azcon-Aguilar, C. (2002) Localised versus systematic effects of arbuscular mycorrhizal fungi on defense response on *Phytophthora* infection in tomato plants. *Journal of Experimental Botany* 53, 525–534.

Ramraj, B., Shanmugam, N. and Reddy, D.A., (1988) Biocontrol of macrophomina root rot of cow pea and fusarium wilt of tomato by using VAM fungi. In: *Mycorrhizae for Green Asia: Proceedings of the First Asian Conference on Mycorrhizae, January 1988, University of Madras, Madras, India.* pp. 29–30, 250–251.

Redecker, D., Mortan, J.B. and Bruns, T.D. (2000) Ancestral lineages of arbuscular mucrorrhizal fungi (Glomales). *Molecular Phylogenetics and Evolution* 14, 276–284.

Reid, C.C.P., Todd, R.L. and Giddens, J.E. (1984) Mycorrhizae: a root–soil interface in plant nutrition. In: Rolston, D.E. and Todd, R.L. (eds) *Microbial–Plant Interaction.* American Society of Agronomy Publication 47, pp. 29–50.

Rosendahl, C.N. and Rosendahl, S. (1990) The role of vesicular arbuscular mycorrhizal fungi in controlling damping-off and growth reduction in cucumber caused by *Pythium ultimum. Symbiosis* 9, 363–365.

Rosendahl, S. (1985) Interaction between the vesicular arbuscular mycorrhizal fungus *Glomus fasciculatum* and *Aphanomyces euteiches* root rot of peas (*Pisum sativum*). *Phytopathology* 114, 31–40.

Ross, J.P. (1972) Influence of endogone mychorrizha in *Phytophthora* root rot of soybean. *Phytopathology* 62, 896–897.

Schenck, N.C., Ridings, W.H. and Cornell, J.A. (1977) Interaction of two vesicular arbuscular mycorrhizal fungi and *Phytophthora parasitica* on two citrus root stocks. (Abst.). In*: Proceedings of 3rd North American Conference on Mycorrhizae*, 23–25 August 1977, University of Georgia, Atlanta, 9 pp.

Schloss, P., Walter, C. and Mader, M. (1987) Basic peroxidases in isolated vacules of *Nicotiana tobacum* I. *Planta* 170, 225–229.

Schonbeck, F. (1979) Endomycorrhiza in relation to plant disease. In: Schipper, B. and Gams, W. (eds) *Soil Borne Plant Pathogens.* Academic, New York, pp. 271–280.

Schonbeck, F. and Dehne, H.W. (1979) Investigations on the influence of endotropic mycorrihza on plant disease of fungal parasites on shoots, *Olpidium brassicae. TMV Zeitschrift fur Pflanzen krank heilen und Pflanzenschutz* 86, 103–112.

Schussler, A. (2005) http//www.tudarmstadt.de/fb/bio/bot/schuessler/amphylo/amphylogeny.html (accessed 19 October 2005).

Schussler, A., Schwarzott, D. and Walker, C. (2001) A new fungal phylum, the Glomeromycota, phylogeny and evolution. *Mycological Research* 105, 1413–1421.

Sharma, A.K. (1990) Nature of interaction between *Rhizobium* and *Glomus caledonium* in chickpea *(Cicer arietinum l.). National Academy of Science* 66(b), 81–85.

Sharma, A.K. and Johri, B.N. (2002) *Arbuscular Mycorrhizae, Interactions in Plants, Rhizosphere and Soils.* Oxford and IBH Publication Co, New Delhi, 311 pp.

Sharma, S., Singh, A. and Vasudevan, P. (2002) Interaction of AM fungi with nitrogen fixing and phosphorus solubilizing microorganisms: a review. *International Journal of Ecology and Environmental Sciences* 28, 115–131.

Siddiqui, Z.A. and Singh, L.P. (2004) Effects of soil inoculants on the growth transpiration on wilt disease of chickpea. *Journal of Plant Diseases and Protection* 111, 151–157.

Sivaprasad, P. Joseph, P.J. and Sulochana, K.K. (2006) Effect of arbuscular mychorrhizal fungi on foot rot of black pepper and rhizome rot of ginger under field conditions. In: Prakash, A. and Mehrothra, V.S. (eds) *Mycorrhiza.* Scientific Publishers, Jodhpur, India, pp. 123–126.

Smith, G.S., Roncadori, R.W. and Hussey, R.S. (1986) Interaction of endomycorrhizal fungi, superphosphate and *Meloidogyne. Journal of Nematology* 20, 539–544.

Smith, S.E. and Gianinazzi-Pearson, V. (1988) Physiological interactions between symbionts in vesicular-arbuscular mycorrhizal plants. *Annual Review of Plant Physiology and Plant Molecular Biology* 39, 221–244.

Smith, S.E. and Read, D.J. (1997) *Mycorrhizal Symbiosis*, 2nd edn. Academic, London.

Sreenivasa, M.N. and Bagyaraj, D.J. (1989) Use of pesticide for mass production of vesicular arbuscular mycorrizhal inoculum. *Plant and Soil* 119, 127–132.

Strullu, D.G., Perrin, R., Plenchette, C. and Garbaye, J. (1991) *Les mycorrhizes – arbres et des plantes cultivees.* Lavoisier, Paris.

Sundaresan, P., Ubalthoose, R. and Gunashekaran, P. (1993) Induction and accumulation of phytoalexins in cowpea roots infected with a mycorrhizal fungus *Glomus fasciculatum* and their resistance to *Fusarium* wilt disease. *Journal of Biosciences* 18, 291–301.

Tello, J.C., Vares, F., Notario, A. and Lacasa, A. (1987) Mycorrhizae: a potential against plant disease. *ITEA (Informacion techica economica agrarian)* 18(73), 40–64.

Torres-Barragan, A., Zavaleta-Mejial, E., Gonzalez-Chavez, C. and Ferrera-Cerrato, R. (1996) The use of arbuscular mycorrhizae to control onion white rot (*Sclerotium cepivorum* Berk.) under field conditions. *Mycorrhiza* 6, 253–257.

Trotta, A., Varese, G.C., Gnavi, E., Fusconi, A., Sampo, S. and Berta, G. (1996) Interaction between soil borne root pathogen *Phytophthora nicotianae* var. *parasitica* and the arbuscular mycorrhizal fungus *Glomus mossae* in tomato plants. *Plant and Soil* 185, 199–209.

Utkhede, R.S., Li, T.S. and Smith, E.M. (1992) The effect of *Glomus mosseae* and *Enterobacter aerogens* on apple seedlings grown in apple replant disease soil. *Journal of Phytopathology* 135, 281–288.

Vidhyasekaran, P. (2004) *Encyclopedia of Plant Pathology.* Viva Books Private Limited, New Delhi, 619 pp.

Vyas, A. (2005) Arbuscular mycorrhizal fungi in biocontrol of soilborne diseases of medicinal plants in Indian Thar desert. In: Khetarpal, R.K., Gupta, K., Chalam, V.C., Gour, H.N., Thakore, B.B.L. and Bhargava, S. (eds) *Second Global Conference on Plant Health–Global Wealth*, 25–29 November 2005. Indian Society of Mycology and Plant Pathology, Udaipur, India, 201 pp.

Wallace, H.R. (1973) *Nematode Ecology of Plant Diseases.* Alden Press, Oxford, UK, 228 pp.

Wilson, J., Ingleby, K., Mason, P.A., Ibrahim, K. and Lawson, G.J. (1994) Long- term changes in vesicular arbuscular mycorrhizal spore populations in *Terminalia* plantations in Cote Ivoire. In: Read, D.J., Lewis, D.H., Fitter, A.H. and Alexander, I.J. (eds) *Mycorrhizae in Ecosystems.* CAB International, Wallingford, UK, pp. 268–275.

Yao, M., Tweddell, R. and Desilets, H. (2002) Effect of two vesicular arbuscular mycorrhizal fungi on the growth of micropropagated potato plantlets and on the extent of disease caused by *Rhizoctonia solani. Mycorrhiza* 12, 235–242.

Zambolin, L. and Schenck, N.C. (1983) Reduction of the effect of pathogenic root rot infecting fungi on soyabean by the mycorrhizal fungus *Glomus mosseae. Phytopathology* 73, 1402–1405.

15 Role of Fungal Endophytes in Plant Protection

S.K. Gond, V.C. Verma, A. Mishra, A. Kumar and R.N. Kharwar
Mycopathology and Microbial Technology Laboratory, Centre of Advanced Study in Botany, Banaras Hindu University, Varanasi, India

Abstract

Endophytes are the microorganisms that reside inside healthy plant tissues without causing any detectable disease symptoms to the host. Often, each and every plant harbours either one or a battery of endophytic microorganisms. The study of endophytes is now on a voyage of interest, not only because of their role in filling the divide between discovered and undiscovered microbial diversity, but also due to their harbouring a great potential to produce novel natural products. Other than soil, higher plants also act as an alternative resource to isolate potential microorganisms. Natural compounds ranging from crop protection to human welfare have been isolated from this alternative source of endophytes. Several anticancer, antibiotic, antimycotic, antiviral, antioxidant, nematicide, insecticide and immunosuppressive compounds have been reported from endophytes, such as cytochalasines, ambuic acid, oocydin, jesterone, cryptocandin, lolitrem B, and 3-hydroxypropionic acid and taxol, etc. Many of them produce some toxic alkaloids and protect their hosts from herbivores. They also improve the growth and yield of crops under various stressed conditions. Endophytic fungi have been emerging as a new tool in genetic engineering, the pharmaceutical industry and in crop protection as well. In this chapter, the ability and role of endophytic fungi to ward off pests and environmental stresses on plants is discussed.

Introduction

The use of agrochemicals as a single control measure in the field to protect crops from their pests has been generating resistance in these pests, and also represents a high risk to field workers and consumers. Most of these chemicals are non-biodegradable and are responsible for polluting the environment. The control of phytopathogens has relied mostly on chemical control agents such as methyl bromide (Jarvis, 1993) but after the Montreal Protocol (1991), the manufacturing of and trade in methyl bromide was phased out in 2005. The utilization of biological materials is an alternative and safe way to protect plants from phytopathogens. The control of plant pathogens by phylogenetically diverse microorganisms acting as natural antagonists has been demonstrated repeatedly over the past 100 years. The antifungal ability of *Trichoderma* sp. has been well known since the 1930s and extensive efforts have been made since then to use them seriously for plant disease control (Harman, 1996).

Although the term 'endophyte' was used much earlier in 1866 by German scientist,

Heinrich Anton De Bary, the presence of endophytes in plant was only recorded in 1904 when Freeman (1904) described an entire plant's endophyte life history in the seeds of darnel (*Lolium perenne* sub sp. *temulentum* = *L. temulentum*). This is a topographical term and includes bacteria, fungi, actinomycetes and algae which spend their whole life, or a period of their life cycle, inside healthy plant tissue without causing any disease symptoms. Among all endophytes, after bacteria, fungi are dominant in higher plants (Figs. 15.1 and 15.2). It seems that other microbial forms almost certainly exist in plants as endophytes such as mycoplasmas (pleuro pneumonia-like organisms – PPLO), rickettsia and archaebacteria; however, no evidence of them has yet been observed. On the basis of their nature, endophytes may be categorized in three groups: (i) pathogens of another host that are non-pathogenic in their endophytic relationship; (ii) non-pathogenic microbes; and (iii) pathogens that have been rendered non-pathogenic but still capable of colonization by selection methods or genetic alteration (Backman and Sikora, 2008). Endophytes are recorded from lower plant to higher plant hosts (Stone *et al.*, 2000). Each and every plant is a reservoir of one or a suite of endophytes. In angiosperms, *Poaceae* members are studied more for their endophytes.

Endophytic fungi are now attracting great interest from researchers as an alternative source in controlling plant and human pathogens. Some of the earlier workers before the 1970s documented the endophytic fungi residing inside the plant, exploring the biodiversity of hidden fungi. The period of 1981 to 1985 can be considered a historical one in the study of endophytes, as plant protection against herbivore insects was demonstrated by endophytic microorganisms. Webber (1981) demonstrated for the first time the role of endophytic *Phomopsis oblonga* in the protection of elm trees against the beetle *Physocnemum brevilineum*. This report generated interest in the role of endophytes in plant protection. Now, their beneficial role to plants as well as to humans is being considered. In this regard, a large number of antimicrobial compounds have been isolated

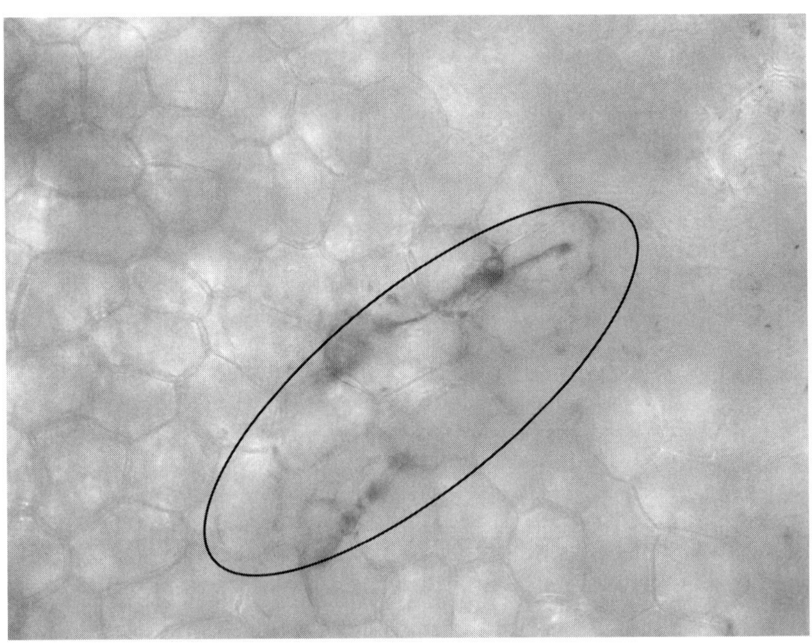

Fig. 15.1. Endophytic fungal mycelia and spores within plant tissue stained with aniline-blue.

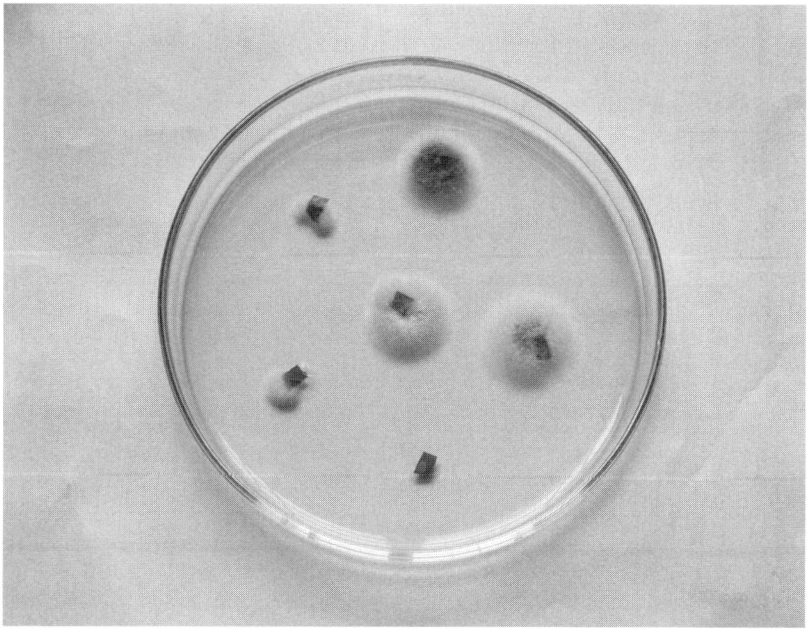

Fig. 15.2. Leaf pieces in Petri plate (21 days old) showing emergence of endophytic fungal mycelia.

from these endophytic microorganisms (Strobel, 2002, 2003; Zhang *et al.*, 2006; Kharwar *et al.*, 2009). Endophytic fungi are now recognized as a new tool in the production of antimicrobials and pharmaceutical compounds. In the search for bioactive compounds, several endophytic fungi have been reported from the medicinal plants of North India (Gond *et al.*, 2007; Verma *et al.*, 2007; Kharwar *et al.*, 2008).

Antimicrobials and their Activities Produced from Endophytes

Antifungal activity of endophytes

Fungi are major causal organisms of various diseases in plants. Many synthetic fungicides are available on the market, but they are giving resistance to pathogens and are also assisting in increasing the hazards to human health. Data show that 52.3% of endophytic fungal fermentation broths display growth inhibition to at least one pathogenic fungus, such as *Neurospora* sp., *Trichoderma* sp. and *Fusarium* sp. (Huang *et al.*, 2001). In a similar study, fermentation broths of 9 (4.8%) out of 187 endophytic fungi isolated from mainly woody plants were highly active against *Phytophthora infestans* in tomato plants (Park *et al.*, 2005). Induced resistance against *Fusarium* wilt by endophytic *F. oxysporum* was generated in tomato plants (Duijff *et al.*, 1998). *Sclerotinia sclerotiorum* is a common root, crown and stem rot causing pathogen to several hosts such as cabbage, common bean, citrus, celery, coriander, melon, squash, soybean, tomato, lettuce, cucumber, etc. Cyclosporine is characterized as a major antifungal substance against *S. sclerotiorum* from the fermentation broth of endophytic *F. oxysporum* (Rodriguez *et al.*, 2006). Out of 510 isolates of endophytic fungi, 64 isolates gave antifungal activities against *Candida albicans*, *C. glabrata*, *C. krusei*, *Cryptococcus neoformans*, *Aspergillus fumigatus*, *A. flavus*, *Rhizopus oryzae*, *Trichophyton rubrum* and *Microsporum canis* (Anke *et al.*, 2003).

Narisawa *et al.* (2000) found that the root endophytic hyphomycete, *Heteroconium chaetospira*, suppressed *Verticillium* sp. in

Chinese cabbage in the field. *Verticillium* wilt is one of the most destructive diseases of aubergine. Eleven out of 123 isolates of endophytic fungi, especially *H. chaetospira*, *Phialocephala fortinii*, *Fusarium*, *Penicillium*, *Trichoderma* and *Mycelium radicis atrovirens* (MRA), after being inoculated on to axenically reared aubergine seedlings, almost completely suppressed the pathogenic effects of a post-inoculated, virulent strain of *V. dahliae* (Narisawa *et al.*, 2002). Out of 39 endophytes of *Artemisia annua*, 21 showed *in vitro* antifungal activity against a number of fungal pathogens (Liu *et al.*, 2001). The extracts of endophytic *Alternaria* sp., isolated from medicinal plants of the Western Ghats of India, inhibited the growth of *C. albicans* (Raviraja *et al.*, 2006). *Colletotrichum gloeosporioides* was isolated as an endophyte from healthy leaves of *Cryptocarya mandioccana*, giving antifungal activity against phytopathogenic fungi *Cladosporium cladosporioides* and *C. sphaerospermum* (Inacio *et al.*, 2006). Fungal endophytes *Chaetomium* and *Phoma* sp., isolated from asymptomatic leaf of wheat, reduced the number and the area of pustules of *Puccinia recondita* f. sp. *tritici*. A study showed 40%, 65% and 27% antagonistic interaction by endophytic morphospecies *in vitro* against cacao pathogens, *Moniliophthora roreri*, *P. palmivora* and *Crinipellis perniciosa*, respectively, while in the field the endophytic *C. gloeosporioides* produced a significant decrease in pod loss (Mejia *et al.*, 2008).

Cryptosporiopsis quercina is an endophytic fungus of a medicinal plant, *Tripterigium wilfordii*. It was observed that *C. quercina* produced an antimycotic compound, cryptocandin, which was active against a number of human and plant pathogenic fungi, including *C. albicans*, *S. sclerotiorum* and *Botrytis cinerea* (Strobel *et al.*, 1999a). A number of antifungal compounds have been identified by Yue *et al.* (2000) from the cultures of *Epichloe* and *Neotyphodium* species which showed activity against chestnut blight fungus, *Cryphonectria parasitica*. These compounds were indole derivatives, indole-3-acetic acid and indole-3-ethanol, a sesquiterpene and a diacetamide. Consequently, a tetramic acid, cryptocin, has also been isolated from the cultures of *C. quercina*, which exhibits strong antifungal activity against *Pyricularia oryzae*, the causal agent of blast of rice, as well as some other plant pathogenic fungi (Li *et al.*, 2000).

Colletotrichum gloeosporioides was isolated from *A. mongolica*, which produced antifungal metabolite colletotric acid, against the fungus *Helminthosporium sativum* (Zou *et al.*, 2000). Another *Colletotrichum* sp., isolated from *A. annua*, produced bioactive metabolites that were fungistatic to several plant-pathogenic fungi (Lu *et al.*, 2000). *Pestalotiopsis microspora* is a commonly isolated and well-identified fungus from every rainforest plant and, as a single endophytic species, it contributes a high percentage to the total mass of fungal endophytes in any host. *Pestalotiopsis* is observed to produce many antimicrobial secondary metabolites. One such secondary metabolite is ambuic acid, an antifungal agent which has been described from several isolates of *P. microspora* (Li *et al.*, 2001). *P. jesteri*, isolated from the Sepik River area of Papua New Guinea, produced jesterone and hydroxyjesterone which exhibited antifungal activity against a variety of plant-pathogenic fungi (Li and Strobel, 2001). Two new metabolites, ethyl 2,4-dihydroxy-5,6-dimethylbenzoate and phomopsilactone, have been isolated from *P. cassiae*, an endophytic fungus in *Cassia spectabilis*, with strong antifungal activity against the phytopatogenic fungi, *C. cladosporioides* and *C. sphaerospermum* (Silva *et al.*, 2005).

An aquatic plant, *Rhyncholacis penicillata*, is known worldwide to harbour a potent antifungal microbe, *Serratia marcescens*, which produces an antioomycetous compound named oocydin A (Strobel *et al.*, 1999b). Oocydin A provides the plants with a strong protection against several water moulds.

Antibacterial activity of endophytes

The antimicrobial activity of endophytic fungi has been observed in a range of bacteria representing pathogens to plants and humans. The broths of 16 endophytic fungi isolated from the medicinal herb, *Cynodon dactylon* (*Poaceae*), were identified as having potent anti-*Helicobacter pylori* activity. The most active endophyte, identified as *Aspergillus* sp. (strain number: CY725), produced four active fractions and was identified as: (i) helvolic acid; (ii) monomethylsulochrin; (iii) ergosterol; and (iv) 3β-hydroxy-5α, 8α-epidioxy-ergosta-6, 22-diene with corresponding MICs of 8.0, 10.0, 20.0 and 30.0 μg/ml against *H. pylori*, respectively (Li *et al.*, 2005). Bioactivity of endophytic fungi of *Coffea arabica* and *C. robusta* was screened against *Salmonella choleraesuis*, *Staphylococcus aureus*, *Pseudomonas aeruginosa* and four different *Escherichia* spp. Out of these endophytic fungi, *T. harzianum*, *Guignardia* sp. and *Phomopsis* sp. have inhibited four to five bacterial species successfully (Sette *et al.*, 2006). Out of 377 isolates of endophytic fungi from *Garcinia* plants, 18.6% isolates displayed antimicrobial activity against at least one pathogenic microorganism, such as *S. aureus*, a clinical isolate of methicillin-resistant *S. aureus*, *C. albicans* and *C. neoformans* (Phongpaichit *et al.*, 2006).

Epicoccum purpurascens and *Truncatella hartigii* were found to have significant action against human pathogenic bacteria. *E. purpurascens* expressed a good antibacterial effect on *S. aureus* and *P. aeruginosa* and a very good antibacterial effect on *E. coli*, while *T. hartigii* exhibited a significant antibacterial effect on *Enterococcus faecalis* (Janes *et al.*, 2007). *Fusarium* was the most frequently isolated endophyte from the Chinese traditional medicinal plant, *Dioscorea zingiberensis*, and *F. redolens* showed the most potent antibacterial activities against *B. subtilis*, *S. haemolyticus*, *E. coli* and *X. vesicatoria* (Xu *et al.*, 2008).

Two antibacterial cerebrosides, one new and another known, were isolated from *Fusarium* sp., an endophytic fungus found in *Quercus variabilis*. The new cerebroside was named fusaruside with structure (2*S*,2′*R*, 3*R*,3′*E*,4*E*,8*E*,10*E*)-1-O-*β*-d-glucopyranosyl-2-*N*-(2′-hydroxy-3′-octadecenoyl)-3-hydroxy-9-methyl-4,8,10-sphingatrienine. Both of them were active against *B. subtilis*, *E. coli* and *P. fluorescens* (Shu *et al.*, 2004). Periconicins A and B were isolated from endophytic fungus *Periconia* sp. of *Taxus cuspidata* and exhibited antibacterial activity against many pathogenic bacteria. The minimum inhibitory concentration (MIC) of periconicin A was even less (3.12 μg/ml) than that of gentamicin (12.5 μg/ml) against *Klebsiella pneumoniae* (Kim *et al.*, 2004). The endophytic fungus, *Xylaria* sp., isolated from *Ginkgo biloba*, showed strong antibacterial activity *in vitro* against *S. aureus* (MIC 16 μg/ml), *E. coli* (MIC 10 μg/ml), *S. typhae* (MIC 20 μg/ml) and *S. typhimurium* (Liu *et al.*, 2008). Recently, some bioactive nitronaphthalenes have been isolated from endophytic fungus, *Coniothyrium* sp. (Krohn *et al.*, 2008). Javanicin, an antibacterial naphthaquinone, has been isolated from neem endophyte, *Chloridium* sp., which was significantly active against *Pseudomonas* spp. (Kharwar *et al.*, 2009).

Antiviral activity of endophytes

Viruses are an important causal agent of various diseases in plants and animals. Endophytes can induce plant resistance against viral diseases, but there is a contradiction and Guy (1992) found no correlation between virus infection and the incidence of endophyte in perennial ryegrass (*L. perenne*), whereas other correlative studies have revealed that some endophyte-infected tall fescue (*Festuca arundinaceum*) seem to be more resistant to barley yellow dwarf virus (BYDV) than the others (Mahmood *et al.*, 1993; Guy and Davis, 2002). Lehtonen *et al.* (2006), when releasing the viruliferous aphid vectors to endophyte-infected and endophyte-free *L. pretense* plants in a common garden, found the number of aphids and the percentage of BYDV infections were lower in endophyte-infected plants compared to endophyte-free plants. Human cytomegalovirus (hCMV) is a ubiquitous opportunistic pathogen. Two novel human cytomegalovirus protease inhibitors, cytonic acids A and B, have been isolated from the solid-state fermentation of the endophytic fungus, *Cytonaema* sp. (Guo *et al.*, 2000).

Nematicidal activity of endophytes

Endophytic fungi are known to produce some compounds which are toxic to nematodes. The first report on antagonistic activity of endophytic fungi against plant parasitic nematodes was observed in tall fescue (*F. arundinacea*) infected by *Pratylenchus scribneri*. The nematode population was found to be comparatively less in the soil surrounding endophyte-infected plants. Since the root of tall fescue (*F. arundinacea*) was infected by *Acremonium coenophialium*, it was considered that the presence of *A. coenophialium* deterred the nematode population. The colonization of fungal endophyte, *F. oxysporum*, in the roots of tomato plant reduced 60% infection of *Meloidogyne incognita* successfully. Endophyte-free perennial ryegrass plants are shown to have a larger number of *M. incognita* population in roots than endophyte-containing plants (Ball *et al.*, 1997). Pregaliellalactone and structurally related lactones were isolated with nematicidal activity from non-graminaceous endophytes and related saprophytic ascomycetes (Kopcke *et al.*, 2002a,b). Another endophytic microbe, *Burkholderia ambifaria*, isolated from corn root, produced some toxic metabolites which inhibited egg hatching and mobility of second-stage juveniles of *M. incognita* (Li *et al.*, 2002).

Diedhiou *et al.* (2003) demonstrated the successful nematicidal activity of an arbuscular mycorrhiza, *Glomus coronatum*, and an endophytic fungus, *F. oxysporum*, against the *M. incognita* in tomato plant. Several endophytic fungi isolated from above-ground plant organs produced 3-hydroxypropionic acid (HPA) by bioactivity-guided fractionation of extracts and showed selective nematicidal activity against the plant-parasitic nematode, *M. incognita*, with LD_{50} values of 12.5–15 μg/ml (Schwarz *et al.*, 2004). *Radopholus similis* is an important parasitic nematode on banana and other plants. It is suggested that the dual inoculations of endophytic fungal isolates reduce a large number of the *R. similis* population (Felde *et al.*, 2006).

Insecticidal activity of endophytes

Fungi are known to produce a large number of insecticidal metabolites such as destruxins, ibotenic acid, pantherine, tricholomic acid, etc. Endophytic fungi are also known to deter insect pests (Clay, 1989; Carroll, 1991, 1995; Azevedo *et al.*, 2000). Several toxins are produced by endophytic fungi and these substances confer host protection against different herbivores. The endophytic fungus, *P. oblonga*, was responsible for reducing the spread of Dutch elm disease causal agent, *Ceratocystis ulmi*, by controlling its vector beetle (*P. brevilineum*) (Webber, 1981). In 1985, Claydon and his co-workers confirmed that endophytic fungi belonging to the Xylariaceae family synthesized secondary metabolites in host *Fagus* sp. and that these substances affected the beetle larvae. Susceptible and resistant cultivars of perennial rye grass (*L. perenne* L.) against sod webworms (*Crambus* spp.) were analysed for the presence of an endophytic fungus. All resistant cultivars were found to have a high infection of endophytic fungi. Several highly infected ryegrass species with endophytic fungi consequently have shown less attack frequency of Argentine stem weevils (*Listronotus bonariensis*) (Gaynor and Hunt, 1983). Barker *et al.* (1984) and Prestidge *et al.* (1984) also observed that the same grass infected with endophytic *Acremonium* sp. was more resistant to stem weevils in New Zealand. In the white spruce, *Picea glauca*, the death rate of the Homoptera, *Adelges abietis*, was considerably higher when galls were infected with the endophytic fungus, *C. sphaerosperum* (Lasota *et al.*, 1983). In *L. perenne* and a few members of genus *Cyperus*, insect-pest *Spodoptera frugiperda* was affected adversely by endophytic fungus like *Balansia cyperi* (Clay *et al.*, 1985a,b). Ahmad *et al.* (1985) showed that endophytic *Acremonium* sp. deterred the grasshopper, *Acheta domesticus*. Patterson *et al.* (1992) observed the production of alkaloids by endophytic *Acremonium* in plants *Lolium* and *Festuca* that reduced the attack of the Japanese beetle, *Popilla japonica*. *Muscodor vitigenus*, an endophytic fungus of *Paullinia paullinioides*, from the Peruvian Amazon, is known to produce naphthalene, which effectively repels the adult stage of the wheat stem sawfly, *Cephus cinctus* (Daisy *et al.*, 2002). Endophyte-mediated resistance was reported in strong creeping and chewings fescue species against red thread (Bonos *et al.*, 2005). *Beauveria bassiana* is a highly effective entomopathogen of a wide range of insects. Grass varieties infected by *Neotyphodium* endophyte have affected the feeding performance and preference of newly hatched nymphs of the hairy chinch bug, *Blissus leucopterus hirtus*, a common turfgrass pest in north-eastern USA (Steeve *et al.*, 2007). Akello *et al.* (2007) incorpotated *B. bassiana* as an artificial endophyte in banana plants to combat the banana weevil, *Cosmopolites sordidus*. The endophytic fungi, *B. bassiana* and *Clonostachys rosea*, isolated from coffee plant, showed strong antagonistic activity against coffee berry borers (Vega *et al.*, 2008).

Two new insecticidal compounds, 5-hydroxy-2-(1′-oxo-5′-methyl-4′-hexenyl) benzofuran and 5-hydroxy-2-(1′-hydroxy-5′-methyl-4′-hexenyl) benzofuran were isolated via bioassay-directed fractionation of culture extracts of an unidentified endophytic fungus obtained from wintergreen, *Gaultheria procumbens* (Findlay *et al.*, 1997). These compounds exhibited toxicity to spruce budworm (*Choristoneura fumiferana* Clem.) cells. Peramine and lolines, potent insecticides, are produced in endophyte-infected perennial ryegrass and protect them from the Argentine stem weevil, *Listronotus bonariensis* (Rowan and Latch, 1994; Tanaka *et al.*, 2005). Nodulisporic acids, novel indole diterpenes, have potent insecticidal properties against the larvae of the blowfly by activating insect glutamate-gated chloride channels. *Nodulisporium*, an endophytic species from the plant, *Bontia daphnoides*, produces such nodulisporic compounds (Demain, 2000). A strain of endophytic *Penicillium* sp., isolated from the fresh roots of *Derris elliptica*, produces some insecticidal compound analogues to rotenone against the adult turnip aphid, *Lipaphis erysimi* (Hu *et al.*, 2005).

Plant Protection in Abiotic Stresses

Endophytes are also involved in the protection of plants in various abiotic stresses like drought, temperature, pH, heavy metals, etc. (Rodriguez *et al.*, 2004). Water stress tolerance was observed in epacrids and their endophytic partners in south-west Australia (Hutton *et al.*, 1996). In drought conditions, water content of some endophyte-associated, field-grown tall fescues may be maintained at higher levels than those of endophyte-free plants (Elbersen and West, 1996; Buck *et al.*, 1997). This phenomenon may be explained by enhanced accumulation of solutes in tissues of endophyte-infected plants as compared to non-infected plants, or by reduced leaf conductance and a slowdown of the transpiration stream, or due to thicker cuticle formation (Malinowski and Belesky, 2000).

The endophytic mutants and wild-type *C. magna* confer drought tolerance that allows symbiotic tomato and pepper plants to survive desiccation for 24 and 48 h longer than non-symbiotic plants, respectively (Redman *et al.*, 2001). Endophytic colonization was observed to increase the minimum leaf conductance in *Theobroma cacao*, a measure of leaf water loss after maximal stomatal closure under drought stress (Arnold and Engelbrecht, 2007). However, no evidence for endophyte-mediated drought tolerance was observed in *Acremonium*-infected tall fescue (White *et al.*, 1992). It is suggested that endophyte-mediated drought resistance may be due to alterations in drought avoidance.

Malinowski and Belesky (1999) observed that the pH of a limed, acidic soil increased faster as a result of the root activity of endophyte-infected tall fescue compared with non-infected plants under phosphate-deficient conditions. Liu *et al.* (1996) observed that aluminium tolerance in endophyte-infected fine fescues (*Festuca* spp.) was greater as compared to non-infected plants. In an experiment, endophyte-infected clone grew significantly better in high aluminium soils relative to the endophyte-free clone (Zaurov *et al.*, 2001). *L. perenne*, symbiotic with *N. lolii*, showed higher values of total dry weight and tiller number compared to non-symbiotic plants in Zn stress (Monnet *et al.*, 2001). In low NaCl salt stress condition, the endophyte *Piriformospora indica*-infected barley plants showed higher biomass than non-infected plants (Waller *et al.*, 2005). The mechanism of endophyte-conferred salt tolerance has not been investigated so far.

In the USA, a plant species, *Dichanthelium lanuginosum*, has been found growing in the geothermal soils of Yellowstone National Park (YNP) and Lassen Volcanic National Parks (LVNP) at temperatures as high as 57°C (Stout and Al-Niemi, 2002). Redman *et al.* (2002) observed that those plants colonized by an endophytic fungus, *Curvularia protuberata*, were able to tolerate the higher temperature and, thus, we might conclude that endophytes supported the plant to withstand heat or drought stresses. However, an in-depth investigation by Marquez and his colleagues (2007) showed that this was not only because of plant–fungus symbiosis but it also included a virus as a third partner, which parasitized on *C. protuberata*. Thus, it is a complex tripartite symbiosis and the heat tolerance ability of the fungus is, in fact, related to the virus. That mycovirus is called a *Curvularia* thermal tolerance virus (CThTV).

Two mechanisms are involved in the endophyte-conferred biotic and abiotic stress tolerance: (i) rapid activation of host stress response systems in exposure to stress (Redman *et al.*, 1999); and (ii) synthesis of anti-stress biochemicals in the host, either by endophytes or through endophyte induction (Bacon and Hill, 1996). Endophyte-produced anti-stress biochemicals are mostly alkaloids. In addition to anti-stress biochemicals, plant and fungal mutualism has been maintained over an evolutionary time by the ability of fungi to control the activation of host stress response systems and, in core, act as 'biological triggers' (Rodriguez *et al.*, 2004). When a plant interacts with environmental biotic and abiotic stresses, it produces several damaging reactive oxygen species (ROS). Therefore, it is hypothesized that endophytes inside plants scavenge these ROS rapidly and protect their host (Rodriguez and Redman, 2005; Tanaka *et al.*, 2006).

Indian Contributions to Fungal Endophyte Research

The past history of endophytic research in India, especially with fungi, is not so encouraging. It seems that workers who started this research in India are still actively involved in advancing their research manifesto with this 'under-studied' group of microbial population and have not advanced to the fields and forests of the countryside looking for novel microbe/plant associations. Prof Suryanarayanan and his group (Chennai) have initiated biodiversity and distribution patterns of fungal endophytes with some medicinal plants in India and have published several papers along this line. He has also isolated some bioactive compounds and melanin from endophytic fungi (Suryanarayanan *et al.*, 2004). Several research groups have started paying more attention to various aspects of endophytic fungi. No more than a dozen research groups at various locations in India are vigorously involved in either biodiversity or natural product discovery from this untapped and alternative resource (Table 15.1).

It has become obvious to many workers throughout the world that endophytic microbes have enormous potential to solve many of mankind's problems. Thus, with the discovery of new compounds, we can protect our agriculture and medicine industries, as well as plant health. After more than 20 years of effort, the total number of publications from Indian researchers, including some fairly recent ones (Shankar *et al.*, 2003; Seena and Sridhar, 2004; Amna *et al.*, 2006; Tejesvi *et al.*, 2007; Gangadevi and Muthumarry, 2008), is relatively small.

Due to the great variation in plant biodiversity and seasonal changes in India, we may have a better opportunity to collect/isolate various types of promising endophytic fungi, especially from rainforests and mangrove swamps, which may be able to produce an enormous variety of potential bioactive natural compounds. An increasing population of AIDS and immunocompromised patients in India compels us to bear them in mind when searching for safe drugs.

Table 15.1. List of Indian workers involved in endophytic fungal research.

Sr. No.	Name of group leader	Place of work	Work specialization	E-mail addresses
1.	Dr T.S. Suryanarayanan	Department of Botany, Vivekanand College, Chennai	Endophytic fungal diversity and natural product discovery	tssury@md3.vsnl.net.in ts_sury2002@yahoo.com
2.	Dr K.R. Sridhar	Department of Biosciences, Mangalore University, Mangalagangotri, Mangalore	Endophytic fungal diversity and bioactive molecules	sirikr@yahoo.com
3.	Dr H.S. Prakash	Department of Botany, University of Mysore, Manasagangotri	Endophytic fungal diversity and bioactive molecules	hasriprakash@gmail.com
4.	Dr D.J. Bhat	Department of Botany, Goa University, Panji, Goa	Endophytic fungal diversity	bhatdj@rediffmail.com
5.	Dr J. Muthumarry	Department of Botany, University of Madras, Chennai	Endophytic fungal diversity and bioactive fungal compounds	mm_j@rediffmail.com
6.	Dr Arun Arya	Department of Botany, MSU, Baroda	Endophytic fungal diversity	aryaarunarya@rediffmail.com
7.	Dr R. Uma Shaankar	University of Agriculture, Karnataka	Endophytic fungal diversity and bioactive molecules	umashaanker@gmail.com
8.	Dr Absar Ahmad	National Chemical Laboratory, Pune	Synthesis of silver and gold particles from endophytes	aahmad@dalton.ncl.res.in
9.	Dr S.K. Singh	Agharkar Research Institute, Pune	Endophyte diversity	singhsksingh@rediffmail.com
10.	Dr S.C. Puri/ R.K. Khajuria	Regional Research Laboratory, Kanal Road, Jammu Tawi	Natural product development	khajuriark@yahoo.com
11.	Dr R.N. Kharwar	Department of Botany, B.H.U., Varanasi-221005	Endophytic fungal diversity and natural product discovery	rnkharwar@yahoo.com kharwar1@rediffmail.com

India really needs a variety of novel antimicrobial compounds of biological origin, so that we can solve the problems of eco-friendly farmers and the weaker sections of society in which the above-mentioned diseases are prevalent. The fungi, as a group, hold enormous potential as sources of antimicrobials. Observations prove that this group of organisms resides inside healthy plant tissues as endophytes without causing any detectable symptoms. Therefore, we feel strongly that India needs to gear up and exact its research to exploit the maximum potential of the promising endophytes for natural product discovery, which could at least facilitate some of the existing problems of its huge population.

Conclusions

In the study of mycodiversity, we often forget the endospheric fungi as researchers focus their attention on the phyllospheric and rhizospheric fungi. The endosphere is a special niche where endophytic microorganisms reside and, in response, produce a variety of metabolites, which are mostly toxic to plant and human pathogens. In this aspect, plant pathogens interact with the plant itself, as well as the plant's endophytes. The role of endospheric or so-called endophytic fungi in plant protection is quite clear in the above-mentioned examples. Besides protecting plants from biotic and abiotic stresses, endophytes also improve the health and yield of plants by producing some growth-regulating phytohormones. Although endophytes are still poorly investigated microorganisms, they have shown that they are going to play a prominent part in the discovery of many bioactive natural compounds. Bioactive natural products of endophytic origin can change the scenario of existing agropesticides because of their easy and sustainable production. Many scientists throughout the world are engaged in the search for bioactive compounds from endophytes. There is a gap in the knowledge on the genetic and biochemical communications between the plant and endophytic symbionts. We have to minimize this gap for better utilization of endophytic microorganisms.

Acknowledgements

The authors are thankful to the Head of the Department of Botany, BHU, Varanasi, for providing the necessary facilities. They also extend their thanks to the CSIR, New Delhi, for financial support.

References

Ahmad, S., Govindarajan, S., Funk, C.R. and Johnson-Cicalese, J.M. (1985) Fatality of house crickets on perennial ryegrass infected with a fungal endophyte. *Entomologia Experimentalis et Applicata* 39, 183–190.

Akello, J., Dubois, T., Coyne, D. and Hillnhütter, C. (2007) *Beauveria bassiana* as an artificial endophyte in tissue-cultured banana plants: a novel way to combat the banana weevil *Cosmopolites sordidus* (http://www.banana2008.com/cms/details/Paper1.doc, accessed 26 July 2008).

Amna, T., Khajuria, R.K., Puri, S.C., Verma, V. and Qazi, G.N. (2006) Determination and quantification of Camptothecin in an endophytic fungus by liquid chromatography-positive mode electrospray ionization tandem mass spectrometry. *Current Science* 91, 208–211.

Anke, H., Weber, R.W., Heil, N., Pauls, S., Teinert, M., Pauluat, T., Kuenzel, E., Huth, F., Eckard, P. and Kappe, R. (2003) Antifungal compounds from endophytic fungi. *Interscience Conference on Antimicrobial Agents and Chemotherapy, 14–17 September 2003.* Abstract No. F-1236. American Society for Microbiology, Chicago, Illinois.

Arnold, A.E. and Engelbrecht, B.M.J. (2007) Fungal endophytes nearly double minimum leaf conductance in seedlings of a neotropical tree species. *Journal of Tropical Ecology* 23, 369–372.

Azevedo, J.L., Maccheroni, W., Pereira, J.O. and Araujo, W.L. de (2000) Endophytic microorganisms: a review on insect control and recent advances on tropical plants. *Electronic Journal of Biotechnology* 3, 1–36.

Backman, P.A. and Sikora, R.A. (2008) Endophytes: an emerging tool for biological control. *Biological Control* 46, 1–3.

Bacon, C.W. and Hill, N.S. (1996) Symptomless grass endophytes: products of coevolutionary symbioses and their role in the ecological adaptations of grasses. In: Redkin, S.C. and Carris, L.M. (eds) *Endophytic Fungi in Grasses and Woody Plants.* APS Press, St Paul, Minnesota, pp. 155–178.

Ball, O.J.P., Bernard, E.C. and Gwinn, K.D. (1997) Effect of selected isolates of *Acremonium* endophytes on adult black beetle (*Heteronychus arator*) feeding. *Proceedings of the 47th N.Z. Plant Protection Conference.* New Zealand Plant Protection Society, pp. 227–231.

Barker, G.M., Pottinger, R.P., Addison, P.J. and Prestidge, R.A. (1984) Effect of *Lolium* endophyte fungus infection on behavior of adult Argentine stem weevil. *New Zealand Journal of Agricultural Research* 27, 271–277.

Bonos, S.A., Wilson, M.M., Meyer, W.A. and Funk, C.R. (2005) Suppression of red thread in fine fescues through endophyte-mediated resistance. *Applied Turfgrass Science* July 25, 1–7.

Buck, G.W., West, C.P. and Elbersen, H.W. (1997) Endophyte effect on drought tolerance in diverse *Festuca* species. In: Bacon, C.W. and Hill, N.S. (eds) *Neotyphodium/Grass Interactions.* Plenum Press, New York, pp. 141–143.

Carroll, G. (1991) Fungal associates of woody plants as insect antagonists in leaves and stems. In: Barbosa, P., Krischik, V.A. and Jones, C.G. (eds) *Microbial Mediation of Plant–Herbivore Interactions.* John Wiley and Sons, New York, pp. 253–271.

Carroll, G. (1995) Forest endophytes: pattern and process. *Canadian Journal of Botany* 73, S1316–S1324.

Clay, K. (1989) Clavicipitaceous endophytes of grasses: their potential as biocontrol agents. *Mycological Research* 92, 1–12.

Clay, K., Hardy, T.N. and Hammond, A.M. Jr (1985a) Fungal endophytes of *Cyperus* and their effect on the insect herbivore. *American Journal of Botany* 72, 1284–1289.

Clay, K., Hardy, T.N. and Hammond, A.M. Jr (1985b) Fungal endophytes of grasses and their effects on an insect herbivore. *Oecologia* 66, 1–6.

Claydon, N., Grove, J.F. and Pople, M. (1985) Elm bark beetle boring and feeding deterrents from *Phomopsis oblonga. Phytochemistry* 24, 937–943.

Daisy, B., Strobel, G.A., Castillo, U., Ezra, D., Sears, J., Weaver, D.K. and Runyon, J.B. (2002) Naphthalene, an insect repellent, is produced by *Muscodor vitigenus*, a novel endophytic fungus. *Microbiology* 148, 3737–3741.

De Bary, A. (1866) Morphologic und physiologie der plize, Flechten, und Myxomyceten. In: De Bary, A., Irmisch, T., Pringsheim, N. and Sachs, J. (eds) *Handbuch der Physiologischen Botanik (Hofmeister's Handbook of Physiological Botany)*, Volume 2, Leipzig.

Demain, A.L. (2000) Microbial natural products: a past with a future. In: Wrigley, S.K., Hayes, M.A., Thomas, R., Chrystal, E.J.T. and Nicholson, N. (eds) *Biodiversity: New Leads for Pharmaceutical and Agrochemical Industries.* The Royal Society of Chemistry, Cambridge, UK, pp. 3–16.

Diedhiou, P.M., Hallmann, J., Oerke, E.C. and Dehne, H.W. (2003) Effects of arbuscular mycorrhizal fungi and a non-pathogenic *Fusarium oxysporum* on *Meloidogyne incognita* infestation on tomato. *Mycorrhiza* 13, 199–204.

Duijff, B.J., Poulain, D., Olivain, C., Alabouvette, C. and Lemanceau, C. (1998) Implication of systemic induced resistance in the suppression of *Fusarium* wilt of tomato by *Pseudomonas fluorescens* WCS417r and by non-pathogenic *Fusarium oxysporum* Fo47. *European Journal of Plant Pathology* 104, 903–910.

Elbersen, H.W. and West, C.P. (1996) Growth and water relations of field-grown tall fescue as influenced by drought and endophyte. *Grass and Forage Science* 51, 333–342.

Felde, A.Z., Pocasangre, L.E., Monteros, C.A.C., Sikora, R.A., Rosales, F.E. and Riveros, A.S. (2006) Effect of combined inoculations of endophytic fungi on the biocontrol of *Radopholus similis. InfoMusa* 15, 1–2.

Findlay, J.A., Buthelezi, S., Li, G., Seveck, M. and Miller, J.D. (1997) Insect toxins from an endophytic fungus from Wintergreen. *Journal of Natural Products* 60, 1214–1215.

Freeman, E.M. (1904) The seed fungus of *Lolium temulentum* L., the darnel. *Philosophical Transactions of the Royal Society London B* 196, 1–27.

Gangadevi, V. and Muthumarry, J. (2008) Taxol, an anticancer drug produced by an endophytic fungus *Bartalinia robillardoides* Tassi, isolated from a medicinal plant, *Aegle marmelos* Correa ex Roxb. *World Journal of Microbiology and Biotechnology* 24, 717–723.

Gaynor, D.L. and Hunt, W.F. (1983) The relationship between nitrogen supply, endophytic fungus and Argentine stem weevil resistance in ryegrass. *Proceedings of the New Zealand Grassland Association* 44, 257–263.

Gond, S.K., Verma, V.C., Kumar, A., Kumar, V. and Kharwar, R.N. (2007) Study of endophytic fungal community from different parts of *Aegle marmelos* Correae (Rutaceae) from Varanasi (India). *World Journal of Microbiology and Biotechnology* 23, 1371–1375.

Guo, B., Dai, J.R., Ng, S., Huang, Y., Leong, C., Ong, W. and Carté, B.K. (2000) Cytonic acids A and B: novel tridepside inhibitors of hCMV protease from the endophytic fungus *Cytonaema* species. *Journal of Natural Products* 63, 602–604.

Guy, P.L. (1992) Incidence of *Acremonium lolii* and lack of correlation with barley yellow dwarf viruses in Tasmanian perennial ryegrass pastures. *Plant Pathology* 41, 29–34.

Guy, P.L. and Davis, L.T. (2002) Variation in the incidence of barley yellow dwarf virus and in the ability of *Neotyphodium* endophytes to deter feeding by aphids (*Rhopalosiphum padi*) on Australasian tall fescue. *Australasian Plant Pathology* 31, 307–308.

Harman, G.E. (1996) *Trichoderma* for biocontrol of plant pathogens: from basic research to commercialized products. Cornell Community Conference on Biological Control, 11–13 April, Cornell (http://www.nysaes.cornell.edu/ent/bcconf/, accessed 12 July 2008).

Hu, M.Y., Zhong, G.H., Sun, Zh.T., Sh, G., Liu, H.M. and Liu, X.Q. (2005) Insecticidal activities of secondary metabolites of endophytic *Pencillium* sp. in *Derris elliptica* Benth. *Journal of Applied Entomology* 129, 413–417.

Huang, Y., Wang, J., Li, G., Zheng, Z. and Su, W. (2001) Antitumor and antifungal activities in endophytic fungi isolated from pharmaceutical plants *Taxus mairei*, *Cephalataxus fortunei* and *Torreya grandis*. *FEMS Immunology and Medical Microbiology* 31, 163–167.

Hutton, B.J., Sivasithamparam, K., Dixon, K.W. and Pate, J.S. (1996) Pectic zymograms and water stress tolerance of endophytic fungi isolated from western Australian heaths (Epacridaceae). *Annals of Botany* 77, 399–404.

Inacio, M.L., Silva, G.H., Teles, H.L., Trevisan, H.C., Cavalheiro, A.J., Bolzani, V. da S., Young, M.C.M., Pfenning, L.H. and Araujo, A.R. (2006) Antifungal metabolites from *Colletotrichum gloeosporioides*, an endophytic fungus in *Cryptocarya mandioccana* Nees (Lauraceae). *Biochemical Systematics and Ecology* 34, 822–824.

Janes, D., Kreft, S., Jurc, M., Seme, K. and Strukelj, B. (2007) Antibacterial activity in higher fungi (mushrooms) and endophytic fungi from Slovenia. *Pharmaceutical Biology* 45, 700–706.

Jarvis, W.R. (1993) *Managing Diseases in Greenhouse Crops*. American Phytopathological Society Press, S. Paul, Minnesota.

Kharwar, R.N., Verma, V.C., Strobel, G. and Ezra, D. (2008) The endophytic fungal complex of *Catharanthus roseus* (L.) G. Don. *Current Science* 95, 228–233.

Kharwar, R.N., Verma, V.C., Kumar, A., Gond, S.K. and Strobel, G. (2009) Javanicin, an antibacterial naphthaquinone from an endophytic fungus of neem – *Chloridium* sp. *Current Microbiology* 58, 233–238.

Kim, S., Shin, D.S., Lee, T. and Oh, K.B. (2004) Periconicins, two new fusicoccane diterpenes produced by an endophytic fungus *Periconia* sp. with antibacterial activity. *Journal of Natural Products* 67, 448–450.

Kopcke, B., Weber, R.W.S. and Anke, H. (2002a) Galiellalactone and its biogenetic precursors as chemotaxonomic markers of the Sarcosomataceae (Ascomycota). *Phytochemistry* 60, 709–714.

Kopcke, B., Johansson, M., Sterner, O. and Anke, H. (2002b) Biologically active secondary metabolites from the ascomycete A111-95.1 production, isolation and biological activities. *Journal of Antibiotics* 55, 36–40.

Krohn, K., Kouam, S.F., CludiusBrandt, S., Draeger, S. and Schulz, B. (2008) Bioactive nitronaphthalenes from an endophytic fungus, *Coniothyrium* sp. and their chemical synthesis. *European Journal of Organic Chemistry* 21, 3615–3618.

Lasota, J.A., Waldvogel, M.G. and Shetlar, D.J. (1983) Fungus found in galls of *Adelges abietis* (L.) (*Homoptera adelgidae*): identification within tree distribution and possible impact on insect survival. *Environmental Entomology* 12, 245–246.

Lehtonen, P.T., Helander, M., Siddiqui, S.A., Lehto, K. and Saikkonen, K. (2006) Endophytic fungus decreases plant virus infections in meadow ryegrass (*Lolium pratense*). *Biology Letters* 2, 620–623.

Li, J.Y. and Strobel, G.A. (2001) Jesterone and hydroxy-jesterone antioomycetcyclohexenenone epoxides from the endophytic fungus *Pestalotiopsis jesteri*. *Phytochemistry* 57, 261–265.

Li, J.Y., Strobel, G.A., Harper, J.K., Lobkovsky, E. and Clardy, J. (2000) Cryptocin, a potent tetramic acid antimycotic from the endophytic fungus *Cryptosporiopsis cf. quercina*. *Organic Letters* 2, 767–770.

Li, W., Roberts, D.P., Dery, P.D., Meyer, S.L.F., Lohrke, S., Lumsden, R.D. and Hebbar, K.P. (2002) Broad spectrum antibiotic activity and disease suppression by the potential biocontrol agent *Burkholderia ambifaria* Bc-F. *Crop Protection Journal* 21, 129–135.
Li, Y., Song, Y.C., Liu, J.Y., Ma, Y.M. and Tan, R.X. (2005) Anti-*Helicobacter pylori* substances from endophytic fungal cultures. *World Journal of Microbiology and Biotechnology* 21, 553–558.
Liu, H., Heckman, J.R. and Murphy, J.A. (1996) Screening fine fescues for aluminum tolerance. *Journal of Plant Nutrition* 19, 677–688.
Liu, C.H., Zou, W.X., Lu, H. and Tan, R.X. (2001) Antifungal activity of *Artemisia annua* endophyte cultures against phytopathogenic fungi. *Journal of Biotechnology* 88, 277–282.
Liu, X., Dong, M., Chen, X., Jiang, M., Lv, X. and Zhou, J. (2008) Antimicrobial activity of an endophytic *Xylaria* sp.YX-28 and identification of its antimicrobial compound 7-amino-4-methylcoumarin. *Applied Microbiology and Biotechnology* 78, 241–247.
Lu, H., Zou, W.X., Meng, J.C., Hu, J. and Tan, R.X. (2000) New bioactive metabolites produced by *Colletotrichum* sp., an endophytic fungus in *Artemisia annua*. *Plant Science* 151, 67–73.
Mahmood, T., Gergerich, R.C., Milus, E.A., West, C.P. and Darcy, C.J. (1993) Barley yellow dwarf viruses in wheat, endophyte-infected and endophyte-free tall fescue, and other hosts in Arkansas. *Plant Disease* 77, 225–228.
Malinowski, D.P. and Belesky, D.P. (1999) Endophyte infection enhances the ability of tall fescue to utilize sparingly available phosphorus. *Journal of Plant Nutrition* 2, 835–853.
Malinowski, D.P. and Belesky, D.P. (2000) Adaptations of endophyte-infected cool-season grasses to environmental stresses: mechanisms of drought and mineral stress tolerance. *Crop Science* 40, 923–940.
Marquez, L.M., Redman, R.S., Rodriguez, R.J. and Rossinck, M.J. (2007) A virus in a fungus in a plant: three-way symbiosis required for thermal tolerance. *Science* 315, 513–515.
Mejia, L.C., Rojas, E.I., Maynard, Z., Arnold, E.A., Hebbar, P., Samuels, G.J., Robbins, N. and Herre, A.E. (2008) Endophytic fungi as biocontrol agents of *Theobroma cacao* pathogens. *Biocontrol* 46, 4–14.
Monnet, F., Vaillant, N., Hitmi, A., Coudret, A. and Sallanon, H. (2001) Endophytic *Neotyphodium lolii* induced tolerance to Zn stress in *Lolium perenne*. *Physiologia Plantarum* 113, 557–563.
Narisawa, K., Ohki, T. and Hashiba, T. (2000) Suppression of clubroot and *Verticillium* yellows in Chinese cabbage in the field by the endophytic fungus, *Heteroconium chaetospira*. *Plant Pathology* 49, 141–146.
Narisawa, K., Kawamata, H., Currah, R.S. and Hashiba, T. (2002) Suppression of *Verticillium* wilt in eggplant by some fungal root endophytes. *European Journal of Plant Pathology* 108, 103–109.
Park, J.H., Choi, G.J., Lee, H.B., Kim, K.M., Jung, H.S., Lee, S.W., Jang, K.S. and Cho, K.Y. (2005) Griseofulvin from *Xylaria* sp. strain F0010, an endophytic fungus of *Abies holophylla* and its antifungal activity against plant pathogenic fungi. *Journal of Microbiology and Biotechnology* 15, 112–117.
Patterson, C.G., Potter, D.A. and Fannin, F.F. (1992) Feeding deterrency of alkaloids from endophyte-infected grasses to Japanese beetle grubs. *Entomology Experimental and Applied* 61, 285–289.
Phongpaichit, S., Rungjindamai, N., Rukachaisirikul, V. and Sakayaroj, J. (2006) Antimicrobial activity in cultures of endophytic fungi isolated from Garcinia species. *FEMS Immunology and Medical Microbiology* 48, 367–372.
Prestidge, R.A., Zijpp, S. and Badan, D. (1984) Effects of Argentine stem weevil on pastures in Central Volcanic Plateau. *New Zealand Journal of Experimental Agriculture* 12, 323–331.
Raviraja, N.S., Maria, G.L. and Sridhar, K.R. (2006) Antimicrobial evaluation of endophytic fungi inhabiting medicinal plants of the Western Ghats of India. *Engineering in Life Sciences* 6, 515–520.
Redman, R.S., Freeman, S., Clifton, D.R., Morrel, J., Brown, G. and Rodriguez, R.J. (1999) Biochemical analysis of plant protection afforded by a non-pathogenic endophytic mutant of *Colletotrichum magna*. *Plant Physiology* 119, 795–804.
Redman, R.S., Dunigan, D.D. and Rodriguez, R.J. (2001) Fungal symbiosis: from mutualism to parasitism, who controls the outcome, host or invader? *New Phytologist* 151, 705–716.
Redman, R.S., Sheehan, K.B., Stout, R.G., Rodriguez, R.J. and Henson, J.M. (2002) Thermotolerance conferred to plant host and fungal endophyte during mutualistic symbiosis. *Science* 298, 1581–1581.
Rodriguez, M.A., Cabrera, G. and Godeas, A. (2006) Cyclosporine A from a non-pathogenic *Fusarium oxysporum* suppressing *Sclerotinia sclerotiorum*. *Journal of Applied Microbiology* 100, 575–586.
Rodriguez, R. and Redman, R. (2005) Balancing the generation and elimination of reactive oxygen species. *Proceedings of the National Academy of Sciences* 102, 3175–3176.
Rodriguez, R.J., Redman, R.S. and Henson, J.M. (2004) The role of fungal symbiosis in the adaptation of plants to high stress environments. *Mitigation and Adaptation Strategies for Global Change* 9, 261–272.

Rowan, D.D. and Latch, G.C. (1994) Utilization of endophyte infected perennial ryegrass for increased insect resistance. In: Bacon, C.W. and White, J.F. (eds) *Biotechnology of Endophytic Fungi of Grasses*. CRC Press, Boca Raton, Florida, pp. 169–183.

Schwarz, M., Kopcke, B., Weber, R.W.S., Sterner, O. and Anke, H. (2004) 3-Hydroxypropionic acid as a nematicidal principle in endophytic fungi. *Phytochemistry* 65, 2239–2245.

Seena, S. and Sridhar, K.R. (2004) Endophytic fungal diversity of two sand dune wild legumes from southwest coast of India. *Canadian Journal of Microbiology* 50, 1015–1019.

Sette, L.D., Passarini, E.M.R.Z., Delarmelina, E.C., Salati, E.F. and Duarte, E.M.C.T. (2006) Molecular characterization and antimicrobial activity of endophytic fungi from coffee plants. *World Journal of Microbiology and Biotechnology* 22, 1185–1195.

Shankar, S.S., Ahmad, A., Pasricha, R. and Sastry, M. (2003) Bioreduction of chloroaurate ions by geranium leaves and its endophytic fungus yields gold nanoparticles of different shapes. *Journal of Material Chemistry* 13, 1822–1826.

Shu, R.G., Wang, F.W., Yang, Y.M., Liu, Y.X. and Tan, R.X. (2004) Antibacterial and xanthine oxidase inhibitory cerebrosides from *Fusarium* sp. IFB-121, and endophytic fungus in *Quercus variabilis*. *Lipids* 39, 667–673.

Silva, G.H., Teles, H.L., Trevisan, H.C., Bolzani, V. da S, Young, M.C.M., Pfenning, L.H., Eberlin, M.N., Haddad, R., Costa-Neto, C.M. and Araujo, A.R. (2005) New bioactive metabolites produced by *Phomopsis cassiae*, an endophytic fungus in *Cassia spectabilis*. *Journal of Brazilian Chemical Society* 16, 1463–1466.

Steeve, B., Jacques, B. and Yves, C. (2007) Endophyte-grass complexes and the relationship between feeding preference and performance in a grass herbivore. *Entomologia Experimentalis et Applicata* 124, 221–228.

Stone, J.A., Bacon, C.W. and White, J.F. Jr (2000) An overview of endophytic microbes: endophytism defined. In: Bacon, C.W. and White, J.F. Jr (eds) *Microbial Endophytes*. Dekker, New York, pp. 3–30.

Stout, R.G. and Al-Niemi, T.S. (2002) Heat-tolerance flowering plants of active geothermal areas in Yellowstone National Park. *Annals of Botany* 90, 259–267.

Strobel, G.A. (2002) Rainforest endophytes and bioactive products. *Critical Reviews in Biotechnology* 22, 315–333.

Strobel, G.A. (2003) Endophytes as sources of bioactive products. *Microbes and Infection* 5, 535–544.

Strobel, G.A., Miller, R.V., Miller, C., Condron, M., Teplow, D.B. and Hess, W.M. (1999a) Cryptocandin, a potent antimycotic from the endophytic fungus *Cryptosporiopsis* cf. *quercina*. *Microbiology* 145, 1919–1926.

Strobel, G.A., Li, J.Y., Sugawara, F., Koshino, H., Harper, J.K. and Hess, W.M. (1999b) Oocydin A, a chlorinated macrocyclic lactone with potent anti-oomycete activity from *Serratia marcescens*. *Microbiology* 145, 3557–3564.

Suryanarayanan, T.S., Ravishanker, J.P., Venkateshan, G. and Murali, T.S. (2004) Characterization of the melanin pigment of a cosmopolitan fungal endophyte. *Mycological Research* 180, 974–979.

Tanaka, A., Tapper, B.A., Popay, A., Parker, E.J. and Scott, B. (2005) A symbiosis expressed non-ribosomal peptide synthetase from a mutualistic fungal endophyte of perennial ryegrass confers protection to the symbiotum from insect herbivory. *Molecular Microbiology* 57, 1036–1050.

Tanaka, A., Christensen, M.J., Takemoto, D., Park, P. and Scott, B. (2006) Reactive oxygen species play a role in regulating a fungus–perennial ryegrass mutualistic interaction. *Plant Cell* 18, 1052–1066.

Tejesvi, M.V., Kini, K.R., Prakash, H.S., Subbiah, V. and Shetty, H.S. (2007) Genetic diversity and antifungal activities of *Pestalotiopsis* isolated as endophyte from medicinal plant. *Fungal Diversity* 24, 37–54.

Vega, F.E., Posada, F.J., Aime, M.C., Pava-Ripoll, M., Infante, F. and Rehner, S.A. (2008) Entomopathogenic fungal endophytes. *Biological Control* 46, 72–82.

Verma, V.C., Gond, S.K., Kumar, A., Kharwar, R.N. and Strobel, G.A. (2007) Endophytic mycoflora from leaf, bark, and stem of *Azadirachta indica* A Juss. from Varanasi India. *Microbial Ecology* 54, 119–125.

Waller, F., Achatz, B., Baltruschat, H., Fodor, J., Becker, K., Fischer, M., Heier, T., Huckelhoven, R., Neumann, C., Wettstein, D.V., Franken, P. and Kogel, K.H. (2005) The endophytic fungus *Piriformospora indica* reprograms barley to salt-stress tolerance, disease resistance, and higher yield. *Proceedings of the National Academy of Sciences* 20, 13386–13391.

Webber, J. (1981) A natural control of Dutch elm disease. *Nature* 292, 449–451.

White, R.H., Engelke, M.C., Norton, S.J., Johnson-Cicalease, J.M. and Ruemmele, B.A. (1992) *Acremonium* endophyte effects on tall fescue drought tolerance. *Crop Science* 32, 1392–1396.

Xu, L., Zhou, L., Zhao, J., Li, J., Li, X. and Wang, J. (2008) Fungal endophytes from *Dioscorea zingiberensis* rhizomes and their antibacterial activity. *Letters in Applied Microbiology* 46, 68–72.

Yue, Q., Miller, C.J., White, J.F. and Richardson, M.D. (2000) Isolation and characterization of fungal inhibitors from *Epichloe festucae*. *Journal of Agricultural and Food Chemistry* 48, 4687–4692.
Zaurov, D.E., Bonos, S., Murphy, J.A., Richardson, M.D. and Belanger, F.C. (2001) Endophyte infection can contribute to aluminum tolerance in fine fescues. *Crop Science* 41, 1981–1984.
Zhang, H.W., Song, Y.C. and Tan, R.X. (2006) Biology and chemistry of endophytes. *Natural Product Reports* 23, 753–771.
Zou, W.X., Meng, J.C., Lu, H., Chen, G.X., Shi, G.X., Zhang, T.Y. and Tan, R.X. (2000) Metabolites of *Colletotrichum gloeosporioides*, an endophytic fungus in *Artemisia mongolica*. *Journal of Natural Products* 63, 1529–1530.

Part V

Managing Fungal Pathogens Causing Leaf Damage

16 The Rust Fungi: Systematics, Diseases and Their Management

M.S. Patil[1] and Anjali Patil[2]
[1]Department of Botany, Shivaji University, Kolhapur, India; [2]Department of Botany, Rajaram College, Kolhapur, India

Abstract

The rust fungi (*Uredinales*) consist of 7000 species belonging to 163 genera in 14 families and comprise about 10% of all described species in the Kingdom Fungi. All the rust fungi are ecologically obligate parasites on ferns, gymnosperms and angiosperms. There are six vital processes in plants and, correspondingly, six ways in which rusts affect their hosts adversely. Rusts as pathogens damage foliage, the main organ of photosynthesis, destroy seedlings, impair growth and interfere in the metabolism of the hosts. Management of any disease begins with correct identification of the pathogen; hence, some important concepts in rust systematics are discussed, along with detailed information about rust diseases of some economically important crops. Of course, discussion on plant diseases would not be complete without recent management strategies. The discussion includes the following;

1. Rust systematics, including characteristic features of rust fungi, their occurrence and geographical distribution, vegetative and reproductive propagules, pleomorphism, autoecious and heteroecious nature and host range, etc.
2. Rust diseases of crops, including field crops – medicinal, ornamental, cereals, pulses, millets, oilseeds, fruit and plantation crops, etc. – nature of disease, epiphytotics, disease development index; $X = Xo^{eRT}$, assessment of crop losses.
3. Management strategies citing food crisis, need for another green revolution, crop losses, famines, social impact of rust diseases, e.g. change in coffee-drinking habit due to coffee rust, management methods – Sharvelle's strategy (1961): (i) protective, (ii) preventive; and (iii) corrective (physiological disorders), cultural, chemical, biological, breeding, biotechnology – transgenic plants are described in detail.

Introduction: Rust Systematics

The rust fungi (*Uredinales*) consist of 7000 species belonging to 163 genera in 14 families and comprise *c.*10% of all described species in the Kingdom Fungi (Kirk *et al.*, 2001; Ono, 2002). The *Uredinales* are believed to be monophyletic taxa and recent molecular–phylogenetic analysis (Swann and Taylor, 2001) supports this perspective. Very recently, a rust, *Uredo vetus* Henne, has been reported for the first time on *Selaginella* sp. The parasitism of rust fungi to the host plant is highly specific; however, this specialization varies with species, for example, two well-known rusts of soybean, namely *Phakopsora pachyrhizi* Syd. and Syd. and *P. meibomiae* Arthur (American rust) (Ono *et al.*, 1992), occur on a large number of species of family *Leguminosae*. The rust fungi

 Management of Fungal Plant Pathogens (eds A. Arya and A.E. Perelló)

are unique in having complex life cycle patterns with elaborate spore forms. Accordingly, many types of life cycles are known (Laundon, 1973) with modified spore types with immense functional diversities; for example, aecioid teleutospores in endoform rusts like *Endophyllum*, *Monosporidium* and *Kulkerniella* or uredinoid teleutospores in *Hemileia vastatrix* Berk. & Br., which Rajendren (1967) described as 'Kamat Phenomenon'. The spermogonial–aecial host/s and uredinial–telial host/s are closely associated eco-geographically. Heteroecious life cycle is widespread in the uredinales. If all spore forms are produced in unidirectional order, the life cycle is said to be macrocyclic (*Puccinia graminis*, *P. helianthi*, the heteroecious and autoecious rusts of wheat and sunflower, respectively). There is a tendency to omission of spore form in life cycles and thus many different patterns exist, for example, demi-cyclic, in which the uredial stage is absent. In rust fungi, a widespread assumption is that parasitism and host specialization are acquired at an early stage of rust fungus evolution. Nine types with 11 variations are found in nuclear cycles associated in metabasidium development of microcyclic rust fungi (where only a telial stage with or without a spermogonial stage is formed on a plant throughout the season). Thus, rust species that produce only teleutospores that germinate without dormancy to initiate new generations repeatedly in a single growing season would be highly adaptive, e.g. *P. pampeana* Speg. on chillies (*Capsicum* spp.), *P. alyxiae* Arthur on *Alyxia* spp., *P. xanthii* Schw., *M. machili* (Hennings) T. Sato, *E. acacia* Hodges and Gardner. Microcyclic species exhibit two or more patterns of nuclear cycles and different metabasidium development, indicating that microcyclic lineages might have evolved independently and repeatedly from macrocyclic parental species.

Families of Rust Fungi Based on Teleutospores

Rust fungi comprise one of the largest and best described groups of fungi and include many economically important pathogens of vascular plants. Dietel (1900) divided the order Uredinales into four families based on sessile or pedicillate teleutospores as follows:

Pedicillate teleutospores -------*Pucciniaceae*
Sessile teleutospores
In a single layer --------*Melampsoraceae*
In 1–2 layers forming waxy crust --------
--------------------------*Coleosporiaceae*
In chains --------------------*Cronartiaceae*

Families	Genera
Pucciniastraceae (Arthur) Gaumann	06
Coleosporiaceae Dietel	02
Cronartiaceae Dietel	01
Melampsoraceae Schroeter	02
Phakopsoraceae (Arthur) Cummins & Hiratsuka	11
Mikronegeriaceae Cummins & Hiratsuka	01
Chaconiaceae Cummins & Hiratsuka	10
Uropyxidaceae (Arthur) Cummins & Hiratsuka	10
Pileolariaceae (Arthur) Cummins & Hiratsuka	03
Raveneliaceae (Arthur) Leppik	14
Phragmidiaceae Corda	10
Sphaerophragmidiaceae Cummins & Hiratsuka	06
Pucciniaceae Chevalier	15
Puccinosiraceae (Dietel) Cummins & Hiratsuka	09
	99
Unassigned genera	08
Total	107

However, this dependence on teleutospore morphology has brought many unrelated genera into the same family. Some of the larger genera are:

1. *Puccinia* Pers. ex Pers. (*c.*3000–4000 spp.).
2. *Uredo* Pers. ex Pers. (*c.*3000 spp.).
3. *Uromyces* (Link) Unger (*c.*600–700 spp.).
4. *Ravenelia* Berkeley (*c.*200 spp.).

5. *Melampsora* Castagne (*c*.100 spp.).
6. *Hemileia* Berk. & Br. (*c*.50 spp.).
7. *Coleosporium* Lev. (*c*.80 spp.).
8. *Phragmidium* Link (*c*.60 spp.).

In some cases, the original function has been changed irrespective of its basic nature or structure; for example, the species of *Endophyllum*, *Monosporidium* or *Kulkerniella* produce aeciospores morphologically in aecial cups, but they function like teleutospores producing promycelium bearing basidiospores and are thus called aecioid teleutospores. In *H. vastatrix*, urediniospores occasionally function like teleutospores and are called uredinoid teleutospores. Of course, in the first example, no teleutospores are produced, while in the second example, teleutospores normally are produced in the life cycle. In the course of the evolution of rust fungi, there is a tendency to eliminate spores, narrowing the host range and surviving in spite of unfavourable environmental conditions or non-availability of the required host. There are numerous examples in which a rust survives or continues its life cycle by producing one type of spore, e.g. *Aecidium*, *Uredo*, *Peridermium*, *Caeoma*, etc.

Spore morphology

In groundnut rust, the uredinospores are one-celled, spherical to oval or angular, stalked, mostly brown coloured, faint or dark, thick-walled with spiny, verrucose, or a modification of these two, rarely smooth, bearing visible areas in the walls through which germination takes place. Tulasne and Tulasne (1847) observed for the first time, pores/oscules varying from 2 to 20 in number. The germ pores may be distributed equatorially, zonal or scattered. Cummins (1936) also recognized their phylogenetic significance in the rust taxonomy. There appears to be some correlation between the arrangement of pores and the shape of urediniospores; globoid spores usually have scattered pores, while ellipsoid, oblong or asymmetrical spores usually have zonate pores. Urediniospores are bi-nucleate, repeating like conidia, germinate very easily within 24 h at high temperatures and relatively higher humidity, but their viability is lost at very high temperatures. At lower temperatures, spores remain viable for a long time. They germinate mostly by germ tubes, except in coffee rust, where they behave like uredino teleutospores.

Urediniospores are most significant in disease development and spread on epiphytotic scale (van der Plank, 1963, 1968). Hence, this spore state is also considered as the conidial state during sporulation and the release of spores form spore clouds in the air and serves as secondary and tertiary inoculum within the crop in a favourable season. Rust diseases, due to their repeating nature, are known as compound interest diseases. Log $e(x/1 - x)$ where x is a proportion of infected susceptible tissue, if the pathogen is systemic. Urediniospores are light, airborne and travel long distances at various heights; for example, wheat rust urediniospores travel from Mexico to Canada; in Indian wheat rust, the rust originates in South India from the Nilgiri and Pulney Hills and travel via Central India from the plateau of Mahabaleshwar and Panchgani to North India.

Rust physiology – urediniospore germination

Spore liberation is active and the terminal velocity of fungal spores in the air is 0.05–2.5 cm/sec. In calm weather, only 0.05% spores travel more than 100 m from their source of origin. Spores of black stem rust fall from a height of 1.6 km to the ground at a speed of 12 mm/sec and travel from place to place at 11–32 km/h (Gregory, 1973). Urediniospores of different rust species have a different period of viability as they are affected by environmental factors such as RH, light intensity, as well as their own structural characteristics, namely wall thickness, etc. In northern India, urediniospores are killed by high temperatures in the field and cannot serve as a source of inoculum the following year.

Viability of urediniospores in rust fungi

Rust urediniospores show different periods of viability at 20–40% RH and 23°C, e.g. *P. graminis tritici* 36 days, *P. recondita* 63 days, *P. coronata* 87 days, *P. menthae* 173 days, *P. helianthi* 185 days and *U. pisi* 75 days. Viability of urediniospores decreases at higher or lower humidity. Spores germinate generally at 90–100% RH, while the temperature requirement varies greatly in different species. The period required for sporulation (urediniospores) in black stem rust of wheat is found to be 5 days at 24°C. If the temperature is lowered to 0°C, then sporulation occurs after 85 days. Spore longevity depends on light, temperature, relative humidity, species of rust and type of spore. Basidiospores and pycniospores are delicate and have least viability. But if the spores are kept at a low temperature, viability lasts for 18 days. In the case of sunflower rust, relative humidity is more important than temperature. At 80% RH, only 5% aeciospores remain viable after 56 days.

Teleutospores spores are produced at the end of a rust fungus's life cycle, i.e. spores terminating the life cycle of rusts. They are produced in telia in or on the host, are innate or erumpent, covered or exposed in telial sori, in the leaves or in the stem. There is a tendency in rusts to eliminate spore states showing progressive reduction either due to non-availability of host or climatic conditions, for example, rusts in temperate regions on the family *Liliaceae*. Rust systematics, a dynamic science, is far from perfect; hence, there is still a lot of work to be done on their taxonomy and pathology. Study of their host's behaviour during development, varieties, races, physiological forms and pathotypes is beyond the scope of taxonomists.

Rust Diseases of Some Economically Important Crops

Plant pathology originated in Europe and migrated to North America, where it flourished and spread to different parts of the world. Among all known living organisms, three groups are dominant, namely insects, flowering plants and fungi. Among all these plant pathogens, fungi are the most dominant and successful plant pathogens and are estimated to produce more than 25,000 diseases. Among these, 8000 diseases of cultivated and plantation crops are extremely damaging in the field every year. Kuhn, (1858) wrote a book entitled *Diseases of Cultivated Plants*. Rusts are complex; hence, it is difficult to understand how they damage standing crops in the field qualitatively and quantitatively, creating problems of food crisis and insecurity; a global problem today.

Epidemiological Studies

Epidemiology is the science of epidemics or diseases in plant population. Types of epiphytotics are:

1. Based on the rate of disease development:
   (i) Tardiv (slow epiphytotics);
   (ii) Explosive (rapid epiphytotics).

2. Area covered and time of development:
   (i) Pandemic – developing on a continental scale;
   (ii) Sporadic – seasonal and irregular incidence.

There are also secondary epiphytotics known. Epiphytotics is also defined as 'a host–pathogen system, out of genetic balance in favour of the pathogen'. Such epiphytotics of crop plant diseases are known, in the history of plant pathology, to be followed by food famines:

1. Wheat rust epidemics occurred in 1916 in America and Canada; 1935 and 1937 in America; 1951 in Europe; and 1827, 1907, 1947, 1949–1950, 1957, 1971–1972 in India.
2. Coffee rust epidemics occurred in 1867 and 1875 in Sri Lanka; 1891 in the Philippines; 1891–1892 in Java; 1911–1913 in Central Africa; 1871–1878 in South Africa; and 1970–1971 in Brazil.

The coffee rust famines influenced the coffee-drinking habit, which then changed to tea.

Rusts are compound interest diseases and an increase of infection at a compound interest rate exponentially/logarithmically increases the rate of compound interest of disease by primary and secondary infection. The compound interest equation can be given as:

$$X = Xo^{eRT}$$

where X = the amount of disease at time T, Xo = initial amount of disease at O time, R = infection rate, which is variable, and $e = 2.718$ for cereal rusts.

The rate (R) of increase % per unit of time is a fundamental concept in epidemiology, e.g. 12.5%/day in *P. recondita* and 57%/day in *Phytophthora infestans*. Development of epiphytotics is basically a transport problem to get enough inoculum to the right place at the right time. Plant–pathogen–environment is a triple interaction and may be complicated by vectors and humans; according to van der Plank (1963, 1968), the pathogen must be virulent. To express virulence quantitatively, the disease reaction type is expressed in numerical values as:

R (resistant) = 01, MR (moderately resistant) = 02, S (susceptible) = 03

Aggressiveness corresponds to disease severity on a 0–9 score scale:

0 = absent, 9 = more than 75% leaf area in 12 days after inoculation

therefore

VI (virulent index) = [1 + (virulence × aggressiveness) × latent period]

$$VI = [1 + VAL - 1]$$

where VI = virulent index, A = aggressiveness and L = latent period.

Crop Losses

Conservative estimates of total annual losses in crop production by diseases, insects and weeds worldwide are 220 billion US$ corresponding to 31–42% of all losses, of which diseases are 14.1%, insects 10.2% and weeds 12.2%, while 6–12% losses are postharvest losses.

The estimated annual crop losses worldwide (Agrios, 2005) are:

	US$
Attainable crop production (2002 prices)	1–5 trillion
Actual crop production	995 billion
Production without crop protection	445 billion
Losses prevented by crop protection	415 billion
Actual annual losses to world crop production	550 billion
Losses caused by disease	220 billion

Rusts damage plants and plant products, causing economic losses. Crop protection measures result in increased prices of primary products to consumers and pollution of the environment. Rusts are also pathogenic to animals and humans. Diseases are responsible for minor aesthetic losses – in domestic gardens, avenues and forests. There are six vital processes in plants and, correspondingly, six ways in which rusts affect their hosts adversely. Rusts as pathogens damage foliage, the main organ of photosynthesis, destroy seedlings, impair growth and interfere in the metabolism of the hosts. Rusts keep hosts alive and active for their own growth, development and spread. Hence, pathogens and hosts have coevolved. At the same time, rusts are a useful means of controlling weeds as their infection results in thinning of plants in the field. Artificial infection of rust fungi in fodder grasses brings about an increase in protein content.

Green Revolution and Grain Production in India

In India, a green revolution began in 1960. In 1965, hybrid varieties were introduced, followed by an increased consumption of fertilizers (N, K, P). From the 53% of total area under cereal cultivation, hybrid cultivars have been introduced in 16% of the area. These innovations in agricultural practices revolutionized grain production in India from 1900 to 1971. From 1900/01 to 1910, grain production was 67.6 Mt. It then remained stable until 1948, in the pre-independence era. During 1948–1949, there was a wheat famine. During the first three of the Five Year Plans (1950–1965), the rate of grain production increased by 2.5%. In the fourth Five Year Plan, grain production was approximately 100 Mt, i.e. grain production increased by 5%. The food crisis provided not only a warning but also an opportunity for new thinking. Still, there is some hope as the International Grains Council has forecast a 7% increase in global wheat production. India is the world's second largest producer of wheat and rice and the expected wheat harvest this year is 76.8 Mt and rice production is 95.7 Mt. Among the major cereal crops, wheat is the most suitable due to its superior quality of grain, coupled with its wide adaptability for cultivation under varied conditions; humans and wheat will survive in any environment. Even today, the wheat rust management mission is still incomplete and awaiting novel solutions to increase the yield of quality grains. Lord John Boyd Orr, the first Director of the FAO, said in 1948, 'a lifetime of poor nutrition and actual hunger is the fate of at least 2/3rd of the world's population'. This is still true, even today.

Diseases of Crop Plants and Associated Pathogens

Grain crops

1. Wheat rusts (*Triticum* spp.)
(i) Black stem rust: *P. graminis* Pers. *tritici* Eriks. and Hennen;
(ii) Brown rust: *P. recondita* Rob. ex Desm.;
(iii) Yellow or stripe rust: *P. striiformis* West.

2. Leaf rust of rye (*Secale cerealis* L.): *P. graminis* Pers. *secalis*.

3. Leaf or crown rust of oat (*Avena sativa* L.): *P. coronata* Corda and *P. graminis* Pers. *avenae* Fraser & Ledingham.

4. Rust of jowar (*Sorghum bicolor* (L.) Moench):
(i) *P. purpurea* Cke.;
(ii) *P. levis* Arthur;
(iii) *P. nakanishiki* Dietel.

5. Rust of pearl millet or bajara (*Pennisetum glaucum* (L.) R. Br.):
(i) *P. substriata* Ell. and Barth var. *indica* Ramachar and Cummins, India;
(ii) *P. substriata* Ell. and Barth. var. *decrospora* Eboh, Nigeria;
(iii) *P. substriata* Ell. and Barth. var. *penicillaris* Ramachar and Cummins.

6. Rust of maize (*Zea mays* L.):
(i) *P. sorghi* Schw. (common corn rust);
(ii) *P. polysora* Underw. (southern corn rust);
(iii) *Physopella zeae* (tropical corn rust).

Oilseed crops

1. Soybean rust (*Glycine* max (L.) Merr.): *Malupa sojae* (P. Henn.) Ono, Y. *et al.* or *Malupa* state of *P. pachyrhizi* H. and P. Sydow.

2. Groundnut rust (*Arachis hypogaea* L.): *P. arachidis* Speg.

3. Sunflower rust (*Helianthus annuus* L.): *P. helianthi* Schwein.

4. Safflower rust (*Carthamus tinctorius* L.):
(i) *P. carthami* Corda;
(ii) *P. caleitrapae* var. *centaureae* (DC.) Cummins.

5. Linseed/flax rust (*Linum usitatissimum* L.): *M. lini* (Ehrb.) Lev.

Plantation crops

1. Coffee leaf rust (*Coffea arabica* L. and other spp.):
(i) *H. vastatrix* Berk. and Br.;
(ii) *H. coffeicola* (reported only from Cameroon, West Africa).

2. Mulberry rusts (*Morus alba* L. and other spp.):
(i) *A. mori* Barclay;
(ii) *Cerotelium fici* (Butler) Arthur.

3. *Dalbergia* rust (*Dalbergia* spp.): *Sphaerophragmium dalbergiae* Dietel = *U. dalbergiae*

P. Henn. (1895) = *U. sisso* Syd. & Butl. (1906).

4. Teak rust (*Tectona grandis* L.): *Olivea tectonae* (Ramkr., T.S. and K.) Mulder = *Chaconia tectonae* Ramkr., T.S. and K.

Pulses and vegetables

1. Green gram rust (*Cicer arietinum* L.): *Uromyces ciceris arietini* (Gron.) Jack.
2. Rust of *Phaseolus* sp: *U. appendiculatus* var. *appendiculatus* (Pers.) Unger.
3. Cowpea rust (*Vigna* sp.): *U. vignae* Barclay.
4. Bean rusts (*Pisum, Vicia, Lens, Lathyrus* spp.):
(i) *U. viciae-fabae* (Pers.), Schroeter, autoecious rust in Europe and America;
(ii) *U. pisi* (Pers.) Wint., heteroecious rust in Europe, rarely in India.
5. Chilli rust (*Capsicum annuum* L.): *P. pampaeana* Speg. (Mexico, Peru, Brazil, Columbia and Gautemala).

Forage crops (Fabaceae)

1. Bersim/clover rust (*Trifolium* spp.).
2. Lucerne (*Medicago* spp.).
3. Alfalfa rust (*M. sativus* L.): *U. striatus* Schroeter.
4. Clover rusts (*Trifolium* spp.):
(i) *U. trifolii-repentis* Liro;
(ii) *U. fallens* (Arthur) Barth.

Fibre crops

1. Cotton rusts (*Gossypium* spp.):
(i) *P. gossypi* (Arthur) Hiratsuka = *U. gossy* = *C. desmium*;
(ii) *A. gossypi*, an aecial state of *P. cacabata* Arthur & Holway.
2. Linseed/flax rust (*Linum usitatissimum* L.): *M. lini* (Ehrb.) Lev.

Medicinal plants

1. *Adhatoda zeylanica* Nees. rust: *Chrysocelis butteri* (Dietel and Sydow) Laundon.
2. Rust of *Vitex* spp.:
(i) *O. fimbriata* (Mains) Cumm. & Hirat.;
(ii) *O. scitula* H. Sydow;
(iii) *O. viticis* Ono, Y. and Hennen.
3. Rust of *Vinca major* L.: *P. vincae* Berk.

Ornamental plants

1. Rose rusts (*Rosa* spp.): *Phragmidium* spp. (10 spp.).
2. Gladiolus rust (*Gladiolus* spp.): *P. gladioli* (Duby) Cast.
3. Tulip rust (*Tulipa* spp.): *P. prostii* Moug. (on wild species).
4. Canna rust (*Canna* spp.): *P. thalie* Dietel.
5. Chrysanthemum rust (*Chrysanthemum* spp.): *P. chrysanthemi* Roze.
6. Saxifraga rust (*Saxifraga* spp.): *P. saxifragae* Schlecht.

The key to species and varieties of *Puccinia* and *Uromyces* producing rust diseases can be found in the Appendix at the end of this chapter.

Rust of wheat (*Triticum* spp.)

In India, wheat is cultivated on 16m ha, but the average yield is very low, that is, 810–1000 kg/ha as compared to the wheat yield in other countries, namely Argentina 1210 kg/ha, America 1610 kg/ha, Belgium 3700 kg/ha and Denmark 4000 kg/ha. The five major wheat-growing areas in India are in the north-western zone, the north-eastern zone and the central, peninsular and northern hilly zones. Gene exchange of the new allopolyploids inevitably would have resulted through hyphal fusion, nuclear exchange and genetic recombination in urediniospore population. This results in widening the host range of the hybrid rusts because of gene diversity and increases the value of survival and nutritional status of the rust. Continuous introduction of hybrid cultivars in the field through plant breeding serve as new hosts to rust. Three species of wheat and their cultivars are used mainly for cultivation of durum wheat (*T. durum* Desf.) or macaroni wheat, which covers about an 85% area. Bread wheat (*T. aestivum* L.) covers a 14%

area and emmer wheat only about a 1% land area. Rusts of wheat have many physiological races and wheat cultivars are recommended by plant breeders in specific regions for cultivation to avoid rust disease development. Epidemiological studies of three rusts have shown that collateral hosts have a restricted role, while alternate hosts virtually have no role at all. The survival of these rusts in India is primarily through urediniospores that survive on self-growing plants or volunteer plants. The airborne urediniospores favoured by wind and rain due to tropical cyclones in the months of October and November get dispersed and deposited over Central India from the Nilgiri Hills in South India. Brown and black stem rusts become established there and then subsequently spread to the eastern and northern states of India over the Indo-Gangetic Plain. Brown rust appears first in the Himalayan foothills, eastern Uttar Pradesh and north Bihar in the month of January.

The western lines associated with disturbances and rain spread the pathogen to the north-western states of India, along with yellow rust. The Nilgiri and Pulney Hills are the primary focal point providing the source of inoculum, that is urediniospores migrating upward with air currents towards the north via Central India, periodically trapped and studied by Mehta (1929, 1952). The Mahabaleshwar and Panchgani plateaus also serve as a focal point for the secondary source of inoculum (Joshi *et al.*, 1986).

Rust of coffee

Coffee rust is a disease of the coffee plantation crop, namely *Coffea arabica* L., *C. libarica* and *C. canephora*, cultivated for berries to produce coffee, a well known non-alcoholic drink like tea, in Ethiopia, Yemen, Sri Lanka, South and Central Africa, Cameroon, Bahamas, Brazil and India. The world production of coffee is 3.16 Mt/year. In India, coffee is cultivated mainly on the hill slopes of Karnataka, Tamil Nadu and Kerala. Coffee production is estimated to be 964,000 t/year on a worldwide basis and its production in India is 230,000 t. The pathogen of the coffee plant is *H. vastatrix* Berk. and Br., which develops very serious disease on foliage, leading to defoliation.

Rust of groundnut (*Arachis hypogaea* L.)

Groundnut is the world's second largest source of edible oil and ranks 13th in production among world food crops. India is the largest producer of groundnut. Groundnut is cultivated in 26m ha of land worldwide and produces 34.5 Mt/year. Groundnut is cultivated in India on 7.6m ha and produces 7.8 Mt/year. The major states in India cultivating groundnut are Gujarat, Maharashtra, Tamil Nadu and Andhra Pradesh. This crop is attacked by 55 pathogens. Among all the diseases, three diseases, namely groundnut rust and early and late leaf spot diseases, are more serious. They generally develop simultaneously and pod yield decreases by up to 10–70% (Ghewande and Savalya, 1999). Rust is caused by *P. arachidis* Speg.; this perpetuates by uredinia, the only spore state throughout the world except teleutospores, which were recorded in Paraguay only once. It is not known how groundnut rust perpetuates and it occurs regularly every year in India without having a telial state, alternate or collateral host. It is said that groundnut rust develops first in South India and then migrates to North India. This disease, along with early and late leaf spot disease, renders the crop uneconomical in the rainy season, which is the major period of groundnut cultivation in India.

Rust of jowar (*Sorghum* spp.)

The genus *Sorghum* Moench has 23 species (Simon, 1993). The crop is damaged by four different fungal diseases: seed and seedling disease, foliage disease, head disease and root and stalk disease. *Sorghum* rust is a foliage disease which infects almost all species of *Sorghum*. High temperatures (75–80°F) and humid weather is favourable for disease development. The species of *Puccinia* that infect *Sorghum* (Cummins, 1971) are *P. purpurea*, *P. levis* and *P. nakanishiki*. The most prevalent rust of jowar throughout the world is *P. purpurea* Cooke, which is heteroecious and its aecial host is *O. corniculata* L. But the aecial stage plays a negligible role in rust disease. The rust infection and host reaction results in the formation of bright purple-coloured spots on the leaves. There are 32 races in cultivated *Sorghum* distributed in South-east Asia (11) and Africa (21).

Rust of maize (*Zea mays* L.)

Maize rust or leaf rust of maize, *P. sorghi* Schw., is an American rust. However, maize is susceptible to two more rusts, i.e. *P. polysora*, southern corn rust, and *P. zeae*, tropical corn rust. It is a heteroecious rust and its aecia are produced on species of *Oxalis*, namely *O. stricta*, according to Arthur (1929). Aecia are more common in this rust than in *Sorghum* rust. However, Mishra (1962) has claimed that the alternate host of maize rust is *O. corniculata*, on which aecia were collected from Nepal. The rust infects all types of maize with a varying degree of severity.

Rust of bajra/pearl millet (*Pennisetum* spp.)

The genus *Pennisetum* Rich. has *c.*80 species and is distributed throughout the tropics. Pearl millet (*P. glaucum* (L.) R. Br.) is a staple food crop of the semi-arid tropical parts of the world, mainly Asia and Africa. There are about 14 rusts reported on bajra (Cummins, 1971). However, only two are well-known, namely *P. substriata* Ell. and Barth. var. *indica* Ramachar and Cummins in India and *P. substriata* Ell. & Barth. var. *decrospora* Eboh., recently reported from Nigeria. The first rust is predominant in India and is heteroecious. The alternate aecidial host is the species of *Solanum*. Rust infection produces pustules on both sides of the leaf with necrotic spots, due to which premature drying of leaves may result. Occasionally, pustules also develop on leaf sheaths and stem.

Rust of flax/linseed (*Linum usitatissimum* L.)

This crop is cultivated mainly for oil and fibre. It is affected by rust in most of the linseed-growing areas of the world such as Asia, America and Europe. The rust appears in India in February. This rust, *M. lini* (Ehrb.) Lev., also infects wild species of *Linum*. It is an autoecious rust and infects all the green parts of the plant. Telia develop late on stems and form crusts covered by epidermis. Aecia of this rust are caeomoid. Flor (1956) studied this rust and differentiated 179 races from America alone. Eighteen races are reported from India. *L. mysorense* L., a wild host, has been reported to harbour the rust from India.

Rust of pea (*Pisum sativum* L.)

Pulse crops are affected by two rusts, namely *U. pisi* (Pers.) Wint., a heteroecious rust reported from Europe only and rarely in India, while *U. viciae-fabae* (Pers.) Schroeter, an autoecious rust, is found in Europe, America and Asia. It was found that aeciospores played a major role in the dissemination of lentil rust during the active growing season. It was also suggested that secondary aecia are produced at low temperatures (17–22°C), while higher temperatures induced uredinia. Teleutospores are dormant spores, survive for 2 years and remain viable at low temperatures (3–18°C).

Rust of gram (*Cicer arietinum* L.)

Gram rust is caused by *U. ciceris-arietini* (Gron.) Jack. It is a heteroecious rust, but no alternate host or pycnia and aecia have been collected. The same rust has been collected on wild species of *Trigonella polycerata*, a weed of *Fabaceae* growing at higher altitudes. It is claimed to be the source of urediniospores, the primary inoculum.

Rust of bean

Beans belong to different genera of the family *Fabaceae*, namely *Phaseolus*, *Vicia*, *Lathyrus*,

Pisum, *Dolichos* and *Vigna*. Their commercial cultivars are a source of vegetables, pulses and forage crops. They are cultivated extensively all over the world as kharif and rabi crops. These crops are infected in the field by many rusts. The pathogens are *U. appendiculatus* (Pers.) Unger var. *appendiculatus*, *U. vignae* Barclay, *U. viciae-fabae* (Pers.) Schroeter and *U. pisi* (Pers.) Wint. The first three pathogens are autoecious and the fourth is heteroecious. Many races of these pathogens are known. Heavy infection of leaves results in defoliation and poor productivity.

Rust of rose (*Rosa* spp.)

All rusts of roses belong to the genus *Phragmidium* and ten species infect roses worldwide. However, seven species are very common. The most common species is *P. disciflorum*, which perpetuates on hybrids of *R. canina* and *R. gallica*, but is less likely to attack climbing or rambling roses like *R. multiflora*. The rusts are autoecious and all spore types, except pycniospores, are equally harmful to the foliage. As a result of infection in some seasons, severe defoliation ensues and plants are greatly weakened. Keeping a garden clean is the most effective method of keeping roses healthy.

Rust of cotton (*Gossypium* spp.)

The genus *Gossypium* is known by 4–5 species. Use of cotton fibre in India is found in 'Rig Veda'. Cotton cultivation is mainly for fibre and oil from seeds. Cotton is cultivated as a cash crop in 80 countries of the world. The cotton crop is infected heavily by a large number of pathogens, including rusts. Among all the rusts, *P. gossypi* (Arthur) Hiratsuka is the most troublesome to cotton, not only in cultivated varieties but also in perennial cotton. The rust disease seriously damages the cotton crop and is responsible for *c.*20–70% financial loss to cotton growers annually.

Rust of soybean (*Glycine max* (L.) Mill.)

Soybean as an oilseed crop is cultivated all over the world as kharif. About 25 diseases are known on soybean crop. Among these, there are 19 predominant fungal diseases. In fungal pathogens, rust is the most serious in India. The rust entered India in 1970 from the New World to Japan via Nepal in northern India and spread to the south-western parts of India up until 1995. The rust was first reported from Taiwan. The fungus inciting soybean in Asia was first described as *U. sojae* P. Henn. from Japan in 1903. It was subsequently described and renamed by Sydow, as *U. sojae* H. and P. Sydow. However, this was erroneous due to the host not being soybean but *Mucuna* spp. (Butler and Bisby, 1931). Moreover, *U. sojae* P. Henn. is not considered as an anamorph of *U. mucunaei* Rabenh. The anamorph and teliomorph connection was first proved by Sawada (1931) and named as *P. sojae* Sawada. In the two rusts of soybean, namely *P. pachyrhizi* and *P. meibomaiae* (Arthur) Arthur, the former species shows wide geographical distribution in Asia, Australia and Africa, while the latter is restricted to America. In India, soybean rust does not produce telia and perpetuates only by uredinia, possibly due to environmental factors.

Rust of fig (*Ficus carica* L.)

Rust of fig and other species of *Ficus* is produced by *C. fici* (Butler) Arthur. It is a common rust found throughout tropical and subtropical parts of the world. Telia have been observed only in India on *F. glomerata* Roxb., an evergreen shade tree. In the commercial fig, *F. carica* L., this rust appears late in the season (monsoon) and does not affect the quality of the fruits, but defoliation exposes them to sunburn. It also reduces host vigour.

It is possible to have a cell-free culture using a complex culture medium to culture rust urediniospores. The method was first used successfully in black stem rust, *P. graminis tritici*, for sporulation. Nowadays, compounds like kinetin and benzimidazole, which exert a cytokinin effect, are used to culture detached leaves in solution in test tubes for up to a month to determine the races of rusts.

Rust Disease Management Strategies

There is a need for effective disease forecasting and warning systems, as well as a disease calendar for each crop. Crop protection/management varies widely with different crops, due to different factors such as:

- varietal susceptibility of crops
- soil types
- agronomic practices and
- cropping patterns.

The study of diseases, disease development, disease outbreak, pathogens, varieties, races, biotypes, ecotypes, pathotypes, specialization, host plants, their hybrids, cultivars, fluctuating factors like soil, water, fertilizers, pesticides, host–pathogen complex, etc. is very vast and difficult. The outbreak of disease on an epiphytotic scale resulting from the interaction of pathogen, host and environment can be represented through disease progress curves (DPC). The problem becomes more serious due to a pathogen having high pathogenicity, which includes:

- virulence and
- aggressiveness (vigorous races).

Today, virulence, based on evidence, can often be considered to be oligogenic, in which a few genes are involved, as suggested by van der Plank (1968), while aggressiveness is generally polygenically inherited. Virulence is conditioned due to gene diversity and aggressiveness by variation in the doses of enzymes. Hence, disease reaction is a chemical process which entails changes in the host and parasite cell metabolism. Black stem rust is found to be more severe on wheat than barley and yellow rust on barley than wheat. To avoid heavy losses, early sowing was recommended by Pliny. In Greek and Roman civilizations, the appearance of plant diseases was attributed to 'Deities – rust gods, Robigan and Robigus'. Thomas Knight, the English plant physiologist, gave experimental proof of the ability of aeciospore to infect cereals and after that same year, voluntary eradication started in Denmark.

Strategy of management of plant diseases (Sharvelle, Strategy of Plant Disease Control, 1961)

Most control measures either reduce the initial inoculum or rate of spread of plant pathogens (van der Plank, 1963). It is important to reduce and delay the initial infection as much as possible, by disease forecasting in advance so that farmers can protect the crop plants to avoid monetary losses. Sharvelle (1961) classified the strategies for plant disease control into two categories: (i) immunization; and (ii) prophylaxis (to eradicate the pathogen).

Strategy of plant disease control

Immunization

1. Genetical resistance.
2. Induced resistance.

Prophylaxis

1. Protection:
   (i) Chemical prophylaxis;
   (ii) Environmental manipulation.
2. Eradication:
   (i) Crop rotation;
   (ii) Sanitation;
   (iii) Alternate host elimination;
   (iv) Chemical eradication.
3. Legislation:
   (i) Quarantine;
   (ii) Regulatory measures.

Developing New Strategies for Disease Management: Role of Oxidative Burst

The molecular biology of interactions between disease-resistance genes, defence genes and their role in genetic engineered disease-resistance elicitor (signal) molecules has been detected in fungal, bacterial and viral pathogens. These molecules serve as signals to elicit defence mechanism of the host. Host resistance genes may function as receptors of these signals. Only a few disease-resistance genes have been cloned from plants. Analysis of these genes shows the presence of leucine-rich repeats (LRRS), leucine zippers and nuclear localization signals (NLS). LRRS are involved in protein–protein interactions of the signal transduction pathway. Several defence genes are widely found in both resistant and susceptible plants and are involved in the production of antimicrobial compounds, namely phenols, phytoalexins and pathogen-related (PR) proteins; PR-1, 2,3,5,6 and 8, co-enzyme reductase (HMGR), transgenic plant expressing phenylalanine ammonial-yases (PAL) showed enhanced disease resistance (Vidyasekaran, 1997).

Fungicides in the Control of Rust Diseases

The different fungicide dosages recommended for rust disease control are shown in Table 16.1.

Other Methods of Plant Disease Management

Biological methods

Biological control using different microbes is the most popular and ecologically safe method, but is not used practically due to many constraints. The following are some potential and promising parasites of rusts which can be used in biological control in the future.

1. *Aphanoderma album* (Preuss.) W.Gams
This *Hyphomycetes* is characterized to produce a metabolite which switches off sporulation of urediniospores to teleutospores, thus terminating the life cycle of rust.

2. *Cladosporium* spp.
(i) *C. aecidiicola* Theum.;
(ii) *C. exobasidii* Jaap;
(iii) *C. uredinicola* Speg. on *P. recondita* (UK).

3. *Verticillium* spp.
(i) *V. hemileiae* Steyaert;
(ii) *V. lecanii* (Zimm.) Viegas.

V. hemileiae and *V. lecanii*, as a virtue of their growth on uredinia of coffee rust in a moist environment, produce a chitinase enzyme to weaken the wall of the spores, as a result of which the spores burst. Even the cultural filtrates are effective (Ellis, 1971, 1976).

4. *Sphaerellopsis filum* (Biv. – Bern. ex. Fr.) B.C. Sutton = *Darluca filum* (Biv.) Cast. This hyperparasite was considered by Tarr (1972) as an ecologically balanced mycoparasite on rust fungi, especially the *uredinia* of the species of *Puccinia* and *Uromyces* of grasses. It is distributed in the tropical and subtropical moist regions of the world. However, it has not been used commercially.

5. *Tuberculina costaricana* H. Sydow.

6. *Olpidium uredinis*, an endoparasite in urediniospores.

Plant breeding

1. Cultivation of hybrid cultivars recommended by plant breeders in different regions for different rusts. The hybrids have high resistance coupled with good quality and high productivity. Some rust-resistant cultivars are 'Maris Ranger', 'Heines VII', 'Fenman', 'Hybrid 46', 'Minster', 'Opal', CS 2D/2M, *T. spelta* 391, *T. spelta* G652 (ICARDA), CIM 25, 'Dove' (CIMMYT), HD 4502, 'Arkan', 'Blueboy II', 'Centurk', 'Chris' (USA), 'Banks' and 'Egret' (Australia).

2. Use of defence activators – spray of salicylic acid (SA), ferric chloride ($FeCl_3$) and dipotassium hydrogen phosphate (K_2HPO_4).

3. Adult plant resistance (APR) – there has been an increasing interest in adult plant resistance, especially against leaf rust or brown rust pathogens of wheat in North India, because of its widespread occurrence in germplasm and its durability.

Table 16.1. The various fungicide dosages recommended for different diseases.

I – Sulphur fungicides			
Sulphur	2–4 kg/ha	Spray or dust 2–3 times	Rust of beans – *Uromyces appendiculatus* var. *appendiculatus*, *U. fabae*
Lime – sulphur	0.75 g/100 g	Sprays	Bean rust – *U. fabae*
Ziram	0.2% solution	Spraying at the outbreak of disease and repeated at weekly intervals	Bean – *U. appendiculatus* or *U. phaseoli* Mint – *Puccinia menthae* Sunflower – *P. helianthi*
Ferbam	0.15% solution	Sprays as required	Rose – *Phragmidium mucronatum*
	0.3% solution	Seed treatment	Safflower – *P. carthamii*
Thiram	0.15–0.2% spray	Sprays	Apricot – *Tranzschelia discolor* Plum – *T. discolor*
	0.2–0.3% solution	Dry seed dressing	Safflower – *P. carthamii*
	0.5% solution	Two sprays at 10–15-day intervals	Beet – *U. betae*
Zineb	0.2% solution	Applied at 4-week intervals from petal fall	Almond and apricot – *T. discolor*
	0.2% solution	3 sprays at 14-day intervals	Barley – *P. striiformis*
	0.15% solution	Sprays at disease outbreak and repeat after 10–15-day intervals	Chrysanthemum – *P. chrysanthemi*
	0.2% solution	5 sprays at 10-day intervals	Garlic – *P. allii*
	0.2% solution	Applied at 4-week intervals from petal fall	Peach – *T. discolor*
	0.2% solution	3 sprays at 15-day intervals starting in last week of September before disease outbreak	Phalsa – *Dasturiella grewiae*
	0.2% solution	5 weekly sprays in the growing season	Soybean – *Phakopsora pachyrhizi*
	0.2% solution	Spray at disease outbreak and repeat at *c.*10-day intervals as necessary	Wheat – *P. recondita* and *P. graminis tritici*
Maneb/ mancozeb	0.2% solution	Sprays at disease outbreak and repeat at 7–10-day intervals	Bean – *U. appendiculatus* var. *appendiculatus*
	0.25% solution	Sprays at 14-day intervals along with benomyl or carbandazim	Groundnut – *P. arachidis*
	0.2% solution	4 sprays at 10-day intervals from disease outbreak. Add triton at rate of 2 ml/l suspension	Peas – *U. fabae*
	0.3% solution	Seed treatment	Safflower – *P. carthamii*
	0.2% solution	4 sprays at 12-day intervals	Sorghum – *P. purpurea*
	400 ppm	Sprays	Sunflower – *P. helianthi*
	2–3 kg/ha	Dusting 3–6 applications at 10-day intervals	Wheat – *P. graminis tritici* and *P. recondita*
Nabam	0.15–0.25% solution	Sprays 4–5, beginning when disease appears	Wheat – *P. graminis tritici* and *P. recondita*

continued

Table 16.1. *continued.*

II – Copper fungicides			
Bordeaux mixture	5:5:1/2:50	Sprays, 5–10 applications, add urea and $ZnSO_4$	Coffee – *Hemileia vastatrix*
Copper oxychloride	1.8 kg/400 l	Sprays	Beans – *U. appendiculatus*
	Colloidal copper 1500 ml/400 l	Sprays	Beans – *U. appendiculatus*
	0.35% solution (kocide 110) and 0.5% (cupravit)	Sprays in March–April 300 l/ha	Coffee – *H. vastatrix*
III – Mercury fungicides			
Agrosan GN	0.3% solution	Seed treatment	Safflower – *P. carthamii*
IV – Heterocyclic nitrogenous compounds			
Captafol	0.3% solution	Seed treatment	Safflower – *P. carthamii*
	0.2% solution	3 sprays at 10-day intervals as soon as disease appears	Brown rust of wheat: *P. recondita*
V – Systemic fungicides			
Benzimidazole	1.5 kg/ha	Dust or spray with carbendazim or carbendazim (0.07%) plus mancozeb (0.15%)	Groundnut – *P. arachidis* Speg. and early and late leaf spot disease
	1 g/kg	Seed dressing with benomyl	Safflower – *P. carthamii*
Oxathiin-carboxin	2.5 g/kg	Seed treatment/foliar spray and soil treatment	Bean – *U. appendiculatus*
Oxycarboxin	2630 g/100 kg	Seed treatment	Stripe rust of wheat – *P. striiformis*
	1.12 kg/ha	Granules in soil	Leaf rust wheat – *P. recondita*
	1.68 kg/ha	Spray	
	3 kg/ha	Spray – single in Europe or 2–3 sprays common	Black stem rust – *P. graminis tritici* and *P. recondita*
	15 kg/ha	Soil treatment	Safflower (seedling) stage – *P. carthamii*
			Fig rust – *Cerotelium fici*
		Soil treatment with benomyl spray	Peanut – *P. arachidis*
		Soil treatment with benomyl as spray	Sunflower – *P. helianthi*
VI – Benzanilide derivatives			
Benzanilide	1.87 kg/ha	Foliar spray	Stripe rust of wheat – *P. striiformis* and barley – *P. hordei*
Benodanyl/ Vitavax	500 mg/ml	2 sprays or soil drenching	Effective against most of the rusts of cultivated plants
	0.75 l/ha	Spray/seed treatment	Stripe rust of wheat – *P. striiformis*
	0.3%	4 sprays at 15-day intervals	Leaf rust of wheat – *P. recondita*
	250 ml/l	2 sprays or soil drenching	Sunflower – *P. helianthi*
Oxycarboxin	0.1125%	Sprays at 6–9-day intervals	Sunflower – *P. helianthi*
	100 mg/l	2 sprays or soil drenching	Sunflower – *P. helianthi*
Plantavax w.p.	10 g/kg	Spray/seed treatment	Stripe rust of wheat – *P. striiformis*
	0.2%	Spray/seed treatment	Black stem rust – *P. graminis tritici*

Integrated disease management

1. Cultural methods include early sowing.
2. Spraying of micronutrients such as $Na_2B_4O_7$, $CuSO_4$ increases resistance in plants.

Acknowledgement

The authors express their gratitude to The Principal, Agriculture College, Kolhapur, India, for providing the facilities of their library.

References

Agrios, G.N. (2005) *Plant Pathology*, 5th edn. Academic Press, New York.

Arthur, J.C. (1929) *The Plant Rust (Uredinales)*. John Wiley and Sons, Inc, New York, 446 pp.

Butler, E.J. and Bisby, G.R. (1931) *The Fungi of India*. Science Monograph I. Indian Council of Agricultural Research, India, 65 pp.

Cummins, G.B. (1936) Phylogenetic significance of the pores in the *Uredinales*. *Mycologia* 23, 1–102.

Cummins, G.B. (1971) *The Rust Fungi of Cereals, Grasses and Bamboos*. Springer-Verlag, Berlin, 570 pp.

Dietel, P. (1900) *Uredinales*. In: Engler, A. and Prantl, H. (eds) *Naturlichen Pflangen Families*, 1(1), pp. 28–88, 546–553.

Ellis, M.B. (1971) *Dematiaceous Hyphomycetes*. Commonwealth Mycological Institute, Kew, UK.

Ellis, M.B. (1976) *More Dematiaceous Hyphomycetes*. Commonwealth Mycological Institute, Kew, UK.

Flor, H.H. (1956) The complementary genetic systems in flax and flax rust. *Advances in Genetics* 8, 29–55.

Ghewande, M.P. and Savalya, S.D. (1999) Integrated management of groundnut diseases in S.D. India. (Abstract). In: Kambale, S.Y. (ed.) *Proceedings of National Seminar on Recent Advances in Plant Pathology*, Department of Botany, University of Pune, Pune, India, pp. 4–5.

Gregory, P.H. (1973) *The Microbiology of the Atmosphere*, 2nd edn. Leonard Hill Books Aylesbury, UK, 377 pp.

Joshi, L.M., Srivastava, K.D. and Singh, D.V. (1986) Monitoring of wheat rust in Indian subcontinent. *Proceedings of Indian Academy of Sciences (Plant Science)* 84, 387–406.

Kirk, P.M., Cannon, P.F., David, J.C. and Stalper, J.A. (eds) (2001) *Dictionary of Fungi*, 9th edn. CAB International, Wallingford, UK.

Kuhn, J.G. (1858) *Diseases of Cultivated Plants*. 312 pp.

Mehta, K.C. (1929) Annual recurrence of rust on wheat in India. Presidential Address (Botany section) at Madras. *Proceedings of Indian Science Congress*, pp. 199–223.

Mehta, K.C. (1952) Further studies on cereal rust in India-II. *ICAR Science Monograph* 18, India, 368 pp.

Mishra, D.P. (1962) Natural occurrence of the aecial stage of *Puccinia sorghi* Schw. on *Oxalis corniculata* L. in Nepal. *Indian Phytopathology* 16, 8–9.

Ono, Y. (2002) The diversity of nuclear cycles in microcyclic rust fungi (Uredinales) and its ecological and evolutionary implications. *Mycoscience* 43, 421–439.

Ono, Y., Burticia, P. and Hennen, J.F. (1992) Delimitation of *Phakopsora* , *Physopella* and *Cerotelium* and their species on Leguminosae. *Mycological Research* 96, 825–850.

Rajendran, R.B. (1967) Atypical and typical germination of uredinoid teliospores of *Hemileia vastatrix*. *Mycologia* 59(5), 918–921.

Sawada, K. (1931) Materials of the Formosan fungi. *Transactions of the Natural History Society of Taiwan* 21, 227–235 (in Japanese).

Sharvelle, E.G. (1961) *The Nature and Uses of Modern Fungicides*. Burgess Publishing Co, Minneapolis, Minnesota, 30 pp.

Simon, B.K. (1993) *A Key to Australian Grasses*. Queensland Department of Primary Industries, Brisbane, 206 pp.

Swann, E.C. and Taylor, J.W. (2001) Urediniomycetes. In: McLaughlin, D.J., McLaughlin, E.G. and Lemke, P.A. (eds) *et al. The Mycota Volume VII: Systematics and Evolution, Part B*. Springer, Berlin, pp. 37–56.

Sydow, H.P. (1914) Beitrg zur kenntnis der parasitischen Pilz der insel Formosa. *Ann. Mycologia* 12, 105–112.

Tarr, S.A.J. (1972) *The Principles of Plant Pathology*. The McMillan Press, London.

Tulasne, L.-R. and Tulasne, C. (1847) Memoire sur les ustilaginees compares Uredinees- *Annales des Sciences Naturelles Botanique, Série* 3, (7), 12–27.

van der Plank, J.E. (1963) *Plant Disease Epidemics and Control*. Academic Press, New York, 349 pp.

van der Plank, J.E. (1968) *Disease Resistance in Plants*. Academic Press, New York.

Vidhyasekaran, P. (1997) *Fungal Pathogenesis in Plants and Crops*. Marcel Dekker, New York.

Appendix 1

Key to species and varieties of *Puccinia* and *Uromyces* producing rust diseases to cultivated crops which are very common but confusing to identify

Wheat rusts

I. Uredinial infection produces chlorotic streaks with halos on leaves ----------*P. striformis*
I'. Uredinial infection does not produce chlorotic streaks with halos on leaves ----------II
II. Urediniospores with 5–6 germ pores arranged equatorially ------------*P. graminis tritici*
II'. Urediniospores with 4–5 germ pores, scattered ---------------------------------- *P. recondita*

Jowar rusts

I. Uredinia aparaphysate --*P. levis*
I'. Uredinia paraphysate ---II
II. Urediniospores with 5–8 germ pores, scattered --------------------------------*P. nakanishiki*
II'. Urediniospores with 3–5 germ pores, equatorial --------------------------------- *P. purpurea*

Bajra rusts

I. Infection mostly hyphophyllous, teleutospores measure 21–49 µm long, 2-celled, pedicel coloured and short --------- *P. substriata* Ell. & Barth. var. *indica* Ramachar & Cummins
I'. Infection amphigenous, teleutospores large, up to 5-celled --------------------*P. substriata* Ell. & Barth. var. *decrospora* Eboh.

Maize rusts

I. Teleutospores stalked and 2-celled ---II
I' Teleutospores sessile, in chain and innate -- *P. zeae*
II. Telia exposed/erumpent, urediniospores 26–31 µm long (aecia on *Oxalis stricta* L.)------ -- *P. sorghi*
II'. Telia covered, urediniospores 29–30 µm long (aecia not known) ------------- *P. polysora*

Rusts of Phaseolus *and* Vigna *spp.*

I. Urediniospores measure 24–29 × 17–19 µm, germ pores 4, equatorial (on *Phaseolus* spp.) --- *U. appendiculatus* (Pers.) Unger var. *appendiculatus*
I'. Urediniospores measure 29–32 × 20–22 µm, germ pores 2, equatorial (on *Vigna* spp.) --- *U. vignae*

Rusts in forage crops

I. Urediniospores have 4–7 germ pores, scattered (on red clover) -----------------*U. fallens*; *T. pratens* (only uredia and telia)
I'. Urediniospores have 2–4 germ pores, equatorial (on white clover) -------------------------- ------------------------ *U. trifolii-repentis T. repens* and other spp. autoecious and macrocyclic

Rusts on vegetable and pulse crops

I. Autoecious, occurs in America and Europe ----------------------------------- *U. viciae-fabae*
I'. Heteroecious, occurs only in Europe (Arthur, 1929) on *P. sativum* L. ------------- *U. pisi*

17 Etiology, Epidemiology and Management of Fungal Diseases of Sugarcane

Ayman M.H. Esh
Biotechnology and Tissue Culture Laboratories, Sugar Crops Research Institute, Agricultural Research Center, Giza, Egypt

Abstract

Sugarcane (*Saccharum officinarum* L.) is one of the most important commercial crops in many countries of the world. It contributes nearly 70% of world sugar and provides the base materials essential for many other industries. Sugarcane crop is attacked by numerous foliar and root pathogens. Some of these diseases cause serious quantitative and qualitative losses which have negative effects on sugarcane production, as well as in the sugar industry.

About 56 diseases of sugarcane have been reported so far from different parts of the world. Of these, 40 are caused by fungi, several of which can cause economic losses. The major sugarcane fungal diseases in different tropical and subtropical regions are: smut disease (*Ustilago scitaminea*); rust disease (*Puccinia melanocephela*); red rot (*Glomerella tucumanensis* [*Colletotrichum falcatum*]); eye spot disease (*Bipolaris sacchari*), pokkah boeng disease (*Fusarium moniliforme*); and pineapple disease (*Ceratocystis paradoxa*). This chapter includes the major fungal diseases of sugarcane and the various control practices used against them.

Introduction

Sugarcane (*S. officinarum* L.) is a monocotyledonous plant from the family *Poaceae* of the subfamily *Andropogoneae* (Cox *et al.*, 2000) and is considered as one of the oldest cultivated crops known to man. Sugar, along with honey, is the oldest natural sweetener (Peng, 1984; Naik, 2001). Sugarcane is grown in the tropical and subtropical regions of the world and is cultivated in nearly 60 countries as a commercial crop, with Brazil, India, China, Cuba, Thailand and Pakistan as the major sugarcane-growing countries (FAO, 2005).

Due to its wide range of adaptability, it supplies more than 60% of world sugar demand and basic raw material in many industries, which makes it one of the most important cash crops that plays an enormous role in the economy. Various biotic and abiotic factors are responsible for yield reduction and economic losses. Among these factors, fungal diseases are the major cause. Over 100 fungi, 10 bacteria, 10 viruses and about 50 species of nematodes are pests of sugarcane in different parts of the world (Singh and Waraitch, 1981). Sugarcane, being a long duration crop (10–12 months), remains in the field for several years (ratoons

 Management of Fungal Plant Pathogens (eds A. Arya and A.E. Perelló)

remain in the field for up to more than 6 years). A serious drawback of this practice, however, is that pathogens may build up within the field and be disseminated during propagation of new seed cane. The pathogen within the seed pieces is transmitted easily into young plants, which in turn serve as sources of inoculum for secondary infections in adjoining healthy plants. Several management strategies have been developed as a result of research and development work. The endless struggle between varieties and the complexity of disease have led correspondingly to the development of a variety of approaches for control. The role of fungicides in modernizing and changing the condition of agriculture is quite significant (Mehta, 1971; McFarlane *et al.*, 2006).

Sugarcane Smut Disease

Sugarcane smut, caused by the *Basidiomycetes* fungus *Ustilago scitaminea* Syd., is cosmopolitan in distribution and has been an important disease in nearly every sugarcane-producing country of the world. It can reduce crop yields by over 50% and make ratoon crops unprofitable to maintain. It is highly infectious and even developed countries have been unable to stay smut free with the use of appropriate quarantine measures (Antony, 2008). The disease was first noted in South Africa in 1877, then in the early 1930s it caused severe problems in India and other countries in Asia. Years later, the disease started to establish and cause serious problems in different parts of the world: 1943, Argentina, (Cross, 1960); 1950, Brazil and Paraguay; 1957, Bolivia; 1960 and 1971, Hawaii (Byther *et al.*, 1971); and 1974, Guyana (James, 1976). By 1981. the disease had been found in most of the Caribbean and North, South and Central America (Ferreira and Comstock, 1989). In 1998, the disease was reported for the first time in the Ord River area of Western Australia (Riley *et al.*, 1999). Australia is a major exception since the disease is present only in Western Australia. The sugar industries of eastern Australia, Fiji and Papua New Guinea are still free of the disease (Braithwaite *et al.*, 2004b).

Causal agents

The fungus belongs taxonomically to Phylum: *Basidiomycota*; Class: *Ustilaginomycetes*, (Bisby *et al.*, 2007). Classification of *U. scitaminea* H. & P. Sydow, the causal agent of sugarcane smut disease, is based mainly on differences in spore morphology and the characteristics of germinating spores (Lee-Lovick, 1978).

Smut races have been reported subsequently based both on observations and inoculation studies (Gillaspie *et al.*, 1983). Usually, races are suggested when a cultivar succumbs to smut after being grown for several years without being infected (James, 1976). *U. scitaminea* races have been reported in Hawaii, Pakistan, the Philippines and Taiwan; the presence of the actual number of races and their prevalence are unknown (Ferreira and Comstock, 1989). Pathogenic races of sugarcane smut have been observed in several countries, including two races A and B from Hawaii (Comstock and Heinz, 1977) and three races (1, 2, 3) reported in Taiwan (Leu and Teng, 1972; Lee *et al.*, 1999). However, Ferreira and Comstock (1989) considered the true prevalence of races to be controversial. Many claims are based on the reaction of the same cultivar in different countries, but the interpretation of these claims is confused by test-to-test variation and the use of different inoculation methods. Two international collaborations have attempted to standardize race typing. Gillaspie *et al.* (1983) performed race typing under glasshouse conditions to standardize the environment and six races were identified. Grisham (2001) coordinated a race typing study in nine countries using local isolates tested against a standardized set of 11 differential cultivars. On the molecular level, many researchers have studied the genetic diversity among *U. scitaminea* isolates, either between local isolates or between isolates collected from different parts of the world (Braithwaite *et al.*, 2004a,b; Xu *et al.*, 2004; Singh *et al.*, 2005). Genetic variation estimated from 12 AFLP primer combinations showed that, overall, there was little variation in the smut population across the world. However, isolates from the Philippines, Taiwan and Thailand form a distinct cluster; it is therefore suggested that genetic variation is limited between the isolates and the phylogeny of *U. scitaminea* is poorly understood (Braithwaite *et al.*, 2004b).

Disease symptoms

Smut-infected plants are distinguished by the emergence of a 'smut whip'. The whips are the flowering structures of the pathogen which produce teliospores. The flowering structures transform into a whip-like sori that grows out between the leaf sheaths. At first, it is covered by a thin silvery peridium (this is the host tissue), which peels back easily when desiccated to expose the sooty black-brown teliospores. Whips begin emerging from infected cane by 2–4 months of age, with peak whip growth occurring at the 6th or 7th month. Spindle leaves are erect before the whip emerges. Affected sugarcane plants may tiller profusely, with the shoots being more spindly and erect with small narrow leaves (i.e. the cane appears 'grass-like'). Less common symptoms are leaf and stem galls and bud proliferation (Ferreira and Comstock, 1989; Agnihotri, 1990).

Pathogenesis

Ustilago scitaminea produces diploid spores called teliospores. When teliospores germinate, they undergo meiosis, which gives rise to a septate promycelium bearing four haploid sporidia (basidiospores). *U. scitaminea*, like most parasitic *Heterobasidiomycetes*, has a diallelic bipolar mating system (Alexander and Srinivasan, 1966; Leu, 1978; Moosawi-Jorf *et al.*, 2006) in which only sporidia of opposite mating types conjugate. Of the four initial sporidia or basidiospores from each teliospore, two have a positive mating allele and two have a negative mating allele. *U. scitaminea* can thus both self- and outcross, but the frequency of natural selfing versus outcrossing is unknown. A dikaryotic mycelium develops after fusion of compatible sporidia. This dikaryotic mycelium is infectious, penetrates behind bud scales and invades the meristematic zone of the bud. Entry into the meristem in the bud occurs between 6 and 36 h after the teliospores are deposited on the surface (Alexander and Ramakrishnan, 1980). Finally, the apical meristem of smut-infected cane produces a long whip-like structure bearing billions of teliospores (i.e. sorus).

Sugarcane smut is spread by spores which have an aerial dispersal mode. The whip serves as a source of spores that release approximately one billion spores/whip/day into the air to infect the buds of the standing sugarcane. The infected buds remain dormant until the cane is cut for seed. The spores mixed with the soil of cropped or newly prepared fields also become a source of infection to the disease-free seed pieces. Under normal soil moisture, the spores only survive for a short time in the soil. On the other hand, several species of insects have been associated consistently with smut whips; this suggests insects could play a role in spore dispersal (Ferreira and Comstock, 1989; Agnihotri, 1990).

Disease control

The best control method is to use resistant cultivars. There is a strong genetic basis for resistance and resistant varieties have been readily available and used to control outbreaks of smut in several countries (Churchill *et al.*, 2006). Disease-free planting material usually can be obtained by subjecting seed to hot water treatment. Hot water treatment, however, may not be practical on a large scale and its effectiveness may be subject to varietal differences (McFarlane *et al.*, 2007).

Several fungicides (triadimefon, fludioxonil:mefenoxam:azoxystrobin, mancozeb, metalaxyl + carboxin + furathiocarb, pyroquilon, benomyl and chlorothalonil) have been used to control sugarcane smut when used as pre-planting fungicidal dips of planting setts (Wada *et al.*, 1999; Wada, 2003).

Sugarcane Rust Disease (Common and Orange Rust)

Throughout the world, the important leaf rust disease causes severe losses in sugarcane fields (Magarey *et al.*, 2008). In 2000, sugarcane rust was once considered a minor pathogen in the Australian sugar industry. In 2000, it devastated most plantations of the cultivar Q124 in Australia, causing yield losses of up to 40% (Apan *et al.*, 2003; Braithwaite *et al.*, 2004a; Magarey *et al.*, 2008). In the USA, the yield loss caused by a rust epidemic due to cultivar CP 72-1210 in 1987 was 20% (Raid and Comstock, 2000).

Sugarcane rust is caused by two species belonging to the genus *Puccinia*, *P. melanocephala* and *P. kuehnii*, the former causes common rust disease, while the later causes orange rust. Common rust caused by *P. melanocephala* H. & P. Syd. was first reported on sugarcane in 1949 in the Deccan area in India (Patel *et al.*, 1950). The disease has been also reported from: Japan (Ohtsu, 1975 [cited by Muta, 1987]); the Philippines (Serra *et al.*, 1983); Australia (Egan and Ryan, 1979); Taiwan (Hsieh *et al.*, 1977); Dominican Republic (Presley *et al.*, 1978); Jamaica (Burgess, 1979); Puerto Rico (Liu, 1979); Cuba (Sandoval *et al.*, 1983); Carribean and Central America (Purdy *et al.*, 1983); Hawaii (Comstock *et al.*, 1982); and Angola, Kenya, Madagascar, Tanzania, Uganda, Zambia, Zimbabwe, South Africa, Mozambique and Malawi (Egan, 1980; Sivanesan and Waller, 1986).

Orange rust caused by *P. kuehnii* Butl. was first reported on sugarcane in Java in 1890 (Ryan and Egan, 1989). The disease has been reported from Japan (Ito, 1909), Australia, Indonesia, the Philippines, Taiwan, Pacific Islands, Sri Lanka, Malaysia, Thailand, New Caledonia, China (Egan, 1980; Sivanesan and Waller, 1986) and India (Mukerji and Bhasin, 1986).

Disease symptoms

The initial symptoms of common rust are small, elongated yellowish spots, which are visible on both leaf surfaces. The spots increase in size up to 1.5 mm in diameter and usually turn brown to orange-brown or red-brown. The lesions occur irregularly and typically range from 2 to 10 mm in length, but occasionally reach 30 mm. The spots are raised and are surrounded by a pale yellow halo (Raid and Comstock, 2000). The raised pustules are formed predominantly on the undersurface of the leaves and the urediospores formed therein are orange to orange-brown. On a highly susceptible variety, considerable numbers of pustules may occur on a leaf, coalescing to form large, irregular, necrotic areas. High rust severities may even result in premature death of young leaves. Severe rust has caused reductions in both stalk mass and stalk numbers (Rao *et al.*, 1999; Raid and Comstock, 2000).

Causal agent

Puccinia melanocephala Syd. and P. Syd. (common rust) and *P. kuehnii* Butler (orange rust) are reported to cause rust diseases on sugarcane (Butler, 1914; Cummins and Hiratsuka, 1983; Shine *et al.*, 2005; Ido *et al.*, 2006; Comstock *et al.*, 2008; Ovalle *et al.*, 2008).

The two obligate parasitic fungi belong to Phylum: *Basidiomycota*; Class: *Urediniomycetes*, Order: *Uredinales*. The causal agents of common and orange rust cannot be clearly distinguished based on colour of lesions and uredinia and the size of urediniospores. However, they are distinguishable based on the presence or absence of abundant capitate paraphyses in uredinia, echinulation, colour and wall thickness of urediniospores, colour of the telia and colour and wall thickness of teliospores. *P. melanocephala* has abundant capitate paraphyses in uredinia and urediniospores with dense echinulation, darker brown and uniformly thick walls. They also have dark brown to blackish telia with brown to dark brown teliospores with apically thickened walls. *P. kuehnii* has morphologically indistinct paraphyses in uredinia and urediniospores with moderate echinulation, lighter brown and sometimes apically or uniformly thickened walls (Virtudazo *et al.*, 2001).

Pathogenesis

The life cycle of sugarcane rust is simple, with the urediniospore being the only known infectious spore. These are produced in, and are released from, pustules that develop on the underside of sugarcane leaves. The development of substomatal vesicles, infectious hyphae, haustoria and subsequent infection processes are similar to other *Puccinia* spp. Urediniospore production occurs 8–18 days after the initial urediniospore lands on a leaf, depending on varietal susceptibility and environmental conditions (CABI CPC, 2006).

Spread of rust disease occurs primarily by wind and water-splash movement of urediniospores. The movement of diseased vegetative parts of sugarcane, contaminated equipment and workers from one location to another may also provide a means of spread. The expression of the disease is influenced by the interaction of genetic, environmental (primarily air temperature and leaf wetness) and physiological (age of infected plants) factors. The infection may occur within the temperature range of 5–34°C; however, the optimal temperatures for spore germination are between 15° and 30°C. Heavy rains tend to remove spores from the atmosphere, rendering them infective if they land on the soil (Egan, 1964; Comstock and Ferreira, 1986). On the other hand, it has been found that rains favour the development of orange rust but inhibit the development of common rust (Croft *et al.*, 2000). Sugarcane plants appear to be most susceptible at 3–6 months old (Ryan and Egan, 1989).

Disease control

The best control of sugarcane rust is use of resistant sugarcane varieties. The development of resistant cultivars has decreased the economic losses caused by this disease (Ryan and Egan, 1989). Nevertheless, existing resistant cultivars are threatened by the establishment of new races of the pathogen. For example, cultivar CP 78-1247 was considered to be resistant or moderately resistant until 1988, and then it exhibited extremely high rust susceptibility throughout south Florida (Raid, 1989). However, resistance has not been stable or durable on certain varieties, presumably because of rust variants. For this reason, it is highly recommended that growers should diversify their varietal holdings (Raid and Comstock, 2000). Chemicals like propiconazole/mancozeb, cyproconazole, triadimefon and triadimenol have been used for the control of sugarcane rust. Several soil factors influence rust infection levels on sugarcane significantly. Studies have shown that rust levels are higher on sugarcane grown on low pH soils, high soil moisture and high levels of phosphorus and potassium nutrients present in the soil (Johnson *et al.*, 2007).

Red Rot Disease of Sugarcane

Red rot is one of the oldest known diseases of sugarcane. It occurs in most cane-growing countries. The disease was first described from Java by Went (1896) and then the disease was reported from Australia, India, Hawaii and the USA. It is clear that the disease was widely distributed before the knowledge of its impact on sugarcane crop (Singh and Singh, 1989).

Symptoms

The pathogen, *Colletotrichum falcatum* Went, can attack any part of the sugarcane plant – stalk, leaf, buds or roots – but it is usually considered a stalk and a seed-piece disease. *C. falcatum* completes its life cycle on the sugarcane leaf and usually the damage to the leaf does not pose a serious threat to cane or cause much harm to the plant (Singh and Singh, 1989; Raid, 2006).

The most damaging phase of this disease occurs when the pathogen attacks the stalk. Depending on the age of the stalk, time of infection and susceptibility of the cane genotype, it produces different types of symptoms. The typical stalk symptoms, that is, presence of white spots in otherwise rotten (dull red) internodal tissues and nodal rotting, appear when the crop is at the fag end of the grand growth phase in subtropical areas. These white patches are specific to the disease and are of significance in distinguishing red rot from other stalk rots. At a later stage, some discoloration of rind often becomes apparent when internal tissues have been badly damaged and are fully rotten (Singh and Singh, 1989; Raid, 2006; Duttamajumder, 2008).

In susceptible varieties, the red colour, sometimes along with some grey colour, may be seen throughout the length of the stalk. The infection is confined largely to the internodes in resistant varieties. On the leaves, the pathogen may produce elongated red lesions on the midribs, reddish patches on the leaf sheaths and, infrequently, small dark spots on the leaf blades. Eventually, the lesions may develop a straw colour in the centre. In seed pieces, the entire seed piece may become rotten and the internal tissues turn various shades of red, brown or grey (Singh and Singh, 1989).

Causal agent

The fungus causing red rot of sugarcane is commonly known by its imperfect state, i.e. *C. falcatum* Went (*Glomerella tucumanensis*). The perfect state of the fungus belongs to Phylum: *Ascomycota*; Class: *Ascomycetes*; Family: *Glomerellaceae*; Genus: *Glomerella* (Bisby *et al.*, 2007).

Conidia are falcate (but not markedly so), fusoid, apices obtuse, 15.5 (25–26.5)–48 μm × 4 (5–6)–8 μm and contents are granular and sometime contain oil globules. At least two races have been identified. The variations in the asexual state of the fungus (*Colletotrichum* state) may originate through: (i) heterokaryosis; (ii) by recombination through parasexual mechanism; and (iii) by the universal mechanism of mutation, selection and adaptation in response to the changes in the host environment. Heterokaryosis is the mechanism through which the fungus collects and consolidates two or more genetically different nuclei in the hypha and derives the benefit of the introduced genetic material. These newly gathered nuclei also multiply in tandem with the native nuclei (Duttamajumder, 2008).

Pathogenesis

The pathogen mainly infects the stalks through the nodes. Once the infection is established in the stalk, the fungal mycelium grows intracellularly and is sparse in the reddened areas. The dead cells are packed in white patches with profuse hyphae. The size and number of these white areas are correlated with the susceptibility of the variety. The lesions become dark red, narrow and sharp margins, with a few white spots in resistant varieties, while in susceptible varieties the lesions become wide, light red and ill-defined margins with prominent white spots (Singh and Singh, 1989).

One of the major sources of inoculum is midrib lesions. Also, diseased stalks and crop debris and infected plant material are important sources of inoculum and cause secondary infections. Wind, rain, heavy dews and irrigation water play a role in the dispersal of the inoculum. The pathogen spores washed into the soil may produce infection in planted seed pieces. Climatic factors affect both the spread and severity of red rot. In newly-planted cane, the disease is favoured by excessive soil moisture, drought conditions and low temperatures.

Disease control

The use of resistant varieties is the most effective method of prevention and control of sugarcane red rot disease (Singh and Singh, 1989; Raid, 2006; Singh *et al.*, 2008b). Management of the disease by the use of disease-free seed canes for planting is impractical due to the difficulty in diagnosing dormant infections of the fungus in seed canes under field conditions (Viswanathan and Samiyappan, 2002).

It is difficult to manage red rot through chemotherapy because the impervious nature of rinds and fibrous nodes at cut ends does not allow sufficient absorption in setts (Agnihotri, 1990). However, better crop stands have been achieved from enhanced germination obtained by treating seed pieces with a fungicide before planting (Raid, 2006).

Thermotherapy (moist hot air or hot water) is thought useful for inactivation of red rot pathogen, but it is difficult to remove deep-seated infections. It is limited in checking secondary infections (Singh, 1973; Singh and Singh, 1989; Raid, 2006). In India, extensive studies about the possibilities of using biological control to control sugarcane red rod disease have been carried out (Mohanraj *et al.*, 1999). Seventy-five per cent of canes may be protected against secondary infection of red rot by dipping the setts for 15 min in 2.5% culture filtrate of *Trichoderma harzianum* (Th 38) and also by applying *Trichoderma* multiplied culture in press mud 20 kg/ha beneath the setts in furrows. Besides the biological control of red rot, the growth in improved resulting is to enhanced yield by 15.4 t/ha (Singh *et al.*, 2008a,b).

Sugarcane Eye Spot Disease

Eye spot has been reported in many sugarcane-growing areas of the world. The disease was first described by van Breda de Haan in Java (1892, cited in Comstock and Lentini, 2005). The disease is prevalent and is found in 66 sugarcane-growing countries (Agnihotri, 1990). Generally, the disease has a minor economic impact on sugarcane yield in most areas because of the use of resistant varieties (Comstock and Lentini, 2005). In India, in 1976, the disease was in epidemic form and affected 1600 ha sugarcane crop in Mandya district of Karnataka only (Kumaraswami and Urs, 1978). The disease can reduce sugarcane yield by 15–20% (Sharma *et al.*, 2004).

Disease symptoms

Typical mature eye spot symptoms are characterized by a reddish-brown elliptical lesion (0.5–4.0 mm long, 0.5–2.0 mm wide) with yellowish-brown margins. Reddish-brown to yellowish-brown streaks, sometimes called 'runners', extend upward from individual lesions toward the leaf tip. These streaks are 3–6 mm wide and 30–90 cm long. The entire leaf eventually may become necrotic (Comstock and Steiner, 1989; Comstock and Lentini, 2005).

Causal agent

Eye spot disease is caused by *Bipolaris sacchari* (Butler) Shoemaker. The fungus belongs taxonomically to Phylum: *Ascomycota*; Class: *Ascomycetes*; Order: *Pleosporales*. *B. sacchari* is the teleomorphic stage of *Helminthosporium sacchari* Butler. The name *H. sacchari* is still used occasionally.

Pathogenesis

Sugarcane eye spot fungus *B. sacchari* (*H. sacchari*) produces a host-specific toxin (HST). HSTs are a group of structurally complex and chemically diverse metabolites produced by plant pathogenic strains of certain fungal species and function as essential determinants of pathogenicity or virulence. HSTs are referred to as 'host selective' because they are typically active only toward plants that serve as hosts for the pathogens that produce them and disease never occurs in the absence of toxin production (Wolpert *et al.*, 2002). The HST is a mixture of three isomers (A, B and C) (Lesney *et al.*, 1982; Livingston and Scheffer, 1984) and the HST produced by the sugarcane red rot pathogen is responsible for the disease symptoms (Steiner and Byther, 1971). The fungus causes eye-shaped lesions on the leaves, followed by the development of reddish brown streaks or 'runners' extending from the lesions toward the tip of the leaf. The toxic compound from the fungus causes the runner (Steiner and Byther, 1971).

Eye spot spores, which are produced abundantly on leaf lesions, are dispersed by wind and rain. High humidity and dew formation are the favoured conditions for spore germination. The disease is not transmitted by seed pieces and mechanical transmission by equipment and by humans is unimportant.

Disease control

The only practical and efficient method of control of eye spot disease is with resistant clones. Chemical control using foliar fungicides is not practical (Comstock and Lentini, 2005).

Pokkah Boeng Disease

Pokkah boeng, which is a potentially destructive disease of sugarcane, is caused by *Gibberella moniliformis* (Sheldon) Wineland. There have been many reported outbreaks of the disease which have been severe, like the Java outbreak in 1896, but they have caused little economic loss (Martin *et al.*, 1989).

Disease symptoms

In the early stages of infection, the symptoms of the disease are chlorotic areas at the base of young leaves, distortion (wrinkling and twisting) and shortening of the infected leaves and finally stalk death in severe cases. The infected leaves can be distinguished by their narrow base. Irregular reddish stripes and specks develop in the chlorotic parts which appear in mature leaves.

The infection is present in the stalk and dark reddish streaks may be found extending through several internodes. Also, in the internodes, the infection may form long lesions with cross-depressions that give them a ladder-like appearance. These lesions sometimes break through the surface of the rind, causing curvature and distortion of the stalk. Exaggerated versions of these depressions may look like neatly made 'knife-cuts' in the stalk. In the most advanced stage of pokkah boeng, the entire top (growing point) of the plant dies (referred to as 'top rot'). The ladder-like lesions are due to rupturing of the diseased cells that cannot keep up with the growth of the healthy tissue (Martin *et al.*, 1989; Raid, 2009a).

Causal agents

The disease is caused by the fungi, *Fusarium moniliforme* (*G. fujikuroi*) and *F. moniliforme* var. *subglutinans* (*G. subglutinans*). The perithecia of *G. fujikuroi* occur only on dead plant material, while the perithecia of *G. subglutinans* are rarely formed in nature; thus, the perithecia of these two pathogens are rarely associated with infected sugarcane plants (Martin *et al.*, 1989). The pathogens have a wide host range, i.e. rice, corn, sorghum and many other grasses. These fungi also cause other diseases, such as seedling blight, scorch, stalk rot, root rot and stunting in different crops (Martin *et al.*, 1989; Raid, 2009a).

Pathogenesis

The pathogens of pokkah boeng disease are transmitted by the movement of spores from one locality to another by air currents (Martin *et al.*, 1961; Raid, 2009a,b). Pokkah boeng disease of sugarcane may also spread from seeds contaminated with the fungus (Martin *et al.*, 1961). It appears to be favoured by dry climatic conditions being followed by a wet season. Cane that is 3–7 months old and growing vigorously appears to be most susceptible (Martin *et al.*, 1989).

The pathogen spores enter the spindle along the margin of a partially unfolded leaf, then germinate and grow into the inner tissue of the spindle leaves. The conidia germinate and the mycelium can pass through the soft cuticle of young leaves to the inner tissues because the epidermal tissues are still fragile and not protected by the plant system (Dillewijn, 1950). The mycelium spreads to vascular bundles of the immature stem and blocks the vessels, which eventually leads to growth distortions and rupture, and this development shows the ladder-like lesions (Holliday, 1980; Martin *et al.*, 1989).

Disease control

The only efficient control for pokkah boeng disease is the use of resistant varieties. Sugarcane resistance to pokkah boeng has been shown to be highly heritable (Martin *et al.*, 1989).

Pineapple Disease of Sugarcane

Pineapple disease is an economically important sugarcane disease that is widely distributed in almost all the regions where sugarcane is grown (Wismer and Bailey, 1989). In India, the disease has been noticed on different sugarcane varieties (Singh *et al.*, 1990). The disease is caused by the fungus *Ceratocystis paradoxa*, which induces seed-piece decay following planting. The affected setts emit a smell resembling that of the mature pineapple fruit (Went, 1896).

Disease symptoms

The disease affects sugarcane setts in the first weeks after planting. The fungus spreads rapidly through the parenchyma and colonizes all the internal tissue of the seed piece, which turns red and eventually black. The black coloration results from the production of fungal spores within the seed piece. Nodes act as partial barriers to the spread of rotting, but with susceptible varieties, entire seed pieces may become colonized by the fungus (Wismer and Bailey, 1989). The pineapple odour resulting in the decayed seed pieces is due to ethyl acetate, formed by the metabolic activity of the pathogen. The ethyl acetate content in the infected tissue may rise up to 1%, which is sufficient to inhibit the germination of buds (Kuo *et al.*, 1969). As a result, gappy stands are evident and young crops have a patchy and uneven appearance. In the early stages of rotting, the disease may be diagnosed by a strong odour of overripe pineapple. Although pineapple disease is not considered important in standing cane, infection may occur if the stalks are physically damaged or stressed (Wismer and Bailey, 1989).

Causal agents

Pineapple disease is caused *C. paradoxa*. The fungus belongs to Phylum: *Ascomycota*; Class: *Ascomycetes*. The fungus produces two types of imperfect spores, conidiospores 6–24 um × 2–5.5 μm (thin-walled cylindrical conidia) and chlamydospores 10–25 um × 7.5–20 μm (thick-walled and oval). These spores are produced intensively on the internal tissues of the infected seed pieces. They are released into the soil on seed piece decay. The spores may survive for several years in the soil, serving as a source of inoculum for the next crop (Wismer and Bailey, 1989). The perfect stage of the fungus has been reported and it occurs naturally on cacao (Dade, 1928) and sugarcane (Kuo *et al.*, 1969). The pathogen also causes diseases of pineapple, banana, cacao, coconut and oil palm (Wismer and Bailey, 1989).

Disease control

Control is a priority, especially where the soil inoculum level is high through the amelioration of conditions that favour the germination of buds and the emergence of young shoots (e.g. good quality cuttings, adequate irrigation, right planting time and depth). The ends of cuttings are dipped in a fungicidal solution, either as a cold dip or the fungicide may be incorporated into the water tank at the time of the hot water treatment (Antoine, 1956). Other fungicides used to control pineapple disease include benomyl, propiconazole carbendazim, etc. (Autrey, 1974). Systemic fungicides are found more efficient than non-systemic fungicides (Vijaya *et al.*, 2007).

Since pineapple disease is a soilborne disease, crop rotation or a fallow period between cane crops may prove to be of some benefit in reducing its impact. Seed-piece infection by pineapple disease frequently proceeds from the exposed cut ends to the centre of the seed piece. Therefore, the use of seed pieces containing at least three nodes increases the likelihood that buds closer to the centre will germinate (Wismer and Bailey, 1989).

For disease control, sett treatment with chemical fungicides before planting is an effective method and is widely followed. Systemic fungicides, benomyl and carbendazim, were found more efficient than non-systemic fungicides such as captan and mancozeb (Vijaya *et al.*, 2007).

References

Agnihotri, V.P. (1990) *Diseases of Sugarcane and Sugarbeet.* Oxford and IBH, Pub. Co. Pvt. Ltd., New Delhi, 483 pp.

Alexander, K.C. and Ramakrishnan, K. (1980) Infection of the bud, establishment in the host and production of whips in sugarcane smut (*Ustilago scitaminea*) of sugarcane. *Proceedings of the International Society of Sugarcane Technology* 17, 1452–1455.

Alexander, K.C. and Srinivasan, K.V. (1966) Sexuality in *Ustilago scitaminea* Syd. *Current Science* 23, 603–604.

Antoine, R. (1956) Cane diseases. Pineapple disease. Mauritius Sugar Industry Research Institute Annual Report 1956, 59 pp.

Antony, G. (2008) *Sugarcane Smut: The Political Economy of Biosecurity.* In: Plunkett, H. (ed.) *Proceedings of The 52nd Annual Conference of the Australian Agricultural and Resource Economics Society*, Canberra, Australia 5–8 February 2008, Volume 20, 12 pp (http://ageconsearch.umn.edu/bitstream/6048/2/cp08an01.pdf).

Apan, A., Held, A., Phinn, S. and Markley, J. (2003) Formulation and assessment of narrow band vegetation indices from EO-1 Hyperon imagery for discriminating sugarcane disease. In: Lees, B. (ed.) *Proceedings of the Spatial Sciences Conference 2003*, National Convention Centre Canberra, 17–23 September 2003. Spacial Sciences Institute of Australia, pp. 1–13.

Autrey, L.J.C. (1974) Control of *Ceratocystis paradoxa* (Dade) C. Moreau in sugarcane by the systemic fungicide benomyl. Special study, BSc Degree Botany, University of London, London, 39 pp.

Bisby, F.A., Roskov, Y.R., Ruggiero, M.A., Orrell, T.M., Paglinawan, L.E., Brewer, P.W., Bailly, N. and Hertum, J. van (eds) (2007) Species 2000 and ITIS Catalogue of Life: 2007 Annual Checklist (www.catalogueoflife.org/annual-checklist/2007/, accessed 15 May 2009). Species 2000, Reading, UK.

Braithwaite, K., Croft, B. and Brumbley, S. (2004a) Genetic diversity within collections of the sugarcane orange rust fungus. In: Hogarth, D.M. (ed.) *Proceedings of 2004 Conference of the Australian Society of Sugar Cane Technologists* held at Brisbane, Queensland, Australia, 4–7 May 2004, Volume 26. PK Editorial Services Pty Ltd, Brisbane, Australia, p. 82.

Braithwaite, K.S., Bakkeren, G., Croft, B.J. and Brumbley, S.M. (2004b) Genetic variation in a worldwide collection of the sugarcane smut fungus *Ustilago scitaminea. Proceedings of the Australian Society of Sugar Cane Technology* 26, 48–56.

Burgess, R.A. (1979) An outbreak of sugarcane rust in Jamaica. *Sugarcane Pathologists' Newsletter* 22, 4–5.

Butler, E.J. (1914) Notes on some rusts in India. *Annals of Mycology* 12, 76–82.

Byther, R.S., Steiner, G.W. and Wismer, C.A. (1971) New sugarcane diseases reported in Hawaii. *Sugarcane Pathologists' Newsletter* 7, 18–21.

CABI CPC (2006) *Crop Protection Compendium on Internet* (http://www.cabi.org/compendia/CPC/, accessed 30 April 2009).

Churchill, J.H., Croft, B.J., Kokot, S. and O'Shea, M.G. (2006) Determining plant ratings without field trials II – evaluation and prediction of clonal smut resistance status by morphological and spectroscopic examinations of sugarcane bud scale surfaces. In: Hogarth, D.M. (ed.) *Proceedings of the 2006 Conference of the Australian Society of Sugar Cane Technologists*, 2–5 May 2006, Mackay, Queensland, Australia. Australian Society of Sugar Cane Technologists, pp. 232–240.

Comstock, J.C. and Ferreira, S.A. (1986) Sugarcane rust: factors affecting infection and symptom development. In: *Proceedings of the International Society of Sugarcane Technology*, Congress Jakarta, Indonesia, pp. 403–410.
Comstock, J.C. and Heinz, D.J. (1977) A new race of culmicolous smut of sugarcane in Hawaii. *Sugarcane Pathologists' Newsletter* 19, 24–25.
Comstock, J.C. and Lentini, R.S. (2005) Sugarcane eye spot disease. In: Gilbert, R.A. (ed.) *Florida Sugarcane Handbook* (SS-AGR-203). Agronomy Department, Cooperative Extension Service, Institute of Food and Agricultural Sciences, University of Florida, Florida (http://edis.ifas.ufl.edu/SC006, accessed 30 April 2009).
Comstock, J.C. and Steiner, G.W. (1989) Eye spot. In: Ricaud, C., Egan, B.T., Gillaspie, A.G. Jr and Hughes, C.G. (eds) *Diseases of Sugarcane, Major Diseases.* Elsevier, Amsterdam, pp. 123–133.
Comstock, J.C., Tew, T.L. and Ferreira, S.A. (1982) Sugarcane rust in Hawaii. *Plant Disease* 66, 1193–1194.
Comstock, J.C., Sood, S.G., Glynn, N.C., Shine, J.M., McKemy, J.M. and Castlebury, L.A. (2008) First report of *Puccinia kuehnii*, causal agent of orange rust of sugarcane, in the United States and Western Hemisphere. *Plant Disease* 92, 175.
Cox, M., Hogarth, M. and Smith, G. (2000) Cane breeding and improvement. In: Hogarth, M. and Allsopp, P. (eds) *Manual of Cane Growing.* Bureau of Sugar Experimental Stations, Indooroopilly, Australia, pp. 91–108.
Croft, B., Magarey, R. and Whittle, P. (2000) Disease management. In: Hogarth, M. and Allsopp, P. (eds) *Manual of Cane Growing.* Bureau of Sugar Experiment Stations, Brisbane, Australia, pp. 263–289.
Cross, W.E. (1960) Studies of the sugarcane smut in Tucuman. *International Sugar Journal* 62, 239–242.
Cummins, G.B. and Hiratsuka, Y. (1983) *Illustrated Genera of Rust Fungi.* Revised edn. American Phytopathological Society, St Paul, Minnesota.
Dade, H.A., (1928) *Cerastomella paradoxa* the perfect stage of *Thielaviopsis paradoxa* (de Seynes) von Hohnel. *Transactions of the British Mycological Society* 13, 184–194.
Dillewijn, C. van (1950) Fusarium Pokkahboeng. *Proceedings of the International Society of Sugarcane Technology* 7, 473–498.
Duttamajumder, S.K. (2008) *Red Rot of Sugarcane.* Indian Institute of Sugarcane Research Lucknow, India, 115 pp
Egan, B.T. (1964) Rust. In: Hughes, C.G., Abbott, E.V. and Wismer, C.A. (eds) *Sugarcane Diseases of the World, Volume 2.* Elsevier Publishing Co., Amesterdam, pp. 61–68.
Egan, B.T. (1980) A review of the world distribution of *Puccinia* spp. attacking sugarcane. *Proceedings of the International Society of Sugarcane Technology* 17, 1373–1381.
Egan, B.T. and Ryan, C.C. (1979) Sugarcane rust, caused by *Puccinia melanocephala,* found in Australia. *Plant Disease Reporter* 63, 822–823.
FAO, (2005) The statistic division (http://www.fao.org/es/ess/top/ commodity. html?lang=en&item=156&year=2005, accessed 30 April 2009).
Ferreira, S.A. and Comstock, J.C. (1989) Smut. In: Ricaud, C., Egan, B.T., Gillaspie, A.G. Jr and Hughes, C.G. (eds) *Diseases of Sugarcane.* Elsevier, Amsterdam, pp. 211–229.
Gillaspie, A.G. Jr, Mock, R.G. and Dean, J.L. (1983) Differentiation of *Ustilago scitaminea* isolates in greenhouse tests. *Plant Disease* 67, 373–375.
Grisham, M.P. (2001) An international project on genetic variability within sugarcane smut. *Proceedings of the International Society of Sugarcane Technology* 24, 459–461.
Holliday, P. (1980) *Fungus Diseases of Tropical Crops.* Cambridge University Press, Cambridge, UK.
Hsieh, W.H., Lee, C.S. and Chan, S.L. (1977) Rust disease of sugarcane in Taiwan. Taiwan Sugar Research Institute Annual Report, Tainan, pp. 38–39.
Ido, O.T., da Lima Neto, C.V., Daros, E., Possamai, J.C., Zambon, J.L.C., Weber, H. and Oliveira, R.A. (2006) Rust incidence and severity in sugarcane clones at Parana State, Brazil. *Pesquisa Agropecuaria Tropical* 36, 159–163.
Ito, S. (1909) On the Uredineae parasitic on the Japanese *Gramineae. Journal of the College of Agriculture, Tohoku Imperial University, Sapporo, Japan* 3, 244–245.
James, G.I. (1976) Sugarcane smut infection in Guyana and Martinique. *Sugar Journal* 38, 17–18.
Johnson, R.M., Grisham, M.P. and Richard, E.P. Jr (2007) Relationship between sugarcane rust severity and soil properties in Louisiana. *Phytopathology* 97, 748–755.
Kumaraswami, S. and Urs, S.D. (1978) Eye spot disease of sugarcane – some observations on its outbreak in Mandya district. *Indian Sugar Crops Journal* 5, 40–42.

Kuo, T.T., Chien, M.M. and Li, H.W. (1969) Production of the sexual stage of *Ceratocystis paradoxa* on sugarcane. *Proceedings of the International Society of Sugarcane Technology* 13, 1191–1196.

Lee, C.S., Yuan, C.H. and Liang, Y.G. (1999) Occurrence of a new pathogenic race of culmicolous smut of sugarcane in Taiwan. *Proceedings of the International Society of Sugarcane Technology* 23, 406–407.

Lee-Lovick, G. (1978) Smut of sugarcane. *Ustilago scitaminea. Review of Plant Pathology* 47, 181–188.

Lesney, M.S., Livingston, R.S. and Scheffer, R.P. (1982) Effects of toxin from *Helminthosporium sacchari* on non-green tissues and a reexamination of toxin binding. *Phytopathology* 72, 844–849.

Leu, L.S. (1978) Culmicolous smut of sugar cane in Taiwan VI. New pathogenic strain obtained by artificial hybridization and further studies on compatibility of *Ustilago scitaminea* Sydow. *Annals of the Phytopathological Society of Japan* 44, 321–324.

Leu, L.S. and Teng, W.S. (1972) Pathogenic strains of *Ustilago scitaminea* Sydow. *Sugarcane Pathologists' Newsletter* 8, 12–13.

Liu, L.J. (1979) Rust of sugarcane in Puerto Rico. *Plant Disease Reporter* 63, 256–258.

Livingston, R.S. and Scheffer, R.P. (1984) Selective toxins and analogs produced by *Helminthosporium sacchari. Plant Physiology* 76, 96–102.

McFarlane, K., McFarlane, S.A., Moodley, D. and Rutherford, R.S. (2006) Fungicide trials to determine the effect of brown rust on the yield of sugarcane variety N29. In: *Proceedings of the 80th Annual Congress of the South African Sugar Technologists' Association*, Durban, South Africa, pp. 297–300 (http://www.sasta.co.za/wp-content/uploads/Proceedings/2000s/2006_McFarlane_fungicide%20 trials%20to%20determine%20the.pdf).

McFarlane, K., Moodley, D., Chinnasamy, G. and McFarlane, S.A. (2007) Laboratory, glasshouse and field assessments of fungicides to control *Ustilago scitaminea*, causal organism of sugarcane smut. *XXVI Congress, International Society of Sugarcane Technologists, 30 July–2 August 2007, ICC, Durban, South Africa*, pp. 1031–1035.

Magarey, R.C., Bull, J.I. and Tomasin, W.A. (2008) Yield losses caused by leaf diseases: 1999 and 2003 selection trial analyses. In: Bruce, R.C. (ed.) *Proceedings of the 2008 Conference of the Australian Society of Sugar Cane Technologists, Volume 30*, held at Townsville, Queensland, Australia. Australian Society of Sugar Cane Technologists, pp. 309–321.

Martin, J.P., Hong, H.L. and Wismer, C.A. (1961) Pokkah boeng. In: Hughes, C.G. and Robinson, P.E. (eds) *Sugar-Cane Diseases of the World, Volume 1*. Elsevier, New York, pp. 247–257.

Martin, J.P., Handojo, H. and Wismer, C.A. (1989) Pokkah boeng. In: Ricaud, C., Egan, B.T., Gillaspie, A.G. Jr and Hughes, C.G. (eds) *Diseases of Sugarcane*. Elsevier, Amsterdam, pp. 157–168.

Mehta, P.R. (1971) Role of fungicides in modernizing agriculture in India. *Indian Phytopathology* 24, 235–246.

Mohanraj, D., Padmanaban, P. and Viswanathan, R. (1999) Biological control of sugarcane diseases. In: Gnanamanickam, S.S. (ed.) *Biological Control of Crop Diseases*. Marcel Dekker, New York–Basel, pp. 161–178.

Moosawi-Jorf, S.A., Teymour-Zadeh, M. and Izadi, M. (2006) Sexual incompatibility system in Iranian isolates of *Ustilago scitaminea*, the causal agent of sugarcane smut. In: *Proceedings of the 17th Iranian Plant Protection Congress, 2*. University of Tehran, 265 pp.

Mukerji, K.G. and Bhasin, J. (1986) *Plant Diseases of India: A Source Book*. Tata McGraw-Hill Publishing Company Ltd, New Delhi, pp. 218–219.

Muta, T. (1987) Occurrence of sugarcane rust *(Puccinia rnelanocephala* and *Puccinia* sp.) in Kagoshima Prefecture. *Proceedings of the Association for Plant Protection of Kyushu* 33, 36–38.

Naik, G.R. (2001) *Sugarcane Biotechnology*. Oxford and IBH Publishing Co, New Delhi and Science Publishers, Inc, Enfield, New Hampshire, 153 pp.

Ovalle, W., Comstock, J.C., Glynn, N.C. and Castlebury, L.A. (2008) First report of *Puccinia kuehnii*, causal agent of orange rust of sugarcane, in Guatemala. *Plant Disease* 92, 973.

Patel, M.K., Kamar, M.N. and Padhyem, Y.A. (1950) A new record of *Puccinia* on sugarcane in Bombay. *Current Science* 19, 121–122.

Peng, S.Y. (1984) *The Biology and Control of Weeds in Sugarcane*. Elsevier Science, New York, 336 pp.

Presley, J.T., Perdomo, R. and Ayats, J.D. (1978) Sugarcane rust found in Dominican Republic. *Plant Disease Reporter* 62(10), 843.

Purdy, L.H., Liu, L.J. and Dean, J.L. (1983) Sugarcane rust, a newly important disease. *Plant Disease* 67, 1292–1296.

Raid, N.R. (1989) Physiological specialization in sugarcane rust (*Puccinia melanocephala*) in Florida. *Plant Disease* 73, 183.

Raid, R.N. (2006) Sugarcane red rot disease. In: Gilbert, R.A. (ed.) *Florida Sugarcane Handbook* (SS-AGR-206). Agronomy Department, Cooperative Extension Service, Institute of Food and Agricultural Sciences, University of Florida, Florida (http://edis.ifas.ufl.edu/SC006, accessed 22 April 2009).

Raid, R.N. (2009a) Pineapple disease of sugarcane. In: Gilbert, R.A. (ed.) *Florida Sugarcane Handbook* (SS-AGR-205). Agronomy Department, Cooperative Extension Service, Institute of Food and Agricultural Sciences, University of Florida, Florida (http://edis.ifas.ufl.edu/document_sc005, accessed 30 April 2009).

Raid, R.N. (2009b) Pokkah boeng disease of sugarcane. In: Gilbert, R.A. (ed.) *Florida Sugarcane Handbook* (SS-AGR-204). Agronomy Department, Cooperative Extension Service, Institute of Food and Agricultural Sciences, University of Florida, Florida (http://edis.ifas.ufl.edu/document_sc004, accessed 30 April 2009).

Raid, R.N. and Comstock, J.C. (2000) Common rust. In: Rott, P., Bailey, R. A., Comstock, J. C. and Croft, B. J., and Saumtally, A. S. (eds.) *A Guide to Sugarcane Diseases.* ISSCT/CIRAD, Montpellier, France, pp. 85–89.

Rao, G.P., Bergamin, F., Magarey, Y. and Autrey, L.J.C. (1999) *Sugarcane Pathology.* Science Publishers Inc., Enfield, New Hampshire.

Riley, I.T., Jubb, T.F., Egan, B.T. and Croft, B.J. (1999) First outbreak of sugarcane smut in Australia. In: Singh, V. and Kumar, V. (eds) *Proceedings of the XXIII ISSCT Congress*, New Delhi, India, 22–26 February 1999. STAI, New Delhi, pp. 333–337.

Ryan, C.C. and Egan, B.T. (1989) Rust. In: Ricand, C., Egan, B.T., Gillaspie, A.G. Jr and Hughes, C.G. (eds) *Diseases of Sugarcane, Major Diseases.* Elsevier, Amsterdam, pp. 189–210.

Sandoval, I., Picornell, V., Chavez, R. and Ramos, M. (1983) *Puccinia melanocephala.* H. & P. Syd.: biological and ecological aspects. *Proceedings of the International Society of Sugarcane Technology* 18, 845–853.

Serra, R.J., Barredo, F.C. and Tiangco, A.P. (1983) Incidence of sugarcane rust *(Puccinia melanocephala* H. et. P. Sydow) in the Victorias Milling Company District Occidental Negros, Philippines in 1982. *Philippines Phytopathology* 19, 50–53.

Sharma, S.R., Gaur, R.K., Singh, A., Singh, P. and Rao, G.P. (2004) Biological and chemical control of eye leaf spot disease of sugarcane. *Sugar Tech* 6, 77–80.

Shine, J.M. Jr, Comstock, J.C. and Dean, J.L. (2005) Comparison of five isolates of sugarcane brown rust and differential reaction on six sugarcane clones. *Sugar Cane International* 23, 24–29.

Singh, H., Patil, A.S., Hapase, D.G. and Singh, H. (1990) Pineapple disease of sugarcane – a review. *Bharatiya Sugar* 16, 187–192.

Singh, K. (1973) Hot air therapy against red rot of sugarcane. *Plant Disease Reporter* 57, 220–222.

Singh, K. and Singh, R.P. (1989) Red rot. In: Ricaud, C., Egan, B.T., Gillaspie, A.G. Jr and Hughes, C.G. (eds) *Diseases of Sugarcane.* Elsevier, Amsterdam, pp. 169–188.

Singh, N., Somai, B.M. and Pillay, D. (2005) Molecular profiling demonstrates limited diversity amongst geographically separate strains of *Ustilago scitaminea. FEMS Microbiology Letters* 247, 7–15.

Singh, O. and Waraitch, K.S. (1981) Effect of wilt and red rot induced disease stress on quality deterioration of sugarcane. *Sugarcane Pathologists' Newsletter* 27, 25–29.

Singh, V., Srivastava, S.N., Lal, R.J., Awasthi, S.K. and Joshi, B.B. (2008a) Biological control of red rot disease of sugarcane through *Trichoderma harzianum* and *Trichoderma viride. Indian Phytopathology* 61(4), 486–491.

Singh, V., Joshi, B.B., Awasthi, S.K. and Srivastava, S.N. (2008b) Eco-friendly management of red rot disease of sugarcane with *Trichoderma* strains. *Sugar Tech* 10(2), 158–161.

Sivanesan, A. and Waller, J.M. (1986) *Sugarcane Diseases.* Phytopathology Paper No. 29, Commonwealth Mycological Institute, Kew, UK.

Steiner, G.W. and Byther, R.S. (1971) Partial characterization and use of a host-specific toxin from *Helminthosporium sacchari* on sugarcane. *Phytopathology* 61, 691–695.

Vijaya, H.K., Kulkarni, S. and Hegde, Y.R. (2007) Chemical control of sett rot of sugarcane caused by *Ceratocystis paradoxa. Karnataka Journal of Agricultural Science* 20, 62–64.

Virtudazo, E.V., Nojima, H. and Kakishima, M. (2001) Taxonomy of *Puccinia* species causing rust diseases on sugarcane. *Mycoscience* 42,167–175.

Viswanathan, R. and Samiyappan, R. (2002) Induced systemic resistance by fluorescent pseudomonads against red rot disease of sugarcane caused by *Colletotrichum falcatum. Crop Protection* 21, 1–10.

Wada, A.C. (2003) Control of sugarcane smut disease in Nigeria with fungicides. *Crop Protection* 22, 45–49.

Wada, A.C., Mian, A.W., Anaso, A.B., Busari, L.D. and Kwon-Ndung, E.H. (1999) Control of sugarcane smut (*Ustilago scitaminea* Syd) disease in Nigeria and suggestions for an integrated pest management approach. *Sugar Tech* 1, 48–53.

Went, F.A.F.C. (1896) Notes on sugarcane diseases. *Annals of Botany* 10, 583–600.

Wismer, C.A. and Bailey, R.A. (1989) Pineapple disease. In: Ricaud, C., Egan, B.T., Gillaspie, A.G. Jr and Hughes, C.G. (eds) *Diseases of Sugarcane*. Elsevier, Amsterdam, pp. 145–155.

Wolpert, T.J., Dunkle, L.D. and Ciuffetti, L.M. (2002) Host selective toxins and avirulence determinants: what's in a name? *Annual Review of Phytopathology* 40, 251–285.

Xu, L., Que, Y. and Chen, R. (2004) Genetic diversity of *Ustilago scitaminea* in mainland China. *Sugar Tech* 6, 267–271.

18 New and Emerging Fungal Pathogens Associated with Leaf Blight Symptoms on Wheat (*Triticum aestivum*) in Argentina

Analía Edith Perelló
CIDEFI (Centro de Investigaciones de Fitopatología) – CONICET (Consejo Nacional de Investigaciones Científicas y Técnicas), Facultad de Ciencias Agrarias y Forestales de la Universidad Nacional de La Plata, La Plata, Provincia de Buenos Aires, Argentina

Abstract
Regional surveys are being conducted at the CIDEFI to investigate the presence of wheat (*Triticum aestivum* L.) pathogens on leaves and seeds across the Argentinian cropping area. During the past 5 years in the wheat cropping area of Buenos Aires Province, Entre Ríos and Santa Fe Provinces, Argentina, several unusual diseases have been found on wheat leaves. From the symptomatic tissues, the fungi were isolated and identified. To test pathogenicity and fulfil Koch's postulates, inoculations of different wheat cultivars under greenhouse conditions were carried out; disease symptoms and the causal agents are described.

Introduction

Wheat ranks as a primary source of food and livelihood for hundreds of millions of people globally, especially in developing countries. Several serious foliar diseases caused by necrothrophic pathogens occur in this crop in Argentina. Among them, *Septoria tritici* leaf blotch and tan spot caused by *Drechslera tritici-repentis* are the most important. Both can cause serious yield and quality losses under the right conditions. Among biotrophic diseases, leaf rust is a very dangerous one.

Samples were obtained from different wheat cultivars from 2001 in 13 different locations of the main wheat-growing area of Argentina (northern, central and eastern regions of the Buenos Aires Province, Entre Ríos and Santa Fé). Most symptoms were observed on the upper leaves at growth stage 69 (anthesis complete) and at growth stages 80 (early dough) to 85 (soft dough) according to the scale of Zadok *et al.* (1974). Samples of 20–40 plants of each disease area were used for late laboratory identification of leaf-spotting fungi.

The frequency of the pathogens was different among localities and cultivars over the past 10 years. *Pyrenophora tritici-repentis* was predominant, followed by *Mycosphaerella graminicola*. In a total of 193 and 240 infected leaf samples collected in 2006–2007, *D. tritici-repentis* was observed in 63 and 77% of leaf samples, respectively, suggesting an increasing trend in incidence over the years. This may be explained in

 Management of Fungal Plant Pathogens (eds A. Arya and A.E. Perelló)

part due to the adoption of reduced-tillage cropping practices. Surprisingly, a significant increase in the frequency of *Alternaria* spp. isolates was observed, with 41 and 53% of the samples harbouring tan spot in the same period. The isolates were identified as belonging mostly to the *A. infectoria* species group. Moreover, mapping of pathogen distribution during the past years in different agroclimatic zones of Buenos Aires Province shows that *A. infectoria* is a widespread pathogen that is gaining prominence as an emerging wheat pathogen in latter years. Also, it was found, together with *A. alternata*, on almost 70% of wheat seed samples.

Other diseases were sporadic, isolated or minor and included foliar blight or spots caused by the fungus *A. triticina*, *A. infectoria* species group complex, *Bipolaris sorokiniana*, *Cladosporium herbarum*, *Phoma sorguina*, *Ascochyta hordei*, *Pyricularia grisea*, *Cephalosporium gramineum* and others. Some disease symptoms are characteristic and obvious. Other diseases, however, may be difficult to diagnose without microscopic or laboratory analysis. Since different diseases require different control strategies, their accurate diagnosis is essential.

Leaf Blight of Wheat Caused by *Alternaria triticina* in Argentina

Alternaria species are perhaps the most common fungi encountered by mycologists working in plant pathology. As plant pathogens, over 4000 *Alternaria*/host associations are recorded in the USDA Fungal Host Index and the genus ranks 10th among nearly 2000 fungal genera listed based on the total number of host records. In Argentina, this genus has been studied on wheat plants since the last decade (Perelló, 2007; Perelló *et al.*, 1992, 1996, 2002, 2003, 2005a,b, 2008; Perelló and Sisterna, 2005, 2006). Several *Alternaria* species and numerous uncharacterized *Alternaria* taxa have been found associated with leaf blight symptoms. *Alternaria* species are some of the most prodigious producers of toxic secondary metabolites, producing over 70 compounds of varying toxicity (Kumar and Rao, 1979). Some of these metabolites are powerful mycotoxins not yet characterized in Argentina.

During routine investigations across the wheat (*T. aestivum*)-growing area of the Buenos Aires Province, diseased leaf samples were collected from different wheat cultivars. Discoloured oval lesions appear on lower leaves. The disease progresses upwards, lesions enlarge and coalesce to irregular, dark blotches, often with chlorotic margins. Severely infected seeds are discoloured and shrivelled.

Necrotic tissue fragments were surface-sterilized and plated on potato dextrose agar (PDA), from where *Alternaria* specimens were isolated. Morphobiometrical and cultural features of the fungus were examined on potato carrot agar (PCA). Conidia were irregularly oval, ellipsoid conical, gradually tapering into a beak, 15–92 × 8–35 μm, with 1–10 transverse septa and 0–5 longitudinal septa, light brown to dark olive buff, becoming darker with age. All isolates obtained were identified as *A. triticina* following the morphological descriptions by Anahosur (1978) and confirmed by comparison with reference strains of CABI Bioscience (IMI 289962 and IMI 178784) kindly sent by Dr D. Mercado (Université Catholique de Louvain, Unité de Phytopathologie, Belgique). One of the isolates has been lodged in the culture collection of La Plata Spegazzini (LPSC) (accession number 798). Pathogenicity tests were conducted in the greenhouse. Susceptible wheat cultivars were inoculated at tillering and heading stages with a conidial suspension (2×10^5 conidia/ml). After 10 days, typical leaf blight symptoms developed and *A. triticina* was recovered from the lesions. No symptoms appeared on the control plants.

Alternaria triticina causes significant yield losses in wheat on the Indian subcontinent, from where it originates and has spread throughout the world (Agarwal *et al.*, 1993). Although *A. triticina* has been detected previously in Argentina on wheat leaves and seeds (Perelló *et al.*, 1992), it has probably existed as a minor pathogen for many years without being noticed. The recent increase in the severity of leaf blight may be due to new cultural practices such as conservation tillage, nitrogen fertilization, irrigation, use of new germplasm, as well as favourable weather conditions. As *A. triticina* is a quarantine pathogen in many countries, it is important to investigate the incidence and importance of this disease in Argentinean wheat areas. This is the first published record of *A. triticina* on wheat in Argentina and on any host in this country.

Alternaria leaf blight

Alternaria triticina Prasada & Prabhu causes significant yield losses in wheat in the Indian subcontinent, from where it originated and has spread throughout the world. The disease was first reported in 1924 but remained incompletely characterized for several years. Early studies associated *Alternaria* spp. with the disease, but the causal organism, *A. triticina*, was not identified until 1962 (Prasada and Prabhu, 1962). From 1960 to 1964, leaf blight damaged all commercial cultivars on the Indian subcontinent (Prabhu and Prakash, 1973; Bhowmik, 1974; Sokhi, 1974). It developed on plants approaching maturity, caused premature death of the uppermost leaves and heads and reduced yield significantly. Today, durum wheats, their derivatives and introduced Mexican wheats are considered most susceptible (Frisullo, 1982). In addition to wheat, the disease affects triticale in India and other graminaceous hosts in the Middle East and Nigeria (Chaudhuri *et al.*, 1976).

Symptoms

The pathogen may infect all foliar parts. *Alternaria* leaf blight is characterized by small, chlorotic, oval- or elliptical-shaped lesions scattered on lower leaves. As the plant matures, the disease progresses upwards and lesions darken to brown-grey, enlarge and coalesce to irregular, dark blotches, irregular in shape and may have a yellow margin. The chlorotic borders of the lesions may become diffuse and turn light to dark brown in colour. The lesions develop progressively from lower to upper leaves and blighting may extend to heads and leaf sheaths. Symptoms appear on leaves, seeds and spikelets. Severely infected seeds are discoloured and shrivelled (Rault *et al.*, 1983). Under humid conditions, the lesions support visible clusters of dark, powdery conidia. Lesions are difficult to distinguish from those caused by *Helminthosporium* spp. *Alternaria* leaf blight is likely to develop near irrigation ditches, in low areas, or wherever humidity and soil moisture are high. It develops rapidly once wheat plants are 6–8 weeks old, and especially as the crop approaches maturity. Bread and durum wheat, barley and triticale are the primary hosts.

Causal organism

Isolations from leaf lesions routinely yield *Alternaria* spp., many of which are saprophytic and mask the pathogen. *Alternaria triticina* is distinguished by its wheat-specific virulence and cultural characters. The mycelium and conidia of *A. triticina* are initially hyaline and later olive buff. Conidiophores are septate, usually unbranched but occasionally branched, erect, single or fasciculate, emerging through stomata, geniculate, straight, length variable, between septa 17–28 μm, 3–6 μm wide. A chromogenic variant in *A. triticina* was studied (Jain and Prabhu, 1976). Conidia of *A. triticina* are acrogenous, borne singly or in short chains (two to four spores). They vary from 8 to 35 μm in width and 15 to 92 μm in length, are dark, ellipsoid to conical, tapering to a beak.

A. triticina grows on a variety of simple media (Rao and Subrahmanyam, 1974). Colonies on PDA are discrete or effuse, dark blackish brown to black, margin smooth and entire. Growth is optimal between 20° and 24°C, with limits near 5° and 35°C. Physiological specialization of the pathogen exists and six races have been characterized and reported using 15 differentials. Nonspecific phytotoxins produced by the pathogen apparently play a role in wheat pathogenesis.

Alternaria infectoria Complex Associated with Black Point and Leaf Blight Symptoms in Argentina

Among the field fungi found in cereals, *Alternaria* is the dominant genus and, within this habitat, taxa of the *A. infectoria* species group predominate by far. Information available on the *A. infectoria* species group is limited as the taxa it comprises have often been misidentified as other small-spored *Alternaria* species, due to the use of insufficient methods for identification. Members of the *A. infectoria* species group are morphologically distinguishable from other small-spored species of *Alternaria* by their long secondary conidiophores and formation of white or grey colonies on dichloran rose bengal yeast extract sucrose media (DRYES) (Andersen *et al.*, 2002).

Furthermore, this species group is the only one among *Alternaria* where the teleomorph, *Lewia*, has been identified in Argentina. To date, the *A. infectoria* species group comprises the known species *A. arbusti*, *A. conjuncta*, *A. infectoria*, *A. oregonensi*, *A. triticimaculans*, *A. metachromatica*, *A. viburni*, *A. intercepta* and *A. novae-zelandiae*, as well as an unknown number of distinct taxa yet to be described. Members of the *A. infectoria* species group produce a range of unique secondary metabolites that are useful for metabolic profiling and chemotaxonomy of *Alternaria*. Until now, four different profiles have been identified, suggesting that there is a potential risk of *Alternaria* mycotoxins in wheat in Argentina (Pich *et al.*, 2007).

Black point and leaf blight caused by *A. infectoria* species group complex is a new disease of wheat in Argentina. Four hundred and ten isolates collected from 17 different geographical zones of the Argentinian cropping area were tested for their morphological variation. The isolates differed in their morphobiometrical and cultural characteristics on PCA. On this basis, they were categorized into four morphotypes. Further, 20 isolates were characterized according to their pathogenic and biochemical variability. Pathogenicity tests were conducted under greenhouse conditions on wheat plants of the cultivars Buck Arriero, Buck Charrúa, Buck Granar, Buck Poncho, Buck Yatasto, Klein Cacique, K. Estrella, ProINTA Cinco Cerros, ProINTA Elite and Pro INTA. Significant differences between cultivars, isolates and the interaction isolate × cultivar were shown according the ANOVA results. Analysis of severity means (Tukey's test) showed cultivar Buck Charrúa as the one with the best behaviour against all the isolates tested, and cvs. Pro INTA Cinco Cerros and Pro INTA Elite as the most susceptible. Symptoms observed were: chlorosis and/or apical or general necrosis (blight), or elongated necrotic spots surrounded by a chlorotic halo. Additionally, 20 samples of wheat seeds from different localities in the Buenos Aires Province were analysed by blotter test (ISTA) (Neergaard, 1979). After 7 days incubation ($20 \pm 2°C$ and cycles of 12 h light plus NUV light), the microorganisms developed were identified and the *A. infectoria* complex in particular was characterized according to its morphobiometrical features on PCA. A prevalence (samples infected/samples analysed) of 55% and infection values of 37% of *A. infectoria* species group members was registered.

Twenty isolates were tested in a comparative analysis of five isoenzyme patterns (phosphatase, peroxidase, A-esterase, B-esterase, glutaminotransaminase) and total proteins. Mycelium for electrophoresis in polyacrylamide gel was obtained from monosporical cultures of the fungus on PCA over 10 days. The results obtained revealed differences between the strains in the isozyme banding patterns. Each isolate had a characteristic electromorph, showing different main bands of enzyme activity and some minor bands varying in intensity for all the patterns assayed. Isozyme data corroborated the morphological and pathogenic variability observed previously on *A. infectoria* isolates collected from wheat. These results support the usage of isoenzymatic patterns for the characterization of isolates of *A. infectoria* complex associated with black point and leaf blight symptoms on wheat, as a valuable additional tool to aid the traditional taxonomy, base on morphocultural characters only.

Detection of *Lewia infectoria* and its *Alternaria* Anamorph from Wheat in Argentina

Occurrence of *L. infectoria* (Fuckel) Barr & Simmons (teleomorph of *A. infectoria)* developed in culture is described, illustrated and reported for the first time. Monosporic isolates, obtained from infected wheat plants, produced conidia within a week and ascomata with fully mature ascospores within 7 months when stored on slants of PCA at 4 degrees in darkness. The anamorph exhibited the sporulation pattern of *A. infectoria* species group and was identified on the basis of axenic colony morphology and by the prominence of their secondary conidiophore structure. Critical examination of the teleomorph proved it to be *L. infectoria*. The importance in interpreting the teleomorph–anamorph pair is discussed.

A. infectoria species group, causing leaf blight and black point of wheat, was not significant for many years, but currently this constraint has become a new problem (Perelló and Sisterna, 2006). During 2005, wheat samples were collected and typical symptoms of tan–dark brown leaf spot were observed on several cultivars from the cropping area of Buenos Aires Province. Different isolates of a fungus with characteristics of *Alternaria* were obtained from this material. In laboratory conditions on PCA, cultures of this fungus produced conidia within a week. Then, these isolates were stored to maintain a fungal collection. Within 7 months, ascomata with fully mature ascospores of a previous undescribed genus were observed in connection with this anamorphic state. Based on morphological characters (Simmons, 2002), the teleomorph proved to be a *Lewia* species, described here as *L. infectoria* with its *Alternaria* anamorph.

Wheat leaves exhibiting necrotic symptoms were collected during September and October in 2005 from different cultivars of farmers' fields and research stations of eight localities of Buenos Aires. Lesions on leaves were viewed through a stereoscope at × 12 and specific morphological characteristics of pathogens were recorded. Fungi were cultured on PDA. After screening of the cultures, 180 belonged to the *Alternaria* genus. Single spore cultures were obtained on PCA. Based on morphological characters like the conidial sporulation pattern and the prominence of their secondary conidiophore structure, most of the strains were identified as members of the *A. infectoria* species group. The cultures were stored on slants of PCA at 4°C in darkness. Numerous conidia appeared on the surface of the agar within a week and groups of fruiting bodies within 7 months. To determine their stage of development, the fungal fruiting bodies were placed on a microscope slide, stained with 0.25% Trypan blue in lactid acid:glycerol:water (1:1:1) and examined with a light microscope (× 400). After 7 months, groups of fertile ascomata (pseudothecia) with septate hyaline mature ascospores developed in isolates obtained from the cultivar Klein Estrella, from a particular field in the locality of Balcarce (Buenos Aires Province). There were detectable differences between isolates with regard to pseudothecial density and speed of ascospore maturity. In some cases, the production of immature asci was observed.

The morphobiometrical and cultural features of these ascomata on PCA allowed the identification of the teleomorph of *A. infectoria*, *L. infectoria* (Fuckel) Barr & Simmons. Its description is as follows: ascomata ellipsoid, 400–500 × 150 μm, with a short, papillate beak, dark, thin-walled at maturity. Asci 105–125 × 13–16 μm, subcylindrical, straight or somewhat curved. Ascospores 8, 18–22 × 7–8 μm at full development, broadly elliptic, muriform, becoming 5-septate (3 primary septa), only end cells not longitudinally septated, constricted, yellow-brown.

Although most *Alternaria* species do not have teleomorphic affinities, a number of anamorphically defined taxa within the *Pleosporaceae* have recognized teleomorphs and most are not commonly encountered (Simmons, 1986, 2002). These teleomorphs are representative of nearly all major lineages within the *Pleosporaceae*. An evaluation of the teleomorphic characters of well-known *Pleospora* spp. with anamorphs of *Alternaria*, namely *P. infectoria* and *P. scrophularieae*, revealed that *Pleospora* spp. with *Stemphylium* anamorphs were morphologically distinct from *Pleospora* spp. with *Alternaria* anamorphs, particularly in the size of the ascomata and ascospores. This resulted in the designation of the genus *Lewia* for *Pleospora*-like fungi with *Alternaria* teleomorphs (Simmons, 1986). Evidence of production of *Alternaria*-related teleomorphs in axenic culture was previously reported by Bilgrami (1974) in *L. infectoria*, Simmons (1986) in *L. photistica*, Kwasna and Kosiak (2003) in *L. avenicola* and Kwasna *et al.* (2006) in *L. hordeicola*. Other described *Lewia* species, like *L. chlamidosporiformans*, *L. ethzedia*, *L. intercepta*, *L. sauropi*, *L. viburni* and *L. eureka*, usually produce ascomata on tissues of infected plants (Simmons, 1986, 2002; Vieira and Barreto, 2005). During the present study, ascomata of *L. infectoria* were produced *in vitro*, in axenic culture on PCA slants, in connection with the anamorph.

However, the fungus does not often produce both anamorph and teleomorph on the same slant. Crossing between isolates is evidently not necessary for production of the teleomorph since single-ascospore axenic cultures continued to produce ascomata on PCA. These results were similar to observations made by Kwasna and Kosiak (2003) for *L. avenicola*. In addition, its finding in Argentina has provided an important framework for hypothesis testing in advanced studies on *Alternaria*/*Lewia* epidemiology and pathogenicity variability on wheat plants.

Lewia infectoria forms pseudothecia on wheat straw in the field under determined weather conditions. This could play an important role as a source of inoculum in *Alternaria*/*Lewia* disease able to infect wheat and wild grasses as a result of the dispersal of airborne ascospores. The discovery of the sexual stage in nature may have a large influence on localized development of the diseases in different regions of the country.

Occurrence of *Ascochyta hordei* Hara var. *europaea* Punith. on Wheat Leaves in Argentina

Ascochyta leaf spot is often overlooked in association with other leaf spot diseases. It is reported on wheat as a pathogen of minor economic importance but there are some reports pointing out that high humidity conditions could favour the occurrence of outbreaks of the disease (Scharen and Krupinsky, 1971).

During September–October 2002, leaf spot symptoms on wheat cultivar Baguette 10, growing in farmers' fields in Tandil, eastern area of Buenos Aires Province, were commonly observed. The leaves showed symptoms similar to those described for other necrotrophic foliar pathogens (*D. tritici-repentis*) and *Stagonospora nodorum*, suggesting that any of these might have been involved. *Ascochyta* was commonly isolated from affected tissues of the samples collected. *A. tritici* Hori & Enj. is generally accepted as the cause of *Ascochyta* leaf spot, but *A. graminicola* Sacc. is cited in some literature (Zillinsky, 1984). Punithalingam stated the status of *A. tritici* was uncertain but it might be a synonym of *A. hordei* (Farr *et al.*, 1989). Sprague and Johnson (1950) also stated that *A. tritici* was close to *A. hordei* Hara, differing mainly in the symptoms on barley. The identity of the culture of *Ascochyta* isolate A1102 of this study was determined as *A. hordei* Hara var. *europaea* by experts of the Centraalbureau voor Schimmelcultures (CBS), The Netherlands, and deposited in the CBS culture collection under the number 112525.

Ascochyta was not previously reported on wheat and other grasses in Argentina. In this sense, the first occurrence of this fungus as a member of the leaf spotting complex on wheat plants in the Argentinian cropping area is significant. Diseased leaves were collected, stored in paper bags and transported to the laboratory. The pathogen was isolated from typical necrotic symptoms. Morphobiometrical and cultural studies of the fungus were studied.

Inoculation experiments to confirm pathogenicity were performed in the greenhouse at 15–25°C and 80% relative humidity on 16 wheat cultivars: Buck Arriero, Buck Yatasto, Buck Poncho, Buck Charrúa, Buck Halcón, ProInta Granar, ProInta Cinco Cerros, Desimoni Caudillo, ProInta Imperial, ProInta Puntal, ProInta Guazú, ProInta

Colibrí, ProInta Elite, Klein Estrella, Klein Cacique and Klein Dragón. Plants were grown in plastic pots 12 cm in diameter (4 seeds/plot in all samples) with a standard potting mix. Plants were inoculated when they had reached the third expanded leaf stage. Inoculum was prepared from 10-day-old cultures of *A. hordei* var. *europaea* (isolate A1102) growing on PDA and was obtained by flooding each sporulating plate with sterile distilled water and gently scraping the fungal colony with a flame-sterilized scalpel to dislodge conidia. The conidial suspension was filtered once through a single layer of cheesecloth and spore concentration was determined with a haemocytometer. The inoculum consisted of 12×10^6 conidia/ml. Twenty seedlings of each cultivar were used for the inoculation. Leaves were sprayed to runoff with a manually operated sprayer. The inoculated plants and controls were kept in a moist chamber for 48 h. The first symptoms appeared 92 h after inoculation. Between 40 and 60% were necrotic 18 days after inoculation. In natural field infections, it was observed that plants were affected rather severely, especially the basal leaves. Lesions at first are distinct, chlorotic, ellipsoidal or round and 1–5 mm across. Later, they become diffused and grey-brown internally. Pycnidia sometimes form and appear as black dots within necrotic lesions. They are submerged in host tissues, except for a papillate projection. In culture on PDA, pycnidia measured $142.5–225 \times 93.75–206.25$ µm. Conidia (pycnidiospores) are straight, hyaline and oblong, $3.75–5.60 \times 15–18.70$ µm, typically with one median septum.

All wheat plants inoculated with *A. hordei* var. *europaea* in the greenhouse developed symptoms identical to those observed on naturally infected plants in the field. The amount of damage to seedlings was measured as per cent necrotic leaf area from the first leaf of 15 plants per each inoculated cultivar in comparison with controls. Cultivars Buck Arriero and Buck Poncho showed the most conspicuous symptoms 9 days after inoculation, with a disease severity rating of between 8–35% of necrotic leaf area. The rest of the cultivars showed no symptoms or had low disease severity. Buck Arriero, Buck Charrúa, Buck Halcón, Buck Poncho, Buck Yatasto and Klein Dragón showed the most severe symptoms of the disease (between 12–45% of the necrotic leaf area) 20 days after inoculation. The rest of the cultivars showed little evidence of infection with only a 10% necrotic foliar area, except ProInta Imperial and ProInta Puntal, which showed no evidence of infection, indicating that the disease pressure on Buck cultivars was much higher than on the rest of the wheat cultivars examined. These observations are consistent with other reports (Wiese, 1977), which indicates that the fungus is apparently of little economic consequence as a foliar pathogen of wheat. Inoculation studies proved that *A. hordei* var. *europaea* was the cause of this outbreak on wheat in Argentina. New cultural practices (reduced tillage, nitrogen fertilization, irrigation), the use of new germplasm and favourable environmental conditions could have contributed to creating ideal conditions for the increase and spread of inoculum, not only of *A. hordei* var. *europaea* but of the foliar complex of necrotrophic pathogens in general. Other wheat cultivars may also be susceptible to isolates of the pathogen.

Phoma sorghina (Sacc.) Boerema, Dorenbosch & van Kesteren in Wheat Leaves in Argentina

Phoma sorghina (Sacc.) Boerema, Dorenbosch & van Kesteren is plurivorous, ubiquitous and common in the tropics and subtropics, causing diseases of cereals and other *Gramineae* and forage crops (Punithalingham, 1985; Manavolta and Bedendo, 1999; Kumar and Kumar, 2000). Although *P. sorghina* has been found causing leaf spots in different hosts such as *Agave americana*, *Gossypium hirsutum*, *Lycium halimifolium*, *Lycopersicum esculentum*, *Oryza sativa*, *Populus nigra*, *Sorghum* spp. and *Zea mays*, the disease it causes to aerial parts of plants is of minor importance (White and Morgan-Jones, 1983). It causes considerable loss of seedlings of *Macroptilum*, *Stylosanthes* and *Sorghum* through pre- and post-emergence death. The fungus has been found on or associated with sorghum grains in the humid Argentinian Pampa (Gonzalez *et al.*, 1997), but there are no previous reports of its presence on wheat plants in Argentina and therefore the confirmation of this fungus as a foliar pathogen of wheat in Buenos Aires Province is significant. The first occurrence of leaf spot diseases of wheat by *P. sorghina* was in the Buenos Aires Province of Argentina in 2002. The fungus was detected on samples from two localities, Olavarría and Los Hornos, on experimental field plots with wheat cultivars, Buck Poncho and Buck Diamante.

Diseased leaves were collected, stored in paper bags and transported to the laboratory. The pathogen was isolated from typical necrotic lesions on PDA Petri dishes. Fifteen plants of each of the cvs. Buck Arriero, Buck Poncho, Klein Cacique, Klein Estrella, ProInta Cinco Cerros, ProInta Elite, ProInta Granar, ProInta Guazú, ProInta Imperial and ProInta Puntal were grown in plastic pots 12 cm in diameter and containing a potting mix of clay 21.2%, lime 56%, sand 22.8%, soil organic matter (SOM) % = 3.35; C % = 1. Plants were inoculated at the third expanded leaf stage and heading stage. Inoculum was prepared from 10-day-old cultures of *P. sorghina* growing on PDA by flooding each sporulating plate with sterile distilled water and gently scraping the fungal colony with a sterile scalpel to dislodge conidia. The resulting suspension was filtered once through a single layer of cheesecloth and the spore concentration was determined with a haemocytometer. The spore concentration in inoculum was adjusted to 1×10^6 conidia/ml. Control plants were sprayed with sterile distilled water. Leaves were sprayed to runoff with a manually operated sprayer. The inoculated plants and controls were kept in a moist chamber for 48 h and observed daily.

The first symptoms appeared 72 h after inoculation under greenhouse conditions and 15–50% of plants showed necrotic lesions 10 days after inoculation. The wheat cvs. showed different degrees of susceptibility to the pathogen. The cvs. Buck Arriero, Buck Poncho and Klein Cacique became severely infected with up to 40% of their leaf surface spotted. The cvs. ProInta Cinco Cerros, ProInta Elite, ProInta Granar and ProInta Imperial were slightly spotted, with 1–5% of their foliage covered by spots. The cvs. Klein Estrella, ProInta Guazú and ProInta Puntal were free of spots. Elongated necrotic yellowish to light-brown lesions could be observed on the upper surfaces of affected leaves of wheat cvs. Leaf spots later coalesced to form large irregular spots with yellow margins. Under high humidity, pycnidia developed within spots on leaves after 21 days. A gelatinous spore mass was extruded in cirri from pycnidia. No symptoms or spots were seen on the control plants. All wheat plants inoculated with *P. sorghina* in the greenhouse developed symptoms identical to those observed on naturally infected plants in the field. In culture, the fungus developed dark, greyish colonies with dense aerial mycelium with abundant, solitary or sometimes aggregated pycnidia with characteristic beaks. Conidia were globose to ovoid or shortly cylindrical, usually straight, hyaline, unicellular 4–7 × 2 μm. Abundant chlamydospores and dictyochlamydospores were observed. Newly formed chlamydospores quickly became covered with a black coating that obscured their brown colour.

The identification was confirmed by the Centraalbureau voor Schimmelcultures (CBS), Utrecht, The Netherlands. One representative isolate of *P. sorghina* has been lodged in the CBS culture collection with the accession number 112525.

Cephalosporium gramineum Nisikado & Ikata on Wheat Leaves in Argentina

Cephalosporium stripe is a disease of cereals that is sporadic in its distribution and occurrence but can cause severe yield losses when it occurs. The disease is found most consistently in areas where frost heaving, resulting from fluctuating winter temperatures, heavier soils and higher soil moisture damages roots (Bruehl *et al.*, 1976). *Cephalosporium* stripe is caused by *Hymenula cerealis* (synonym *C. gramineum*). This fungus is slow growing in culture and probably in nature, too. It produces tiny conidia on sporodochia in the saprophytic stage on wheat straw, but as a parasite it invades the vascular system, where it interferes with water movement. It is the only true vascular parasite known to attack wheat.

Hosts

C. gramineum attacks most winter cereals, but especially wheat. It invades several grasses (*Bromus*, *Dactylis*, *Poa*) and probably was indigenous to the region in native grasses. Until now in Argentina, it has been detected on *Bromus* and wheat plants only.

Disease symptoms

Cephalosporium stripe is first observed in the spring as distinct yellow stripes on leaf blades, sheaths and stems. The stripes may contain thin brown streaks (necrotic vascular tissues) surrounded by yellow. Frequently, a yellow stripe on the leaf blade continues as a single brown line down the leaf sheath.

Nodes are darker than normal on diseased plants and, when cut lengthwise, the inner nodal tissue is brown in colour. Plants are stunted and the heads are white and sterile. If any seed is set, it is usually shrivelled. Diseased plants have a scorched appearance when hot weather accentuates moisture stress. The fungus survives for as long as 4–5 years in undecomposed infested straw.

In 2004, on a non-tilled wheat assay sown at the Julio Hirschhorn Experimental Station, belonging to the Facultad de Ciencias Agrarias y forestales de la Universidad Nacional de La Plata, Argentina, chlorotic stripes, which became necrotic, were observed on leaves of wheat cultivar Buck Biguá. Samples were collected and remitted to the laboratory at the CIDEFI. Morphocultural and morphobiometrical characteristics allowed identification of the fungus as *C. graminearum*. The fungus was cultured on 2% PDA Petri dishes. Wheat plants of the same cultivar were inoculated with a pathogen conidial suspension by a manual sprayer under greenhouse conditions. Similar symptoms developed from 7 to 21 days after the inoculation.

Occurrence of *Cladosporium herbarum* on Wheat Leaves in Argentina

Cladosporium on wheat was reported to be a common and mild parasite affecting dead or half-dead plant tissues in association with some other fungi. It often appears on the ear heads, causing a greenish black mouldy growth on the affected parts (Wiese, 1987). It was not found to cause severe symptoms on leaves and stems of wheat, but there were some reports pointing out that moist and shady conditions could favour the occurrence of outbreaks of the disease on leaves (Arya and Panwar, 1955).

During the past 5 years, leaf spot symptoms on wheat cultivars Buck Pingo, Buck Biguá, Buck Brasil and Buck Poncho growing in the north-east of the Buenos Aires Province, were commonly observed. Infected leaf samples were collected during September–November in an extensive survey conducted in 2002. Samples were collected from different cultivars in farmers' fields and one experimental research station across the wheat region of Buenos Aires and Entre Ríos Provinces, in five of the eight sites surveyed (Los Hornos, Nogoyá, Olavarría, Tandil and Victoria). In most of the plants, leaves showed symptoms sufficiently similar to those described for the complex of necrotrophic foliar pathogens, i.e. *D. tritici-repentis* (Died.) Shoem., *S. tritici* Rob. in Desm., *A. triticimaculans* Simmons & Perelló and *B. sorokiniana* (Sacc.) Shoem., suggesting that any of these might have been involved. *C. herbarum* was commonly isolated from affected tissues of the samples collected and it has emerged as a widespread and serious foliar disease. The fungus was previously reported in Argentina (Marchionatto, 1948) as *C. herbarum* Link.

var. *cerealinum* Sacc. on leaves and spikes of wheat and other grasses without describing the symptoms in detail. Since then, there have been no other reports of this disease on wheat leaves in Argentina.

Diseased leaves were collected, stored in paper bags and transported to the laboratory. The pathogen was isolated from typical necrotic symptoms. On PDA Petri dishes, morphobiometrical and cultural studies of the fungus were conducted on single spore colonies grown in Petri dishes containing PDA, cultured at 20 ± 2°C under cool-white fluorescent light supplemented with near UV with a 12 h photoperiod.

Inoculation experiments to confirm pathogenicity were performed in the greenhouse at 15–25°C and 80% relative humidity. Fifteen plants of each of the cultivars Buck Biguá, Buck Brasil, Buck Pingo and Buck Poncho were grown in plastic pots (12 cm diameter) with a standard potting mix. Plants were inoculated when they had reached the third expanded leaf stage and heading stage. Inoculum was prepared from 10-day-old cultures of *C. herbarum* (isolates Ch101 and Ch500) growing on PDA and was obtained by flooding each sporulating plate with sterile distilled water and gently scraping the fungal colony with a flame-sterilized scalpel to dislodge conidia. The conidial suspension was filtered once through a single layer of cheesecloth and spore concentration was determined with a haemocytometer. The inoculum consisted of 2×10^5 conidia/ml. Control plants were sprayed with sterile distilled water only. Leaves were sprayed to runoff with a manually operated sprayer. The inoculated plants and controls were kept in a moist chamber for 48 h. The plants were observed periodically. The first symptoms appeared between 72 and 92 h after inoculation. All cultivars showed susceptibility to both of the isolates tested. Between 12 and 75% were necrotic 10 days after inoculation. Reisolation from leaves with lesions was performed and the isolates were compared morphologically with those used for inoculation to fulfil Koch's postulates. In natural field infections, amphigenous, irregular yellowish brown spots were observed that especially affected the basal leaves rather severely. They often confluenced and elongated, developing progressively from lower to upper leaves. The margin of top leaves became brittle when dried and the tissue tore. When lesions spread over the leaf surface, they caused the death of the entire leaf. A velvety olivaceous grey mould of spores and mycelia developed on the surface of the infected tissue, forming dense tufts. Microscopic examination revealed the presence of conidiophores more or less erect, septate, sparsely branched; the spores are often in chains of 2 or 3, subcylindric, pale olive, 1-(2-3) septate, 10–15 × 4–7 µm. The teleomorph, *M. tulasnei* (Jancz.) Rothers was not seen.

All wheat plants inoculated with *C. herbarum* in the greenhouse developed symptoms identical to those observed on naturally infected plants in the field. No differences in degree of infection were noted among the cultivars. Nevertheless, adult plants showed more severe symptoms than younger ones. No symptoms were observed in the control non-inoculated plants. Isolation from symptomatic tissue has consistently yielded cultures of *C. herbarum*. The fungus sporulated on the diseased tissue in the Petri dishes. Comparison of morphological characteristics of *C. herbarum* isolates revealed no differences between field- and glasshouse-produced spores, according to the shape and size of conidia.

The isolates of *C. herbarum* have been lodged in the culture collection of the CIDEFI (Centro de Investigaciones de Fitopatología), Facultad de Ciencias Agrarias y Forestales de la Universidad Nacional de La Plata, Buenos Aires, Argentina, with the accession numbers 111-01, 209-02, 210-02, 212-02 and 215-02.

Inoculation studies proved that *C. herbarum* was the cause of this outbreak on wheat in Argentina. In the past few years, the increased incidence of the disease may be related to new cultural practices (reduced tillage, nitrogen fertilization, irrigation), the use of new germplasm and favourable weather conditions. This contributed to a major spread, not only of *C. herbarum* but also of the foliar complex of necrotrophic pathogens in general.

The fact that other wheat cultivars apart from those checked may also be susceptible to the pathogen shows the importance of conducting thorough research to determine the reactions of those cultivars currently used in the Argentinian cropping area.

Pyricularia grisea on Wheat Leaves in Argentina

During 2006/2007, *P. grisea* (Cooke) Sacc. was detected for first time in the north-east region of Argentina, a non-traditional, marginal culture area. Plants of wheat cv. Klein Chajá presented spots or blight symptoms on all aerial parts. Isolates and pathogenicity tests confirmed the presence of *P. grisea* associated to blight symptoms on leaves, sheets and spikles. *Pyricularia grisea* also affects rice and other gramineous spontaneous species in the region. Simultaneously, during 2007, the fungus was isolated from wheat plants cvs. Cronox, Baguette 11, ACA 304 and BioINTA from Bragado, Baradero, Rojas, Alberti and 9 de Julio localities from the typical wheat area in Argentina.

Conclusion

Wheat (*T. aestivum* L.) cultivars currently grown in Buenos Aires Province, Argentina, are susceptible to different leaf spotting fungi. Surveys conducted over several years in Argentina have determined that the main fungi involved in this disease complex are *M. graminicola* (Fuckel) Schroet. in Cohn (anamorph *S. tritici* Roberge in Desmaz.) (leaf spot), *Cochliobolus sativus* (Ito & Kuribayashi) Drechs. ex Dastur (anamorph *B. sorokiniana* (Sacc.) Shoemaker (spot blotch) and *P. tritici-repentis* (Died.) Drechs. (anamorph *D. tritici-repentis* (Died.) Shoemaker) (tan spot). *A. triticimaculans* Simmons & Perelló was first described on wheat in Argentina in 1996 and commonly observed since then, like others members of the *infectoria* complex. During surveys of wheat commercial fields from 2001 to 2002 to now on the northern, central and southern prairies of the Buenos Aires Province, a dramatic difference was observed between fungal diseases. A high incidence of *D. tritici-repentis* was commonly observed in all locations analysed.

Tan spot, caused by the fungus *P. tritici-repentis* (Died.) Drechs. (anamorph *D. tritici-repentis*) (Died.) Shoem., is a major disease of wheat (*T. aestivum* L.) worldwide (Wiese, 1977; Hosford, 1975). The disease has a fast growth in the Southern Cone region of South America including Argentina, where it was found for the first time affecting wheat crops in the north-central region of the Buenos Aires Province in the early 1980s (Annone, 1985). Subsequently, tan spot has gained predominance among other wheat diseases in most wheat-growing areas in the country (Kohli *et al.*, 1992; Annone, 1997; Carmona *et al.*, 1999; Perelló *et al.*, 2003). Tan spot was often observed throughout the growing season and was the most common leaf disease observed each year, in 72.6% of all wheat fields in 2001 and 90.4% in 2002. Additionally, strains of *Alternaria* spp. from wheat plants with symptoms of leaf blight sufficiently similar to those described for tan spot were collected from 11 localities. All isolations corresponded to the *A. infectoria* species group (Simmons, 1994; Simmons, personal communication, 2001; Perelló *et al.*, 2002). Leaf blight of wheat caused by *Alternaria* spp. isolates was not significant in complex of wheat foliar diseases in Argentina for many years but, currently, this pathogen has become a new problem in the Buenos Aires Province. Symptoms are often difficult to distinguish in the field from those caused by *D. tritici-repentis* (Died.) Shoem.

Other pathogens for wheat, like *C. herbarum*, *P. sorghina*, *C. gramineum*, *Pirycularia oryzae* and *A. tritici* were registered for the first time in Argentina (Perelló, 2007). The pattern of diseases produced by these phytopathogens in some areas of cultivation is changing drastically due, among other causes, to new market trends that induce changes in agricultural practices and the introduction of new crops. Monitoring these changes is important in order to take appropriate and timely action to prevent disease dispersal. Information on the most common leaf spotting fungi would help to identify appropriate benchmarks for selecting for disease resistance in different environments. It would also help breeders to set priorities in the incorporation of disease resistance to the leaf spotting complex into adapted wheat cultivars.

Moreover, widespread occurrence of these fungal diseases in the major wheat-growing region of Argentina described warns regional breeders and pathologists to increase efforts to manage the spotting wheat leaf complex in order to avoid future epidemics.

Acknowledgement

The author is grateful for the financial support of Project 11/A142 'Patógenos fúngicos del trigo y su posibilidad de biocontrol con microorganismos antagonistas en el marco de una agricultura sustentable' of the Programa de Incentivos a la Docencia e Investigación de la Universidad Nacional de la Plata.

References

Agarwal, V.K., Chahal, S.S. and Mathur, S.B. (1993) *Alternaria* leaf blight. In: Mathur, S.B. and Cunfer, B.M. (eds) *Seed-borne Diseases and Seed Health Testing of Wheat.* Danish Government Institute of Seed Pathology for Developing Countries, Hellerup, Denmark, pp. 9–13.

Annone, J.G. (1985) Presencia de la mancha tostada del trigo (*Helminthosporium tritici – repentis*). *Carpeta de Production Vegetal Trigo.* Tomo VII Informe No. 88, EEA, INTA Pergamino, Buenos Aires, pp. 1–9.

Annone, J.G. (1997) Tan spot of wheat in Argentina: importance and disease management strategies. In: Duveiller, E., Dubin, H.J., Reeves, J. and McNab, A. (eds) Helminthosporium *Blights of Wheat. Spot Blotch and Tan Spot. Proceedings of an International Workshop held at CIMMYT, El Batan, Mexico,* 339 pp.

Anahosur, K.H. (1978) *Alternaria triticina. Descriptions of Pathogenic Fungi and Bacteria.* No. 583. Commonwealth Mycological Institute, Association of Applied Biologists, Kew, UK.

Andersen, B., Kroger, E. and Roberts, R.G. (2002) Chemical and morphological segregation of *Alternaria arborescens, A. infectoria* and *A. tenuissima* species-group. *Mycological Research* 106, 170–182.

Arya, H.C. and Panwar, K.S. (1955) Some studies on a virulent strain of *Cladosporium herbarum* (Link) Fr. on Wheat. *Indian Phytopathology* 8, 176–183.

Bhowmik, T.P. (1974) Fungicidal control of *Alternaria* leaf blight of wheat. *Indian Phytopathology* 27, 162–167.

Bruehl, G.W., Murray, T.D. and Allan, R.E. (1986) Resistance of winter wheats to *Cephalosporium* stripe in the field. *Plant Disease* 70, 314–316.

Carmona, M., Reis, E. and Cortese, P. (1999) *Manchas foliares deltrigo. Diagnostico, epidemiologia y nuevos criterios para el manejo.* BASF Argentina S.A. 32 pp.

Chaudhuri, S., Maiti, S.S. and Saha, P. (1976) Leaf blight of *Triticale* caused by *Alternaria triticina. Plant Disease Reporter* 60, 133–134.

Cristensen, K.B., Van Klink, J.W., Weavers, R.T., Larsen, T.O., Andersen, B. and Phipps, R.K. (2005) Novel chemotaxonomic markers of the *Alternaria infectoria* species-group. *Journal of Agricultural and Food Chemistry* 53, 9431–9435.

Frisullo, S. (1982) Parassiti fungini delle piante nell'Italia meridionale. I. *Alternaria triticina* Pras. et Prab. sufrumento duro. *Phytopathologia Mediterranea* 21, 113–115.

Gonzalez, H.H.L., Martinez, E.J. and Resnik, S.L. (1997) Fungi associated with sorghum grain from Argentina. *Mycopathologia* 139, 35–41.

Hosford, R.M. Jr (1975) *Phoma glomerata*, a new pathogen of wheat and Triticales, cultivar resistance related to wet period. *Phytopathology* 65, 1236–1239.

Jain, K.L. and Prabhu, A.S. (1976) Occurrence of a chromogenic variant in *Alternaria triticina. Indian Phytopathology* 29, 22–27.

Kumar, C.S.K.V. and Rao, A.S. (1979) Production of phytotoxic substances by *Alternaria triticina. Canadian Journal of Botany* 57, 1255–1258.

Kumar, P.S. and Kumar, P.R.P. (2000) A new leaf spot disease of *Parthenium hysterophorus* caused by *Phoma sorghina. Indian Phytopathology* 53, 1–115.

Kwasna, H. and Kosiak, B. (2003) *Lewia avenicola* sp. nov. and its *Alternaria* anamorph from oat grain, with a key to the species of Lewia. *Mycological Research* 107, 371–376.

Kwasna, H., Ward, E. and Kosiak, B. (2006) *Lewia hordeicola* sp. nov. from barley grain. *Mycologia* 98(4), 662–668.

Manavolta, V.M.A. and Bedendo, I.P. (1999) Danos devidos a manchas de grãos em arroz causadas pelos fungos *Bipolaris oryzae*, *Microdochium oryzae* e *Phoma sorghina*, em diferentes epocas de infecção. *Summa Phytopatologica* 25(4), 324–330.

Marchionatto, J.B. (1948) *Tratado de Fitopatología*. Ediciones Librería del Colegio, Editorial Sudamericana, Buenos Aires, 537 pp.

Neergaard, P. (1979) *Seed Pathology*. Volume 1 and 2, revised edn. MacMillan Press, London, 1191 pp.

Nema, K.G. and Joshi, L.M. (1971) Symptoms and diagnosis of the 'spot blotch' and 'leaf blight' diseases of wheat. *Indian Phytopathology* 24, 418–419.

Perelló, A. (2007) Enfermedades nuevas y emergentes en trigo. Manejo de bajo impacto ambiental con microorganismos antagonistas. In: *Libro de Resúmenes. Jornada de actualización en enfermedades de trigo. Instituto Fitotécnico de Santa Catalina, Facultad de Cs. Agrarias y Ftales, Universidad Nacional de La Plata, Llavallol, Buenos Aires, 23 November 2007*, pp. 30–34.

Perelló, A. and Moreno, V. (2006) Ocurrencia de *Cephalosporium gramineum* en trigo en la Argentina. *XII Jornadas Fitosanitarias Argentinas, Catamarca*, pp. 270–271.

Perelló, A.E. and Sisterna, M. (2005) Leaf blight of wheat caused by *Alternaria triticina* in Argentina. *Plant Pathology* (NDR 2004-117) (http://www.bspp.org.uk/ndr/ july 2005-35.asp).

Perelló, A. and Sisterna, M. (2006) Leaf blight of wheat caused by *Alternaria triticina* in Argentina. *Plant Pathology* 55(2), 303.

Perelló, A., Cordo, C.Y. and Alippi, H. (1992) Tizón de la Hoja del Trigo (*Triticum aestivum* L) ocasionado por *Alternaria triticina* Pras. & Prab. *VIII Jornadas Fitosanitarias Argentinas. Facultad de Ciencias Agropecuarias, Paraná, Entre Ríos, 7–11 September 1992.*

Perelló, A., Cordo, C. and Simón, M.R. (1996) A new disease of wheat caused by *Alternaria triticimaculans* in Argentina. *Agronomie, INRA, Francia* 16, 107–112.

Perelló, A., Sisterna, M. and Cortese, P. (1998) A scale for appraising the leaf blight of wheat caused by *Alternaria triticimaculans*. *Cereal Research Communications* 26(2), 189–194.

Perelló, A., Sisterna, M.Y. and Cortese, P. (1999) Severidad del tizón de la hoja (*Alternaria triticimaculans*) en diferentes cultivares de trigo bajo condiciones de campo. *X Jornadas Fitosanitarias Argentinas, 7–9 April 1999. San Salvador de Jujuy, Jujuy, Argentina.*

Perelló, A.E., Sisterna, M.N. and Moreno, M.V. (2002) Manchas foliares del trigo causadas por el complejo *Alternaria infectoria* en la Argentina. *XI Jornadas Fitosanitarias Argentinas, 26–28 June 2002, Río Cuarto, Córdoba*, 33 pp.

Perelló, A., Tacaliti, M.S., Moreno, M.V.Y. and Castro, A.M. (2003) Análisis de Patrones Isoenzimáticos en Aislamientos de *Alternaria infectoria* asociados a Tizones Foliares del Trigo. *Boletín de la Sociedad Argentina de Botánica* 38 (suppl), 275.

Perelló, A., Carranza, M. and Larrán, S. (2005a) Patógenos de semillas asociados al manchado del grano de trigo. *XIII Congreso Latinoamericano de Fitopatología y III Taller de la Asociación Argentina de Fitopatólogos, Villa Carlos Paz, Córdoba, Argentina, 19–22 April 2005*, 447 pp.

Perelló, A., Moreno, V., Sisterna, M. and Castro, A. (2005b) Occurrence and characterization on wheat of the *Alternaria infectoria* complex associated with black point and leaf blight symptoms in Argentina. *7th Internacional Wheat Conference. 27 November – 2 December 2005. Mar del plata, Argentina.* Organized by SAGPyA and INTA, 149 pp.

Perelló, A., Moreno, V. and Sisterna, M. (2008) *Alternaria infectoria* species group associated to black point of wheat in Argentina. *Plant Pathology* 57, 379.

Prabhu, A.S. and Prakash, V. (1973) The relation of temperature and leaf wetness to the development of leaf blight of wheat. *Plant Disease Reporter* 57, 1000–1004.

Prasada, R. and Prabhu, A.S. (1962) Leaf blight of wheat caused by a new species of *Alternaria*. *Indian Phytopathology* 15, 292–293.

Punithalingam, E. (1985) *Phoma sorghina. CMI Descriptions of Pathogenic Fungi and Bacteria*. No. 825. Commonwealth Mycological Institute, Kew, UK.

Punithalingam, E. and Holliday, P. (1972) *Phoma insidiosa. CMI Descriptions of Pathogenic Fungi and Bacteria*. No. 333. Commonwealth Mycological Institute, Kew, UK.

Rao, A.S. and Subrahmanyam, P. (1974) A modified medium for preferential isolation of *Alternaria triticina* on wheat. *Indian Phytopathology* 27, 133–134.

Rault, J.G., Guldhe, S.M. and Wangikar, P.D. (1983) Seed-borne infection of *Alternaria triticina* in wheat and its control. *Indian Phytopathology* 36, 274–277.

Scharen, A.L. and Krupinsky, J.M (1971) *Ascochyta tritici* on wheat. *Phytopathology* 61, 675–680.

Simmons, E. (1994) *Alternaria* themes and variations. *Mycotaxon* 50, 409–427.

Simmons, E. (2002) *Alternaria* themes and variations (305–309). *Lewia/Alternaria* revisited. *Mycotaxon* 83, 127–145.

Sokhi, S.S. (1974) *Alternaria* blight on wheat in India. *Proceedings of the National Academy of Sciences of the United States of America* 20, 55–57.

Sprague, R. and Johnson, A.G. (1950) *Ascochyta* leaf spot of cereals and grasses in the United States. *Mycologia* 42, 523–553.

Vieira, B.S. and Barreto, R.W. (2005) *Lewia chamidosporiformans* sp. nov. from *Euphorbia heterophylla*. *Mycotaxon* 94, 245–248.

Waller, J.M. (1981) The recent spread of some tropical plant diseases. *Tropical Pest Management* 27, 360–362.

White, J.F. and Morgan-Jones, G. (1983) Studies in the genus Phoma. II. Concerning *Phoma sorghina*. *Mycotaxon* 18, 5–13.

Wiese, M.V. (1987) *Compendium of Wheat Diseases*. APS Press, St Paul, Minnesota, 112 pp.

Zadok, J.C., Chang, T.T. and Konzak, C.F. (1974) A decimal code for the growth stages of cereals. *Weed Research* 14, 415–421.

Zillinsky, F.J. (1984) *Guía para la identificación de enfermedades en cereales de grano.* Centro Internacional de Mejoramiento de Maíz y Trigo, CIMMYT, El Batán, México, 64 pp.

19 Diseases of Fenugreek (*Trigonella foenum-graecum* L.) and Their Control Measures, with Special Emphasis on Fungal Diseases

S.N. Acharya,[1] J.E. Thomas,[2] R. Prasad[1,2] and S.K. Basu[1,2]
[1]*Agriculture and Agri-Food Canada Research Centre, Lethbridge, Canada;*
[2]*Department of Biological Sciences, University of Lethbridge, Lethbridge, Canada*

Abstract
Fenugreek (*Trigonella foenum-graecum* L.) is an annual legume crop cultivated in India, the Mediterranean region, China, parts of Africa, Europe and Australia and, in recent years, in North America. Although traditionally used as a spice crop, fenugreek has important medicinal and nutraceutical properties and is also grown as a forage crop in some countries. This multi-use crop has the potential to expand into new areas, as well as increase in the area where it is traditionally grown. Therefore, its reaction to biotic and abiotic factors that can limit its production deserves special attention. Although this review contains a discussion on all fenugreek diseases and insect pests, the main focus is on the causal organisms, symptoms and corresponding control measures for all of the major and minor fungal diseases affecting its productivity. It is interesting to note that only a few diseases have been reported to affect this crop adversely. The two major fungal diseases that affect fenugreek are powdery mildew caused by *Erysiphe polygoni* and *Cercospora* leaf spot caused by *Cercospora traversiana*. However, disease problems may change as this crop is grown more widely and with larger acreages outside of its natural area of adaptation. Ongoing vigilance in disease monitoring and development of new resistant varieties is needed to ensure productivity and usefulness of this crop in the future.

Introduction

Fenugreek (*T. foenum-graecum* L.) is an annual crop belonging to the legume family *Fabaceae*. Although widely cultivated in India, China, northern and eastern Africa, parts of Mediterranean Europe, Argentina and Australia (Acharya *et al.*, 2006a), it was only recently introduced to North America (Acharya *et al.*, 2006b). Fenugreek is a dicotyledonous, self-pollinated plant with trifoliate leaves, branched stems and white flowers, which typically produces golden yellow seeds (Basu *et al.*, 2008). It has two morphological forms of flowering shoots, the common one bearing axillary flowers and an indeterminate growth habit, whereas plants with blind shoots possess both axillary and terminal flowers with a more determinate growth habit (Busbice *et al.*, 1972; Fehr, 1993; Acharya *et al.*, 2008; Basu *et al.*, 2008). Although both closed and open type flowers are reported, the majority of fenugreek flowers belong to the closed

 Management of Fungal Plant Pathogens
(eds A. Arya and A.E. Perelló)

type (Petropoulous, 2002; Acharya *et al.*, 2008).

Fenugreek is capable of fixing atmospheric nitrogen in the soil. The plants require a minimal amount of nitrogen for growth, reducing the need for nitrogen fertilizers to supplement crop growth and facilitating use of the plant in crop rotations. Fenugreek is considered a dryland crop; water requirements are low, making cultivation of fenugreek an increasingly attractive alternative to producers in regions with limited water supply. Use of fenugreek in arid and semi-arid environments can reduce the cost of irrigation, reduce the potential for eutrophication of surface water and limit contamination of groundwater sources (Basu *et al.*, 2004; Acharya *et al.*, 2008). According to Acharya *et al.* (2004), dryland adaptation of fenugreek was a major consideration for introducing it as a forage crop for growth in the temperate climates of western Canada. Well-drained, loamy soils are most favourable for the crop (Rosengarten, 1969; Acharya *et al.*, 2008), while heavy and wet soils are known to restrict fenugreek growth (Petropoulos, 1973; Acharya *et al.*, 2008) and increase susceptibility of the plant to disease.

Although fenugreek is known primarily as a spice crop used especially in India and Mediterranean regions for cooking, the species name *foenum-graecum* refers to 'Greek hay', highlighting its use as a forage crop in early years (Acharya *et al.*, 2006b). Fenugreek also has been referred to as a medicinal herb, both in Indian Ayurvedic and traditional Chinese medicines (Tiran, 2003). Its leaves and seeds have been used extensively to prepare extracts and powders for medicinal use like wound healing and promotion of lactation in weaning mothers (Basch *et al.*, 2003; Tiran, 2003; Acharya *et al.*, 2007a). The medicinal value of fenugreek comes mainly from three chemical constituents; i.e. steroidal sapogenins, galactomannans and isoleucine (Acharya *et al.*, 2006a, 2007a, 2008). Steroidal sapogenins are often used as a raw precursor for the production of steroidal drugs and hormones such as testosterone, glucocorticoids and progesterone (Fazli and Hardman, 1968) and are also effective in the treatment of hypercholesterolaemia (McAnuff *et al.*, 2002). Fenugreek galactomannans appear to aid in the control of type 2 diabetes in both animals (Raju *et al.*, 2001; Tayyaba *et al.*, 2001; Puri *et al.*, 2002; Vats *et al.*, 2002, 2003) and humans (Sharma *et al.*, 1996; Puri *et al.*, 2002). The amino acid isoleucine is a precursor of 4-hydroxyisoleucine, which is known to regulate the secretion of insulin in animals (Broca *et al.*, 2000) and also making it potentially useful in the control of diabetes.

Since fenugreek can be used for multiple purposes (e.g. as a spice, forage crop, eco-friendly dryland crop and for medicinal and nutraceutical applications), there is interest in cultivation of the plant in new biogeographical areas of the world (Acharya *et al.*, 2008; Basu *et al.*, 2008). Acharya *et al.* (2006b) have described fenugreek as a traditional 'Old World' crop with significant potential for use in the 'New World'. 'Tristar' is the first North American variety of fenugreek released by a research group in Canada (Acharya *et al.*, 2007c). As a result of its increased economic and industrial importance, those involved in fenugreek production need to become more aware of the diseases that can affect both yield and quality of the plant adversely. In many fenugreek-growing areas, infectious and non-infectious diseases are becoming an important production constraint because of their ability to cause variation in crop yield and quality (Basu *et al.*, 2006a). When fenugreek is introduced to a new biogeographical area, new diseases may emerge that can cause a reduction in productivity, and even crop losses (McCormick and Hollaway, 1999; Fogg *et al.*, 2000). This review looks at major and minor diseases of fenugreek reported worldwide, as well as emerging fungal diseases that are increasing in importance, to use of the plant commercially.

Diseases of Fenugreek

Fenugreek production is affected by both biotic and abiotic agents. Abiotic diseases or disorders are non-infectious and are often caused by a deficiency in nutrients, extremes in temperature, moisture, soil acidity or alkalinity, an excess of certain micronutrients within the soil and toxic impurities in the atmosphere (Petropoulos, 2002; Acharya *et al.*, 2008). For example, Sinskaya (1961) reported yellowing of some fenugreek plants under field conditions due to mineral deficiencies in boron, mag nesium, manganese or potassium. Physiological diseases resulting from abiotic agents can lead to premature death of the plant and loss of forage and seed yield. In western Canada, exposure of fenugreek crops to very dry and hot conditions has resulted in stunted growth and yellowing, with occasional loss of leaves from the plant.

Diseases caused by living or biotic agents (pathogens) are often infectious (Acharya *et al.*, 2008). The most important diseases of fenugreek are caused by plant pathogenic fungi. Bacterial diseases are next in degree of importance, followed by viral diseases (AAFRD, 1998; Fogg *et al.*, 2000; Prakash and Sharma, 2000; Petropoulos, 2002; Weiss, 2002; Jongebloed, 2004). While information on specific pests and diseases damaging fenugreek is limited, in general, insects and pathogenic organisms that attack other common legume crops grown in the vicinity of fenugreek, such as alfalfa, can also attack fenugreek (Basu *et al.*, 2006b).

Viral diseases

Bean yellow mosaic virus, alfalfa mosaic virus, cowpea mosaic virus, soybean mosaic virus, pea mosaic virus, potato virus A and Y and clover vein mosaic virus are all common viral infections of fenugreek (Petropoulos, 2002). Bhasker and Summanwar (1982) reported mosaic wilt on fenugreek. Flexuous rod-shaped viruses like bean yellow mosaic potyvirus (Singh, 1969) and pea streak carlavirus (Hagedorn and Walker, 1949) have also been reported on fenugreek. These viral diseases have been associated with moderate losses of fenugreek seed and forage yield (Table 19.1).

Bacterial diseases

McCormick and Hollaway (1999) found that infection of fenugreek with *Pseudomonas syringae* resulted in bacterial blight. First reported in Victoria, Australia, infection with these bacteria caused small isolated patches, to entire crop loss in the field. Fogg *et al.* (2000) reported the same disease on fenugreek in New Jersey, USA. It has also been suggested that the bacterium *Xanthomonas alfalfa* can infect fenugreek (Petropoulos, 2002), leading to loss in productivity (Table 19.1).

Nematode diseases

Various nematodes, typically not identified as a problem for other crops, can damage fenugreek roots (Weiss, 2002; Jongebloed, 2004). The soilborne nematode *Meloidogyne incognita* has been shown to cause root rot and the death of immature fenugreek plants in Australia (Jongebloed, 2004) (Table 19.1). However, it is interesting to note that fenugreek has also been reported to have some anti-nematicidal properties. Zia *et al.* (2003) reported that decomposed seeds of fenugreek caused a marked reduction in population densities of the soil nematode *M. javanica*, which causes root-knot development in mungbean. Decomposed seed and aqueous extracts of fenugreek were also able to enhance plant height and shoot fresh weight in mungbean.

Insect pests

In Australia, insects such as thrips, podborers and *Heliothis* can cause serious damage to forage yield in fenugreek (Lucy, 2004). Basu *et al.* (2006b) reported that in southern Alberta (Canada), a low level of insect pests such as *Lygus* bugs and, to a lesser extent, alfalfa plant bugs and aphids had been observed in fenugreek fields. In addition, the researchers reported that western flower thrips (especially severe under greenhouse conditions), alfalfa looper, alfalfa weevil

Table 19.1. The major non-fungal diseases of fenugreek reported worldwide.

Disease groups	Causal organisms	Country reported	Tolerant varieties/ genotypes	References
Viral diseases	Bean yellow mosaic virus	England	Fluorescent, Ethiopian	Brunt, 1972; Petropoulos, 1973, 2002
	Potato virus A	NA*	NA	Schmelzer, 1967; Anonymous, 1968
	Cowpea mosaic virus	NA	NA	Vidamo and Conti, 1965; Anonymous, 1968
	Potato virus Y	NA	NA	Schmelzer, 1967
	Tobacco etch virus	NA	NA	Petropoulos, 2002
	Pea streak virus	NA	NA	Hagedorn and Walker, 1949; Anonymous, 1968
	Pea mosaic virus	NA	NA	Petropoulos, 2002
	Soybean mosaic virus	NA	NA	Quantz, 1968; Schmelzer and Wolf, 1971
	Alfalfa mosaic virus	NA	NA	Quantz, 1968; Schmelzer and Wolf, 1971; Latham and Jones, 2001
	Tomato black ring virus	NA	NA	Quantz, 1968; Schmelzer and Wolf, 1971
	Clover vein mosaic virus	NA	NA	Quantz, 1968; Schmelzer and Wolf, 1971
Bacterial diseases	*Pseudomonas syringae* pv. *syringae*	Australia	NA	McCormick and Hollaway, 1999; Fogg *et al.*, 2000
	Xanthomonas alfalfa		NA	Petropoulos, 2002
Nematode diseases	*Meloidogyne incognita*	Australia	NA	Jongebloed, 2004
Insect-mediated diseases	*Lygus keltoni*, *L. elisus*, *L. borealis* and *L. lineolaris*	Canada	Tristar	Basu *et al.*, 2006a
	Adelphocoris lineolatus	Canada	Tristar	Basu *et al.*, 2006a
	Acyrthosiphon pisum	Canada	Tristar	Basu *et al.*, 2006a
	Frankliniella occidentalis	Canada	Tristar	Basu *et al.*, 2006a
	Sitona sp.	Canada	Tristar	Basu *et al.*, 2006a
	Hypera postica	Canada	Tristar	Basu *et al.*, 2006a
	Autographa californica	Canada	Tristar	Basu *et al.*, 2006a
	Aphis craccivora	India, West Asian countries	NA	Weiss, 2002
	Myzodes persicae	India, West Asian countries	NA	Weiss, 2002
	Scirtothrips dorsalis	Australia, India, the Mediterranean region	NA	Weiss, 2002; Lucy 2004
	Tetranychus cucurbitae	India	NA	Weiss, 2002
	Pachymerus pallidus	Sudan	NA	Weiss, 2002
	Diacrisia oblique	India	NA	Weiss, 2002
	D. orichalcea	India	NA	Weiss, 2002
	Prodenia litura	India	NA	Weiss, 2002
	Maruca testulalis	India	NA	Weiss, 2002

Note: *NA = not available.

and *Sitona* sp., were attracted to standing fenugreek crops under field conditions in western Canada. *Aphis craccivora* and *Myzodes persicae* have caused damage to fenugreek crops from west Asia to India, while various *Thysanoptera* (thrips), including *Scirtothrips dorsalis*, have been found on almost all fenugreek crops grown from the Mediterranean to India (Potropoulos, 2002; Weiss, 2002). There have also been reports of mite (*Tetranychus cucurbitae*) attacks on fenugreek in India (Weiss, 2002). *Pachymerus pallidus*, a seed beetle, which attacks a wide range of crops, is a major pest of fenugreek in the Sudan (Weiss, 2002). A number of polyphagous caterpillars belonging to the order Lepidoptera, including *Diacrisia oblique*, *D. orichalcea* and *Prodenia litura*, and especially the mung moth (*Maruca testulalis*), have been reported to affect fenugreek in India (Weiss, 2002) (Table 19.1).

Common Fungal Diseases of Fenugreek

The two most common fungal diseases infecting fenugreek are *Cercospora* leaf spot and powdery mildew (AAFRD, 1998). Powdery mildew on fenugreek, caused by *E. polygoni*, can seriously reduce crop yield (Prakash and Sharma, 2000; Jongebloed, 2004) and has the potential to affect biomass and seed yield in crops grown under moist agroclimatic conditions in North America. In Australia, yield of fenugreek was seriously affected by blight caused by *C. traversiana* and wilt caused by *Fusarium oxysporum* and *Rhizoctonia solani* (Jongebloed, 2004). The pathogen *C. traversiana* is spread by contaminated seed and is now found in many countries, ranging from India to Europe, eastern Africa including Ethiopia and in several countries in South America; it is slowly becoming a major fenugreek disease concern (Weiss, 2002). Other well-known fungal diseases observed to be associated with fenugreek are collar rot, leaf spot and pod spot diseases (Petropoulos, 2002) (Table 19.2).

In India, 27 species of fungi have been isolated from fenugreek seeds (Prabha and Bohra, 1999). Time of sowing can influence the damage caused by an infection. For example, in Haryana, India, seed sown in mid October as compared with the end of November exhibited a 30% reduction in damage caused by *E. polygoni* and *Leveillula taurica* (Sharma, 1999). Downy mildew caused by *Peronospora trifoliorum* and spring black stem and leaf spot caused by *Phoma pinodella* have recently become more common (Lakra, 2002, 2003; Bretag and Cunnington, 2005). Several leaf diseases causing varying degrees of damage generally or in specific seasons, including rust due to *Uromyces anthyllidis*, have been reported in India (Weiss, 2002). Root and collar rots caused by *Rhizoctonia* spp., typically *R. solani* and *Alternaria* spp., often *A. alternata*, can damage individual crops (Weiss, 2002).

Antifungal activity for fenugreek has also been reported in the primary literature (El-Gizawy *et al.*, 2000). Lupin and fenugreek seed extracts significantly suppressed *Pythium* damping-off of cucumber and tomato seedlings, as well as radish damping-off caused by *R. solani*. Moreover, application of seed extracts had a significant positive effect on seedling growth of the vegetables tested (El-Gizawy *et al.*, 2000). A detailed description of major and minor fungal diseases of fenugreek reported all across the globe and their prescribed control measures are outlined individually in the following sections.

Cercospora leaf spot

Cercospora leaf spot is a seedborne fungal disease, considered to be one of the most serious threats to fenugreek. This disease is capable of causing considerable economic loss (Leppik, 1959, 1960; Khare *et al.*, 1981; Zimmer, 1984; Ryley, 1989). The *Cercospora* leaf spot of fenugreek has been reported all across the world and is most common in Australia, several eastern European countries, South America, North America, in the Near East and India (Voros and Nagy, 1972; Cook, 1978; Khare *et al.*, 1981; Ryley, 1989).

Table 19.2. The major fungal diseases of fenugreek reported worldwide.

Name of the disease	Pathogenic fungal species	Country reported	Tolerant varieties/ genotypes	References
Cercospora leaf spot	*Cercospora traversiana*	India, Australia, Canada	NA	Leppik, 1959, 1960; Khare *et al.*, 1981; Zimmer, 1984; Ryley, 1989
Collar rot	*Rhizoctonia solani*	India	TG-18, UM-20, Pusa Early Bunching	Hiremath *et al.*, 1976; Hiremath and Prasad, 1985; Raian *et al.*, 1991; Petropoulos, 2002; Datta and Chatterjee, 2004
Leaf spot	*Ascochyta* sp.	UK	Fluorescent, Ethiopian	Walker, 1952; Petropoulos, 1973
Powdery mildew	*Oidiopsis* sp.	Israel, Ethiopia, England	Fluorescent	Palti, 1959; Rouk and Mangesha, 1963; Petropoulos, 1973
Downy mildew	*Peronospora trigonellae*	India	HM-346, HM-350, HM-444	Lakra, 2002, 2003; HAU, 2008
Powdery mildew	*Leveillula taurica*	Israel	NA*	Palti, 1959
Leaf spot	*Pseudoperiza medicaginis*		NA	Glaeser, 1961
Spring black stem and leaf spot	*Phoma pinodella*	Australia	NA	Bretag and Cunnington, 2005
Powdery mildew	*Erysiphe polygoni*	Israel, Ethiopia, India, Canada	HM-350, HM-444, Fluorescent	Petropoulos, 1973; Zimmer, 1984; Prakash and Saharan, 2000; Basu *et al.*, 2006a; HAU, 2008
Rust	*Uromyces trigonellae*	Israel	NA	Ubrizsy, 1965
Pod spot	*Heterosporium* sp.	UK	Kenyan, Moroccan	Petropoulos, 1973
Charcoal rot	*Macrophomina phaseolina*	Pakistan	NA	Haque and Ghaffar, 1992
Root rot	*Sclerotinia trifoliorum*	UK	NA	Petri, 1934
Fusarium wilt	*Fusarium oxysporum*	India, Sudan, Malta	NA	Borg, 1936; Komaraiah and Reddy, 1986; Hashmi and Thrane, 1990; Bansal and Gupta, 2000

Note: *NA = not available.

The causal organism for this disease is *C. traversiana*, a member of the *Ascomycetes* (Agrios, 1997). Several researchers have suggested that *C. traversiana* is the only species of the *Cercospora* infecting fenugreek (Cook, 1978; Ryley, 1989). Conidiophores of *C. traversiana* are dark, paler towards the tip, unbranched, rarely geniculate and rarely septate. These conidiophores develop in fascicles of 3–5 conidiophores per fascicle, with a length of up to 420 μm and width ranging from 3 to 5 μm (Ryley, 1989). The conidia are hyaline, acicular, straight or slightly curved, apex rounded, base truncate and multicellular. The main source of overwintering inocula is plant debris, where sclerotia or stromata can form. Conidia germinate best at a high relative humidity and at a high temperature. They are dispersed mainly by rain-splash and to some extent by wind (Agrios, 1997).

Cercospora leaf spot initially presents itself as circular, sunken lesions that appear bleached in colour, with narrow (1–2 mm) chlorotic halos on the surface of the leaves. These lesions expand rapidly as the infection progresses, producing necrotic areas. Each area of infection is sharply defined, with most lesions surrounded by a characteristic yellowish halo. Lesion size is increased significantly on mature leaves, where sporulation becomes evident, giving the lesions a whitish, velvet-like appearance (Zimmer, 1984). Severely infected plants are reported to have only a few leaves situated towards the apex of the plant (Agrios, 1997). Stem and pods also can become infected. Disease symptoms on pods include discoloured infected areas, as well as severely infected areas that can become shrunken and twisted (Zimmer, 1984). The life cycle of the pathogen is shown in Fig. 19.1.

Control measures

As the pathogen is often seedborne, seed treatment before planting has been an effective control measure in some cases (Leppik, 1960; Khare *et al.*, 1981). However, selection of healthy seeds as planting material may also provide an effective control (Cook, 1978). Rotation with crops outside of the host range for the fungal pathogen *C. traversiana* may also be useful. It appears that prevention of seed contamination by treating plants when the pathogen is first detected will likely be the best approach to limiting spread

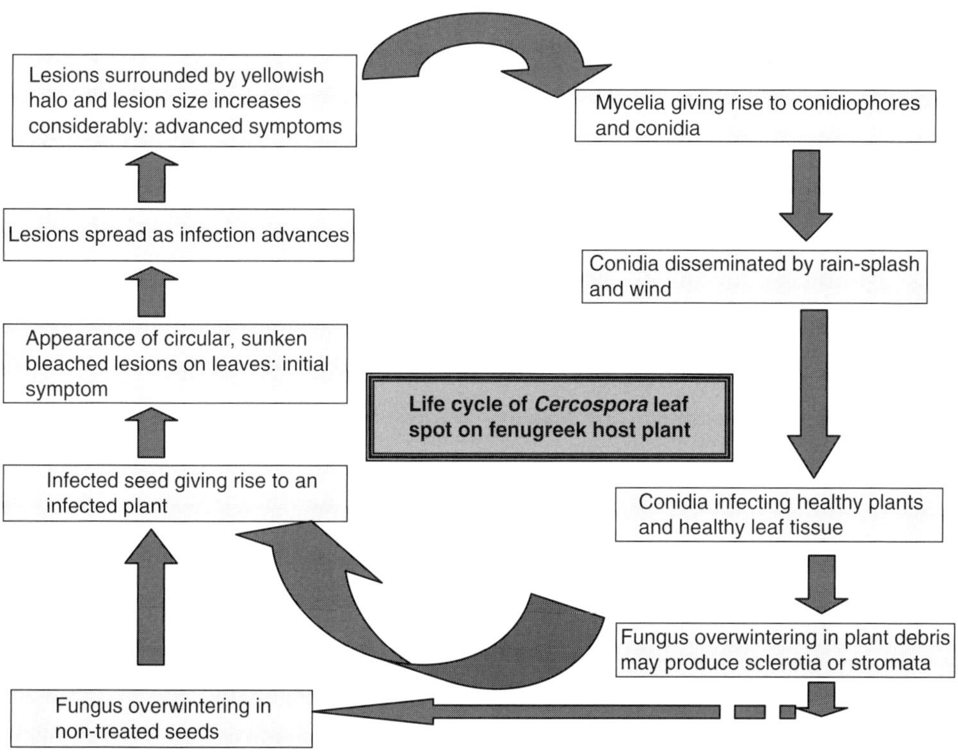

Fig. 19.1. The life cycle of *Cercospora traversiana* on fenugreek host plant.

of this pathogen. Spraying the plants with fungicides such as benomyl, chlorothanolin, Bordeaux mixture, mancozeb and maneb has been suggested as an effective chemical control measure (Agrios, 1997).

Collar rot

Collar rot is another important fungal disease of fenugreek and has been reported in all parts of India (Hiremath *et al.*, 1976; Hiremath and Prasad, 1985; Raian *et al.*, 1991). The causal organism for this disease is a member of the *Basidiomycetes*. *R. solani* reduces yield of fenugreek causing foot-rot and damping-off where freshly emerged seedlings fall over and die (Petropoulos, 2002). The vegetative mycelium of *R. solani* is colourless when young but turns brown on maturity. The mycelium consists of hyphae partitioned into distinct individual cells by a septum consisting of a doughnut-shaped pore (Ogoshi, 1987; Alexopoulos *et al.*, 1996). *R. solani* survives as sclerotia in the soil and on plant tissue, and as mycelia by colonizing soil organic matter as a saprophyte. Sclerotia and/or mycelia present in the soil and/or on plant tissue germinate to produce fungal hyphae that can attack the subsequent year's crop (Alexopoulos *et al.*, 1996). The pathogen primarily attacks below-ground plant parts such as the root system, but is also capable of infecting other parts such as green foliage, seeds and hypocotyls. The most common symptom of the disease is damping-off (Petropoulos, 2002). Most of the severely infected seedlings may die at pre- or post-soil emergence stages. The infected seedlings may develop reddish-brown cankers on roots and stems at or near ground level (Anderson, 1982; Adam, 1988; Agrios, 1997).

Control measures

Cultivation of resistant varieties has been suggested as the best control measure for the disease (Prasad and Hiremath, 1985). According to Haque and Ghaffar (1992), seed dressing and soil drenching with *Rhizobium meliloti*, *Trichoderma banatum*, *T. harzianum* and *T. pseudokonongii* can control the infection effectively. The Gram positive bacterium *Bacillus subtilis* can also be used effectively as a biological control agent for *R. solani* (Tschen and Kou, 1985; Tschen, 1987). Prasad and Herimath (1985) demonstrated that carbendazim could be used as a seed and dry soil mix fungicide and that captan also could be used to drench the soil and kill the fungus.

Leaf spot

Leaf spot is another seedborne disease of fenugreek that is caused by fungal pathogens of the *Ascochyta* sp. belonging to the *Ascomycetes* (Walker, 1952; Petropoulos, 2002). The fungus attacks the leaves, stems and pods of fenugreek, reducing both yield and quality severely. It can survive in the soil, on infected seed and on crop residues. The pathogen is disseminated by both wind and rain-splash (Agrios, 1997; Petropoulos, 2002). Irregular brown to black spots with distinct margins are detected on infected leaves. As the disease progresses, the leaves on the plant may die and fall off. Infected seeds have round, dark brown lesions. Seedlings from infected seeds start rotting from the point of seed attachment and rotting advances towards the stem and taproot; subsequently, the young seedlings die (Petropoulos, 2002). Cool, moist weather is favourable for rapid dissemination and growth of the fungus (Anonymous, 1970; Agrios, 1997).

Control measures

Cultivation of tolerant genotypes is a good idea to avoid rapid infestation of the fungus (Agrios, 1997). To protect fenugreek plants from primary infection, seeds can be treated effectively with benlate, while to prevent secondary infection, use of a frequent foliar spray containing benlate is also recommended (Petropoulos, 2002).

Fusarium wilt

Fusarium wilt of fenugreek is caused by the fungus *F. oxysporum*, an *Ascomycete* that has been reported by several investigators across the world (El-Bazza *et al.*, 1990; Borg, 1936; Hashmi, 1988; Bansal and Gupta, 2000; Petropoulous, 2002). The pathogen *F. oxysporum* is both seed and soilborne (Komaraiah and Reddy, 1986; Hashmi and Thrane, 1990; Bansal and Gupta, 2000; Pierre and Francis, 2000). The pathogen can remain in infested soils for up to 10 years. Dissemination of the pathogen occurs through seed, soil and infested plant parts (Pierre and Francis, 2000). *Fusarium* wilt first appears as a slight clearing in veins found on the outer portion of younger leaves, followed by downward drooping of the mature leaves. At the seedling stage, plants infected by *F. oxysporum* may wilt and die soon after the symptoms appear. In mature plants, vein clearing and downward drooping of the leaf are often followed by stunting, yellowing of the lower leaves and subsequent wilting of leaves and young stems. Marginal necrosis of the infected leaves, rapid defoliation and finally death of the entire plant typically follow (Agrios, 1997). Browning of the vascular tissue is strong evidence of *Fusarium* wilt infestation. Furthermore, symptoms become more apparent on mature plants during the period between blossoming and fruit maturation (Jones *et al.*, 1982; Smith *et al.*, 1988).

Control measures

Some effective means of controlling *F. oxysporum* include disinfection of the soil and planting of the seeds with thiram or captan, crop rotation with non-hosts of the fungus, or use of resistant cultivars (Singh, 2001).

Pod spot

Petropoulos (1973) first investigated and described this disease in fenugreek and identified *Heterosporium* sp., an *Ascomycete*, as the causal agent. *Heterosporium medicaginis* is the only species of *Heterosporium* that has been reported to be pathogenic to legumes (Karimov, 1956). The fungus overwinters on dead leaves. Spores spread from old plant debris to initiate new plant infections. Fungal spores produced within leaf spots during the growing season are spread by splashing rain (Petropoulos, 1973). Symptoms of the disease become visible at the third stage of pod development and can be seen as dark brown to black spots on the pods that extend to produce a dark olive, velvet-like cover. Initially, localized spots of infection elongate transversely to the pod axis but with time spread over the pod surface and transform into more rounded to oblong lesions. These spots are also visible on the stems, but are rarely found on the plant leaves. Petropoulos (2002) suggests that the fungus does not enter into the seeds as the mycelium of the fungus is not buried deeply in the epidermis of the pod and that contamination of fenugreek seeds by this fungus takes place specifically during the threshing process.

Control measures

Hot water treatment of the seeds before planting is efficient to remove the fungus from the seeds (Pirone *et al.*, 1960). Resistant cultivars tolerant to that fungus have been suggested as the best way to restrict rapid dissemination of the disease on a standing crop effectively (Petropoulos, 2002).

Spring black stem and leaf spot

This disease of fenugreek has been reported in Australia by Bretag and Cunnington (2005). These investigators also identified *P. pinodella*, an *Ascomycete* previously known as *A. pinodella* (Jones, 1927), as the causal agent of the disease. This observation is supported by another investigation conducted by Boerema *et al.* (2004). *Phoma* sp. has been isolated from the seeds of fenugreek in Egypt, India, Nepal, Pakistan, Sri Lanka, Sudan and Syria (Hashmi, 1988), suggesting that the organism is not new to fenugreek crops. The pycnidia of the fungus are more or less globose, glabrous, ovoid to ellipsoid and usually aseptate (Punithalingam and Gibson, 1976; Bretag and Cunnington, 2005). Onfroy *et al.* (1999) reported that the length of conidia could range from 7.3–9.6 μm. The pathogen persists as pycnidia and mycelia in plant debris. It is dispersed mainly by splashing rain and to some extent by wind. Numerous small, irregular-shaped, dark brown to black leaf lesions surrounded by small chlorotic areas often appear as disease symptoms on the leaves, petioles and stems of the growing plant. Elongated black lesions may also develop on the taproot (Bretag and Cunnington, 2005). Infected plants become stunted with a mild chlorosis. In cases of severe infection, most of the leaves turn completely yellow, wither and the taproot system becomes completely girdled with sharp lesions (Bretag and Cunnington, 2005).

Control measures

Bretag *et al.* (2006) suggested practising crop rotation, destruction of infected plant portions and chemical seed treatments to control primary infections by the disease efficiently. When the fungus was restricted effectively at the primary infection level, it did not spread to advanced stages in most disease trials conducted (Bretag *et al.*, 2006), suggesting that restricting primary infection is the key to control of the disease.

Powdery mildew

Powdery mildew is one of the most common and serious fungal diseases of fenugreek, affecting both biomass and yield (Petropoulos, 2002; Basu *et al.*, 2006a). Powdery mildew is most commonly found in hot and humid tropical and subtropical areas, as well as in temperate to subtemperate regions (Palti, 1959; Rouk and Mangesha, 1963; Prakash and Saharan, 2000; Basu *et al.*, 2006a). On the basis of recent observations, Basu *et al.* (2006a) suggested that powdery mildew could become a serious disease problem in North America, where fenugreek is a recent crop introduction. Although some investigators have reported *Oidiopsis* sp. as the causal organism (Rouk and Mangesha, 1963; Petropoulos, 1973), the majority of investigators from across the globe consider *E. polygoni* (an *Ascomycete*) as the causal organism for the disease (Zimmer, 1984; Prakash and Saharan, 2000; Bretag and Cunnington, 2005; Basu *et al.*, 2006a). The conidiophores of the fungus are simple and erect and the corresponding conidia are unicellular, hyaline in colour, ellipsoidal to cylindrical in shape (Agrios, 1997; Nyvall, 1999; Basu *et al.*, 2006a) (Figs 19.2 and 19.3). The conidiophores vary in size from 32.5 to 65.6 μm × 9.7 to 13.6 μm, whereas dimensions of the conidia are 22.6–48.4 μm × 12.4–20.8 μm (Basu *et al.*, 2006a) (Fig. 19.2). The pathogen survives mostly by developing cleistothecia in diseased plant debris. They survive in soil until the next season. Ascospores are released after the disintegration of the wall of the asci. The ascospores first infect the lower and older leaves in the next season. The spores are carried by the

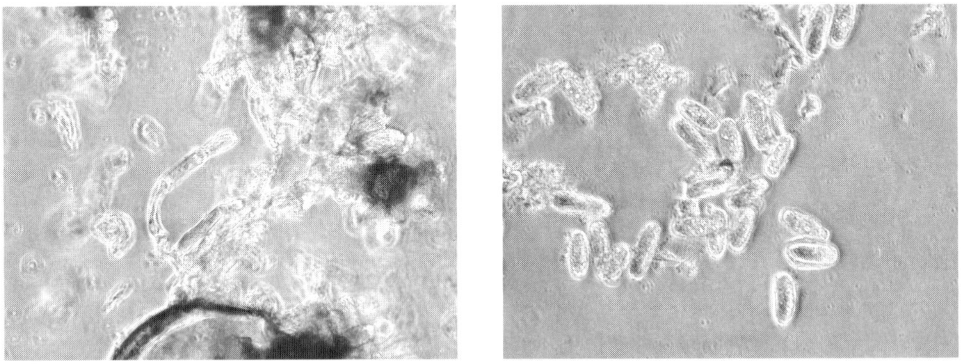

(a) (b)

Fig. 19.2. Light microscopy images of *Erysiphe polygoni* conidiophores (a) and conidia (b).

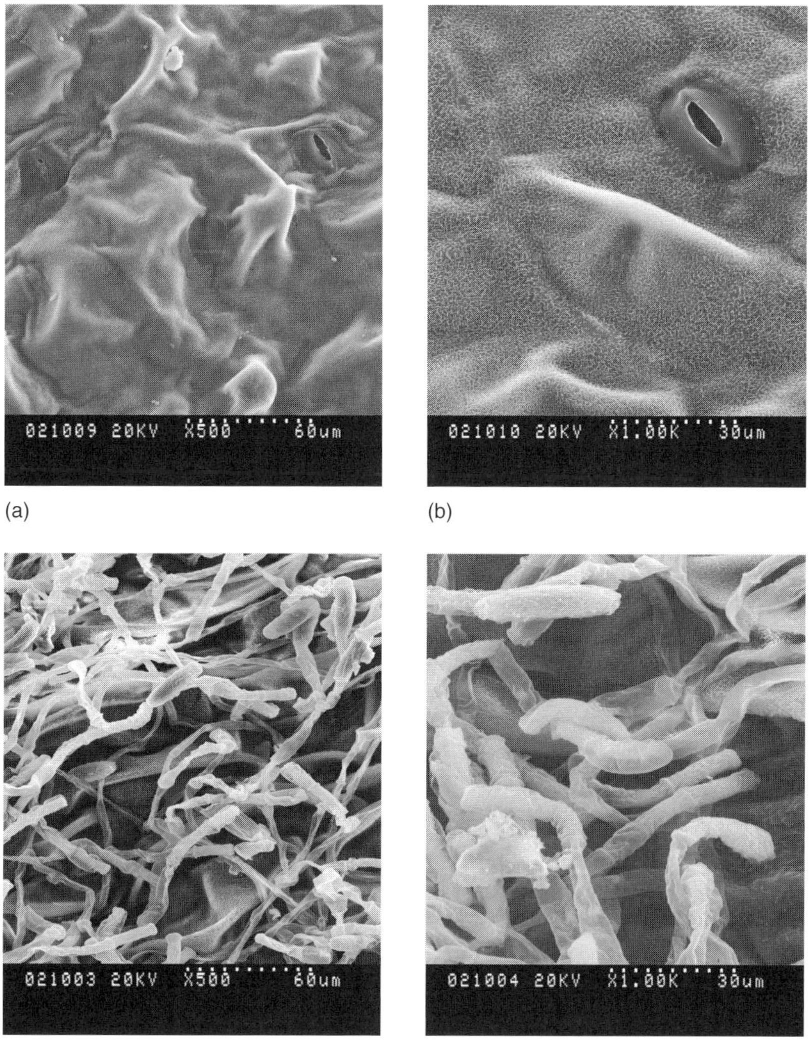

Fig. 19.3. Scanning electron microscopy (SEM) images of healthy fenugreek upper leaf surfaces [Top, (a) and (b)] compared to powdery mildew (caused by *Erysiphe polygoni*) infected upper leaf surface [Bottom, (c) and (d)]. Left images were magnified 500×, while the right images were magnified 1000×.

wind to new hosts. The pathogen is also known to survive as a mycelium (Sharma, 2005).

Powdery mildew is one of the easier diseases to identify on plants as its symptoms are quite distinctive. The disease can be identified easily by the presence of white to grey powdery masses or distinct circular to ellipsoidal patches on both the upper and lower surfaces of the leaves (Fig. 19.4), on pods but rarely on flowers, and by the strong odour emitted by the infected plants. During the initial stages of an infection, fungal patches appear isolated or in scattered patches which coalesce as the infection progresses. At first, leaves near ground level are infected, after which the whole plant can become covered with the fungus over a

Fig. 19.4. Comparison of healthy fenugreek upper leaflet surfaces (centre) with infected leaflets from the same plant.

short period of time (Fig. 19.5). The upper surface of the leaves typically bears more fungal structures and spores than the lower surface (Fig. 19.4).

Severely infected leaves become irregular in shape, dry and shrivelled, resulting in stunted growth of the whole plant (Basu *et al.*, 2006a) (Figs 19.4 and 19.5). Although Zimmer (1984) first identified powdery mildew infecting fenugreek in North America, Basu *et al.* (2006a) reported the first major in-depth investigation of powdery mildew as a major disease of fenugreek in North America based on trials that were conducted at different locations and under different physico-geographic conditions and variable climatic factors on the west and east coast of North America and the mid interior of Canada. The life cycle of the pathogen is presented in Fig. 19.6.

Control measures

Petropoulos (1973) and Avtar *et al.* (2003) reported variation in the sensitivity of fenugreek genotypes to powdery mildew; hence, use of resistant varieties has been strongly recommended to avoid disease infestation. Basu *et al.* (2006a) demonstrated that application of tilt 250E-propiconazole or milgo-ethrinol (28% at 2.5 ml/l) and captan-captane (50%) or benlate-benomyl (50% at 2.0 g\l) could control the disease at a satisfactory level, whereas Petropoulos (1973) showed that spraying with dinocap (8–10 oz a.i/acre in 100 gals) could also control the disease.

Conclusions

Fenugreek is affected mostly by seedborne fungal diseases. From our experience, and other reports of fenugreek disease, it is clear that powdery mildew and *Cercospora* leaf spot are the two most important diseases currently affecting this crop. These diseases can reduce the production and quality of fenugreek crops significantly all across the globe. Other minor fungal diseases of fenugreek, namely collar rot, leaf spot, *Fusarium*

Fig. 19.5. Spread of powdery mildew infection on a fenugreek potted plant in the greenhouse at Lethbridge Research Center, Agriculture and Agri-Food Canada, Lethbridge, Alberta, Canada.

wilt, pod spot, spring black stem and leaf spot, and downy mildew, have the potential to become major fenugreek diseases in many areas, including subtemperate climatic zones. Use of resistant cultivars and application of suitable chemical agents are suggested by most research groups as potential control measures against infection and spread of fungal diseases. Although certain Internet sites do make widely optimistic claims about effective biological control of fenugreek fungal diseases, they do not have strong evidence from multi-location and multi-year trials to support their claims and so are not included in this review.

Fenugreek is being cultivated in many new areas as it becomes more widely recognized as a multiple-use crop. Development of new fenugreek cultivars and improvement of existing cultivars with disease resistance using conventional plant breeding methods (Acharya *et al.*, 2007b) and advanced plant biotechnological approaches (Laroche, 2007) could be a good strategy to prevent emergence of new fungal diseases for this crop. Susceptibility of a plant to disease is determined by the genetic relationship between the plant and the pathogen. The relationship between genes of the host and the pathogen can determine disease expression in the host. Genetic resistance in plants is considered a major form of biological control of disease and is possibly the most cost-effective and environmentally friendly way to control crop diseases. Resistant cultivars have been used effectively to control diseases in many crops. However, development of resistant cultivars takes time and so work should continue in the interim to find chemical and other biological control agents to protect the crop from disease and other pest damage. It should also be noted that disease control measures should not only be cost-effective but also need to be environment friendly and socially acceptable.

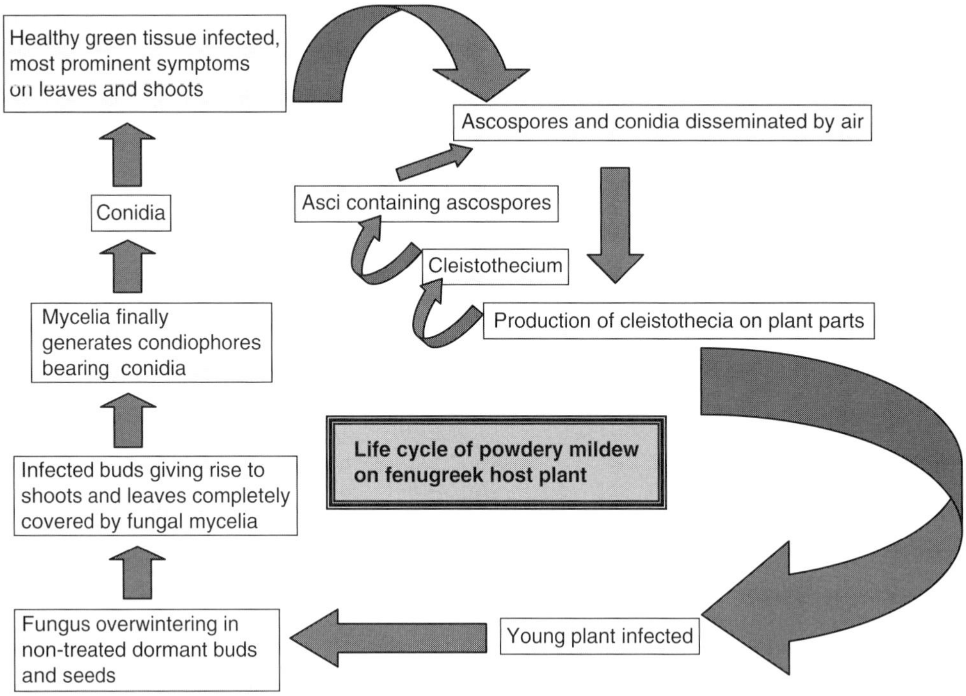

Fig. 19.6. The life cycle of *Erysiphe polygoni* on fenugreek host plant.

Acknowledgements

The authors express their sincere thanks to Mr Byron Lee, Research Technician, Electron Microscopy and Image Analysis Laboratory, Lethbridge Research Centre (LRC), Agriculture and Agri-Food Canada (AAFC) for his help in taking all the light microscopy and SEM images, Mr Doug Friebel, Technician, Forage Lab, AAFC, LRC for his help with field trials. The authors also extend their gratitude to the School of Graduate Studies, University of Lethbridge, Alberta Agriculture Research Institute (AARI) and AAFC matching grant initiatives for graduate student assistantships and project funding, respectively.

References

AAFRD (1998) Fenugreek. *Agrifax Agdex*. 147/20-25. Alberta Agriculture, Food and Rural Development, Alberta.

Acharya, S., Srichamoren, A., Basu, S., Ooraikul, B. and Basu, T. (2004) Improvement in the nutraceutical properties of fenugreek (*Trigonella foenum-graecum* L.). In: *Proceedings of the World Congress on Clinical Nutrition*, 10th, 30 November–3 December 2004, Phuket, Thailand. International College of Nutrition (ICN), Canada in collaboration with Prince Songlanakarin University, Phuket, Thailand, 45 pp.

Acharya, S., Srichamoren, A., Basu, S., Ooraikul, B. and Basu, T. (2006a) Improvement in the nutraceutical properties of fenugreek (*Trigonella foenum-graecum* L.). *Songklanakarin Journal of Science and Technology* 28(1), 1–9.

Acharya, S.N., Thomas, J.E. and Basu, S.K. (2006b) Fenugreek: an 'old world' crop for the 'new world'. *Biodiversity* 7, 27–30.

Acharya, S.N., Basu, S.K. and Thomas, J.E. (2007a) Medicinal properties of fenugreek (*Trigonella foenum-graecum* L.): a review of the evidence-based information. In: Acharya, S.N. and Thomas, J.E. (eds) *Advances in Medicinal Plant Research*. Research Signpost (Publisher), Trivandrum, India, pp. 81–122.

Acharya, S.N., Thomas, J.E. and Basu, S.K. (2007b) Improvement in the medicinal and nutritional properties of fenugreek (*Trigonella foenum-graecum* L.). In: Acharya, S.N. and Thomas, J.E. (eds) *Advances in Medicinal Plant Research*. Research Signpost (Publisher), Trivandrum, India, pp. 491–512.

Acharya, S.N., Blade, S.F., Mir, Z. and Moyer, J.R. (2007c) Tristar fenugreek. *Canadian Journal of Plant Science* 87, 901–903.

Acharya, S.N., Thomas, J.E. and Basu, S.K. (2008) Fenugreek, an alternative crop for semi-arid regions of North America. *Crop Science* 48, 841–853.

Adam, G.C. (1988) *Thanatephorus cucumeris* (*Rhizoctonia solani*) a species of wide host range. In: Sidhu, G.S. (ed.) *Advances in Plant Pathology, Volume 6. Genetics of Plant Pathogenic Fungi*. Academic Press, New York, pp. 535–552.

Agrios, G.N. (1997) *Plant Pathology*. 4th edn. Academic Press, San Diego, California.

Alexopoulos, C.J., Mims, C.W. and Blackwell, M. (1996) *Introductory Mycology*. 4th edn. John Wiley and Sons, New York, 869 pp.

Anderson, N.A. (1982) The genetics and pathology of *Rhizoctonia solani*. *Annual Review of Phytopathology* 20, 329–347.

Anonymous (1968) Review of applied mycology. In: *Plant Host–Pathogen Index. Volumes 1–40*. Commonwealth Mycological Institution, Kew, Surrey, UK.

Anonymous (1970) *Short-term leaflet* 60. Ministry of Agriculture, Fisheries and Food, USA.

Avtar, R., Rathi, A.S., Jatasra, D.S. and Joshi, U.N. (2003) Changes in phenolics and some oxidative enzymes in fenugreek leaves due to powdery mildew infection. *Acta Phytopathologica et Entomologica Hungarica* 38(3–4), 237–244.

Bansal, R.K. and Gupta, R.K. (2000) Evaluation of plant extracts against *Fusarium oxysporum*, wilt pathogen of fenugreek. *Indian Phytopathology* 53(1), 107–108.

Basch, E., Ulbricht, C., Kuo, G., Szapary, P. and Smith, M. (2003) Therapeutic application of fenugreek. *Alternative Medicine Review* 8, 20–27.

Basu, S., Acharya, S., Bandara, M. and Thomas, J. (2004) Agronomic and genetic approaches for improving seed quality and yield of fenugreek (*Trigonella foenum-graecum* L.) in western Canada. In: *Proceedings of Science of Changing Climates – Impact on Agriculture, Forestry and Wetlands*, 20–23 July 2004. University of Alberta, Edmonton, Alberta, Canada. Canadian Society of Agronomy, Canadian Society of Soil Sciences and Canadian Society of Animal Sciences (jointly), Edmonton, Alberta, Canada, 38 pp.

Basu, S.K., Acharya, S.N. and Thomas, J.E. (2006a) A report on powdery mildew infestations caused by *Erysiphe polygoni* D.C. in North American grown fenugreek. *Journal of Mycopathological Research* 44(2), 253–256.

Basu, S.K., Acharya, S.N., Cárcamo, H.A. and Thomas, J.E. (2006b) Study on the potential insect pests of fenugreek (*Trigonella foenum-graecum* L.) in North America with particular emphasis on the Western flower thrips (*Frankliniella occidentalis Pergande*) in the greenhouse and plant bugs (*Lygus* and *Adelphocoris*, Miridae, Hemiptera) in the field. *Journal of Environment and Sociobiology* 3, 1–7.

Basu, S.K., Acharya, S.N. and Thomas, J.E. (2008) Genetic improvement of fenugreek (*Trigonella foenum-graecum* L.) through EMS-induced mutation breeding for higher seed yield under western Canada prairie conditions. *Euphytica* 160, 249–258.

Bhaskar, R.B.L. and Summanwar, A.S. (1982) Physical properties and host range of math mosaic virus. *Indian Phytopathology* 35, 688–689.

Boerema, G.H., Gruyer, J. de, Noordeloos, M.E. and Hamers, M.E.C. (2004) *Phoma Identification Manual: Differentiation of Specific and Intraspecific Taxa in Culture*. CAB International, Wallingford, UK.

Borg, P. (1936) Report of the plant pathologist. *Report of the Institute of Agriculture, Malta* 35, 53–61.

Bretag, T.W. and Cunnington, J.H. (2005) First report of black stem and leaf spot in fenugreek (*Trigonella foenum-graecum*) caused by *Phoma pinodella* in Australia. *Australasian Plant Pathology* 34, 619–620.

Bretag, T.W., Keane, P.J. and Price, T.V. (2006) The epidemiology and control of ascochyta blight in field peas: a review. *Australian Journal of Agricultural Research* 57(8), 883–902.

Broca, C., Manteghetti, M., Gross, R., Baissac, Y., Jacob, M., Petit, P., Sauvaire, Y. and Ribes, G. (2000) 4-Hydroxyisoleucine: effects of synthetic and natural analogues on insulin secretion. *European Journal of Pharmacology* 390, 339–345.

Brunt, A. (1972) Data results. Official report to Bath University, Glasshouse Crops Research Institute, Virology Department, Littlehampton, UK.
Busbice, T.H., Hill, R.R. Jr and Carnahan, H.L. (1972) Genetics and breeding procedures. In: Hanson, C.H. (ed.) *Alfalfa Science and Technology.* ASA Monogr. 15. ASA, Madison, Wisconsin, pp. 283–319.
Cook, A.A. (1978) *Diseases of Tropical and Subtropical Vegetables and Other Plants.* Hafner Press, New York.
Datta, S. and Chatterjee, R. (2004) Performance of fenugreek (*Trigonella foenum-graecum* L.) genotypes under new alluvial zone of West Bengal. *Journal of Spices and Aromatic Crops* 13(2), 132–134.
El-Bazza, Z.E., Mahmoud, M.I. and Mohamed, Z.G. (1990) Production of aflatoxin by fungi isolated from some medicinal plants used as drinks in Egypt. *Egyptian Journal of Microbiology* 25, 323–334.
El-Gizawy, A.M., Al-Rokaibah, A.A. and Abdalla, M.Y. (2000) Efficacy of *Lupinus termis* and *Trigonella foenum-graecum* seed extracts on *Pythium* and *Rhizoctonia* diseases of cucumber, tomato and radish seedlings. *Agricultural Science (Riyadh)* 12(1), 51–59.
Fazli, F.R.Y. and Hardman, R. (1968) The spice fenugreek (*Trigonella foenum-graecum* L.). Its commercial varieties of seed as a source of diosgenin. *Tropical Science* 10, 66–78.
Fehr, W.R. (1993) *Principles of Cultivar Development: Theory and Technique.* Volume 1. Macmillan Publishing Company, New York.
Fogg, M.L., Kobayashi, D.Y., Johnston, S.A. and Kline, W.L. (2000) Bacterial leaf spot of fenugreek: a new disease in New Jersey caused by *Pseudomonas syringae* pv. *syringae.* Publication No. P-2001–0012-NEA. In: *Northeastern Division Meeting Abstracts*, 1–3 November 2000, Cape Cod. The American Phytopathological Society (APS), Cape Cod, North Falmouth, Massachusetts.
Glaeser, G. (1961) Common leaf spot an autumn disease of Lucerne. *Pflanzenarat* 14(10), 88–89.
Hagedorn, D.J. and Walker, J.C. (1949) Wisconsin pea streak. *Phytopathology* 39, 837–847.
Haque, S.P. and Ghaffar, A. (1992) Efficiency of *Trichoderma* sp. and *Rhizoctonia meliloti* in the control of root rot of fenugreek. *Pakistan Journal of Botany* 24(2), 217–221.
Hashmi, M.H. (1988) Seedborne mycofloras of *Trigonella foenum-graecum* L. *Pakistan Journal of Botany* 20, 233–237.
Hashmi, M.H. and Thrane, U. (1990) Mycotoxins and other secondary metabolites in spices of *Fusarium* isolated from seeds of capsicum, coriander and fenugreek. *Pakistan Journal of Botany* 21, 106–116.
HAU (Haryana Agricultural University) (2008) Research highlights. Varieties released/identified. Haryana Agricultural University, Hisar, India (http://hau.ernet.in/vegetable_crops.htm, accessed 1 December 2008).
Hiremath, P.C. and Prasad, C.K.P.S. (1985) Vertical distribution of *Rhizoctonia solani* in soil and effect of culture filtrate on the rhizosphere mycoflora of fenugreek. *Journal of Soil Biology and Ecology* 5(2), 126–128.
Hiremath, P.C., Anilkumar, T.B. and Sulodmath, V.V. (1976) Occurrence of collar rot of fenugreek in Karnataka, India. *Current Science* 45, 405.
Jones, J.P., Jones, J.B. and Miller. W. (1982) Fusarium wilt on tomato. In: *Plant Pathology Circular No. 237.* Florida Department of Agriculture and Consumer Services, Division of Plant Industry.
Jones, L.K. (1927) Studies of the nature and control of blight, leaf and pod spot, and foot-rot of peas caused by species of *Ascochyta. New York State Agricultural Experiment Station Bulletin* 547, 1–45.
Jongebloed, M. (2004) Coriander and fenugreek. In: Salvin, S., Bourke, M. and Byrne, T. (eds) *The New Crop Industries Handbook.* Rural Industries Research and Development Corporation (RIRDC), Australian Government, Australia, pp. 229–235.
Karimov, M.A. (1956) Survey of fungal diseases of Lucerne (*Medicago sativa*). *Not. Syst. Sect. Crypt. Inst. Sci., USSR* 11, 118–131.
Khare, M.N., Agrawal, S.C. and Sharma, N.D. (1981) A seed-borne fungus on *Trigonella foenum-graecum. Indian Phytopathology* 34, 71–72.
Komaraiah, M. and Reddy, S.M. (1986) Influence of humidity on seed deterioration of methi (*Trigonella foenum-graecum*) seeds by some seed-borne fungi. *Indian Journal of Mycology and Plant Pathology* 16, 77–79.
Lakra, B.S. (2002) Role of temperature and humidity in oospore formation of *Peronospora trigonella* causing downy mildew of fenugreek. *Plant Disease Reporter* 17(2), 339–340.
Lakra, B.S. (2003) Effect of date of sowing and crop geometry on downy mildew incidence and seed yield of fenugreek. *Forage Research* 29(2), 65–67.
Laroche, A. (2007) Biotechnology and medicinal plants. In: Acharya, S.N. and Thomas, J.E. (eds) *Advances in Medicinal Plant Research.* Research Signpost (Publisher), Trivandrum, India, pp. 339–356.

Latham, L.J. and Jones, R.A.C. (2001) Incidence of virus infection in experimental plots, commercial crops, and seed stocks of cool season crop legumes. *Australian Journal of Agricultural Research* 52, 397–413.

Leppik, E.E. (1959) World distribution of *Cercospora traversiana. FAO Plant Protection Bulletin* 8, 19–21.

Leppik, E.E. (1960) *Cercospora traversiana* and some other pathogens of fenugreek new to North America. *Plant Disease Reporter* 44(1), 40–44.

Lucy, M. (2004) Fenugreek (www.dpi.qld.gov.au/fieldcrops/9050.html, accessed 12 September 2007).

McAnuff, M.A., Omoruyi, F.O., Morrison, E.Y.S.A. and Asemota, H.N. (2002) Plasma and liver lipid distributions in streptozotocin-induced rats fed sapogenin extract of the Jamaican bitter yam (*Dioscorea polygonoides*). *Nutrition Research* 22, 1427–1434.

McCormick, K.M. and Hollaway, G.J. (1999) First report of bacterial blight in fenugreek (*Trigonella foenum-graecum*) caused by *Pseudomonas syringae* pv. *Syringae. Australasian Plant Pathology* 28, 338.

Nyvall, R.F. (1999) *Field Crop Diseases.* 3rd edn. Iowa State University Press, Ames, Iowa, pp. 3–787.

Ogoshi, A. (1987) Ecology and pathogenicity of anastomosis and intraspecific groups of *Rhizoctonia solani* Kunn. *Annual Review of Phytopathology* 25, 125–143.

Onfroy, C., Tivoli, B., Coubiere, R. and Bouznad, Z. (1999) Cultural, molecular and pathogenic variability of *Mycosphaerella pinodes* and *Phoma medicaginis* var. *pinodella* isolates from dried pea (*Pisum sativum*) in France. *Plant Pathology* 48, 218–229.

Palti, J. (1959) Oidiopsis diseases of vegetable and legume crops in Israel. *Plant Disease Reporter* 43(2), 221–226.

Petri, L. (1934) Review of phytopathological records noted in 1933. In: *Review of Applied Mycology,* Volume 13. Commonwealth Mycological Institute, Kew, Surrey, UK.

Petropoulos, G.A. (1973) Agronomic genetic and chemical studies of *Trigonella foenum-graecum* L. PhD thesis, Bath University, UK.

Petropoulos, G.A. (2002) *Fenugreek -The Genus Trigonella.* Taylor and Francis, London and New York, pp. 120–127.

Pierre, D. and Francis, R. (2000) *Detection and Isolation of Soil Fungi.* Science Publishers, Inc., Enfield, New Hampshire.

Pirone, P., Dodge, B. and Rickett, H. (1960) *Diseases and Pests of Ornamental Plants.* 3rd edn. Constable and Company Ltd., London.

Prabha, P. and Bohra, A. (1999) Seed mycoflora associated with some important spice seeds. *Advances in Plant Sciences* 12, 195–198.

Prakash, S. and Saharan, G.S. (2000) Conidial germination of *Erysiphe polygoni* causing powdery mildew of fenugreek. *Indian Phytopathology* 53(3), 318–332.

Prasad, C.K.P.S. and Hiremath, P.C. (1985) Varietal screening and chemical control of foot-rot and damping-off caused by *Rhizoctonia solani. Pesticides* 19(5), 34–36.

Punithalingam, E. and Gibson, I.A.S. (1976) *Phoma medicaginis* var. *pinodella. CMI Descriptions of Pathogenic Fungi and Bacteria.* No. 518. Commonwealth Mycological Institute, Kew, UK.

Puri, D., Prabhu, K.M. and Murthy, P.S. (2002) Mechanism of action of a hypoglycaemic principle isolated from fenugreek seeds. *Indian Journal of Physiology and Pharmacology* 46, 457–462.

Quantz, L. (1968) Leguminosen. In: Klinkowsky, M. (ed.) *Pflanzliche viologie,* II, 2. Akademie Verlag, Berlin.

Raian, F.S., Vedamuthu, P.G.B., Khader, M.P.A. and Jeyarajan, R. (1991) Management of root disease of fenugreek. *South Indian Horticulture* 39(4), 221–223.

Raju, J., Gupta, D., Rao, A.R., Yadav, P.K. and Baquer, N.Z. (2001) *Trigonella foenum-graecum* (fenugreek) seed powder improves glucose homeostatis in alloxan diabetic rat tissues by reversing the altered glycolytic, gluconeogenic and lipogenic enzymes. *Molecular and Cellular Biochemistry* 224, 45–51.

Rosengarten, F. (1969) *The Book of Spices.* Livingstone, Wynnewood, Pennsylvania.

Rouk, H.F. and Mangesha, H. (1963) Fenugreek (*Trigonella foenum-graecum* L.). Its relationship, geography and economic importance. *Experimental Station Bulletin of the Empirical Ethiopian Cooperative of Agriculture and Mechanical Arts,* No. 20.

Ryley, M.J. (1989) *Cercospora traversiana* on fenugreek (*Trigonella foenum-graecum*) in Queensland. *Australasian Plant Pathology* 18(3), 60–63.

Schmelzer, K. (1967) Hosts of potato virus Y and potato etch virus outside of Solanaceae. *Phytopathology* 60(4), 301–315.

Schmelzer, K. and Wolf, P. (1971) *Wirtspflanzen der Viren und Virosen Europas.* Barth verlag, Leipzig, Germany.

Sharma, P.D. (2005) *Fungi and Allied Organisms.* Alpha Science International Ltd, Oxford, UK.

Sharma, R.D., Sarkar, A., Hazra, D.K., Misra, B., Singh, J.B., Sharma, S.K., Maheshwari, B.B. and Maheshwari, P.K. (1996) Use of fenugreek seed powder in the management of non-insulin dependent diabetes mellitus. *Nutrition Research* 16(8), 1331–1339.

Sharma, S. (1999) Effect of sowing dates on powdery mildew on fenugreek. *Journal of Mycology and Plant Pathology* 29(1), 144–145.

Singh, R.P. (1969) Natural infection of fenugreek by two legume viruses. *Plant Disease Reporter* 53, 297–304.

Singh, R.S. (2001) *Plant Disease Management.* Science Publishers, Inc, Enfield, New Hampshire, pp. 127–137.

Sinskaya, E. (1961) Flora of the cultivated plants of the USSR. XIII. In: *Perennial Leguminous Plants. Part I: Medicago, Sweet Clover, Fenugreek.* Israel Programme for Scientific Translations, Jerusalem.

Smith, I.M., Dunez, J., Phillips, D.H., Lelliott, R.A. and Archer, S.A. (1988) *European Handbook of Plant Diseases.* Blackwell Scientific Publications, Oxford, UK, 583 pp.

Tayyaba, Z., Hasnian, S.N. and Hasan, S.K. (2001) Evalution of the oral hypoglycaemic effect of *Trigonella foenum-graecum* L. (methi) in normal mice. *Journal of Ethnopharmacoligy* 75, 191–195.

Tiran, D. (2003) The use of fenugreek for breast-feeding women. *Complementary Therapies in Nursing and Midwifery* 9, 155–156.

Tschen, J.S.M. (1987) Control of *Rhizoctonia solani* by *Bacillus subtilis*. *Transactions of the Mycological Society of Japan* 28, 483–493.

Tschen, J.S.M. and Kou, W.L. (1985) Antibiotic inhibition and control of *Rhizoctonia solani* by *Bacillus subtilis. Plant Protection Bulletin (Taiwan, R.O.C.)* 27, 93–101.

Ubrizsy, G. (1965) *Novenykortan II.* Akademiai Kiado, Budapest.

Vats, V., Grover, J.K. and Rathi, S.S. (2002) Evaluation of anti-hyper-glycaemic and hypoglycaemic effect of *Trigonella foenum-graecum* Linn, *Ocimum sanctum* L. and *Pterocarpus marsupium* L. in normal and alloxanized diabetic rats. *Journal of Ethnopharmacology* 72, 95–100.

Vats, V., Yadav, S.P. and Grover, J.K. (2003) Effect of *Trigonella foenum-graecum* on glycogen content of tissues and the key enzymes of carbohydrate metabolism. *Journal of Ethnopharmacology* 28, 1–6.

Vidamo, C. and Conti, M. (1965) Aphid transmission of a cowpea mosaic virus, isolated from cowpea in Italy. Atlantic Academy of Sciences, Torino, Italy, pp. 1041–1050.

Voros, J. and Nagy, F. (1972) *Cercospora traversiana* Sacc., a new destructive pathogen of fenugreek in Hungary. *Acta Phytopathologica Academiae Scientiarum Hungaricae* 7, 71–76.

Walker, J.C. (1952) *Diseases of Vegetable Crops.* McGraw-Hill Book Co, Inc, London.

Weiss, E.A. (2002) Fenugreek. In: *Spice Crops.* CAB International, Wallingford, UK, pp. 77–85.

Zia, T., Siddiqui, I.A., Shaukat, S.S. and Nazarul-Hasnain, S. (2003) *Trigonella foenum-graecum* (fenugreek)-mediated suppression of *Meloidogyne javanica* in mungbean. *Archives of Phytopathology and Plant Protection* 36(1), 23–31.

Zimmer, R.C. (1984) Cercospora leaf spot and powdery mildew of fenugreek, a potential new crop in Canada. *Canadian Plant Disease Survey* 64(2), 33–35.

20 Fungal Diseases of Oilseed Crops and Their Management

S.S. Adiver and Kumari
Oilseeds Scheme, Main Agricultural Research Station, University of Agricultural Sciences, Dharwad, Karnataka, India

Abstract
During the past few years, pathogens in oilseed crops have been recognized as major forces causing economic losses, with identification of certain important ones based on their symptoms, etiology and also ecological zones. Recent research has helped by developing new resistant varieties and other effective management strategies. This chapter describes the causal organisms, symptoms and management of diseases of oilseed crops like castor, groundnut, safflower, sesame and sunflower. Cultural practices for managing certain diseases have been pinpointed. Critical stages for growth of some foliar diseases, namely rust, early and late leaf spot of groundnut, blight and mildew of sunflower, fusarial wilt of safflower and castor, have been identified. Recommendations are given on controlling various diseases by chemical, botanical and other effective and eco-friendly methods. Oil is an essential household commodity required for food and daily use. Certain oils are used as therapeutic agents and are in much demand for their conversion into energy or potential biodiesel. Losses to the tune of 20% in certain oilseed crops need our utmost attention. Various fungal diseases of groundnut, sunflower, safflower, sesame and castor are described. Disease management with fungicides and other available methods are illustrated.

Groundnut

Groundnut, known as poor man's almond, contributes about 38% to the oilseed pool of India. India is the second largest producer of groundnut after China. The crop is subjected to attack by numerous pests and pathogens. Among foliar fungal diseases, early and late leaf spots, commonly called 'Tikka' disease, and rust are economically important.

Early leaf spot caused by *Cercospora arachidicola* Hori

The perfect stage of the fungus is *Mycosphaerella arachidis*. In India, losses in yield due to leaf spots have been estimated to be in the range of 15–59%. Besides the loss in pod and kernel yield, the value of fodder is also affected adversely. Lesions are subcircular in shape and measure 1 to over 10 mm. On the upper surface of the quadrifoliate

 Management of Fungal Plant Pathogens (eds A. Arya and A.E. Perelló)

leaves, the lesions appear dark brown, while on the lower surface they are a lighter shade of brown. The early leaf spot usually has a light to dark brown centre and a yellow halo. These are oval to elongate in shape and have more distinct margins than the late leaf spot lesions. The early leaf spot pathogen survives through conidia on affected plant debris in soil, or through conidia being carried on the pod shell.

Disease management

Tolerant varieties like GPBD-4, ICGV-86590, ICGS-44, M-335, BG-3 and M-522 can be grown wherever early leaf spot is severe. Intercropping pearl millet/sorghum with groundnut (1:3) is useful in reducing the intensity of early leaf spot. Foliar spraying of carbendazim (0.05%) + mancozeb (0.2%), chlorothalonil (0.2%), difenaconazole (0.1%), tebuconazole (0.1%) or hexaconazole (0.1%) is recommended 2–3 times at 2- to 3-week intervals starting from the initiation of the disease (Adiver *et al.*, 1995).

Late leaf spot caused by *Phaeoisariopsis personata* (Burk. and Curt.) Van Arx

The perfect stage of the fungus is *M. berkeleyii* W.A. Jenkins. Late leaf spot is more severe in the southern and central parts of India. Dark brown to black, circular to subcircular lesions, measuring 1–6 mm diameter appear on the lower surface of the quadrifoliate, where most sporulation occurs. The lesions are black in colour and fruiting structures occur in concentric rings on the lower leaflet surface, giving lesions a slightly rough appearance. The ambient temperature required is between 25 and 30°C. Prolonged leaf wetness hours and high relative humidity (> 80%) favour infection and disease development. Conidia are disseminated by the wind and insects, leading to secondary infection.

Disease management

Deep burying of crop residues in the soil and removal of volunteer groundnut plants are important measures in reducing the primary source of infection. Resistant/tolerant varieties like Girnar-1, RG-141, IGV-87160, ICGV-86590, ICGV-86325, R-8808, GPBD-4, Kadari-4, Co-3,4, M-335 and BG-3 can be grown wherever late leaf spot is severe. Foliar spraying of carbendazim (0.05%) + mancozeb (0.2%), chlorothalonil (0.2%), difenaconazole (0.1%) or hexaconazole (0.1%) is recommended 2–3 times at 2- to 3-week intervals, starting from the initiation of the disease (Adiver *et al.*, 1995).

Rust caused by *Puccinia arachidis* Speg.

Rust of groundnut is prevalent throughout India; however, it is more severe in the southern states. In India, losses in yield due to rust alone have been reported in the range of 10–52%, depending on the variety. Rust can be recognized readily as orange-coloured pustules (uredinia) that appear on the lower leaflet surface and rupture to expose masses of reddish-brown urediniospores. Pustules appear first on the lower surface and, in highly susceptible cultivars, the original pustules may be surrounded by colonies of secondary pustules.

Disease management

Early sowing in the first fortnight of June is recommended to avoid incidence. Use of resistant/tolerant varieties like Girnar-1, ICGV-87160, ICGV-86590, DRG-12, ALR-2,3, Co-4, ALR-1, ICGS-5 and DRG-17 is recommended. Spraying mancozeb (0.2%), tridemefon (0.1%), chlorothalonil (0.2%), difenaconazole (0.1%), tebuconazole (0.1%), hexaconazole (0.1%) or cyproconazole (0.1%) 2–3 times at 2- to 3-week intervals starting from the initiation of the disease helps to control the disease.

Seed and seedling diseases of groundnut

Pre-emergence seed rot and post-emergence seedling mortality are of common occurrence. The disease develops either from the fungi already present in the seed or result from direct invasion of seeds and seedlings by soil fungi. Among seedling diseases, collar rot, root rot and stem rot are of economic importance and are known to reduce yields by 25–50%.

Collar rot caused by *Aspergillus niger* van Tiegh

In India, collar rot, also known as crown rot or seedling blight, is prevalent in almost all groundnut-growing states, causing 28–50% losses. The diagnostic symptoms are pre-emergence rotting of seeds and rotting of hypocotyls, but the most common cause of loss is early post-emergence seedling blight. The first symptom in emerged seedlings is usually a rapid withering of the entire plant or its branches. Lesions develop on the stem below the soil and spread upwards along the branches. The dead and dried branches are easily detached from the disintegrated collar region.

Disease management

Avoiding deep sowing (not more than 5 cm), mixed cropping with moth bean in alternate rows, deep tillage and early sowing of crop is recommended. Soil application of neem cake/castor cake at 500 kg/ha, seed treatment with *Trichoderma harzianum*/*T. viride* at 4.0 g/kg seed, bacterization of groundnut seeds with strains of fluorescent pseudomonads or seed treatment with carbendazim (1.0 g/kg), mancozeb (2.0 g/kg seed) or chlorothalonil/captan (2.0 g/kg) is suggested.

Stem rot caused by *Sclerotium rolfsii* Sacc.

The stem rot pathogen has a very wide host range. In India, stem rot, also known as *Sclerotium* wilt, occurs in all groundnut-growing states and is particularly severe in Maharashtra and Gujarat. In India, 27% or more yield loss has been reported. *S. rolfsii* also causes indirect losses such as reduction in both dry weight and oil content of groundnut kernels. The first symptom is partial or complete wilting of the stem or branch that is in contact with the infected soil. The leaves turn brown and wilt, but remain attached to the plant. The pathogen has a wide host range. *S. rolfsii* can colonize either living plant tissue or plant debris. Deeply buried sclerotia survive a year or less, while those near the soil surface remain viable for many years. Disease development occurs when soil moisture is 40–50%. Generally, when the temperature remains between 29 and 32°C during the day and seldom drops below 25°C during the night, the disease develops more favourably.

Disease management

Deep ploughing, early sowing and close planting is recommended and rotation of groundnut with cotton, maize, sorghum and pearl millet. Seed treatment with *T. viride*/*T. harzianum* at 4.0 g/kg seed or seed treatment with carbendazim/captan at 2–3 g/kg seed is suggested.

Sunflower

Sunflower (*Helianthus annuus* var. *macrocarpus* (DC) Cockerell) is an important edible oilseed crop. It belongs to the family *Asteraceae*. The sunflower head is composed of about 1000–2000 individual flowers. The fertile disc florets bear the seed, which is white, black or striped grey and black. The seeds contain 40–50% oil and 50–55% meal, which contain high protein (35%), calcium, phosphorus, iron, potassium and vitamin E. The sunflower is a native of North America, where it is used in dyes, food preparations and medicines.

Alternaria leaf spots and blight of sunflower

Different species of *Alternaria*, namely *A. alternata*, *A. helianthi*, *A. zinniae*, *A. helianthiacola*, *A. leuconthemi* and *A. tenuissima* have been reported to cause the disease. Among these, *A. helianthi* (Hansf.) Tubaki and Nishihara are economically important.

The disease has been reported to cause a huge grain yield loss in Australia, where yield potential of 1.25 t/ha of the crop was reduced to 0.1 t/ha (Allen *et al.*, 1981). In Karnataka, India, the disease occurred in epidemic form in 1987, with a disease incidence of 95–100% (Hiremath *et al.*, 1990). The disease is caused by *A. helianthi* (Hansf.) Tubaki and Nishihara. The mycelium of the fungus is septate, rarely branched, brown and 2.5–5.0 μm in breadth. Conidiospores are cylindrical and yellow to black grey, with one to 11 transverse septa and a few longitudinal septa. The conidia measure in the range of 40–110 × 8–28 μm, with an average of 74 × 19 μm.

Disease management

Summer deep ploughing reduces the inoculum level in the soil, which is present in plant debris as dormant mycelium; altering the date of sowing in order to reduce disease pressure and sowing during August–September onwards is suggested. Following spacing 60 × 30 cm under irrigated and 45 × 20 cm under rainfed conditions is recommended. Use of resistant varieties like GP-145, AH-303, BSH-1 is suggested. Foliar sprays of mancozeb (0.2%), chlorothalonil (0.2%), difenaconazole (0.1%) or tebuconazole (0.1%) can prevent the crop from *Alternaria* blight.

Rust of sunflower

The pathogen *Puccinia helianthi* Schw. is a macrocyclic, autoecious fungus and it produces all the five stages on sunflower only. The disease has been reported to cause variable yield losses in the crop, depending on variety, environmental conditions and time of the outbreak of the disease in the crop season. Early infection of the variety 'Sunrise' and S-37-338 showed 17% and 68% less yield, respectively. Under field conditions, the disease usually starts at flowering stage, when the crop attains a dense canopy. The disease appears in the form of small cinnamon brown-coloured uredia on the lower surface of the lower leaves. In severe conditions, younger leaves, stems, petioles and floral parts are also infected. When the crop reaches physiological maturity, most of the uredia are converted to telia and are dark brown in colour.

Disease management

Altering the date of sowing reduces disease pressure. Removal of self-sown plants, crop rotation for at least 3 years and deep summer ploughing are recommended to reduce the inoculum level in the soil. Use of resistant varieties like SH-41, SH-187, PH-1, 2, 3, 4, 7 and 8, ICI-306, 331, PAC-36, 9128 and systemic fungicides containing triazoles, namely hexaconazole and cyproconazole (0.1%), are found suitable under field conditions.

Downy mildew of sunflower

Downy mildew causes heavy yield losses in sunflower-growing countries of the world. A serious outbreak (80–90%) of the disease was recorded in the Red River Valley of North Dakota and Minnesota (USA) during 1970, resulting in a reduction of about 50% yield, with a loss of about US$0.5m. Later, it spread to many European countries, then to Asia. This spread was mainly through the seed trade. In India, the disease first appeared during 1984, in experimental plots of the Regional Research Station, Latur, particularly during September–October. Later, it spread to many areas of Maharashtra (Mayee, 1989), Karnataka and Madhya Pradesh (Agarwal *et al.*, 1991). Causal organisms are *Plasmopara halstedii*, *P. perennis* and *P. patens*. The sporangiophores, measuring 150–750 μm, are monopodially branched almost at right angles and bear zoosporangia singly at the tips of the branches. Zoosporangia produced from leaves are elliptical with an apical papilla and measure 17–30 × 15–21 μm. The zoosporangia from roots are uniform, pyriform to oval with 1–3 papillae and 36–66 × 39–40 µm. The sporangia germinate at 5–28°C, with the optimum temperature being 16–18°C. Zoosporangia formed at 27°C show 86–95% germination. Oospores are formed in the intercellular spaces of roots, stem and seeds and measure 27–32 µm. The fungus causes damping off, systemic infection and local lesions on leaves and basal root or stem galls, depending on the stage of infection during plant growth. Damping-off occurs either as pre- or post-emergence under damp and cool weather at seedling stage and gives poor plant stand (Goosen and Sackston, 1964). Systemically infected plants remain stunted with chlorotic leaves.

Disease management

Regulatory measures: the pathogen is seedborne and exhibits races; therefore, a quarantine measure has been imposed to check the movement of virulent races from endemic areas to other countries. Use of the resistant variety LDMRSH-1 and seed treatment with metalaxyl MZ 72 WP at 5–6 g/kg or apron 35 SD at 5–6 g/kg seed is suggested.

***Rhizopus* head rot of sunflower**

The disease is caused by three different *Rhizopus* spp., namely *R. nigricans*, *R. arrhizus* and *R. oryzae*. The fungal colony is cottony-white to brown in *R. arrhizus*, while it is cottony-white turning brownish-grey to blackish-grey in *R. oryzae* and *R. nigricans*. The optimum temperature for the growth of *R. arrhizus*, *R. oryzae* and *R. nigricans* is reported to be 37°C (thermophyllic), 30°C and 22°C, respectively. The disease causes severe yield losses, particularly in wet weather conditions. The disease has no effect on seed size but it reduces seed weight. Affected seeds become scurfy with discoloration of the hull and partial to complete discoloration of the nut meal and the quality of the oil is affected because of off-flavours. The disease first appears as brown, water-soaked irregular spots on the back of the ripening head, usually adjacent to the flower stalk. The spots enlarge and turn soft and pulpy, with superficial whitish to blackish mycelium on the head. Under severe conditions, rotting spreads to the flower stalk and the head drops off. Sometimes, the seeds from the rotted head shed and those that remain on the head have a bitter taste.

Disease management

Management of insects by spraying endosulphan or diazinon at the onset of bloom and spraying of fungicide, i.e. carbendazim (0.1%), on completion of the flowering stage is effective in controlling the disease.

Safflower

Safflower (*Carthamus tinctorius* L.) is one of the *rabi* season oilseed crops cultivated in medium to heavy textured soils, mainly in Maharashtra, Karnataka and Andhra Pradesh, India. Being a crop mostly of the poor smallholder, it came to be recognized as an edible oilseed crop because of its superior role over animal fats and other vegetable oils, resulting in a boom in the cultivated area under the crop.

***Alternaria* leaf blight of safflower**

Leaf blight caused by *A. carthami* Choudhary is the most destructive disease of safflower in India, appearing in a severe form wherever the crop is grown and causing up to 90% reduction in crop yield and oil content of affected seeds. However, the pathogen is reported to increase significantly the level of free fatty acids in the seeds (Heaton *et al.*, 1978). Mycelium of the pathogen *A. carthami* is septate, inter and intracellular and dark coloured on maturity. Conidiophores are septate, unbranched, erect and brown to olivaceous brown, pale near the apex, measuring 15–85 µm × 6–10 µm, arising through the epidermis or stomata singly or in clusters. Conidia are light brown to translucent in shade, with/without a long beak, showing constrictions at the septa and borne singly or in short chains. The disease appears in seedlings on hypocotyls and on cotyledons as dark necrotic lesions up to 5 mm in diameter, which may sometimes result in damping-off. Spots, having concentric rings up to 2 cm in diameter, light to dark brown with the centre lighter in colour, are observed in mature plants on leaves and frequently coalesce into large irregular lesions.

Disease management

The disease can be managed by using seeds from early sown dry land crops and treating them with thiram and TPTH, captan 0.3% (Siddaramaiah *et al.*, 1980). Hot water treatment at 50°C for 30 min is also found useful (Sastry, 1996). Bulb extract (1.0% w/v) of *Allium sativum* also shows promise in checking the disease. Varieties like EC-32012, JLA-1753, C-2603, Co-1, C75-7218, HUS-524, 476, 305, 260, SSF-112, CTC-251, 248, 252, etc., are reported to exhibit a variable degree of tolerance to *A. carthami* infection.

***Fusarium* wilt of safflower**

Wilt of safflower is caused by *Fusarium oxysporum* f.sp. *carthami*. Fusarial mycotoxins capable of causing mycotoxicoses have been reported as being produced in sufficient quantities on infested seeds of safflower in storage (Ghosal *et al.*, 1977). The disease manifests at all growth stages. It may cause pre-emergence death or delayed germination of seeds. Symptoms on seedlings during post-emergence are blackening at the collar region; chlorotic, small brown spots appear on cotyledonary leaves, which then shrivel, become brittle, sometimes get rolled and droop downwards; finally, the seedlings bend and die. Plants grown from infected seeds rarely survive beyond the seedling stage. In mature plants, lateral branches on one side may be killed, while the other half of the plant shows no disease symptoms. Such plants show partial recovery, but symptoms may reappear later. Sporodochial production on stems may also be visible. Flower head size is reduced in severely affected plants, less seeds are formed and many of them are small, distorted, black and chaffy (Chakrabarti, 1980; Sastry and Jayaraman, 1993). The mycelium of the pathogen *F. oxysporum* f.sp. *carthami* Klisiewicz and Houston is septate and branched conidia, straight or curved, often pointed at the tip with a rounded base and measure up to 10–36 μm × 3–6 μm. Microconidia are oval to elliptical, one-celled and measure up to 5–16 μm × 2.2–3.5 μm in size. Chlamydospores are single celled, smooth, faintly coloured, single or in chains and 5–13 μm × 10 μm in size. Four biotypes of the pathogen have been identified on the basis of the reaction of safflower differentials to its isolates (Sastry and Chattopadhyay, 1999).

Disease management

Seed treatment with carbendazim (1.0 g/kg), captan (2.0 g/kg), thiram (2.0 g/kg) or *Trichoderma* (4.0 g/kg) helps to avoid infection of the plant by the pathogen. Crop rotation with legumes like chickpea, cowpea and pigeon pea helps to manage the disease (Sastry and Jayaraman, 1993). Use of tolerant varieties HUS-3234, 3123, 305, BSF-3, CTV-53, etc., is recommended.

***Phytophthora* root rot of safflower**

The mycelium of the pathogen *Phytophthora drechsleri* is hyaline, aseptate, branched and 4.5 μm wide. Sporangia are hyaline to faint in colour, thin-walled, non-papillate, pyriform to ovate, 34–38 μm × 15–24 μm in size and having zoospores measuring 10–20 μm in diameter. Oospores are spherical, smooth, thick-walled, yellow to bright brown and are 16–45 μm in diameter (Klisiewicz, 1977).

Root rot of Safflower caused by *P. drechsleri* Tuck is reported to cause about 3% losses on average, although 80% losses have been observed in a few instances, particularly when grown under surface irrigation (Sastry, 1996). Safflower is affected by *Phytophthora* root rot at any stage from pre-emergence to maturity. Symptoms on seedlings of 2–3 weeks of age appear as water-soaked lesions with softening and collapse of cortical tissue of the lower stem, whereon the plants lodge, shrivel and die.

Disease management

Draining out excess water from beds after irrigation and avoidance of monocropping may help to control the disease (Kolte, 1965). Use of resistant varieties US-10, Gila, Frio and VFR-1 is recommended.

Rust of safflower

Uredosori of the obligate, autoecious, heterothallic, macrocyclic pathogenic fungus *P. carthami* (Hutz) Corda contain numerous globoid or broadly ellipsoid echinulate, light chestnut brown uredospores measuring 21–27 μm × 21–24 μm, thick-walled and 3–4 equatorial germpores (Singh, 1998). Safflower rust caused by *P. carthami*, is an important disease in India. It causes a stand loss of 55–97% in susceptible varieties with considerable yield loss, particularly if the infection starts early in the crop growth. The first pathological phase of the safflower rust is seen in the seedling stage of the crop, when orange to yellow spots representing pycnia appear on cotyledons, which ultimately leads to drooping and wilting of the plants. With the development of uredospores and teliospores, the colour of the spots later changes to brownish black. The second pathological phase of the rust is uredia development on leaves, flowers and fruits, where teliospores are formed later towards crop maturity when the atmospheric temperature rises (Schuster and Christiansen, 1952).

Disease management

Destruction of the infected host, crop debris and collateral host *Carthamus oxycantha* (Pohli weed) and crop rotation checks the disease to some extent. Three sprays of tridemorph (0.5%), thiophanate methyl (0.15%) or tridimefon (0.1%) are effective against safflower rust (Singh *et al.*, 1997). Use of resistant varieties APPR-1 and APPR-3 is suggested.

Sesamum

Sesame (*Sesamum indicum* L.) is an important oilseed crop of India. It is grown mainly in light soils during *kharif* and in heavy soils during the *early rabi* season. It occupies an area of 17.50 hundred thousand ha in India, with production of 587.1 thousand t. The overall productivity of this crop in India is 335 kg/ha. About 72 fungi have been reported on this plant in India (Vyas *et al.*, 1984).

Phytophthora blight of sesame

Phytophthora blight is caused by *P. parasitica* var. *sesame.* It was first reported from India by Butler (1918). Now, it has become an important disease of sesame and has been reported from the Dominican Republic (Ciferri, 1930) and Argentina (Frezzi, 1950). In India, it was severe in Madhya Pradesh Rajasthan, Uttar Pradesh and Gujarat (Vasudeva, 1961; Verma, 2002). This disease has caused 66% losses in Gujarat (Kale and Prasad, 1957) and 79.8% in Central Madhya Pradesh (Singh *et al.*, 1976). It may cause even 100% loss under the most favourable conditions for infection to occur severely at seedling stage. Disease occurs on all the aerial plant parts. The symptoms of the disease appear as brown, water-soaked spots on the leaves of seedlings at a very early stage. Gradually, the spots increase in size. Under favourable weather conditions, the whole leaf rots and becomes black. Rotting progresses further and the whole stem is rotted. Frequently, the attack on the seedling starts at the collar region and gives damping-off like symptoms. The cottony-white growth of the fungal mycelia appears on the lower side of the leaves and on pods under humid condition.

Disease management

Intercropping with soybean, castor, maize, sorghum and pearl millet in the ratio of 1:3 or 3:1 shows a low incidence of the disease, with a higher yield. Application of FYM alone or neem cake with inorganic fertilizer (N_{60}, P_{40}, K_{20}) reduces the disease as compared to without FYM. Application of the species of *Pseudomonas*, *Bacillus* and *Streptomyces*, which are most active at 25–27°C at field capacity moisture level, can be suppressive to *Phytophthora* species in soil. Seed treatment with vitavax (1.0 g/kg) and captan (2.0 g/kg) controls seedling disease effectively. Captan 75D is the best fungicide for reducing the disease, followed by thiram 75D.

Fusarium wilt of sesame

Fusarium wilt of sesame is quite serious wherever the crop is grown. In India, it has been reported from all the sesame-growing areas, such as Madhya Pradesh, Maharashtra, Andra Pradesh, Rajasthan, Haryana, Punjab, etc. The disease is quite serious when it starts in the early stages of crop growth. The causal organism is *F. oxysporum* f.sp. *sesami*. The fungus produces profuse light pink mycelial growth on PDA. Microconidia are hyaline, ovoid to ellipsoid, unicellular and produce abundantly even on the medium and are about 8.5×3.25 μm in size. The macroconidia are produced abundantly in sporodochia and size ranges from 35 to 49×4.5 μm. The chlamydospores are globose to subglobose, smooth or wrinkled and about 7–16 μm in diameter. The pathogen grows at a temperature range of 10–25°C, with an optimum temperature of 26°C and a pH of 5.6. The initial symptoms of the disease appear as yellowing of the leaves, which later droop and desiccate. On the infected plant, the leaves may show inward rolling of the edge and eventually may dry up. If the disease appears at the later stages of crop growth, the symptoms may appear on one side of the plant, resulting in partial wilting. Discoloration of the vascular system is conspicuous in the roots.

Disease management

Seed treatment with benlate (1.0 g/kg) and vitavax (1.0 g/kg) is most effective against wilt. Application of conidial dust of *Gliocladium virens* gave better disease control. Similarly, application of *T. harzianum* and *T. viride* in the field also reduced the incidence of wilt significantly. Soil drenching with antibiotic KB-8A isolated from *B. polymyxa* at a concentration of 13 μm/ml inhibited *F. oxysporum* f.sp. *sesami* completely (Hyun *et al.*, 1999).

Alternaria leaf spot of sesame

The pathogen is *A. sesame* (Kawamura) Mohanty and Behera. The conidiophores of the pathogen are pale brown, cylindrical, erect, not rigid and arise singly with a size of $30–54 \times 4–7$ μm. Conidiophores produce conidia at the apex, which are in chains of one to two. The conidia are straight or slightly curved, obclavate, yellowish brown to dark brown in colour and measure $30–120 \times 9–30$ μm. The disease affects all the aboveground plant parts. The initial symptoms appear as small, brown, round to irregular spots on the leaf blade. Later, the spots enlarge and turn dark with concentric rings. On the lower surface of the leaves, spots are light brown in colour. The appearance of the disease at the seedling stage can cause post-emergence damping-off. On capsules, small, brown spots appear which result in the formation of shrivelled and deformed seeds.

Disease management

Application of Bordeaux mixture (0.1%) and zineb (0.1%) has been reported to be effective. Application of mancozeb (0.2%) at the time of disease initiation is effective in managing the disease.

Powdery mildew of sesame

This disease is common, especially in South India. It has been reported that powdery mildew of sesame is caused by *Oidium erysiphoides*, *Leveillula taurica* (Lav.) Trnaud, *Sphaerotheca fuliginea* (Schlecht) Pollacci and *Erysiphe cichoracearum* DC. The disease causes considerable losses in yield, depending on the time of its appearance, as well as the intensity of the disease. Powdery mildew causes a loss of 42%; every 1% increase in disease intensity results in a yield loss of 5.63 kg/ha. Four different fungi have been reported to cause powdery mildew, but in India *E. cichoracearum* is predominantly prevalent. Both conidia and ascospores on germination give rise to an abundant superficial mycelium of uninucleate cells, which form a white coating on the leaf and send haustoria into the host. The disease normally appears after 45–60 days. The initial symptoms appear as dirty whitish fungal patches on the upper surface of the leaves. Later, these specks coalesce to cover the entire leaf and result in premature defoliation. Generally, it affects the leaves but in severe cases, the disease spreads to petioles and other plant parts. In severe infection, pods or capsules are shrivelled and produce smaller seeds.

Disease management

Two sprays of wettable sulphur (0.3%), dinocap (0.1%) or hexaconazole (0.1%) at 15-day intervals can help to control the disease.

***Cercospora* leaf spot/white leaf spot of sesame**

Mycelium of *Cercospora sesami* Zimmerman is yellowish-white in colour and produces profuse conidiophores in culture. The conidiophores are olivaceous, septate, usually single but sometimes up to 10, epiphyllous, nodulase, thickened towards the tip, conidia with 7–10 septa and measure about 90–135 × 3–4 μm. Generally, the symptom of the disease appears at the time of flowering, but the disease may also appear after 30–40 days after sowing. The initial symptoms of the disease are circular spots scattered on both leaf surfaces. These spots enlarge rapidly and become up to 5 mm in diameter. The spots are initially brown in colour with a whitish centre, but later they may be brown to dark brown in colour. The symptoms on petioles are visible as elongated lesions, whereas on capsules they are more or less circular and brown to dark brown in colour.

Disease management

Three sprays either of carbendazim (0.05%) and topsin M-70 (0.2%) at 10-day intervals were best in controlling the disease. Resistant varieties recommended are BIC-7-2, Sidhi-54, Rewa-114 and Seoni Malwa.

Castor

Castor (*Ricinus communis* L.), belonging to the family *Euphorbiaceae*, is the most important non-edible oilseed crop of arid and semi-arid regions of India. Castor oil finds its application in the manufacture of a wide range of ever expanding industrial products, such as nylon fibres, jet engine lubricants, hydraulic fluids, dyes, detergents, soaps, ointment, greases, paints, varnishes, cosmetics and perfumes, etc. (Pathak, 2003).

Castor is grown in tropical and subtropical climates; the major growing countries are India, China and Brazil. India occupies about 57% of the world castor acreage, but produces about 62% of world production. The major castor-growing states in India are Gujarat, Andhra Pradesh, Tamil Nadu and Orissa. Productivity is highest in Gujarat state because more that 90% of the cultivated area is covered by castor hybrids under irrigation. There are a number of diseases occurring on castor and the important ones are explained below.

***Alternaria* blight of castor**

This is caused by *A. carthami* (Yoshii) Hansford. The disease appears on leaves, stem, inflorescence and capsules. At seedling stage, light brown spots first appear on cotyledonary leaves, which become angular with age. Severe infection results in the death of young seedlings or foliar blight. Symptoms on adult plant leaves are brown, zonate and variable in size and usually surrounded by yellow halos. In the case of severe infection, premature defoliation occurs. Sunken spots develop on capsules on one side, which gradually enlarge to cover the whole capsule with fungal growth. Such capsules are smaller in size and have underdeveloped or wrinkled seeds with little oil content. In heavily infected field crop, all the young racemes and even flower primordia are killed.

Disease management

Foliar application of mancozeb (0.2%) at intervals of 15 days starting from the appearance of the disease is beneficial. Judicious use of nitrogenous fertilizers also reduces the development of the disease.

Botrytis **grey rot of castor**

This is a very serious disease of castor as it affects the flowers and capsules directly and the entire crop may be lost if there are continuous rains during capsule formation. The disease is confined to only a few states in India and is serious in Andhra Pradesh and Tamil Nadu. It is caused by *Botrytis ricini* Godfrey. The disease is confined to spikes or racemes. Generally, pale to olive grey coloured woolly growth of the fungus is observed on flowers or capsules. The disease appears initially as small blackish spots, exuding a drop of yellow liquid. Fungal infection from these spots further spreads to racemes. The infected flowers appear soft due to the profuse growth and sporulation of the pathogen. This later turns to grey masses covered with dusty powder, resulting in the rotting of capsules. The unripe seed becomes soft and mature ones hollow, resulting in a discoloured seed coat and loss in seed weight.

Disease management

Adoption of wider spacing with varieties having open racemes reduces the severity of the disease. Two prophylactic sprays of carbendazim (0.05%), one at 50% flowering and the other soon after the appearance of the disease, reduces incidence of the disease effectively.

Macrophomina **root rot of castor**

Macrophomina phaseolina (Tassi) Goid is reported to cause different symptoms on castor, namely seedling blight, dieback due to aerial infection, spike blight, stem blight, collar rot, root rot and twig blight. The disease appears at different phases as collar rot, stem blight and root rot. Initially, the infected plant shows signs of water shortage. Within a week, the leaves and petiole droop and finally, within a fortnight, the entire plant dries up and can be pulled up easily. Collar rot phase is observed 30–40 days after sowing. Dark black discolorations are seen at the collar region of the plant, which gets sunken and later becomes abnormal. The affected tissue becomes shredded and weak and finally shows sign of wilting. Stem blight symptoms appear slightly later, due to aerial infection, as straw-coloured or brown depressed small lesions on the stem, usually at the nodes. The lesions increase in size by both upward and downward extension of the infection, resulting in a 2–20 cm oval-shaped necrotic area. The surface of the infected stem shrinks at this region and the plant breaks easily at this point. The affected spikes are discoloured, turn black and dry up in the course of time. Infected capsules become discoloured and drop off easily. In the case of the root-rot phase, the taproot shows signs of drying and the root bark shreds off easily. Rotting sometimes spreads partly above the ground. At an advanced stage, sclerotial bodies may be seen as minute black dots on the surface of woody tissues and in the pith region.

Disease management

Crop rotation with non-host crops and mixed cropping with moth bean can be helpful in reducing the disease. Infected plant material should be collected and burnt. Application of thiram (2.0 g/kg) or carbendazim (1.0 g/kg) as seed dresser along with spray and soil drench is recommended. Topsin M-70 has also been found effective for controlling root-rot disease in castor.

Wilt of castor

Wilt of castor is caused by *F. oxysporum* f. sp. *ricini* Nanda and Prasad. The extent of disease incidence has been up to 80% in

Russia (Moshkin, 1986). Losses in yield were realized in all cultivated castor hybrids in Gujarat and up to 85% incidence of the disease has been reported in North Gujarat (Dange *et al.*, 1997). Young seedlings at the two- to three-leaf stage exhibit discoloration of hypocotyls and loss of turgidity, with or without change in colour. The mycelium penetrates the vascular system of the roots, stems and leaves causing necrosis, which leads to wilting and finally death of the plant. At the time of flowering and spike formation stages, the disease is characterized by a gradual yellowing and shrivelling, with marginal and interveinal necrosis of leaves. Infected plants rarely bear seeds and such seeds are deformed and light in weight. Roots of wilted plants show blackening and necrosis, while in the case of partial wilted plants, only one side of the root system is observed as being blackish and necrotic; the other side of the root system remains healthy. When the stem of the wilted plant is split open, a white cottony fungal growth is observed in the pith region, which then becomes blackish.

Disease management

Use of healthy seeds, crop rotation, summer deep ploughing and field sanitation reduce the incidence of the disease. Use of bio-agents like *T. harzianum* and *T. viride* have been screened for their antagonistic activity against castor wilt pathogen. Seed treatment (1.0 g/kg) and pre-sowing soil application of carbendazim at 3.0 kg a.i./ha with thiram (3.0 g/kg seed) is recommended.

References

Adiver, S.S., Anahosur, K.H. and Giri Raj, K. (1995) Triazole for control of foliar diseases of groundnut (*Arachis hypogaea* L.). *Karnataka Journal of Agricultural Science* 8(1), 65–68.

Agarwal, S.C., Gupta, R.K. and Prasad, K.U.V. (1991) A case of downy mildew of sunflower in Madhya Pradesh. *Journal of Oilseeds Research* 8, 12–13.

Allen, S.J., Kochman, J.K. and Brown, J.F. (1981) Losses in sunflower yield caused by *Alternaria helianthi* in Southern Queensland. *Australian Journal of Experimental Agriculture and Animal Husbndry* 21, 98–100.

Butler, E.J. (1918) *Fungi and Diseases in Plants.* Thacker Sprink and Co., Calcutta, India, 547 pp.

Chakrabarti, D.K. (1980) Studies on the *Fusarium* wilt of safflower incited by *Fusarium oxysporum* f. sp. *carthami. Indian National Science Academy* B 46, 120–121.

Ciferri, R. (1930) Phytopathological survey of Santo Damingo, 1925–1929. *Journal of the Department of Agriculture of Porto Rico* 14, 5–44.

Dange, S.R.S., Desai, A.G. and Patel, D.B. (1997) Management of wilt of castor in Gujarat State of India. In: *Proceedings of the International Conference on Integrated Plant Disease Management for Sustainable Agriculture*, 10–15 November 1997. IARI, New Delhi, 107 pp.

Frezzi, M.J. (1950) The species of *Phytophthora* in Argentina. *Revista de Investigaciones Agricolas Buenos Aires* 4, 47–133.

Ghosal, S., Biswas, K., Chakrabarti, D.K. and Basuchoudhary, K.C. (1977) Control of *Fusarium* wilt of safflower by mangiferin. *Phytopathology* 67, 548–550.

Goosen, P.G. and Sackston, W.E. (1964) Biology of *Plasmopara halstedii* on sunflowers. *Proceedings of the Canadian Phytopathological Society* 31, 12 (Abst.).

Heaton, T.C., Knowles, P.F., Mikkelsen, D.S. and Ruckman, J.E. (1978) Production of free fatty acids in safflower seeds by fungi. *Journal of the American Oil Chemists Society* 55, 465–468.

Hiremath, P.C., Kulkarni, M.S. and Lokesh, M.S. (1990) An epiphytotics of *Alternaria* blight of sunflower in Karnataka. *Journal of Agricultural Science* 3, 277–278.

Hyun, J.W., Kim, Y.H., Lee, Y.S. and Park, W.M. (1999) Isolation and evaluation of protective effect against *Fusarium* wilt of sesame plants of antibiotic substance from *Bacillus polymyxa* KB-8. *Plant Pathology* 15, 152–157.

Kale, G.B. and Prasad, N. (1957) *Phytophthora* blight of sesamum. *Indian Phytopathology* 10, 38–47.

Klisiewicz, J.M. (1977) Identity and relative virulence of some heterothallic *Phytophthora* species associated with root and stem rot of safflower. *Phytopathology* 67, 1174–1177.

Kolte, S.J. (1965) *Diseases of Annual Edible Oilseed Crops. Volume III: Sunflower, Safflower, and Niger Seed Diseases*. CRC Press, Inc, Florida, 135 pp.
Mayee, C.D. (1989) Downy mildew of sunflower in India: problem and approaches. In: Raychaudhari, S.P. and Verma, J.P. (eds) *Review of Tropical Plant Pathology* Vol 5, 181–192.
Moshkin, V.A. (1986) *Castor.* Amarind Publishing Co Pvt Ltd, New Delhi.
Pathak, H.C. (2003) Emerging trends in castor seed development. In: *Proceedings of National Seminar on Castor Seed, Castor Oil and Its Value Added Products*, 22nd May 2003. Solvent Extract Association of India, Ahmedabad, India, pp. 54–62.
Sastry, K.R. (1996) Symptoms of wilt disease – clues for use in resistance breeding. In: Hegde, D.M., Raghavaiah, C.V. and Patil, D. (eds) *Proceedings of Training Programme on Breeding Approaches for Improving Productivity of Safflower and Group Meeting as Heterosis Breeding in Safflower.* Directorate of Oilseeds Research, Hyderabad, India, pp. 25–32.
Sastry, K.R. and Chattopdhyay, C. (1999) Development of *Fusarium* wilt resistant genotype in safflower. *Journal of Mycology and Plant Pathology* 29, 276–277.
Sastry, K.R. and Jayaraman, J. (1993) Eradication of *Fusarium oxysporum* f. sp. *carthami* from heavily infected safflower seed. *Journal of Oilseeds Research* 10, 277–281.
Schuster, M.L. and Christiansen, D.W. (1952) A foot and root disease of safflower caused by *Puccinia carthami. Phytopathology* 42, 211–212.
Siddaramaiah, A.L., Desai, S.A., Bhat, R. and Hegde, R.K. (1980) Eradication of *Alternaria carthami* Chowdhary, a seed borne pathogen of safflower. *Pesticides* 14, 22–23.
Singh, B.P. Shukla, B.N. and Kaushal, P.K. (1976) Evaluation of sesamum varieties for their susceptibility to *Phytophthora parasitica* Dastur at Jabalpur, (M.P.). *Jawaharlal Nehru Krishi Vishwa Vidyalaya Research Journal* 10, 76–77.
Singh, R. (1998) Spore stages and disease cycle of safflower rust (*Puccinia carthami*). *Journal of Mycology and Plant Pathology* 28, 168–170.
Singh, R., Khare, M.N., Vyas, S.C. and Niranjan, V.K. (1997) Chemical control of safflower rust. *Indian Phytopathology* 50, 69–75.
Vasudeva, R.S. (1961) Disease of sesamum. In: Joshi, A.B. (ed.) *Sesamum Monograph*. Indian Central Oilseeds Committee, Hyderabad, India, 287–291.
Verma, M.L. (2002) *Fungal and Bacterial Disease of Sesame and Their Management Challenges for the Millenium.* Jyoti Publishers, New Delhi, pp. 161–192.
Vyas, S.C., Kotwal, I., Prasad, K.V.V. and Jain, A.C. (1984) Note on seed borne fungi of sesamum and their control. *Seed Research* 12, 93–94.

21 Occurrence of *Pyrenophora tritici-repentis* Causing Tan Spot in Argentina

M.V. Moreno[1,2] and A.E. Perelló[2,3]

[1]*Laboratorio de Biología Funcional y Biotecnología, CEBB, Facultad de Agronomía de Azul, Universidad Nacional del Centro de la Provincia de Buenos Aires, Buenos Aires, Argentina;* [2]*Consejo Nacional de Investigaciones Científicas y Técnicas (CONICET);* [3]*Centro de Investigaciones de Fitopatología (CIDEFI), Facultad de Ciencias Agrarias y Forestales, Universidad Nacional de La Plata, La Plata, Provincia de Buenos Aires, Argentina*

Abstract

Wheat (*Triticum aestivum* L.) is currently considered as one of the most important crops worldwide. It can be affected by several diseases. However, only a limited number of them, like 'tan spot' resulting from the fungus produced by *Pyrenophora tritici–repentis*, cause serious problems to the crop and may be given special attention. Tan spot has significant economic consequences. In recent years, the incidence of the disease has increased in many areas where wheat is cultivated, becoming a serious problem by causing losses of up to 70%. It has been found in a lot of countries worldwide: North Dakota, Nebraska and Kansas (USA), Canada, Australia, Asia, Pakistan, Czech Republic, Poland, Ukraine, Hungary, France, Denmark and Belgium.

This disease has increased its incidence, prevalence and severity, particularly in the whole of the South Cone region in the last few years: Argentina, Brazil, Bolivia, Colombia, Ecuador, Peru, Paraguay and Uruguay. Tan spot is one of the most destructive and widespread problems of wheat production in Argentina. In this chapter, we summarize the knowledge of many and diverse contributions and we highlight what is known and unknown about the disease.

Introduction

Wheat is considered one of the most important crops of the world, along with rice, maize and potato. Humans consume around 75% of worldwide production (Wiese, 1987; Rajaram, 2001). In the period between 1970 and 2000, wheat yields rose at an annual rate of 2.3%; however, the area cultivated remained the same. The volume of wheat traded is greater than any other grain (Eikboir and Morris, 2001).

Wheat has been one of the most important crops for the past 100 years in Argentina. Between 2005 and 2006, 2m ha were cultivated, giving a yield of 70–74 Mt (Inf. Económico de Coyuntura No. 261, 2006). The future of this production and South America's participation in the international market depended on how Argentina and

 Management of Fungal Plant Pathogens (eds A. Arya and A.E. Perelló)

Brazil developed the crop. In Argentina, planting in the field to exportation of the crop was dependent on the introduction of new technological developments. These developments include new cultivar composition, management of the crop and the management of future areas to expand yield (Eikboir and Morris, 2001). Fungal pathogens are the result of a combination of these factors (Klein, 2001). Management of these diseases requires specific knowledge and an increased ability to identify the fungus and techniques to reduce crop losses to a minimum (Kohli, 1995).

In the last few years, minimum tillage has been considered advantageous to soil conservation, but it leads to a loss of available nutrients and a potential increase in necrotic pathogens whose saprophytic stage lives in the straw of the crop (Annone, 1985). Establishment of the crop under this management can be affected by pathogens of this type (Table 21.1). In Argentina, the increased incidence in leaf spot since the application of minimum tillage has been a cause for concern (Annone and Kohli, 1996).

Tan Spot

The first time that tan spot was observed on wheat was in the 1920s in Japan (Hosford, 1981). In 1954, the first loss (75%) was reported in Kenya (Gilchrist *et al.*, 1984). In 1940, tan spot was reported for the first time in the USA (Barrus, 1942). At the present time, the name of the fungus causing tan spot is reported with high frequency in the wheat-growing areas of the world (Conners, 1939; Tekauz, 1976; Watkins, *et al.*, 1978; Sim and Willis, 1982; Loughman *et al.*, 1998, Postnifova and Khasanov, 1998; Ali and Francl, 2001a; Sarova *et al.*, 2002).

In South America, tan spot has been observed in Colombia, Ecuador and Peru (Dubin, 1983). It has recently gained predominance among wheat diseases in the Southern Cone region of South America, comprising Argentina, Brazil, Chile, Paraguay and Uruguay (Kohli *et al.*, 1992; Linhares and da Luz, 1994).

The disease began to affect wheat crops noticeably in the north-central region of the Buenos Aires Province in the early 1980s (Annone, 1985, 1996). Since then, tan spot symptoms have been detected in most growing areas of the country. The disease is particularly prevalent and intense in the northern area of the Argentine wheat-producing region (central and northern Buenos Aires, southern Santa Fe, south-eastern Cordoba and Entre Rios Provinces), where highly conducive environmental conditions and increasing use of minimum tillage have created a disease hotspot. In the region, the pseudothecia of the pathogen are formed on wheat residue left on the soil surface at crop sowing and/or early growth stages. Conidia are formed and released soon after the development of the first symptoms on leaves (Annone *et al.*, 1994).

Wright and Sutton (1990) observed that when *P. tritici-repentis* was introduced in an area of wheat, it was dominant over other leaf pathogens. In Argentina, tan spot is one of the most important diseases, along with rust and head blight (Annone, 2006). The massive expansion of minimum tillage in Argentina has encouraged the establishment and development of this disease (Annone and García, 2004).

Importance

Tan spot is frequently observed in most farmers' fields, often affecting the upper leaves at flowering to early grain filling stages. Yield losses of between 9 and 50% have been observed by several authors (Hosford and Busch, 1974; Sharp *et al.*, 1976; Rees *et al.*, 1982; Rees and Platz, 1983). In South America, yield losses of around 40% were observed by Mehta and Gaudencio (1991) in Brazil and Kohli *et al.* (1992) reported wheat yield losses of between 20 and 70% in Paraguay and Argentina. Estimates of losses (10–20%) caused by the disease have been made by comparing fungicides protected with non-protected wheat plots (unpublished data Annone, 1996). Similar results were obtained by Galich and Galich

Table 21.1. Diseases of wheat and the agent cause (Annone and Kohli, 1996).

Pathogen	Disease
Xanthomonas campestris pv. *undulosa*	Bacterial stripe
Septoria tritici	Spot blotch
Drechslera tritici-repentis	Tan spot
Fusarium graminearum	Head blight
Gaeumannomyces graminis var. *tritici*	Take-all

(1994) in Marcos Juarez (Cordoba Province). They determined that losses due to tan spot associated with *Septoria tritici* blotch ranged between 6 and 13.5%. Tan spot is a complex disease that is dependent on its geographic location and the environmental conditions prevailing (de Wolf and Francl, 1998).

The pathogen

The tan spot fungus is an Ascomycota currently known as *P. tritici-repentis* (Ptr) (Died.) Drechs. It is a facultative pathogen whose asexual stage is *Drechslera tritici-repentis* (Dtr) (Died.).

P. tritici-repentis was isolated for the first time from *Agropyron repens* in Germany and it was named *Pleospora trichostoma* by Diecke. In 1928, it was isolated from wheat by Nisikado (Nisikado, 1928), when it was named *Helminthosporium tritici-repentis* (= *Drechslera tritici-repentis*) (Hosford, 1981).

The genera *Pyrenophora* Fr. was used frequently for some ascomycota parasitic on cereals and other grasses (Diaz de Ackerman, 1987). It was described by Fries in 1849 and cited by Shoemaker in 1961 (Shoemaker, 1962). In 1869, Fuckel noted the tendency of *P. phaecomes* to mature only after overwintering and found a *Drechslera* conidial stage of *P. phaecomes* (Shoemaker, 1962). In 1883, Saccardo used the presence of setae on the ascocarp of *Pyrenophora* and the absence of setae on the ascocarp of *Pleospora* to separate these two genera. In 1885, Winter (Shoemaker, 1962) included the species of both genera in *Pleospora* and in 1934, Drechsler (Shoemaker, 1962) agreed with Winter that the presence or absence of setae was not an important enough characteristic to separate the two genera. He emphasized the connection between *Pyrenophora* and the conidial stage *Drechslera*, as found by Fuckel for *P. phaecomes*. Drechsler too determined the connection between *P. teres* and *D. teres*, *P. tritici-repentis* and *D. tritici-repentis*, and *P. bromi* and *D. bromi* (Shoemaker, 1962). At the same time, Ito and Kuribayaski (1931) connected five species of *Pyrenophora* with the conidial stage of *Drechslera*. In 1949, Wehmeyer worked on the distinction in form and size of the *Pleospora* and *Pyrenophora* ascospores (Wehmeyer, 1949).

In 1930, Ito described the genera *Drechslera*. In 1809, Link described the genera *Helminthosporium*, where species of *Drechslera* were included (Ito, 1930). In 1902, Diedicke (Drechsler, 1923) determined *H. tritici-repentis* as formae of *H. gramineum*. In 1923, Drechsler recognized *H. teres*, *H. bromi*, *H. gramineum* and *D. avenae* as unique species. In 1959, Shoemaker (1962) made the distinction between two subgenera, *Cylindro-Helminthosporium*, in which all the species have conidia germinating from all cells and *Eu-Helminthosporium*, in which all the species have fusiform conidia germinating from end cells only. In 1930, Ito (Shoemaker, 1962) proposed the name *Drechslera* for those species with cylindric conidia germinating from all cells, using as a type *D. tritici-repentis*. He used the name *Bipolaris* for those species whose conidia were fusiform, germinating from end cells only. In 1962, Shoemaker considered *D. tritici-vulgaris* as *D. tritici-repentis*. Currently, the teleomorphic nomenclature of the fungus is *P. tritici-repentis* and the anamorph of the fungus is unanimously accepted as *D. tritici-repentis*. Morphological data can be found in Drechler (1923), Shoemaker (1962) and Wehmeyer (1954).

Host–Parasite Interactions

Symptomatology. On susceptible wheat leaves, *P. tritici-repentis*(Ptr) produces characteristic oval to diamond-shaped lesions. However, newly formed tan spot lesions cannot be separated reliably from those caused by other necrotrophic pathogens. Later, lesions elongate and develop a tan colour with a chlorotic halo and a small dark brown infection site. Chlorotic areas tend to coalesce on heavily infected leaves, especially on young plants, a symptom which leads to the disease name, 'yellow leaf spot' (Fig. 21.1). On resistant and partially resistant wheat, lesion size is reduced and chlorosis and necrosis may be absent (de Wolf *et al.*, 1998).

Lamari and Bernier (1989a) identified two different types of symptoms produced by the pathogen: tan necrosis and extensive chlorosis. However, they reported that the pathogen isolates could be characterized by their ability to induce tan necrosis and/or chlorosis. They grouped the isolates into four pathotypes based on the production of different symptoms on different lines. In this system, an unlimited number of isolates were designated as races 1, 2, 3, 4, 5, 6, 7, 8, 9, 10, 11 and 12 (Lamari and Bernier, 1989a,b; Lamari *et al.*, 1995, 1998, 2003, 2005; Lamari and Gilbert, 1998; Ali and Francl, 2001a,b, 2002a,b). Races 9 and 10 have been identified in South America, which indicates that the Ptr population is heterogeneous in this area (Ali and Francl, 2002b). In Argentina, the race population structure is unknown and in 2007, Moreno observed that isolates obtained from Argentina produced three reaction types on cultivars of local and international wheat (Moreno, 2007). Actually, the isolates were inoculated on different wheat sets to determine the races present in Argentina.

Ptr can also infect wheat seed during the grain-filling period (Schilder and Bergstrom, 1994). This disorder is called red smudge, because infected seed has a reddish discoloration (Valder, 1954).

Disease cycle

Dispersal and infection by Ptr can develop between 10° and 30°C with moisture between 6 h and 48 h (Larez *et al.*, 1986; Hosford *et al.*, 1987; Sah, 1994). These conditions are the reason why tan spot can occur all year round and which distinguishes it from the white head disease, but they all depend on environmental conditions (Carmona, 2003).

The disease cycle of tan spot (Fig. 21.2) provides a convenient framework on which

(a) (b)

Fig. 21.1. *Pyrenophora tritici-repentis* produces characteristic oval to diamond-shaped lesions.

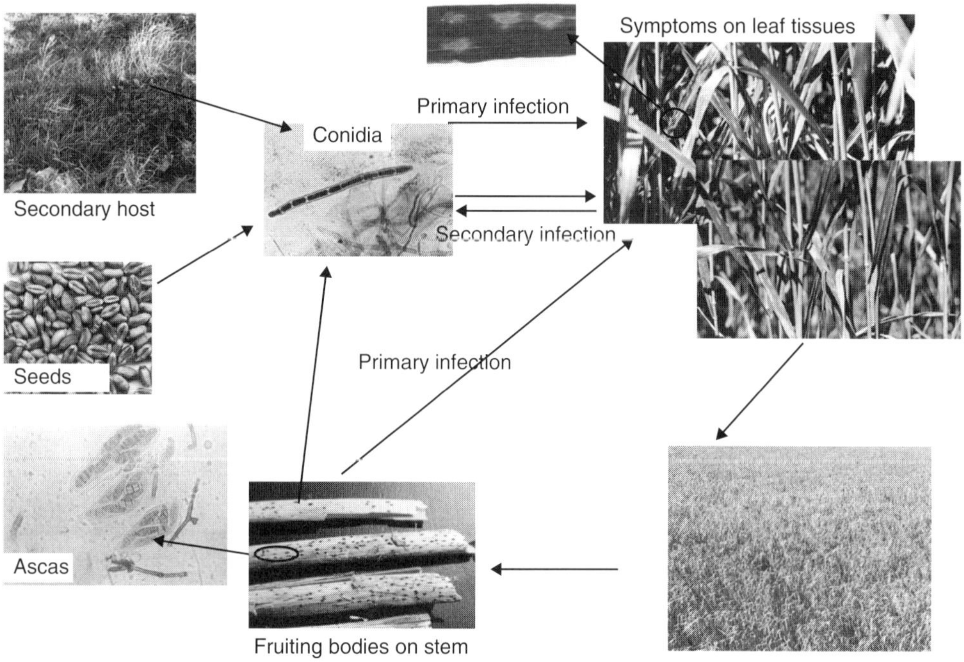

Fig. 21.2. Disease cycle of *Pyrenophora tritici-repentis*, agent cause of tan spot of wheat.

to explain our current understanding of the progress of the disease. The rate of progression through the disease cycle depends on the host and on temporal and environmental components of the pathosystem (de Wolf *et al.*, 1998).

The seeds, straw and collateral hosts are the principal source of inoculum of tan spot. The primary inoculum can travel long distances through the wheat-growing areas and is introduced into new areas by seeds. In the seed, the pathogen lives in the pericarp as mycelium and transmission to the rest of the plant is non-systemic (Schilder and Bergstron, 1994). In Argentina, Barreto (1984, unpublished data) observed infection on 2% of wheat seed. Future investigations are required to establish the sanitatary management of seeds (Carmona, 2003).

Another source of primary inoculum is wheat straw. Several authors consider straw as the principal source of the inoculum of Ptr (Rees and Platz, 1980).

Collateral hosts of Ptr could play an important role as a source of primary inoculum between growing seasons, as a source of genetic variation and as a reservoir of a fungal population genetically different than that prevalent on wheat (de Wolf *et al.*, 1998). The tan spot fungus has been reported on many grass species from different parts of the world, among which are *Agropyron* sp., *Avena fatua*, *A. sativa*, *Echinochloa* sp., *Elymus innovatus*, *Andropogon gerardi*, *Alopecurus arundinaceus*, *Bromus inermis*, *Dactilys glomerata*, *Lolium perenne*, *Phalaris arundinaceae*, *Poa* sp. and *Secale cereale* (Diedicke, 1902; Drechsler, 1923; Conners, 1939; Dennis and Wakefield, 1946; Sprague, 1950; Andersen, 1955; Dickson, 1956; Shoemaker, 1962; Hosford, 1971; Howard and Morral, 1975; Farr *et al.*, 1989; Krupinsky, 1992c; Ali and Francl, 2002b). In Argentina, the host range is unknown.

Ascospores are generated in the pseudothecia that live in the wheat straw. The conidia are formed in the straw containing the pseudothecia and on the leaves of infected plants or the leaves of collateral hosts.

The ascospores of Ptr are dispersed primarily by wind, but the distance an ascospore can travel is limited (Schilder and Bergstrom, 1995). Limitations on ascospore dispersal distance have been attributed in part to short discharge distances from the pseudothecia. However, it is doubtful that the short discharge distance alone can account for these short dispersal distances. Schilder and Bergstrom (1992) proposed that movement was limited during periods of high relative humidity when ascospores were discharged from the ascocarps (de Wolf *et al.*, 1998). Infested residue usually results in significant disease severity at flag leaf emergence and later growth stages due to secondary infections (McFadden and Harding, 1989; Wright and Sutton, 1990; McFadden, 1991).

Following liberation from the host, the conidia of Ptr can be sampled readily during aerial dispersal and differentiated successfully from other fungi (Morral and Howard, 1975; Rees and Platz, 1980; Wright and Sutton, 1990; Krupinsky, 1992b; Maraite *et al.*, 1992; Schilder and Bergstrom, 1992; Wolf and Hoffmann, 1993). Morrall and Howard (1975) reported that conidia numbers reached their highest levels late in the growing season and that the number of conidia has a clear diurnal periodicity. The numbers of conidia of the pathogen decline sharply with dispersal distance. Schilder and Bergstrom (1992) reported that the highest number of conidia occurred within 3 m of the inoculum source and that 60–100% of the recoverable conidia were sampled within 25 m. Only a few conidia could be recovered 100 m away from the inoculum source, but this suggested that longer dispersal distances were possible. When the conidia were deposited on the leaf, their germination was influenced by both temperature and the availability of free moisture (Mihtra, 1934). The conditions that contribute to infection by Ptr in an outdoor environment have also been studied (Ali, 1993; Francl, 1998; de Wolf and Francl, 1997). The precise range of temperatures optimal for disease development varies with cultivar (Luz and Bergstrom, 1986). Leaf age affects the severity of the disease caused by Ptr (Cox and Hosford, 1987; Lamari and Bernier, 1989a,b; Hosford *et al.*, 1990; Perelló *et al.*, 2003a). The growth stage seems to influence tan spot severity and expression of resistance (Hosford *et al.*, 1990; Fernandez *et al.*, 1994; Perelló *et al.*, 2003a).

Then, infecting the wheat leaves, conidia are produced and the pathogen's asexual cycle life develops by infecting new plants of wheat. Even so, conidiogenesis continues in wheat straw. The production of conidia and the development of pseudothecia depend on temperature and water potential. Stem colonization appeared to be the result of the progressive colonization of the leaf sheath and upper internode. No differences in saprophytic colonization were observed among cultivars of varying resistance (de Wolf *et al.*, 1998). Numerous researchers have investigated the factors affecting the initiation and development of pseudothecia in laboratory experiments (Odvody *et al.*, 1982; Pfender and Wootke, 1987; Pfender *et al.*, 1988; Summerell and Burgess, 1988a,b, 1989; Zhang and Pfender, 1993).

The effects of water potential in wheat straw on pseudothecial development have also been studied in an outdoor environment (Fernandes *et al.*, 1991; Zhang and Pfender, 1993). The number of ascocarps per gram of straw in near-soil straw was 32% and 42% of that found in mowed and no-till treatments, respectively. In addition, the number of ascocarps produced in the lower portion of standing stubble of no-till plots was 12% of the number found in the upper portion.

Reports of pseudothecia maturation in an outdoor environment vary from region to region (Rees and Platz, 1980; Odvody *et al.*, 1982; Summerell and Burgess, 1988b, 1989; Wright and Sutton, 1990; Wolf and Hoffmann, 1993). In most regions where wheat is grown, the pseudothecia of Ptr are initiated when the crop has reached full maturity and begins to senesce (Odvody *et al.*, 1982; Wolf and Hoffmann, 1993). However, in colder climates, pseudothecia may not be initiated until the following growing season (Fernandez *et al.*, 1998).

In Argentina, the sexual stage of Ptr has been detected in wheat straw, but it is unknown in which regions and under what conditions development took place.

Physiological Specialization

The terms 'pathogenicity' and 'virulence' are likely to be used to describe the ability of an organism to cause disease. Pathogenicity is regarded as a general attribute of a species, while virulence is an attribute reserved for a particular strain of a pathogen in relation to a particular host genotype (Day, 1960). There exist virulent races of Ptr that interact with wheat hosts in a highly specific manner. This suggests that host-specificity attributes are superimposed on the general pathogenic ability of Ptr.

Variation in virulence in the population of this pathogen is essential in understanding the interaction of the genomes involved in tan spot. Studies of the diversity of virulence within a pathogen population should help in the development of a successful disease management programme, particularly resistant cultivars. Several investigators have described diversity among Ptr isolated from different areas around the world (Christensen and Graham, 1934; Misra and Singh, 1972; Luz and Hosford, 1980; Gilchrist *et al.*, 1984; Krupinsky, 1987, 1992a,b; Diaz de Ackermann *et al.*, 1988; Lamari and Bernier, 1989a; Schilder and Bergstrom, 1990; Ali and Buchenau, 1992; Sah and Ferhmann, 1992; Brown and Hunger, 1993; Moreno, 2007).

In 1971, Hosford observed differences between the reaction type on wheat cultivars produced by isolates of Ptr. Misra and Singh (1972) tested isolates originating from India and they detected significant differences in virulence, based on lesion size. Some results were observed by Gilchrist *et al.* (1984) when they tested isolates collected from Mexico on the wheat cultivar Morocco. Luz and Hosford (1980) grouped the isolates tested into 12 races based on statistical mean separation. However, Díaz de Ackermann *et al.* (1988) did not find any difference in virulence among the isolates tested by Luz and Hosford (1980). Hunger and Brown (1987) tested nine isolates originating from the USA; these isolates showed significant differences on the susceptible cultivar TAM 105. Krupinsky (1987) showed differences in lesion length and percentage of severity among isolates of Ptr obtained from *Bromus inermis*.

Lamari and Bernier (1989a) grouped the isolates of Ptr into three pathotypes on 11 cultivars of wheat based on the type of reaction. Schilder and Bergstrom (1990) tested 70 isolates obtained from Canada on 12 wheat cultivars and detected significant differences among the interaction of isolate × cultivar. Some results were reported by Sah and Ferhmann in 1992 for isolates originating from Brazil, Germany, India, Nepal and the USA. However, Krupinsky (1992a,b) detected variation among levels of aggressiveness but he found no differences in levels of virulence. In 1992, Ali and Buchneau observed physiological specialization based on the reaction type for isolates obtained from the USA. Mehta *et al.* (2004) tested 40 isolates obtained from Parana (Brazil) on six wheat cultivars; they observed low interaction for isolate × cultivar. In 2007, Moreno detected significant differences in isolate × cultivar for isolates of Ptr obtained from wheat-growing areas in Argentina.

Races 1, 2, 3 and 4 of Ptr correspond with those determined by Lamari *et al.* (1995). Races 1 and 2 are predominant in North America (Ali and Francl, 2003). The greater part of isolates identified as race 5 originate from North Africa, North America and Azerbaijan (Ali *et al.*, 1990; Lamari *et al.*, 1995, 1998; Strelkov *et al.*, 2002; Ali and Francl, 2003). Races 6, 7 and 8 were identified from collections originating from Algeria, Caucaso and South America (Ali and Francl, 2002a; Strelkov *et al.*, 2002; Lamari *et al.*, 2003). Finally, races 9 and 10 were identified from isolates originating from South America (Ali and Francl, 2002a,b).

These studies indicate that variation in the pathogen population can be detected by using either quantitative or qualitative rating scales (Table 21.2). Research using quantitative scales generally detected variation in virulence on susceptible lines, but isolates in different studies produced an equal reaction on resistant cultivars (de Wolf *et al.*, 1998).

Table 21.2. Relationships between pathotypes, races and wheat cultivars.

	Cultivars/lines of wheat					
Races	Glenlea	Katepwa	6B662	6B365	Salomouni	M3
1	N (Tox A)	N (Tox A)	R	Cl (Tox C)	R	R
2	N (Tox A)	N (Tox A)	R	R	R	R
3	R	R	R	Cl (Tox C)	R	R
4	R	R	R	R	R	R
5	R	Cl (Tox B)	Cl (Tox B)	R	R	R
6	R	Cl (Tox B)	Cl (Tox B)	Cl (Tox C)	R	R
7	N (Tox A)	N (Tox A) Cl (Tox B)	Cl (Tox B)	R	R	R
8	N (Tox A)	N (Tox A) Cl (Tox B)	Cl (Tox B)	Cl (Tox C)	R	R

Note: R, resistance; N, necrosis; CL, chlorosis; Tox A, presence of Tox A and production of Tox A; Tox B, presence of Tox B and production of Tox B.

In Argentina, the race population structure is unknown. Future research studying physiological specialization in Ptr should consider collections originating in Argentina.

Disease Management Strategies

From the point of view of the disease's development, its management is achieved in different ways: by reducing or delaying the disease early in the growing season or by reducing its rate of development during crop growth (Zadoks and Schein, 1979). This practice has helped to block the life cycle of the pathogens, preventing the introduction of inoculum and susceptible hosts, eliminating certain pathogens (Palti, 1981).

Tan spot is one of a complex of necrotrophic leaf diseases of wheat which overwinter on infested crop residue (Hosford and Busch, 1974; Loughman *et al.*, 1998; Carmona, 2003; Annone, 2006). The occurrence of tan spot with other leaf spots, such as septoria blotch, spot blotch and with rusts and mildews, can complicate disease management practices (de Wolf *et al.*, 1998; Carmona, 2003; Annone, 2006). The management of tan spot is based on integrated management of diseases that use reasonable techniques and resources for sustainable agriculture (Carmona, 2006).

Strategies used for the control of tan spot are the application of fungicides, cultural control and the search for new germplasms and their incorporation in Argentina (Carmona, 2003).

Recently in Argentina, several biological antagonists of Ptr have been identified (Pfender *et al.*, 1989; Li and Sutton, 1995; Perelló *et al.*, 2003b; Annone, 2005).

Genetic resistance

Genetic resistance is complex for diseases such as head blight and leaf spot. The principal limitations are due to the changes made by pathogen populations over the years to challenge new cultivars (Carmona, 2006).

Unfortunately, only a few of the currently grown cultivars have a high level of resistance, while somewhat larger numbers possess a moderate level of resistance (Rees and Platz, 1992). Kohli *et al.* (1992) reported the low presence in South America of cultivars resistant to Ptr. Several studies have been conducted in Argentina to screen breeding material for resistance (Galich and Galich, 1994; Annone, 1995).

In Argentina, cultivars have either a moderate level of resistance or are susceptible to tan spot (Simón, 2006).

Chemical protection

Fungicides offer a complementary tool to the genetic resistance available. Its use in direct seeding crops under-compensates for the lack of genetic protection to facultative parasites. Fungicides are used as seed protection and/or treatment coverage with ground or air equipment.

The majority of literature regarding the use of fungicides to manage tan spot alone or in combination with other leaf diseases has focused on the timing of application and comparative efficiencies. Research results have been mixed, but it appears that in situations where disease pressure is high and conditions favour further development of foliar disease, a single well-timed application of an efficacious fungicide can reduce disease severity, increase yield and improve product quality (Sutton and Roke, 1986; Bockus *et al.*, 1992; Duczek and Jones-Flory, 1994; Stover *et al.*, 1996).

Fungicides such as tebuconazol, frutiafol, fluzilazol propiconazol and prochloraz reduced the intensity of lesions and showed control of from 50 to 70%, depending on the cultivar and the density of wheat straw infested (Annone *et al.*, 1994; Carmona, 1996). The most efficient fungicides are systemic triazoles and estrobirulinas (Carmona, 2003).

Cultural control

Cultural practices alter the development of foliar diseases of wheat, particularly those caused by facultative pathogens. Tan spot, representative of the latter group of diseases, is affected by tillage practices in almost all wheat-growing regions of the world (Mehta and Gaudencio, 1991).

Retention of wheat residue on the soil surface generally results in increased tan spot severity (Gough and Ghazanfani, 1982; Summerell and Burgess, 1988a,b; Schuh, 1990; Bockus and Claasen, 1992; Stover *et al.*, 1996; Carmona and Reis, 1998).

In areas where zero tillage is practised, tan spot and other debris-borne diseases typically increase in incidence and severity (Rees and Platz, 1979; Mehta and Gaudencio, 1991; Kohli *et al.*, 1992). The rotation of crops has a high impact on the sexual stage of this type of pathogen. Because sexual stage viability is minor or low when wheat straw is mineralized, the primary inoculum is therefore low and reduces the severity of tan spot (Carmona *et al.*, 1999).

In relation to crop rotation, of relevance was wheat as an antecessor of barley and oats as an antecessor of wheat and barley as an antecessor of oats (Carmona *et al.*, 2001).

Barley, wheat and oats are common hosts of Ptr and other pathogens, so there are therefore no alternative crops available for crop rotation. Oats were not hosts to leaf spots specific to wheat, so crop rotation is possible in Argentina. However, in Brazil, Paraguay and Uruguay, where *B. sorokiniana* is a relevant pathogen, rotation of these crops should not be authorized (Carmona, 2006).

Biocontrol

Biological control using antagonistic microbes alone or as supplements has become more important in recent years in order to minimize the use of chemicals (Annone, 2005). It is an additional tool available for the design of more sustainable control strategies of wheat diseases. While biological control is a widespread natural phenomenon, it is often inadequate, especially in agricultural ecosystems in which conditions strongly favour pathogens and disease epidemics. It can also fail in natural ecosystems, especially against aggressive alien pathogens that are able to overcome the natural biological buffering of these systems (Sutton, 2005).

Several biological antagonists of Ptr have been identified (Pfender *et al.*, 1989; Li and Sutton, 1995; Perelló *et al.*, 2003b; Perelló *et al.*, 2006, 2009). Luz *et al.* (1998) found that treatment with *Paenibacillus maceruns* or *Pseudomonas putida* reduced transmission of Ptr by seed to levels equivalent to that of a fungicide seed treatment.

Some of the fungi found to be inhibitory to pseudothecia development by Ptr were *Limonomyces roseipellis*, *Myrothecium roridum*, *Acremoniun terricola*, *Stachybotrys* sp. and *Laetisaria arvalis* (Gough and Ghazanfani, 1982; Pfender *et al.*, 1989). Assays in Argentina have demonstrated that some *Trichoderma harzianum* isolates are capable of suppressing growth, the mycelial development of Ptr and the severity of diseases on wheat plants (Perelló *et al.*, 2003b, 2006, 2008, 2009). No previous records of antagonism between isolates of *Trichoderma* spp. and the necrotrophic foliar pathogen have been found. On the other hand, there are increasing economic and social pressures to develop usable biological control strategies in Argentina.

References

Ali, S. (1993) Modeling the progress of tan spot of wheat using weather variables. MSc thesis, South Dakota State University, Brookings, South Dakota, 78 pp.

Ali, S. and Buchenau, G.W. (1992) Incidence of toxin-producing pathotypes of *Pyrenophora tritici-repentis* in South Dakota. *Phytopathology* 82, 1159 (Abstract).

Ali, S. and Francl, L.J. (2001a) First report of tan spot on wheat in Pakistan. *Plant Disease* 85, 1031.

Ali, S. and Francl, L.J. (2001b) Recovery of *Pyrenophora tritici-repentis* from barley and reaction of 12 cultivars to five races and two host-selective toxins. *Plant Disease* 85, 580–584.

Ali, S. and Francl, L.J. (2002a) A new race of *Pyrenophora tritici-repentis* from Brazil. *Plant Disease* 86, 9.

Ali, S. and Francl, L.J. (2002b) Race structure of *Pyrenophora tritici-repentis* isolates obtained from wheat in South America. *Plant Protection Science* 38 (Special Issue 2), 302–304.

Ali, S. and Francl, L.J. (2003) Population race structure of *Pyrenophora tritici-repentis* prevalent of wheat and non-cereal grasses in the Great Plains. *Plant Disease* 87, 418–422.

Ali, S., Francl, L.J. and Wolf, E.D. de (1990) First report of *Pyrenophora tritici-repentis* race 5 from North America. *Plant Disease* 83, 591.

Andersen, H. (1955) Species of *Helminthosporium* on cereals and grasses in Denmark. *Friesia* 5, 80–89.

Annone, J.G. (1985) Presencia de la mancha tostada del trigo (*Helminthosporium tritici-repentis*). *Carpeta de Producción Vegetal. TRIGO.* TOMO VII. Inf. No. 88. EEA INTA, Pergamino, Buenos Aires.

Annone, J.G. (1995) Resistencia genética a agentes fitopatógenos: conceptos sobre hospedante, patógeno, relación hospedante-patógeno y manejo de la resistencia. In: Kohlí, M.M., Annone, J.G. and García, R. (eds) *Curso de manejo de enfermedades del Trigo.* Centro Internacional de Capacitación INTA-CIMMYT, Pergamino, Buenos Aires.

Annone, J.G. (1996) Experiencias de tratamientos con fungicidas en trigo en el norte de la provincia de Buenos Aires. In: *Primera Jornada de Control Químico de Enfermedades del Trigo en Sistemas de Alta Productividad.* CIC "Dr. Norman Borlaog INTA/CIMMYT-CRBAN/INTA-Programa de Trigo de CIMMYT para el Cono Sur. Proyecto IPG/INTA. Bolsa de Cereales de Buenos Aires. 27–28 June 1996, 7 pp.

Annone, J.G. (2005) El desafío del manejo integrado de enfermedades de los cultivos en sistemas agrícolas conservacionistas: el caso de la sanidad del trigo en siembra directa. *13th Congreso Latinoamericano de Fitopatología, 19–22 April 2004, Villa Carlos Paz, Córdoba, Argentina*, pp. 161–164.

Annone, J. (2006) Las principales enfermedades del trigo en Argentina: su importancia relativa en las Regiones productoras Norte y Sur. In: *Congreso A Todo Trigo. Conocimiento y Producción. Federación de Centros y Entidades Gremiales de Acopiadores de Cereales, Hotel Sheraton, 18 and 19 May 2006. Mar del Plata, Argentina*, pp. 53–58.

Annone, J. and García, R. (2004) Cuantificación del efecto supresito de la fertilización nitrogenada sobre la expresión de síntomas de la mancha amarilla en cultivares de trigo de ciclo intermedio-largo y precoz en siembra directa. In: *VI Congreso Nacional de Trigo, IV Simposio Nacional de Cultivos de Siembra Otoño-invernal. Universidad Nacional del Sur/departamento de Agronomía, Bahía Blanca, Argentina*, pp. 193–194.

Annone, J.G. and Kohli, M. (1996) Principales enfermedades del trigo asociadas con la siembra directa en Argentina: importancia y medidas de protección. In: *Trabajos Presentados, IV Congreso Nacional de Siembra Directa: Una estrategia para todos, 27–30 March 1996. AAPRESID, Villa Giardino, Córdioba, Argentina*, pp. 135–159.

Annone, J.G., Botta, G.E and Ivancovich, A. (1994) Ocurrencia de la mancha bronceada del trigo en el área norte de la provincia de Buenos Aires. In: *III Congreso Nacional de Trigo y I Simposio Nacional de Cereales de siembra Otoño-Invernal. Dpto. de Agronomía, Universidad Nacional del sur, Asociación de ingenieros Agrónomos del Norte de la provincia de Buenos Aires, Bahía Blanca, Buenos Aires.* 26–28 October 1994, pp. 205–207.

Barrus, M.F. (1942) Yellow spot disease of wheat in New York State. *Plant Disease Reporter* 26, 246.

Bockus, W.W. and Claasen, M.M. (1992) Effects of crop rotation and residue management practices on severity of tan spot of winter wheat. *Plant Disease* 76, 633–636.

Bockus, W.W., Davis, M.A. and Shroyer, J.P. (1992) Effect of foliar fungicide application on seed size of winter wheat. *Journal of Applied Seed Production* 10, 1–6.

Brown, D.A. and Hunger, R.M. (1993) Production of a chlorosis-inducing, host-specific, low-molecular weight toxin by isolates of *Pyrenophora tritici-repentis*, cause of tan spot of wheat. *Journal of Phytopathology* 137, 221–232.

Carmona, M. (1996) Patógenos de semillas de trigo. Importancia y control químico. In: *Primera Jornada de Control Químico de Enfermedades del Trigo en Sistemas de Alta Productividad.* CIC Dr. Norman Borlaog INTA/CIMMYT-CRBAN/INTA-Programa de Trigo de CIMMYT para el Cono Sur. Proyecto IPG/INTA. Bolsa de Cereales de Buenos Aires. 27–28 June 1996, 7 pp.

Carmona, M. (2003) Manejo Integrado de la mancha amarilla del trigo (*Drechslera triticirepentis*) INTA Rafaela- Información técnica de trigo – Campaña 2003 (http://www.inta.gov.ar/rafaela/info/documentos/misc99/misc99_8.pdf).

Carmona, M. (2006) El manejo integrado de las enfermedades del cultivo de trigo, una asignatura pendiente. In: *Congreso A Todo Trigo. Conocimiento y Producción. Federación de Centros y Entidades Gremiales de Acopiadores de Cereales, Hotel Sheraton, 18 and 19 May 2006. Mar del Plata, Argentina*, pp. 161–165.

Carmona, M. and Reis, E.M. (1998) Las enfermedades y su manejo en siembra directa. *Cuaderno de Actualización Técnica* No. 59 CREA Siembra Directa, pp. 68–81.

Carmona, M., Reis, E.M. and Cortese, P. (1999) Manchas foliares del trigo. Diagnóstico, epidemiología y nuevos criterios para el manejo. Eds de los autores, Buenos Aires, 32 pp.

Carmona, M., Sugia, V., Schmidt, A. and Delfino, S.V. (2001) Control potencial de manchas foliares de trigo, cebada y avena por la rotación de cultivos. *Congresos Nacional de Trigo y III Simposio Nacional de Cereales de invierno. 24–28 September 2001. Villa Carlos Paz, Córdoba, Argentina.*

Christensen, J.J. and Graham, T.W. (1934) *Physiologic specialization and variation in* Helminthosporium gramineum *Rob.* University of Minnesota Agricultural Experiment Station, Minnesota.

Conners, I.L. (1939) Yellow leaf botch. *Canadian Plant Disease Survey* 19, 12–14.

Cox, D.J. and Hosford, R.M. Jr (1987) Resistant winter wheat compared at differing growth stages and leaf positions for tan spot severity. *Plant Disease* 71, 883–886.

Day, P. (1960) Variation in phytopathogenic fungi. *Annual Review of Microbiology* 14, 1–16.

Dennis, R.W.G. and Wakefield, E.M. (1946) New or interesting British fungi. *Transactions of the British Mycological Society* 29, 141–166.

Diaz de Ackermann, M. (1987) *Resistance among winter wheats to* Pyrenophora tritici-repentis *isolates and variability in fungal growth, color, sporulation and pathogenicity.* The North Dakota State University Department of Plant Pathology, Fargo, North Dakota.

Diaz de Ackermann, M., Oxford, R.M. Jr, Cox, D.J. and Hammond, J.J. (1988) Resistance in winter wheats to geographically differing isolates of *Pyrenophora tritici-repentis* and observations of pseudoperithecia. *Plant Disease* 72, 1028–1031.

Dickson, J.G. (1956) *Diseases of Field Crops.* McGraw Hill Book Co, New York, 517 pp.

Diedicke, H. (1902) Uberden zusammenhangwischen *Pleospora* und *Helminthosporium* arten. *Centrablatt fur Bakteriologie und Parasitenkunde Jena*, Abt. 11, 52–59.

Drechsler, C. (1923) Some graminicolous species of *Helminthosporium. Journal of Agricultural Research* 24, 641–740.

Dubin, H.J. (1983) Occurrence of *Pyrenophora tritici-repentis* in the Andean Countries of South America. *Plant Disease* 67, 1040.

Duczek, L.J. and Jones-Flory, L.L. (1994) Effect of timing of propiconazole application on foliar disease and yield of irrigated spring wheat in Saskatchewan from 1991 to 1992. *Canadian Journal of Plant Science* 74, 205–207.

Eikboir, J. and Morris, M. (2001) Perspectivas del Mercado Internacional de Trigo: Implicancias para los países de América del Sur. In: Kohli, M.M., Diaz, M. and Castro, M. (eds) 2003. *Estrategias y Metodologías utilizadas en el mejoramiento de trigo.* Seminario Internacional, La Estanzuela, Uruguay. CIMMYT-INIA, pp. 3–10.

Farr, D.F., Bills, G.F., Chamuris, G.P. and Rossman, A.Y. (1989) *Fungi on Plants and Plant Products in the United States.* APS Press, St Paul, Minnesota, 1252 pp.

Fernandes, J.M.C., Sutton, J.C. and James, T.D.W. (1991) A sensor for monitorig moisture of wheat residues: application in ascospore maturation of *Pyrenophora tritici-repentis. Plant Disease* 75, 1101–1105.

Fernandez, M.R., Clarke, J.M. and De Pauw, R.M. (1994) Response of durum wheat kernels and leaves at different growth stages to *Pyrenophora tritici-repentis. Plant Disease* 78, 597–600.

Fernandez, M.R., McConkey, B.G. and Zentner, R.P. (1998) Tillage and summerfallow effects of leaf spot diseases of wheat in the semiarid Canadian Prairies. *Canadian Journal of Plant Pathology* 20, 376–379.

Francl, L.J. (1998) Genesis and liberation of conidia of *Pyrenophora tritici-repentis. Canadian Journal of Plant Pathology* 20, 387–393.

Galich, M.T.V. de and Galich, A.N. (1994) Enfermedades de trigo en el área sur de Santa Fe y Córdoba correspondientes a la subregión II Norte. Campaña 1993/94. *Informe para Extensión* No. 5. EEA INTA Marcos Juárez, Córdoba, Argentina, 7 pp.

Gilchrist, L., Fuentes, S.F. and Isla de Bauer de la M.L. (1984) Determinación de fuentes de resistencia contra *Helminthosporium tritici-repentis* bajo condiciones de campo y de invernáculo. *Agrociencia* 56, 95–106.

Gough, F.J. and Ghazanfani, J. (1982) Biological antagonists of *Pyrenophora tritici-repentis.* In: Hosford, R.M. Jr (ed.) *Tan Spot of Wheat in Related Diseases Workshop.* North Dakota Agricultural Experiment Station, Fargo, North Dakota, pp. 36–39.

Hosford, R.M. Jr (1971) A form of *Pyrenophora trichostoma* pathogenic to wheat and other grasses. *Phytopathology* 61, 28–32.

Hosford, R.M. Jr (1981) Tan spot. In: Hosford, R.M. Jr (ed.) *Tan Spot of Wheat and Related Diseases Workshop*, North Dakota State University, Fargo, North Dakota, pp. 1–24.

Hosford, R.M. Jr and Busch, R.H. (1974) Losses in wheat caused by *Pyrenophora trichostoma* and *Leptosphaeria avenaria* f. sp. *triticea. Phytopathology* 64, 184–187.

Hosford, R.M. Jr, Larez, C.R. and Hammond, J.J. (1987) Interaction of wheat period and temperature on *Pyrenophora tritici-repentis* infection and development in wheat of differing resistance. *Phytopathology* 77, 1021–1027.

Hosford, R.M. Jr, Jordahl, J.G. and Hammond, J.J. (1990) Effect of wheat genotype, leaf position, growth stage, fungal isolate and wet period on tan spot lesions. *Plant Disease* 74, 385–390.

Howard, R.J. and Morral, R.A.A. (1975) The epidemiology of leaf spot disease in a native prairie I. The progression of disease with time. *Canadian Journal of Botany* 53, 1040–1050.

Hunger, R.M. and Brown, D.A. (1987) Colony color, growth, sporulation, fungicide sensivity and pathogenicity of *Pyrenophora tritici-repentis. Plant Disease* 71, 907–910.

Inf. Económico de Coyuntura (2006) No. 261, Marzo 2006, Año 24. Sector Agropecuario. Consejo Profesional de Ciencias Económicas de la Ciudad Autónoma de Buenos Aires, ISSN: 0327-8174.

Ito, S. (1930) On some new ascigerous stages on the species of *Helminthosporium* parasitic on cereals. *Proceedings of the Imperial Academy of Tokyo* 6, 352–355.

Ito, S. and Kuribayaski, K. (1931) The ascigerous forms of some graminicolous species of *Helminthosporium* in Japan. *Journal of the Faculty of Agriculture Hokkaido University* 79, 85–125.

Klein, O.A. (2001) Trigos adaptados a siembra directa-conociemintos y desconocimientos. In: Kohlí, M.M., Diaz, M. and Castro, M. (eds) *Estrategias y metodología utilizadas en el mejoramiento de trigo: un enfoque multidisciplinario.* Centro Internacional de mejoramiento de Maíz y Trigo (CIMMYT)-Instituto Nacional de Investigación Agropecuaria (INIA), La Estanzuela, Colonia, Uruguay, pp. 37–42.

Kohli, M.M. (1995) Conceptos básicos en el manejo de enfermedades de cultivos. In: Coolí, M.M., Annone, J.G. and García, R. (eds) *Curso de manejo de enfermedades del Trigo.* Centro Internacional de Capacitación INTA-CIMMYT, Pergamino, Buenos Aires.

Kohli, M., Mehta, and Ackermann, M. (1992) Spread of tan spot in the Southern Cone Region of South America. In: Francl, L., Krupinsky, J. and McMullen, M. (eds) *Advances in Tan Spot. Proceedings of the Second International Tan Spot Workshop, North Dakota State University, Fargo, North Dakota*, pp. 86–90.

Krupinsky, J.M. (1987) Pathogenicity on wheat of *Pyrenophora tritici-repentis* isolated from *Bromus inermis. Phytopathology* 77, 760–765.

Krupinsky, J.M. (1992a) Aggressiveness of isolates *Pyrenophora trictici-repentis* obtained from wheat grown under different field cultural practices. In: Francl, L., Krupinsky, J. and McMullen, M. (eds) *Advances*

in Tan Spot. Proceedings of the Second International Tan Spot Workshop, North Dakota State University, Fargo, North Dakota, pp. 50–55.
Krupinsky, J.M. (1992b) Aggressiveness of isolates *Pyrenophora trictici-repentis* obtained from wheat in the northern Great Plains. *Plant Disease* 76, 87–91.
Krupinsky, J.M. (1992c) Grass host of *Pyrenophora tritici-repentis. Plant Disease* 76, 92–95.
Lamari, L. and Bernier, C.C. (1989a) Evaluation of wheat lines and cultivars to tan spot (*Pyrenophora tritici-repentis*) based on type lesions. *Canadian Journal of Plant Pathology* 11, 49–56.
Lamari, L. and Bernier, C.C. (1989b) Virulence of isolates of *Pyrenophora tritici-repentis* on 11 wheat cultivars and cytology of the differential host reactions. *Canadian Journal of Plant Pathology* 11, 284–290.
Lamari, L. and Gilbert, J. (1998) Toward a coherent model of host–pathogen interactions in tan spot of wheat: an epilogue to the Third International Workshop on Tan Spot of Wheat. *Canadian Journal of Plant Pathology* 20, 440–443.
Lamari, L., Sayoud, R., Boulif, M. and Bernier, C.C. (1995) Identification a new race in *Pyrenophora tritici-repentis*: implications for the current pathotype classification system. *Canadian Journal of Plant Pathology* 17, 312–318.
Lamari, L., Gilbert, J. and Tekauz, A. (1998) Race differentiation in *Pyrenophora tritici-repentis* and survey of physiologic variation in western Canada. *Canadian Journal of Plant Pathology* 20, 396–400.
Lamari, L., Strelkov, S.E., Yahyaoui, A., Orabi, J. and Smith, R.B. (2003) The identification of two new races of *Pyrenophora tritici-repentis* from the host center of diversity confirms a one to one relationship in tan spot of wheat. *Phytopathology* 93, 391–396.
Lamari, L., McCallum, B.D. and De Pauw, R.M. (2005) Forensic pathology of Canadian bread wheat: the case of tan spot. *Phytopathology* 95, 144–152.
Larez, C.R., Oxford, R.M. Jr and Freeman, T.P. (1986) Infection of wheat and oats by *Pyrenophora tritici-repentis* and initial characterization of resistance. *Phytopathology* 76, 931–938.
Li, B. and Sutton, J. (1995) Evaluation of leaf associated microorganisms for biocontrol on tan spot in wheat foliage. *Fitopatología Brasileira* 20, 545–552.
Linhares, A.J. and Luz, W.C. da (1994) Detection of *Drechslera tritici-repentis* on wheat seeds from Mato Grosso do Sul. *Fitopatología Brasileira* 19 (Suplemento), 298.
Loughman, R., Wilson, R.E., Roake, J.E., Platz, G.J., Rees, R.G. and Ellison, E.W. (1998) Crop management and breeding for control of *Pyrenophora tritici-repentis*, causing yellow spot of wheat in Australia. In: Duveiller, E., Dubin, H.J., Reeves, J. and McNab, A. (eds) *Heminthosporum Blights of Wheat: Spot Blotch and Tan Spot.* CIMMYT, Mexico, D.F. Mexico, pp. 10–17.
Luz, W.C. da and Bergstrom, G.C. (1986) Effect of temperature on tan spot development in spring wheat cultivars differing in resistance. *Canadian Journal of Plant Pathology* 8, 451–454.
Luz, W.C. da and Hosford, R.M. Jr (1980) Twelve *Pyrenophora trichostoma* races for virulence to wheat in the central plains of North America. *Phytopathology* 70, 1193–1196.
Luz, W.C. da, Bergstrom, G.C. and Stockwell, C.A. (1998) Seed applied bioprotectants for control of seed-borne *Pyrenophora tritici-repentis* and agronomic enhancement of wheat. *Canadian Journal of Plant Pathology* 20, 384–386.
McFadden, W. (1991) The effect of stubble-borne inoculum of leaf spotting pathogens on disease development and yield loss in winter wheat. *Canadian Journal of Plant Pathology* 13, 280 (Abstr).
McFadden, W. and Harding, H. (1989) Cereal stubble as a source of primary inoculum of leaf spotting pathogens of winter wheat. *Canadian Journal of Plant Pathology* 11, 195 (Abstr).
Maraite, H., Berny, J.F. and Goffin, A. (1992) Epidemiology of tan spot in Belgium. In: Francl, L.J., Krupinsky, J.M. and McMullen, M.P. (eds) *Advances in Tan Spot Research.* North Dakota Agricultural Experiment Station, Fargo, North Dakota, pp. 73–79.
Mehta, Y.R. and Gaudencio, R.A. (1991) The effects of tillage practices and crop rotation on the epidemiology of some major wheat diseases. In: Saunders, D.F. (ed.) *Proceedings of International Conference on Wheat for Non-traditional, Warmer Areas.* CIMMYT, Mexico, DF, pp. 226–283.
Mehta, Y.R., Mehta, A. and Riede, C.R. (2004) Molecular and pathogenic variability of *Pyrenophora tritici-repentis* isolates causing tan spot of wheat in the state of Paraná, Brazil. *Summa Phytopathologica* 30, 436–444.
Mihtra, M.A. (1934) Leaf spot disease of wheat caused by *Helminthosporium tritici-repentis* Died. *Indian Journal of Agricultural Science* 4, 692–700.
Misra, A.P. and Singh, R.A. (1972) Pathogenic differences among three isolates of *Helminthosporium tritici-repentis* and the performance of wheat varieties against them. *Indian Phytopathology* 25, 350–353.

Moreno, M.V. (2007) Caracterización cultural, patogénica genética y bioquímico-molecular de *Pyrenophora tritici-repentis* en Argentina. PhD thesis, Facultad de Ciencias Naturales y Museo, Universidad Nacional de La Plata, Argentina, 188 pp.

Morral, R.A.A. and Howard, R.J. (1975) The epidemiology of leaf spot disease in a native praire II. Airborne spore populations of *Pyrenophora tritici-repentis*. *Canadian Journal of Botany* 53, 2345–2353.

Nisikado, Y. (1928) Preliminary notes on yellow leaf spot disease of wheat caused by *Helminthosporium tritici-vulgaris* Nisikado. *Institute of Agricultural Biology* 4, 103–109.

Odvody, G.N., Boosalis, M.G. and Watkins, J.E. (1982) Development of pseudothecia during progressive saprophytic colonization of wheat straw by *Pyrenophora trichostoma*. In: Hosford, R.M. Jr (ed.) *Tan Spot of Wheat in Related Diseases Workshop*. North Dakota Agricultural Experiment Station, Fargo, North Dakota, pp. 33–35.

Palti, J. (1981) *Cultural Practices and Infectious Crop Diseases*. Springer-Verlar, New York, 243 pp.

Perelló, A.E., Moreno, M.V., Simón, M.R. and Sisterna, M. (2003a) Tan spot of wheat (*Triticum aestivum* L.) infection at different stages of crop development and inoculum type. *Crop Protection* 22, 157–169.

Perelló, A., Mónaco, C., Simón, M. and DalBello, G. (2003b) Biocontrol efficacy of *Trichoderma* isolates for tan spot of wheat in Argentina. *Crop Protection* 22(9), 1099–1106.

Perelló, A., Mónaco, C., Moreno, M.V., Simón, M.R. and Cordo, C. (2006) *Trichoderma* spp. isolates as potential biological agents for foliar wheat diseases under field conditions in Argentina. *Biocontrol Science and Technology* 16(8), 803–813.

Perelló, A., Moreno, M.V., Mónaco, C. and Simón, M.R. (2008) Effect of *Trichoderma* spp. Isolates for biological control of tan spot of wheat caused by *Pyrenophora tritici-repentis* under field conditions in Argentina. *BioControl* 53, 895–904.

Perelló, A., Moreno, M.V., Mónaco, C., Simón, M.R. and Cordo, C. (2009) Biocontrol of *Septoria tritici* blotch on wheat by *Trichoderma* sp. under field conditions in Argentina. *BioControl* 54, 113–122.

Pfender, W.F. and Wootke, S.L. (1987) Production of pseudothecia and ascospores by *Pyrenophora tritici-repentis* in response to macronutrient concentrations. *Phytopathology* 77, 1213–1216.

Pfender, W.F., Pacey, C.A. and Zhang, W. (1988) Saprophytic growth and pseudothecia production by *Pyrenophora tritici-repentis* in plant tissue held at controlled water potentials. *Phytopathology* 78, 1205–1210.

Pfender, W.F., Zhang, W. and Nus, A. (1989) Field performance and greenhouse assay of fungi for biocontrol of residue borne *Pyrenophora tritici-repentis*. *Phytopathology* 79, 1152.

Postnifova, E.N. and Khasanov, B.A. (1998) Tan spot in central Asia. In: Duveiller, E., Dubin, H.J., Reeves, J. and McNab, A. (eds) *Heminthosporum Blights of Wheat: Spot Blotch and Tan Spot*. CIMMYT, Mexico, D.F., Mexico, pp. 107–113.

Rajaram, S. (2001) Mejoramiento Internacional de Trigo. Palabras de Apertura. In: Kohli, M.M., Diaz, M. and Castro, M. (eds) 2003. *Estrategias y Metodologías utilizadas en el mejoramiento de trigo. Seminario Internacional, La Estanzuela, Uruguay*. CIMMYT-INIA, 408 pp.

Rees, L.G. and Platz, G.J. (1979) The occurrence and control of yellow spot of wheat in North-Eastern Australia. *Australian Journal of Agricultural Research* 19, 369–372.

Rees, L.G. and Platz, G.J. (1980) The epidemiology of yellow leaf spot of wheat in Southern Queensland. *Australian Journal of Agricultural Research* 31, 259–267.

Rees, L.G. and Platz, G.J. (1983) Effects of yellow spot of wheat in Southern Queensland. *Australian Journal of Agricultural Research* 34, 39–46.

Rees, L.G. and Platz, G.J. (1992) Tan spot and its control. Some experiences. In: *Advances in Tan Spot. Proceedings of the Second International Tan Spot Workshop, North Dakota State University, Fargo, North Dakota*, pp. 1–9.

Rees, L.G., Platz, G.J. and Mayer, R.J. (1982) Yield losses in wheat from yellow spot: comparison of estimates derived from single tillers an plots. *Australian Journal of Agricultural Research* 33, 899–908.

Sah, D.N. (1994) Effects of leaf wetness duration and inoculum level on resistance of wheat genotypes to *Pyrenophora tritici-repentis*. *Journal of Phytopathology* 142, 324–330.

Sah, D.N. and Ferhmann, H. (1992) Virulence patterns of geographically differing isolates of *Pyrenophora tritici-repentis* and sources of resistance in wheat. *Plant Disease* 76, 712–716.

Sarova, J., Hanzalova, N. and Barlos, P. (2002) *Pyrenophora tritici-repentis* one the most important leaf spot pathogens in the Czech Republic. In: Rasmussen, J.B., Friesen, T.L. and Ali, S. (eds) *4th International Wheat Tan Spot and Spot Blotch Workshop, 21–24 July 2002. Bemidji, Minnesota, North Dakota State University, Fargo, North Dakota.*

Schilder, A.M.C. and Bergstrom, G.C. (1990) Variation in virulence within the population of *Pyrenophora tritici-repentis* in New York. *Phytopathology* 80, 84–90.
Schilder, A.M.C. and Bergstrom, G.C. (1992) The dispersal of conidia and ascospores of pyrenophora tritici-repentis. In: Francl, L.J., Krupinsky, J.M. and McMullen, M.P. (eds) *Advances in Tan Spot Research. Proceedings of the Second International Tan Spot Workshop. NDSU Agricultural Experiment Station*, pp. 96–99.
Schilder, A.M.C. and Bergstrom, G.C. (1994) Infection of wheat seed by *Pyrenophora tritici-repentis. Canadian Journal of Botany* 72, 510–519.
Schilder, A.M.C. and Bergstrom, G.C. (1995) Seed transmission of *Pyrenophora tritici-repentis*, causal fungus of tan spot of wheat. *European Journal of Plant Pathology* 101, 81–91.
Schuh, W. (1990) The influence of tillage systems in incidence and spatial pattern of tan spot of wheat. *Phytopathology* 80, 804–807.
Sharp, E.L., Sally, B.K. and McNeal, F.H. (1976) Effect of *Pyrenophora* wheat leaf blight on the thousand kernel weight of 30 spring wheat cultivars. *Plant Disease* 60, 135–138.
Shoemaker, R.A. (1961) *Pyrenophora phaecomes* (Reb. Ex. Fr.) Fr. *Canadian Journal of Botany* 39, 901–908.
Shoemaker, R.A. (1962) *Drechslera* Ito. *Canadian Journal of Botany* 40, 809–908.
Sim, T. and Willis, W.G. (1982) *Kansas Wheat Disease Losses.* Kansas State University, Manhattan, Kansas.
Simón, M.R. (2006) Factores que modifican la expresión de la resistencia a manchas foliares. In: *A Todo Trigo Conocimiento y producción.* Actas de Congreso, 18 and 19 May 2006, Mar del Plata, Argentina, pp. 167–170.
Sprague, R. (1950) *Diseases of Cereals and Grasses in North America.* Ronald Press Co, New York, 538 pp.
Stover, R.W., Francl, L.J. and Jordahl, J.G. (1996) Tillage and fungicide management of foliar diseases in a spring wheat monoculture. *Journal of Production Agriculture* 9, 261–265.
Strelkov, S.E., Lamari, L., Sayoud, R. and Smith, R.B. (2002) Comparative virulence of chlorosis-inducing races of *Pyrenophora tritici-repentis. Canadian Journal of Plant Pathology* 24, 29–35.
Summerell, B.A. and Burgess, L.W. (1988a) Saprophytic colonization of wheat and barley of *Pyrenophora tritici-repentis* in the field. *Transactions of the British Mycological Society* 90, 551–556.
Summerell, B.A. and Burgess, L.W. (1988b) Factors influencing production of pseudothecia by *Pyrenophora tritici-repentis. Transactions of the British Mycological Society* 90, 557–562.
Summerell, B.A. and Burgess, L.W. (1989) Factors influencing survival of *Pyrenophora tritici-repentis*: water potential and temperature. *Mycological Research* 93, 41–43.
Sutton, J.C. (2005) Present and future perspectives of biological disease control in crops. *Proceedings of the 13th Congreso Latinoamericano de Fitopatologı´a, 19–22 April 2005, Villa Carlos Paz, Co´rdoba, Argentina*, 640 pp.
Sutton, J.C. and Roke, G. (1986) Interactive effects of foliar diseases and fungicide sprays in cultivars of winter wheat in Ontario. *Canadian Plant Disease Survey* 66, 37–41.
Tekauz, A. (1976) Distribution, severity and relative importance of leaf spot diseases wheat in Western Canada in 1974. *Canadian Plant Disease Survey* 56, 36–40.
Valder, P.G. (1954) Yellow leaf spot and pink grain in wheat. *The Agricultural Gazette of New South Wales* 65, 36–37.
Watkins, J.E., Odvody, G.N., Boosalis, M.G. and Partridge, J.E. (1978) An epidemic of tan spot of wheat in Nebraska. *Plant Disease Reporter* 62, 132–134.
Wehmeyer, L.E. (1949) Studies in the genus *Pleospora. Mycologia* 41, 565–593.
Wehmeyer, L.E. (1954) Perithecial development in *Pleospora trichostoma. Botanical Gazette* 115, 297–310.
Wiese, M.V. (1987) *Compendium of Wheat Diseases.* The American Phytopathological Society, St Paul, Minnesota, 106 pp.
Wolf, E.D. de and Francl, L.J. (1997) Neural networks that distinguish infection periods of wheat tan spot in an outdoor environment. *Phytopathology* 87, 83–87.
Wolf, E.D. de and Francl, L.J. (1998) Empirical infection period models for tan spot of wheat. *Canadian Journal of Plant Pathology* 20, 394–395.
Wolf, E.D. de, Effertz, R.J., Ali, S. and Francl, L.J. (1998) Vistan of tan spot research. *Canadian Journal of Plant Pathology* 20(4), 349–370.
Wolf, P.F.J. and Hoffmann, G.M. (1993) Biological studies on *Drechslera tritici-repentis* (Died.) Shoem. (teleomorph *Pyrenophora tritici-repentis*) (Died.) Drechsler) the causal agent of a leaf spot disease of wheat. *Zeitschrift fuer Pflanzenkrankheiten und Pflanzenschutz* 100, 33–48.

Wright, K.H. and Sutton, J.C. (1990) Inoculation of *Pyrenophora tritici-repentis* in relation to epidemics of tan spot of winter wheat in Ontario. *Canadian Journal of Plant Pathology* 12, 149–157.
Zadoks, J.C. and Schein, R.D. (1979) *Epidemiology and Plant Disease Management*. Oxford University Press, New York, 427 pp.
Zhang, W. and Pfender, W.F. (1993) Effect of residue management on wetness duration and ascocarp production by *Pyrenophora tritici-repentis* in wheat residue. *Phytopathology* 82, 1434–1439.

22 Epidemiological Studies on *Septoria* Leaf Blotch of Wheat in Argentina

Cristina A. Cordo
Comisión de Investigaciones Científicas de la Provincia de Buenos Aires, Centro de Investigaciones de Fitopatología (CIDEFI) – Facultad de Ciencias Agrarias y Forestales, La Plata, Argentina

Abstract
This chapter introduces the detailed and novel contributions on the epidemiological spread of *Mycosphaerella graminicola* over the wheat field and over time. The within-season and between-crop methods of multiplication, survival and their environmental relations are reviewed. Genetic arguments are given to demonstrate the influence of ascospores as the major source of movement of the pathogen into new fields. Coupled with the evidence that populations worldwide are genetically very similar, it does seem possible that a novel form of the pathogen has been spreading worldwide. This would raise the interesting question as to what epidemiological characteristic confers the new form's invasiveness. There is a clear association between the evolution of the disease and weather conditions. Wheat cultivars exhibit differential responses to infection by *M. graminicola*. Breeding for disease resistance is an important tool in the integrated management of disease. Also, fungicide application and the use of biocontrol organisms alone or in combination with fungicides is mentioned as other integrated action.

Introduction

Cereals and the processed foods derived from them are still the principal sources of nutrition in many parts of the world. Bread wheat (*Triticum aestivum* L.) is the most widely grown and consumed food crop. It is the staple food of nearly 35% of the world's population and the demand for wheat will grow faster than for any other major crop (Rajaram, 1999). The forecasted global demand for wheat in the year 2020 varies between 840 (Rosegrant *et al.*, 1995) to 1050 Mt (Kronstad, 1998). To meet this demand, global production will need to increase by 1.6–2.6% annually from the present production level of 560 Mt. For wheat, the global average yield must increase from the current 2.5 t/ha to 3.8 t/ha. In 1995, only 18 countries worldwide had an average annual rate of growth over 2% between 1961 and 1994 (INTA-CIMMYT, 1996). In Western Europe and North America, the annual growth rate for yield was 2.7% from 1977 to 1985, falling to 1.5% from 1986 to 1995 (Rajaram, 1999). Argentina, with a production of 16.11 Mt in the 2006/07 campaign (Encuesta Agrícola, DIEA-MGAP) and 17.47 Mt in the 2007/08 campaign (OPYPA, 2008), is an important wheat exporter with a volume of 300.00 t and an average yield of 2800 kg/ha (data from OPYPA yield and production). The increased need for wheat export forces producers to improve soil

 Management of Fungal Plant Pathogens (eds A. Arya and A.E. Perelló)

management and plant cultivation strategies, including the application of crop protection agonts to reduce losses, and is associated with the effective use of measures designed to increase yield and ensure quality. With the increased proportion of conservation tillage practice and with the striving for optimal exploitation of the yield and quality potential of cultivars with the aid of appropriate cultivation and fertilization measures, the importance of certain fungal diseases as yield-limiting factors has increased considerably. The occurrence of fungal pathogens can not only limit cereal production in temperate climates, but can also jeopardize the requisite return on capital under conditions of intensive farming. The epidemic development of pathogens, which vary greatly in their ecological requirements, is strongly dependent on the weather and, together with the slightly different cultivation systems, this leads to differences in the type of infection (pathogen species) and in the level of infection (severity of the disease) from one year to the next. Given these circumstances, the level of application of fungicides in cereal cultivation in Argentina has increased since 1996. Its application produced an increased yield of 20–32% in relation to the control test with respect to time of application, fungicide molecular type and wheat variety (Annone *et al.*, 1995).

Chemical crop protection generally has been accepted in Argentina because it should only be used when circumstances make it necessary to achieve the production target and when all other options have been considered. Crop protection measures should be harmonized with the actual infection situation prevailing in the crop and targeted countermeasures must be initiated only where there is a real risk. Cortese *et al.* (1998) and Carmona *et al.* (1999) have fixed the economic damage threshold (UDE) and an action threshold (UDA) for different diseases on wheat. UDA represents the incidence value of the disease to decide the fungicide application on the crop to reduce the cost of application. UDE is based on the formula of Munford and Norton (1984) and represents the value of the disease when the yield losses produced by the pathogen would be equivalent to the cost of fungicide application. The threshold values established should represent infection limits beyond which economic losses are highly likely in the shorter or longer term, and the point at which the pathogen population reach this limits in the field crop determines the time at which fungicide should be used. The pathogen-specific thresholds must be worked out within the framework of exact scientific investigations in relation to the crop management methods used and the environmental conditions prevailing during the growing period. Their development requires extensive case studies (under outdoor conditions) in order to establish how the population dynamics and detrimental effects are influenced by the weather, wheat varieties, use of fertilizer, the preceding crops and the level of inoculum. The purpose of collecting all this information is to control individual pathogens with appropriate products at appropriate application rates, and to do this at a time at which one application is most likely with regard to pathogen development and limitation of damage, using the lowest possible input.

Epidemiology and population genetics are different but related subsets of population biology. Epidemiology focuses on disease progression, the increase in pathogen populations through time and the movement of pathogen populations through space (usually from plant to plant). Most epidemiology studies deal with a short timescale (e.g. 1–2 growing seasons) and small spatial scales (e.g. disease development in a field or a plantation). Epidemiology involves mainly physical processes such as distances of spore movement or effects of weather variables on latent periods. It does not take account of the differences in behaviour or genetically distinct individuals in a collection of individuals.

Population genetics focuses on the processes that lead to genetic changes, or evolution, in populations over time and space. Population genetics deals mainly with genetic processes such as genetic drift, gene flow, mating system, natural selection and mutation. Present study records that epidemiological investigations, based on disease evolution, resistance supply, early detection of the disease, biological crop protection and genetic studies of the pathogen, can be used to orientate the management of disease under natural conditions.

The Disease as a Problem

Septoria tritici blotch (STB) is caused by *M. graminicola* (Fuckel) Schroeter, in Cohn, which is the teleomorphic stage of *S. tritici* Roberger and Desmazieres (anamorph stage). Sanderson (1972) proved the connection between the two stages and the sexual (teleomorph) form has been reported in several countries (Hunter *et al.*, 1999). Cordo and Arriaga (1990) reported the sexual stage in Argentina. It is also known to play a role in the disease's cycle. It causes most of the initial infection in winter wheat crops, during the autumn in the UK (Shaw and Royle, 1989) and USA (Schuh, 1990). In Argentina, an increase in ascospores at harvest time has been reported, suggesting that the sexual stage may be important in initiating the infection in the next growing season (Cordo *et al.*, 1999). Another possible means of spread within a crop during summer is by airborne ascospores, which may play a more major role than previously recognized (Hunter *et al.*, 1999; Cordo *et al.*, 2005).

Studies on the Disease's Evolution

Several control methods, including the use of fungicides and other cultural practices, may reduce the effect of STB, but genetic resistance is the most cost-effective and environmentally safe technique to manage the disease.

In Argentina, an inoculation technique using oat grains covered with the stromatic mycelia of *S. tritici* were presented to check the resistance of the *Septoria* Monitoring Nursery (SMN) set. The international set created by the CIMMYT provides information on the interactions of pathogen × cultivar on different regions of the world. The use of this set allows the generation of extensive information on resistance sources for diverse pathogen populations under variable environmental conditions. Differences among accession reaction were significant due to the rich composition of the selected sources of resistance identified by CIMMYT and a tentative group of differentials proposed by Eyal (Gilchrist *et al.*, 1999) that were incorporated into the SMNs.

The inoculum for grain application was prepared in sterilized 500-ml flasks with 100 g of oat grains and 50 ml of a liquid extract malt medium (Perelló *et al.*, 1997). The grains were soaked with 10 ml of an inoculum suspension (10^7 conidia/ml) of *S. tritici* isolate and incubated for 15–21 days at 23 ± 2°C in darkness and shaken daily to promote good fungal growth. After the incubation, the grains were colonized by a stromatic mycelium and were spread and dried on trays under laboratory conditions. The covered grains were spread on to the soil next to the plants during the tillering growth stage (GS23, Zadoks *et al.*, 1974). Plants in the plots were assessed for *S. tritici* infection at anthesis (GS60) and at the medium milk (GS75) stages.

The accessions (1-BOBWHITE S; 2-TIA. 2/4/CS/TH.CU//GLEN/3/ALD/PVN; 3-CHIRYA.1; 4-CHIRYA 4; 5-CS7TH.CU// GLEN/3/ALD/PVN/4/NANJING; 6-EG-A/H56 7.71//4#EG-A/3/2#CMH79.243; 7-MH86.540-A-1Y-3B-2Y-1B-1B-1B-1Y-1M-1Y;ALD/PVN// YMI#6; 9-SHA5/BOW; 10-ENCOY 1582–1B; 11-BOBWHITE S as the other derivative line; 12-DON ERNESTO INTA; 13-SERI M82; 14-BETHLEHEM; 15-LAKHISH; 16-KAUZ; 17-PENJAMO; 18-ETIT 38; 19-GLENNSON M81) were sown in a factorial design experiment.

The pulverization inoculum was produced using the same isolate as in the previous year. The conidial concentration of the suspension was adjusted to 1×10^7 conidia/ml. A comparison between the pulverization and the grain application methods was made in the field in 2000. The inoculum suspension was sprayed on to the leaves at the tillering stage (GS23). After inoculation, plants were kept moist by sprinkling water several times a day over 3 days. The severity of the infection was registered on the flag leaf at the beginning of the flowering (GS60) and medium milk (GS75) stages using a modified double digit Saari–Prescott scale (Saari and Prescott, 1975). The cut for resistant behaviour was estimated as 5.3 (Gilchrist *et al.*, 1999). Weather variables (daily temperature, relative humidity and rainfall) were recorded from the date of inoculation to anthesis. Plant height was evaluated.

To compare the inoculation techniques, both the necrotic coverage percentage (NCP) and pycnidial coverage percentage (PCP) were scored on the upper three leaves of 15 plants, 21 days after inoculation. The cut-off point between resistant and susceptible response classes was 16.8% NCP following Eyal *et al.* (1985). The comparison between pulverization and grain application showed that, except for the variety Bobwhite 'S' CM 33203-K-10M-7Y-3M-2Y-1M-OM and the line Tia.2/4/CSTH.CU//GLEN/3/ALD/PVN CIGM88.734-1B-3PR-0PR-1M which reacted as in the observations of Gilchrist *et al.* (1999), all genotypes were more susceptible under Argentine conditions. The higher level of virulence of the Argentine isolates and frequency of variation could explain this behaviour (Eyal *et al.*, 1985; Gilchrist *et al.*, 1999; Cordo *et al.*, 2006).

The results of the severity for NLP and PCP in this study are in agreement with previous research (Eyal, 1985; Gilchrist *et al.*, 1999). The advanced resistant lines coming from the crosses with a group of resistant Chinese lines did not show a high level of resistance (Ald/Pvn/YM#6, Milan/Sha#7, Catbird, Talhuen INIA, Sha3/Seri/PSV/Bow and the cultivar with Kavkaz/K4500 sources).

The resistant check Bethlehem was not resistant at CIMMYT or in our conditions. The bread wheat checks SeriM82 and Glennson M81 (with Veery 'S' germplasm) and Lakhish were susceptible, as was expected (Gilchrist *et al.*, 1999). The durum wheat ETIT 38 and the resistant check Bethlehem had the same level of susceptibility as bread wheat checks (SeriM82, Lakhish) and as was scored by Kohli (1995). The disease resistance introduced from Brazilian germplasm was detected on a short, early-maturing resistant line derived from IAS 20 spring wheat and a more susceptible reaction on lines derived from IAS 58. Bobwhite 'S' germplasm and its derivative lines (in Argentina represented by Don Ernesto INTA) showed variable levels of resistance caused by its background with more than one genetic source and the presence of a low number of major genes (Cordo *et al.*, 1994).

Plant height was not associated with the resistant reaction. The negative associations were present when weather conditions were less conducive to the development of the disease. Non-conducive conditions and the further distance between leaves in tall cultivars could have reduced the rain-splash dispersal of pycnidiospores, thus causing this negative association (Arama *et al.*, 1999; Simón *et al.*, 2005; Arraiano and Brown, 2006); it could also depend on the presence of ascospores, which could reduce the effect of plant height on the expression of the disease. In Argentina, the presence of the teleomorphic stage during the whole growing period has been reported (Cordo and Arriaga, 1990; Cordo *et al.*, 1999, 2005).

The modified double digit Saari–Prescott scale was adopted for evaluation of this set (CIMMYT, rules for evaluation, Eyal *et al.*, 1987). The separate analysis of digit 1 and 2 allowed the relative height reached by the disease to be shown simultaneously with the severity of the damage (PCP) (Table 22.1). The differences observed for the first digit in the accession response to the inoculum concentration were attributed to the maximum level of attacked leaf (8th leaf) that was reached with the highest concentration of inoculum: 280 g. In contrast, the second digit did not show differences for either concentration. The lesions were restricted in extension, reaching only a maximum of 20% more of the PCP in the cases of highest susceptibility. This result confirmed that the level of resistance of tested materials was adequate to maintain a low intensity of infection according to the objectives proposed by Eyal and Gilchrist at the beginning of this project (Gilchrist *et al.*, 1999).

Two factors were influencing the expression of the disease on the leaves: the concentration of the grain inoculum (120 g/m^2 was optimum for differentiation between susceptible and resistant accessions) and

Table 22.1. *S. tritici* infection average (digit 1 and 2) for different concentrations of inoculum and different years.

Accession	Inoculum concentration				Years			
	Digit 1[1]		Digit 2[1]		Digit 1[1]		Digit 2	
n	C1	C2	C1	C2	1997	1999	1997	1999
1	6.00 b[2]	6.50 b	2.50 b	1.00 a	6.50 b[2]	6.00 b	2.50 b	1.00 a
2	5.50 b	7.25 d	2.25 a	0.75 a	6.75 b	6.00 b	2.50 b	0.50 a
3	2.50 a	7.50 e	2.75 c	1.75 a	5.50 a	4.50 a	3.00 d	1.50 a
4	7.25 c	7.25 c	0.50 a	1.50 a	8.00 f	6.50 b	1.75 a	0.25 a
5	6.75 b	7.25 d	2.00 a	2.25 a	7.00 c	6.00 b	2.75 c	1.50 a
6	6.25 b	6.75 b	1.25 a	1.75 a	6.75 b	6.25 b	1.75 a	1.25 a
7	6.50 b	8.00 g	1.00 a	0.75 a	7.75 e	6.75 b	1.00 a	0.75 a
8	6.50 b	6.75 b	2.75 c	0.50 a	7.00 c	6.25 b	2.75 c	0.50 a
9	6.50 b	5.75 b	1.50 a	2.00 a	6.25 b	6.00 b	2.50 b	1.00 a
10	6.00 b	7.75 f	1.00 a	1.75 a	7.25 d	6.50 b	1.75 a	1.00 a
11	6.50 b	7.00 c	0.75 a	2.00 a	7.75 e	5.75 a	1.00 a	1.75 a
12	6.25 b	7.00 c	1.00 a	5.00 g	7.25 d	6.00 b	3.25 d	2.75 c
13	6.75 b	7.25 d	0.75 a	1.50 a	7.75 e	6.25 b	0.50 a	1.75 a
14	7.00 c	7.25 c	1.25 a	2.25 a	8.00 f	6.25 b	2.75 c	0.75 a
15	6.25 b	6.50 b	4.00 f	3.00 d	6.75 b	6.00 b	3.25 d	3.75 e
16	7.00 c	7.25 c	3.00 d	3.50 e	8.00 f	6.25 b	2.50 b	4.00 e
17	6.75 b	7.75 f	2.25 a	2.25 a	8.00 f	6.50 b	3.25 d	1.25 a
18	6.50 b	7.75 f	2.00 a	0.50 a	7.750e	6.50 b	1.25 a	1.25 a
19	6.50 b	7.75 f	2.61 c	1.50 a	7.50 d	6.75 b	3.25 d	0.86 a

Note: [1]As assessed by a modified double-digit Saari–Prescott scale (1975). [2]Mean values followed by the same letter are not statistically different. LSD test ($P < 0.01$); C1 = 120 g/m^2; C2 = 280 g/m^2.

the wet environment. For the grain application treatment, the density of the plants was too important for rainfall to produce the infection. If the rain regime was not frequent and intensive, the pycnidiospores could not reach the higher leaves, making it difficult for the inoculum to ascend.

In the pulverization treatment, the surface covered by the inoculum included more than one leaf stratum. A simultaneous proliferation of the pathogen was obtained in all foliage levels, which, in addition to the beneficial structure of the canopy, produced the highest values of severity. The most susceptible varieties at the GS75 stage were those that had a longer period of green leaf during the growth cycle; but something different occurred with ETIT 38 that did not show any difference on NCP and PCP for both growth stages. This could be explained by the quick senescence of the leaves that practically stopped the development of the fungus at the end of GS60.

The higher values of the disease in 1997 compared with those of 1999 were caused by the influence of the climatic conditions. The temperature was not an important factor because there was no statistical difference in 3 years of experiments. In 1998, high humidity (30% more than the following year) and increased rainfall (425.29 mm more than the following year) were responsible for the rapid increase of the disease compared with results of 1999 (data not shown).

Both inoculation techniques were appropriated to monitor the behaviour of the accessions of the SMN set. If the experimental field is under a good rain regimen from tillering to flowering, grain application is recommended. But if it is on a dry irrigated area, pulverization with extra irrigation as a humidity chamber is suggested.

Comparing the efficacy of each inoculation technique, different symptoms produced by each treatment at the beginning of the disease have been associated with environmental conditions (Table 22.2). In relation to the grain application treatment, the generalized necrosis and pycnidial development on the lower leaves could have risen to the upper leaves if irrigation or rainfall had been present at this early stage of infection. The density of the plants was too important for rainfall to produce the infection. The more compact density of the canopy helps to maintain a microclimate for the progress of the disease. The lack of germination that affected seeds of some accessions could have modified the canopy structure and the inner microclimate; consequently, it could have delayed the movement of the pathogen to the upper leaves of the plant (Lovell *et al.*, 1997). It could explain the absence of relation between the increase of inoculum concentration and the decrease of the severity on digit 2. In the pulverization treatment, the surface covered by the inoculum included more than one leaf stratum. A simultaneous proliferation of the pathogen was obtained in all foliage levels, which, in addition to the beneficial structure of the canopy, produced the observed reactions.

The strong differences observed between treatments (pulverization and grain application) could be explained by the different rates in the progress of the disease for each treatment. In pulverization, the inoculum included more canopy levels. In agreement with Lovell *et al.* (1997), the ascending movement of the disease was facilitated, especially in cultivars that, because of their compressed canopy, maintained a more favourable microclimate. On the contrary, the delayed attack with grain application could be explained because only the rain dispersed the conidia from the infected grains. If the rain regimen was not frequent and intensive, the spores could not reach the leaves, making it difficult for the inoculum to ascend.

Although the pulverization and grain application techniques were shown to be effective and could be recommended for field trials, each one demonstrated different advantages. The grain application treatment had the advantage that the incubation period of the disease was maintained with irrigation. In this case, a known weight of inoculated grains was spread on to the soil, between the rows, as primary inoculum. In this technique, pycnidia on stromatic mycelia formed on grains could release pycnidiospores over a long period if wetted. This process may be repeated several times if the grains are dried and wetted again. The splash dispersal effect can increase spore transport from a low to a high level of the crop and from plant to plant.

There was a general correlation between the reaction of some lines and the differential varieties in relation to NCP and PCP. The most susceptible varieties at the GS75 stage were those that had a longer period of green leaf (8 days more); therefore, the pathogen had a higher probability of proliferating in the leaf. Something different occurred with Lakhish. In this differential variety, NLC and PCP did not show any differences for both growth stages. This could be explained by the quick senescence of the leaves that practically stopped the development of the fungus at the end of GS60.

The technique of grain inoculation presented in this research has the advantage of being simple to handle compared with the installation of a barrier to infect wheat plants artificially, as is necessary in the pulverization methods (Sanderson *et al.*, 1986). In the latter, it is necessary to plan the exact date of the previous sowing and the direction of this barrier in relation to the tested wheat rows.

Comparing infectivity of different inoculum concentrations in the grain application treatment, a gradient of infection was obtained and its effect was related to the variation of humidity and rain regime. The differentiation between susceptible and resistant entries of this set was possible using 120 g of oat grains/m^2 covered with stromatic mycelia of *S. tritici*. With this, it is not necessary to use the highest concentration of oat grains. This type of inoculum has a long-lasting effect next to the plants, but the most important characteristics are that the incubation period is produced without the installation of a wet chamber; it is simple to

Table 22.2. Necrotic and pycnidial coverage percentages caused by *Septoria tritici*.

	Inoculum type				Growth Stage			
	Pulverization[a]		Grain application[a]	[a]	GS60[d]		GS75[d]	
	N[b]	P[c]	N	P	N[b]	P[c]	N	P
Accessions				%				
1	41.60 abc[1+]	40.94 [1+]e	28.91 abcd[1+]	28.02 def[1+]	3.21 a[2+]	18.31 cde[2+]	51.28 cde[2+]	50.66 cd[2+]
2	42.59 bc	42.68 de	37.33 defg	37.43 i	12.89 b	12.88 bc	67.03 ijk	67.24 f
3	62.45 fg	61.91 g	32.26 bcde	28.17 ef	30.08 bcd	30.08 fg	64.63 hij	60.00 e
4	27.96 a	11.67 a	25.32 ab	25.32 cd	1.59 a	0.93	54.88 efgh	37.93 a
5	31.92 a	26.98 b	40.91 efg	37.14 hi	20.12 b	12.74 bc	52.71 def	51.37 d
6	46.52 bcd	39.60 d	18.99 a	18.76 ab	21.84 b	21.70 de	43.67 abc	36.66 a
7	38.89 ab	33.34 c	26.14 abc	17.97 ab	22.48 bc	13.16 bc	42.55 abc	38.15 a
8	42.73 bcd	37.68 cd	36.93 def	30.61 fg	17.67 b	16.47 bcd	61.99 ghij	51.83 d
9	70.62 gh	67.48 g	47.88 g	48.56 j	29.93 bcd	30.55 fg	88.58 m	85.49 h
10	51.60 de	47.35 ef	35.07 cde	34.66 ghi	10.83 b	11.18 b	75.84 l	70.82 fg
11	48.73 cd	41.15 de	45.73 fg	40.59 ij	49.17 e	36.89 gh	45.29 bcd	44.86 bc
12	46.77 bcd	47.16 ef	25.63 ab	21.81 bc	11.48 b	42.07 h	60.93 fghi	56.90 de
13	50.64 cde	39.80 d	25.32 ab	20.23 abc	32.39 cd	16.93 bcd	43.58 abc	43.10 b
14	58.97 ef	51.89 f	34.75 cde	31.52 fgh	24.49 bc	17.64 cd	69.22 jkl	75.26 g
15	76.43 hi	63.16 g	44.35 fg	37.24 hi	33.84 d	17.70 cd	86.95 m	82.70 h
16	83.23 i	83.55 h	45.79 g	35.10 ghi	55.20 e	45.17 h	73.82 kl	73.48 g
17	62.21 f	61.88 f	40.54 efg	38.19 i	27.77 bcd	24.81 ef	74.97 l	65.77 f
18	66.24 fg	65.99 g	19.39 a	14.87 a	46.40 e	41.56 e	39.23 ab	39.30 a
19	33.79 a	38.58 cd	20.97 a	25.48 cdf	22.18 bc	23.02 de	32.57 a	41.03 ab
Mean	51.78	47.42	33.27	30.09	25.60	21.15	59.46	56.45

Note: [a]two types of inoculum; [b]mean of necrotic coverage percentage; [c]mean of pycnidial coverage percentage; [d]two growth stages. [1+]Each value is the average of the three upper leaves with two inoculation methods; [2+]each value is the average of the upper three leaves in two growth stages. LSD test ($P < 0.01$).

transport over a long distance and to store for a long period of time (5 days at 5°C).

Climate Influences

After the initial infection or inoculation, the environment is one of the factors conducive to the development of the disease. Different experiments have demonstrated which are the weather conditions for a more favourable expression of the disease (Simón *et al.*, 2005, Cordo *et al.*, 2006). In one experiment conducted in Argentina in 1998 (Simón *et al.*, 2005), the severity of the disease was highest in the early cultivars because precipitation was higher and radiation lower for these cultivars. Precipitation was 53.4 and 18.8 mm and radiation 3511 and 5127 Watt/m^2 for a period of 15 days before evaluation for the earliest and the latest cultivars, respectively. Also, these differences in weather variables in 1998 produced a negative association between pycnidial coverage and days to heading.

In another experiment, Cordo *et al.* (2007) related that the higher values of the disease in 1997 compared with those of 1999 were caused by the influence of the climatic conditions from boot (GS 43) to hard ripening (GS 87) stages, following Zadoks *et al.* (1974). In 1997, high humidity (30% more than the following year) (Fig. 22.1) and increased rainfall (425.29 mm more than the following year) (Fig. 22.2) were responsible for the rapid increase of the disease compared with the results of 1999. For this experiment, temperature was not an important factor in the development of the disease since in 3 years of experiments there were no statistical differences in temperature (mean temperature from the inoculation to the end of the experiment was 17.10°C for the first 2 years and 16.69°C for 1999). Related to the inoculation process that is under discussion, if the experiment is to be carried

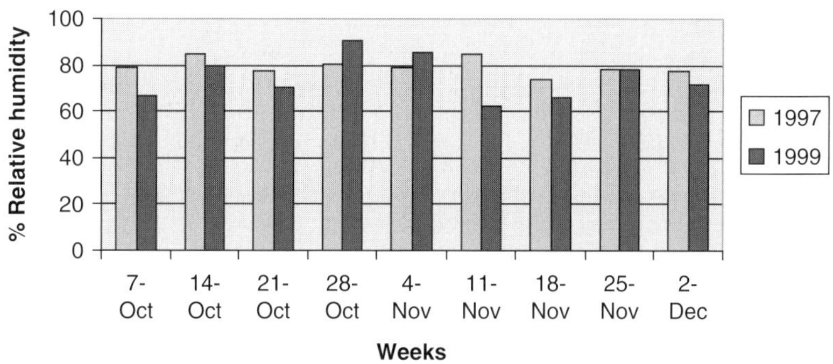

Fig. 22.1. Histogram showing relative humidity during October–December of 1997 and 1999.

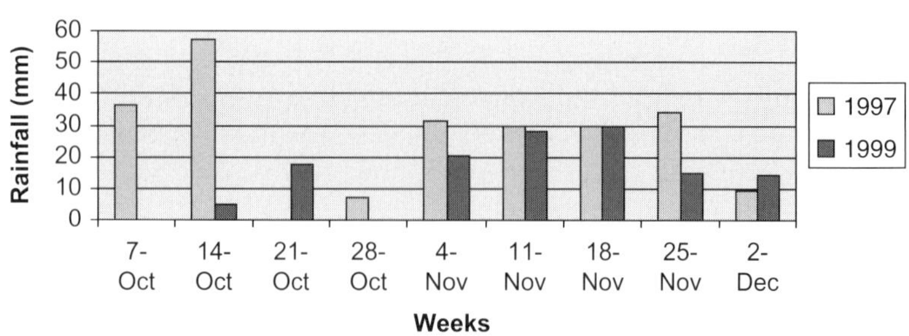

Fig. 22.2. Histogram showing rainfall during October–December of 1997 and 1999.

out on an artificially irrigated area, pulverization with an appropriate suspension of spores and 48 h of extra irrigation in a wet chamber are suggested. However, if the area is under a good rain regime from tillering to flowering stages, the application of grains covered with sporulated mycelia is a feasible option.

Early Detection of the Disease

At the beginning of the wheat-growing season in 1997, Adgen Phytodiagnostic invited the author to participate in a pilot project to identify and quantify *S. tritici* and *S. nodorum* by antibody-based immunoassays and also to compare with the visual method and the sensitive methods for the early detection of *S. tritici*. The objective of this work has been to test the Adgen ELISA kit in a monitoring process on lower to higher leaves during the wheat season in Argentina.

A randomized complete block design with four replicates and a 1.4 × 1 m size subplot was used. Treatment consisted of either an inoculated plot or a control treated with a foliar fungicide spray programme. Plantvax and Tilt were applied at 500 cm^3/ha. Ten main tillers were collected per subplot using a uniform, randomized sampling pattern at GS10.1 (first spikelets just visible 28 October); GS10.3 (heading process 14 November); GS10.5 (flowering 27 November); GS11 (ripening 8 December). Samples for testing consisted of ten leaves bulked for each layer of leaves, each replication and each data of collection. At the same time, the severity of the lesions on the sampled leaves was noted. Samples were homogenized in 50 ml buffer and testing following the protocols described for the DU PONT enzyme-linked immunosorbent assays (ELISA) for *S. tritici*. ELISA results were expressed as the number of *S. tritici* antigen units/ml of homogeneized plant tissue (AgU/ml). AgU/ml values were averaged.

A good correlation was observed between ELISA readings from infected leaves coming from different growth stages (GS10.1, GS10.3, GS10.5) and the visual development on the respective foliar level (C. coefficient = FL 0.68; FL-1 = 0.69). This indicates that the infection increased (correlated with AgU/ml) throughout the different growth stages (Table 22.3). With a variance analysis, the antigenic units registered and the severity of the infection on three wheat cultivars coming from inoculated and protected treatments were compared (Table 22.4). Highly significant differences were observed between inoculated and protected treatments for severity and antigenic units into Flag leaf and Flag leaf-1. A high correlation was calculated (C. coefficient = 0.56) between the average per cent visual attack of a sample and the measured antigenic units. Despite the good correlation, the lower interval of the level of attack scale gave the most confirmable antigenic unit values. This immunoassay has demonstrated to be highly sensitive and quantitative, with antigenic unit concentrations being correlated with the severity of the disease. The infection levels of *S. tritici* in Los Hornos samples could be determined with significant precision. In addition, the specificity of the assay allowed accurate identification of this pathogen, despite the presence of other foliar pathogens. In this assay, only the presence of *Alternaria triticimaculans* gave a cross-reaction.

The Adgen Phytodiagnostic Septoria ELISA kit detected and quantified the amount of *S. tritici* antigen in infected plant tissues. As our experience indicates, the use of this kit can be recommended in a monitoring process for earlier reports of *S. tritici* infection.

Checking the Ascendant Movement of the Inoculum

The ascendant movement of the inoculum was also checked with the diagnostic immunoassay kit for *S. tritici* from the Adgen Company. Increased severity on different levels of the canopy was registered from the latent period of the infection produced by two types of inoculum application (grains covered by pathogen mycelium and pulverization) during 4 weeks from inoculation at tillering stage (GS23).

Table 22.3. Correlation between percentage of lesions covered by pychnidia and antigenic units during the 4 weeks of study.

Susceptible cultivars	Canopy level	Grain application AU[3]	Grain application PCP[4]	Pulverization AU	Pulverization PCP
1°a	Y[1]	42 b	0.0 a	42 c	0.0 a
1°a	B[2]	140 c	0.0 a	300 d	19.16 c
2°b	Y	5.5 a	0.0 a	1.0 a	0.0 a
2°b	B	34 b	3.91 b	120 c	20.8 c
3°c	Y	3.6 a	0.0 a	4.2 a	0.0 a
3°c	B	34 b	3.91 b	82 b	25 c
4°d	Y	5.5 a	0.0 a	22 b	0.0 a
4°d	B	54 b	6.5 b	54 a	6.5 b
Resistant cultivars		AU[3]	PCP[4]	AU	PCP
1°a	Y	2.45 a	0.0 a	3.10 a	0.0 a
1°a	B	65 b	0.0 a	115 b	0.0 a
2°b	Y	9.30 a	0.0 a	3.6 a	0.0 a
2°b	B	8.5 a	0.0 a	180 c	15 b
3°c	Y	65 b	0.0 a	130 b	0.0 a
3°c	B	140 c	4.80 b	195 c	10 b
4°d	Y	42 b	0.0 a	20 b	0.0 a
4°d	B	80 c	1.28 a	80 c	1.28 a

Note: [1]youngest leaf; [2]the leaf below; [3]antigenic units/ml; [4]pycnidial coverage percentage; a, 14 October; b, 9 November; c, 16 November; d, 23 November.

Table 22.4. Information summary for two populations of *Septoria tritici* from Argentina.

	Los Hornos population	Balcarce population
Total isolates	58	62
No. of genotypes	35	39
No. of alleles	24	22
Isolates having fingerprint data	55	58
No. of fingerprint patterns	14	13
Fingerprint pattern types	**A,E,F,G,M,N,O,P,R,S,U,V,W,X**	**A,B,D,E,H,I,K,L,M,P,Q,R,V**

Six leaves per canopy level were sampled per week. Two levels of asymptomatic leaves (the youngest and the leaf below) were chosen from two varieties (Chirya 1 as resistant and Bethlehem as susceptible) belonging to the 8th SMN set. The first sample was taken at GS30 (first node) stage and the following were taken one per week for 3 more weeks. Samples were homogenized in 50 ml buffer and tested following the protocols described for the Adgen ELISA. An analysis of variance was performed with the dates of percentage of lesion covered by pycnidia and antigen units/ml for each week. An LSD test was used to compare treatment means. Correlation between per cent of lesions covered by pycnidia and antigenic units were performed throughout the treatments and during the 4 weeks.

According to the analysis of variance (Table 22.3), highly significant differences were found between each level of the canopy for AU and PCP throughout the weeks and with the two inoculation techniques. There were significant differences for cultivars. A significant correlation was found between PCP and antigen units (C. coefficient = 0.56***) (calculated on 67 dates). The youngest leaves almost had the lowest values of AU. On the leaves below, it had increased. The AU and PCP values were higher in susceptible than in resistant cultivars. In general, on the 2nd or 3rd week after inoculation, the infection was detected in the youngest symptomatic leaf, indicating that the infection was installed in an ascendant level by splashing.

Population Studies of the Pathogen

The population structure and genotypic diversity of *S. tritici* from two crop field populations in Buenos Aires Province separated by 500 km were studied with DNA restriction fragment length polymorphism. From of the 137 isolates from different areas of the Argentine wheat-growing region, only 120 were characterized using the RFLP technique with P32 labelled probes. The pSTL70 fingerprinting probe hybridized many DNA fragments of different sizes in isolates from field populations of both locations. All leaf samples were processed for isolation of the fungus, followed by fungus culture, DNA extraction, Pst1 enzyme digestion, radioactive hybridization and X-ray film detection. Some of the isolates did not yield good quality DNA for the restriction enzyme digestion process. This explains the loss of 17 isolates in the samples of the populations.

In total, 24 alleles were found for the Los Hornos population and 22 alleles for the Balcarce population at the eight RFPL loci (Table 22.4). Despite the difference in the number of alleles, Nei's measure of genetic diversity across all loci was different for both populations (0.2619 for Los Hornos and 0.3161 for Balcarce). Among the 58 isolates of Los Hornos and 62 of Balcarce with complete data from individual RFLP loci, 35 multilocus haplotypes for the first locality and 39 for the second locality were registered. Seven new haplotypes (3a, 20a, 71a, 37a, 47a, 52a, 58a) were added to the list published on the Internet (*S. tritici* RFLP alleles). The haplotype frequency in per cent varied from 1 to 21 times for the Los Hornos population and from 1 to 9 times for the Balcarce population. Genotype diversity was greater in the Balcarce population ($\hat{G}$ = 31.61 or 26.34% of the theoretical maximum of 120) than in the Los Hornos population ($\hat{G}$ = 26.19 or 21.82% of the theoretical maximum of 120). As the mean genetic diversity between populations was high for the 8 loci of RFLP, a significant difference existed between the populations of the two localities.

Fifty-eight multilocus haplotypes and 13 fingerprint patterns were registered for the Los Hornos population and 55 multilocus haplotypes and 14 fingerprint patterns for the Balcarce population when they were hybridized with pSTL70. Many isolates of both populations had from one to several haplotypes for each fingerprint pattern. In the Los Hornos population, the E fingerprint pattern was present on 14 different haplotypes, but it corresponded 3 times with the same 11112110611 haplotype. In the Balcarce population, the same fingerprint was present on 11 different haplotypes, but it corresponded 8 times with the 11101010211 haplotype. This last result showed that there were clones in both populations. Some genotypes were detected as shared across the populations. In other cases, several individuals in the two populations had the same multilocus haplotypes but different DNA fingerprints, indicating that they were not the same clone.

The alleles' frequencies were significantly different from the 8 loci of RFLP. The Argentine population must be compared with other continental populations – Swiss and USA (Oregon) – as independent populations. Over a total of 834 individuals, there was a 40% gene diversity between native populations and the total population differentiation was 11%, showing that differentiation between native and foreign populations exists. The average number of migrants was 3.68. This number meant that 3–4 individuals would need to be exchanged across populations of each generation to maintain the observed level of genetic similarity. Moreover, the amount of gene flow between populations was high when all the populations were compared.

The genetic distance was small when comparing the population of Los Hornos with the other populations, showing a high level of similarity, but the genetic distance of the Los Hornos and Balcarce populations was major compared with the Oregon and Swiss populations. Salamati *et al.* (2000) suggested that the similarity among populations on a regional basis was explained because the gene flow was significant over spatial scales of at least several hundred kilometres. It was found that genetic distances among fields within a region were small, while genetic distances among different continents were larger for the *Rhynchosporium secale* populations.

Genotypic diversity within populations and similarity over regional spatial scale was explained because regular sexual recombination was occurring in *S. tritici* rather than in *R. secalis* (Salamati *et al.*, 2000), *Stagonospora nodorum* (McDonald *et al.*, 1994) and *Phaeosphaeria nodorum* (Keller *et al.*, 1997) populations. This was explained because the ascospores from the teleomorph were dispersed over distances of up to 100 km (Shaw and Royle, 1989; Cordo *et al.*, 1990/1991).

The field populations of the fungus exhibited high degrees of gene and genotype diversity distributed on very small spatial scales. Microgeographical-level observations showed a higher variation of type and number of genotypes for the Balcarce than for the Los Hornos population. In general, different genotypes were often found within a single lesion and most lesions on the same leaf also had different genotypes. This result demonstrated, in coincidence with Boerger *et al.* (1993), that a lesion might result due to coinfection by two or more genotypes.

The genetic distance, for native populations, was very small considering that the geographic distances between them was 500 km; the North American and European populations, separated by to 7000 km, had a low increase of this genetic distance. Then, the high degree of similarity could be caused by the gene flow on a regional scale and between continents (Boerger *et al.*, 1993; Zhan *et al* 2003; Banke *et al.*, 2004; Banke and McDonald, 2005).

The results of this contribution are in agreement with Keller *et al.* (1997), who demonstrated that ascospores were the primary agent for unifying geographically separated populations on a regional scale. Added to this, Cordo *et al.* (2005) showed that ascospores were the most significant component of the *M. graminicola* life cycle in the wheat-producing areas in Argentina. Their release was registered in the vegetative and debris wheat stages for the periods analysed. According to these experiments, the high degree of gene flow among populations would be associated neither with the pycnidiospores presence as dominant in the life cycle of the pathogen nor the infected seeds that could act as a human dispersal mechanism (Keller *et al.*, 1997). The Los Hornos population result was different because the clonal lineages of *S. tritici* probably originated from the inoculations applied for the resistance tests.

If it is assumed that *S. tritici* had not colonized Argentina recently, the high degree of similarity could be explained from the most likely centre of origin for this pathogen. Banke *et al.* (2004) demonstrated that the New World areas (where the South Cone is located) appeared less likely to represent ancestral populations because they had lower diversity, whereas Israel and Europe appeared to be the ancestral populations because they showed the highest genetic diversity. This pattern is related to the fact that wheat has been grown in the Old World for thousands of years, but in the New World for only hundreds of years. Movement of the fungus from Israel into Europe could have been from windblown ascospores or via transport on infected seed or straw. Ascospore movement produced a natural gene flow out of the possible centre of origin and into European populations, which could explain the finding that more haplotypes were found in European than in New World populations.

Another way of dispersion could be an alternate host of *S. tritici* producing pycnidia, which constitute a continuous host population where ascospores (Boerger *et al.*, 1993; Linde *et al.*, 2002) would maintain a uniform source of inoculum that infects the wheat field each autumn. This way of transmission was not demonstrated in Argentina.

Disease Control with Alternative Techniques

The most common approach to biological control consists of selecting antagonistic microorganisms, studying their modes of action and developing a biological control product. Despite progress made in the knowledge of the modes of action of these biological control agents, practical applications often fail to control diseases in the field. Ono of thc rcasons for this failure is that biocontrol products are used in the same way as chemical products. Other methods include the choice of an appropriate crop rotation with the management of crop residues, added to organic amendments and biological disinfestations of soils. In that sense, Cordo *et al.* (2007) evaluated the efficacy and mechanisms of action of *Trichoderma* sp. for controlling leaf blotch in wheat grown under greenhouse conditions.

Because of their capacity to act as biocontrol agents, members of the fungal genus *Trichoderma* have been broadly studied (Barnett and Lilly, 1962; Tronsmo, 1986; Melo, 1991; Harman, 2000; Monte, 2001). Thus, *T. harzianum* and *T. aureoviride* are known to be effective antagonists against phylloplane pathogens (Perelló *et al.*, 1997, 2001, 2003, 2006).

There were significant differences for necrosis and pycnidial coverage percentages for 2 years of experiment and for the behaviour of the 14 antagonists, each treated with pulverization and undercoated seed treatments. Two strains of *Trichoderma* sp. (Th5 and Tk11) were selected. Conversely, trials performed during 2005 examined only plants produced by seeds coated with *Trichoderma* Th5 and Tk11 isolates. The *T. koningii* 11 strain was selected for the third experiment, instead of the Th2 strain, because the necrotic coverage percentage of Tk11 was statistically different with respect to the control and with a higher value than the others (Table 22.5). Moreover, the value of pycnidial coverage percentage was also one of the highest that was statistically different from the control.

This work shows that *T. harzianum*, *T. koningi* and *T. aureoviride* reduce the leaf blotch caused by *S. tritici* in greenhouse-grown wheat. The effect of *T. aureoviride* was considered similar to that of *T. harzianum* in reducing the leaf blotch caused by *S. tritici* because, under the most effective application method (seed coating), the pycnidial coverage percentage was statistically different to the control, but not to that of *T. harzianum*. The positive result of the immunochemical test applied on all asymptomatic leaf intercellular fluid samples demonstrated the presence of *S. tritici* on plants free of *Trichoderma* and plants coming from *Trichoderma*-coated seeds, both inoculated with *S. tritici*.

Effect of *Trichoderma* on Leaf Proteolysis

Plants pretreated with *Trichoderma* Th5 and Tk11 isolates were selected to assess the balance between leaf apoplast proteolysis and protease inhibitory capacity.

Table 22.5. Severity of necrosis and pycnidial coverage percentage in leaves with different *Trichoderma* spp. isolates in 2005.

Trichoderma spp. isolates	Necrotic coverage (%)*	Pycnidial coverage (%)*
Th5	34.23 a	46.91 b
Tk11	41.62 a	57.50 a
Control	46.17 a	59.14 a

Note: *Each value is the mean of two replicates for necrotic and pycnidial coverage percentage. Means followed by the same letter are not significantly different ($P = 0.05$) according to the LSD test.

Compared to controls, leaf proteolytic activity decreased by 40% 12 days after *S. tritici* inoculation. Conversely, it increased in plants produced by Th5-coated seeds. This was visible within 15–22 days after sowing (Table 22.6). Moreover, the increased proteolytic activity coincided with a decreased protease inhibitory capacity. Furthermore, the proteolytic activity remained higher in plants produced by Th5-coated seeds challenged with *S. tritici*. Proteolytic activity did not increase when comparing wheat plants grown without inoculation and plants produced by Tk11-coated seeds.

The genus *Trichoderma* is a soilborne fungus whose survival on the phylloplane environment is difficult (Perelló *et al.*, 1997, 2003). Thus, a *Trichoderma* spp. population applied over wheat leaf decreases rapidly. Conversely, coating wheat seeds with *T. harzianum* is the finest application technique to control the leaf blotch caused by *S. tritici*. To understand how *T. harzianum* acts, the general features of the biological control set up by Viterbo *et al.* (2002) were tested. Leaves of plants formed by pre-coated seeds did not contain *T. harzianum*. This suggests that its biocontrol of leaf blotch is indirect and able to produce morphological or biochemical changes.

As mentioned, inoculation with *S. tritici* decreases the apoplastic serine protease action in plants of susceptible cultivars (Segarra *et al.*, 2002). Conversely, leaf blotch symptoms decrease in susceptible cultivar PRO INTA Molinero plants pretreated with some *Trichoderma* isolates after challenging with *S. tritici*. For this, the protease action in plants pretreated with isolates Th5 and Tk11 with high and low biocontrol capacity, respectively, was tested. The apoplastic protease activity increased only after treatment with Th5. In order to know if this increase ocurred independently of the inoculation with the pathogenic fungus, the kinetics of this phenomenon were analysed in plants pretreated with this isolate. The proteolytic activity was controlled by the leaf germin-like protease inhibitor already described (Segarra *et al.*, 2003).

In plants, the apoplast forms a space the pathogens necessarily must cross to colonize tissues. Therefore, it plays a central role in defence strategies, being a place where not only signals for plant response originate, but also where the proteins for defence mechanisms accumulate: glucanases, chitinases and proteases among others (Bowles, 1990). Within proteases must be mentioned the tomato P-69 induced by the citrus viroid (Vera and Conejero, 1988; Tornero *et al.*, 1996), the tomato aspartil protease that degrades proteins related to pathogenesis (Rodrigo *et al.*, 1988), the specific race join protease

Table 22.6. Effect of *S. tritici* and *Trichoderma* spp. on leaf apoplast proteolytic and inhibitor activity.

Treatment	Days after sowing	Protease activity (%)	Inhibitor activity (%)
T_1	22	100	100
T_2	22	61 +/– 15	150
T_4 (Th5)	7	95 +/– 4	98
T_4 (Th5)	15	167 +/– 25	87
T_4 (Th5	22	140 +/– 20	33
T_6 (Th5)	22	128 +/– 10	60
T_4 (Th11)	22	70 +/– 10	–
T_4 (Th11)	22	98 +/– 12	–

Note: In T_1 (wheat plants without inoculum), the IWF (leaf intercellular washing fluid) was obtained 22 days after sowing. In T_2 (wheat plants inoculated with the pathogen) and T_6 (wheat plants grown from *Trichoderma* spp. pre-coated wheat seeds and inoculated with the pathogen), the plants were inoculated with *S. tritici* 10 days after sowing and the IWF was examined 12 days after inoculation. In the case of plants grown from *Trichoderma* spp. pre-coated wheat seeds (T_4), the IWF was examined 7, 15 or 22 days after sowing. For all treatments, the proteolytic and the protease inhibition activity was considered 100% in non-inoculated plants. Each value is the mean of two replicates.

that processes the AVR9 of the compatible reaction tomato–*Cladosporium fulvum* (de Witt *et al.*, 1985; Schaller and Ryan, 1996), two closely related subtilisin-like proteases that are associated with the defence response of tomato and encoded by the *P69B* and *P69C* genes (Jordá and Vera, 2000) and a unique 33-kDa cysteine protease mobilized in response to caterpillar feeding in maize lines that are resistant to feeding by several lepidopteran species (Pechan *et al.*, 2002).

Conclusions

The relevant advances for *Septoria* leaf blotch of wheat fall into two classes. First, qualitative, as the conditions that allow inoculum transfer, permit infection and encourage sporulation; second, quantitative as, in a given agroecosystem, what factors in practice control pathogen regulation size. It was demonstrated that the distance between leaf layers with and without infection varied greatly according to both the architecture of the wheat cultivar and the latent period of the pathogen on the cultivar. The interaction of these factors, as was observed on the SMN collection, caused great variation in the potential for the spread of pathogens to the upper part of the crop. DNA restriction fragment polymorphism (RFLP) markers labelled with radioactive compounds were used to assess the potential for gene and genetic diversity and for gene flow between geographically separated populations.

The results on the genetic composition of two populations separated by 500 km show shared haplotypes. This has significant implications for wheat-breeding programmes that seek to incorporate resistance to *S. tritici*. In coincidence with Boerger *et al.* (1993), our evidence of gene flow suggests that plant breeders in Argentina are driving the breeding process well. They are testing the resistance of their cultivars at many locations away from the area of local adaptation. The fine scale of patterns with genetic variability suggests that plant breeders should use a wide spectrum of pathogen genotypes when testing wheat cultivars resistant to this pathogen in any location.

Throughout the genetic evidence on the gene flow between continents, it is possible to affirm that lesions of leaf blotch of wheat could arise from seed transmission or could be attributed to a failure of isolation and a stray ascospore. In the face of the high environmental contamination produced by agrochemical products, new ecological alternatives are applied to control diseases in extensive plant cultures. So, biological control is a complementary strategy in the ecological management of wheat cultivation.

These results suggest that the saprophytic fungus *T. harzianum* provokes a biochemical plant defence response, as has been reported previously. Immunochemical tests proved that although these leaves looked asymptomatic, they contained *S. tritici*. Because *T. harzianum* does not meet leaves coming from pre-coated seeds, its stimulation of leaf proteolytic activity might be considered a systemic induced response, which is one of the different biochemical mechanisms of plant defence proposed by Viterbo *et al.* (2002) and Hoitink *et al.* (2006). We conclude that prospects for the biological control of leaf blotch with *T. harzianum* are auspicious. The results encourage trials under field conditions.

References

Annone, J.G., Conta, H., Calzolari, H. and Poliporo, O. (1995) Selección por resistencia a *Septoria tritici* en filiales segregantes avanzadas de trigo pan. In: Gilchrist, L., van Ginkel, M., McNab, A. and Kema, G.H.J. (eds) *Proceedings of a* Septoria tritici *Workshop*, 20–24 September 1993. CIMMYT, El Batán, Mexico, pp. 73–83.

Arama, P.F., Parleviet, J.E. and van Silfhout, C.H. (1999) Heading date and resistance to *Septoria tritici* blotch in wheat non-genetically associated. *Euphytica* 106, 63–68.

Arraiano, L.S. and Brown, J.K.M. (2006) Identification of isolate specific and partial resistance of *Septoria tritici* blotch in 238 wheat cultivars and breeding lines. *Plant Pathology* 55, 726–738.

Banke, S. and McDonald, BA. (2005) Migration patterns among populations of the pathogenic fungus *Mycosphaerella graminicola*. *Molecular Ecology* 14, 1881–1896.

Banke, S., Peschon, A. and McDonald, B.A. (2004) Phylogenetic analysis of globally distributed *Mycosphaerella graminicola* populations based on three DNA sequence loci. *Fungal Genetic and Biology* 41, 226–238.

Barnett, H. and Lilly, V. (1962) A destructive mycoparasite, *Gliocladium roseum*. *Mycologia* 54, 72–77.

Boerger, J.M., Chen, R.S. and McDonald, B.A. (1993) Gene flow between geographic populations of *Mycosphaerella graminicola* (Anamorph *Septoria tritici*) detected with restriction fragment length polymorphism markers. *Phytopathology* 83, 1147–1154.

Bowles, D.J. (1990) Defense-related proteinsin higher plants. *Annual Review of Biochemistry* 59, 873–907.

Carmona, M., Reis, E. and Cortesse, P. (1999) *Manchas foliares del trigo. Diagnóstico, epidemiología y nuevos criterios para el manejo.* BASF Argentina SA, 32 pp.

Cordo, C.A. and Arriaga, H.O. (1990) Variación en patogenicidad entre cepas argentinas de *Mycosphaerella graminicola* (anamorfo *Septoria tritici*). In: Kohli, M.M. and van Beuningen, L.T. (eds) *Conferencia Regional sobre septoriosis del trigo.* CIMMYT, México, D.F., pp. 88–100.

Cordo, C.A., Perelló, A.E., Alippi, H.E. and Arriaga, H.O. (1990/91) Presencia de *Mycosphaerella graminicola* (Fuckel) Schroeter telomorfo de *Septoria tritici* Rob. apud Desm. en trigos maduros de la Argentina. *Revista de la Facultad de Agronomía* 66/67, 49–55.

Cordo, C.A., Perelló, A.E., Arriaga, H.O., Benedicto, G., Avila, V. and Ziglino, I.R. de (1994) Resistencia a la Mancha foliar causada por *Septoria tritici* en el trigo pan (*Triticum aestivum*). *Revista de la Facultad de Agronomía, La Plata* 70, 23–36.

Cordo, C.A., Simón, M.R., Perelló, A.E. and Alippi, H.E. (1999) Spore dispersal of leaf blotch pathogens of wheat (*Mycosphaerella graminicola* and *Septoria tritici*). *Septoria* and *Stagonospora* diseases of cereals. In: *Proceedings of the Fifth International Septoria Workshop.* CIMMYT, México, pp. 98–104.

Cordo, C.A., Simón, M.R., Perelló, A.E., Bayo, D., Kripelz, N.I. and Larrán, S. (2005) Environmental factors affecting the release and dispersal of pycnidiospores and ascospores of *Mycosphaerella graminicola. 7th International Wheat Conference Abstracts, Mar del Plata, Argentina*, 131 pp.

Cordo, C.A., Linde, C., Zhan, J. and McDonald, B. (2006) Genotypic diversity of the wheat leaf blotch pathogen (*Septoria tritici*) in Buenos Aires Province. *Boletin SociedadeAargentine de Botanica* 41(3–4), 293–305.

Cordo, C.A., Mónaco, C.I., Segarra, C.I., Simon, M.R., Mansilla, A.Y., Perello, A.E., Kripelz, N.I., Bayo, D. and Conde, R.D. (2007) *Trichoderma* spp. as elicitors of wheat plant defense responses against *Septoria tritici. Biocontrol Science and Technology* 17(7), 687–698.

Cortese, P., Cordo, C.A. and Zanelli, M. (1998) Aspectos epidemiológicos de la mancha de la hoja del trigo y su influencia sobre los componentes del rendimiento. Trabajo de Tesis para optar Grado de Magíster Scientiae en Protección Vegetal: Universidad Nacional de la Plata. Fsacultad de Ciencias Agrarias y Forestales, 65 pp.

Encuesta Agricola DIEA-MGAP (2007) http//www.mgap.gub.uy/DIEA/Anuario2007/Anuario 2007/ pages/ DIEA-Anuario 2007-cd_001html-2k- (accessed 22/02/2008).

Eyal, Z., Scharen, M.D., Huffman, M.D. and Prescott, J.M. (1985) Global insights into virulence frequencies of *Mycosphaerella graminicola. Phytopathology* 75, 1456–1462.

Eyal, Z., Sharen, A.L., Prescott, J.M. and van Ginkel, M. (1987) *The Septoria diseases of wheat: concepts and methods of disease management.*, CIMMYT, Mexico.

Gilchrist, L., Velázquez, C. and Crossa, J. (1999) Analysis of the *Septoria* Monitoring Nursery. In: Van Ginkel M., McNab, A. and Krupinsky, J. (eds) Septoria *and* Stagonospora *Diseases of Cereals. A Compilation of Global Research. Proceedings of the Fifth International* Septoria Workshop*, 20–24 September 1999.* CIMMYT, El Batán, Mexico, pp. 59–62.

Harman, G. (2000) Myths and dogmas of biocontrol. Changes in perceptions derived from research on *Trichoderma harzianum* T22. *Plant Diseases* 84, 377–393.

Hoitink, H., Madden, L. and Dorrance, A.E. (2006) Systemic resistance induced by *Trichoderma* spp.: interactions between the host, the pathogen, the biocontrol agent, and soil organic matter quality. *Phytopathology* 96, 186–189.

Hunter, T., Coker, R.R. and Royle, D.J. (1999) The teleomorph stage, *Mycosphaerella graminicola*, in epidemics of *Septoria tritici* blotch on winter wheat in the UK. *Plant Pathology* 48, 51–57.

INTA-CIMMYT (1996) Primera jornada de control químico de Enfermedades del Trigo en Sistemas de Manejo para alta productividad. *Bolsa de cereales de Buenos Aires.* 40 pp.

Jordá, L. and Vera, P. (2000) Local and systemic induction of two defense-related subtilisin-like protease promoters in transgenic arabidopsis plants. Luciferin induction of PR gene expression. *Plant Physiology* 124, 1049–1058.

Keller, S.M., Wolfe, M.S., McDemott, J.M. and McDonald, B.A. (1997) High genetic similarity among populations of *Phaeosphaeria nodorum* across wheat cultivars and regions in Switzerland. *Phytopathology* 87, 1134–1139.

Kohli, M.M. (1995) Resistance to *Septoria tritici* blotch in southern cone germplasm. In: Gilchrist, L., van Ginkel, M., McNab, A. and Kema, G.H.J. (eds) *Proceedings of a* Septoria tritici *Workshop*, 20–24 September 1993. CIMMYT, El Batán, México, pp. 62–72.

Kronstad, W.E. (1998) Agricultural development and wheat breeding in the 20th century. In: Braun, H.J., Altay, F., Kronstad, W.E., Beniwal, S.O.S. and MacNab, A. (eds) *Prospects for Global Improvement. Series: Developments in Plant Breeding, Volume 6. Proceedings of the 5th International Wheat Conference, 10–14 June 1996, Ankara, Turkey.*

Linde, C., Zhan, J. and McDonald, B.A. (2002) Population structure of *Mycophaerella graminicola*: from lesions to continents. *Phytopathology* 92, 946–955.

Lovell, D.J., Parker, S.R., Hunter, T., Royle, D.J. and Coker, R.R. (1997) Influence of crop growth and structure on the risk of epidemics by *Mycosphaerella graminicola* (*Septoria tritici*) in winter wheat. *Plant Pathology* 46, 126–138.

McDonald, B.A., Miles, J., Nelson, L. and Pettway, R.E. (1994) Genetic variability in nuclear DNA in field populations of *Stagonospora nodorum. Phytopathology* 84, 250–255.

Melo, Y. (1991) Potencialidades de utilizao *Trichoderma* spp. no controle biológico de dodencas de plantas. In: *Controle biológico de dodencas de plantas.* EMBRAPA, Brasilia, D.F., pp. 388, Cap 9, pp. 135–156.

Monte, E. (2001) Understanding *Trichoderma*: between biotechnology and microbial ecology. *International Microbiology* 4, 1–4.

Munford, S.L. and Norton, G.W. (1984) Economics of decision making in pest management. *Annual Review of Entomology* 29, 157–174.

OPYPA (2008) http//www.mgap.gub.uy/opypa/index.html-2k (accessed 22/02/2008).

Pechan, T., Cohen, A., Williams, P. and Luthe, D. (2002) Insect feeding mobilizes a unique plant defense protease that disrupts the peritrophic matrix of caterpillars. *Proceedings of the National Academy of Sciences of the United States of America* 99, 13319–13323.

Perelló, A., Mónaco, C. and Cordo, C. (1997) Evaluation of *Trichoderma harzianum* and *Gliocladium roseum* in controlling leaf blotch of wheat (*Septoria tritici*) Ander *in vitro* and greenhouse conditions. *Zeitschrift fur Pflanzenkrankheiten und Pflanzenschutz* 104, 588–598.

Perelló, A., Simon, M., Arambarri, A. and Cordo, C. (2001) Greenhouse screening of the saprophytic resident microflora for control of leaf spots of wheat (*Triticum aestivum* L.). *Phytoparasitica* 29, 341–351.

Perelló, A., Mónaco, C., Sisterna, M. and Dal Bello, G. (2003) Biocontrol efficacy of *Trichoderma* isolates for tan spot of wheat in Argentina. *Crop Protection* 22, 1099–1106.

Perelló, A., Mónaco, C., Moreno, V., Cordo, C. and Simón, M.R. (2006) The effect of *Trichoderma harzianum* and *T. koningii* on the control of tan spot (*Pyrenophora tritici-repentis*) and leaf blotch (*Mycosphaerella graminicola*) of wheat under field conditions in Argentina. *Biocontrol Science and Technology* 54, 113–122.

Rajaram, S. (1999) Historical aspects and future challenges of an international wheat program. In: van Ginkel, M., McNab, A. and Krupinsky, J. (eds) *Septoria and Stagonospora Diseases of Cereals: A Compilation of Global Research. Proceedings of the Fifth International Septoria Workshop, 20–24 September.* CIMMYT, México, pp. 1–18.

Rodrigo, I., Verá, P. and Conejero, V. (1988) Degradation of tomato pathogenesis-related proteins by an endogenous 37 kDa aspartyl endoproteinase. *European Journal of Biochemistry* 184, 663–669.

Rosegrant, M.W., Agcaolli-Sombilla, A. and Perez, N. (1995) *Global Food Protections to 2020.* Discussion paper. IFPRI, Washington, DC.

Saari, E.E. and Prescott, J.M. (1975) A scale for appraising the foliar intensity of wheat diseases. *Plant Disease Reporter* 59, 377–380.

Salamati, S., Zhan, J., Burdon, J.J. and MCDonald, B. (2000) The genetic structure of field populations of *Rynchosporium secalis* from three continents suggests moderate gene flow and regular recombination. *Phytopathology* 90, 901–908.

Sanderson, F.R. (1972) A *Mycosphaerelle* species as the ascogenous state of *Septoria tritici* Rob. and Desm. N. Z. *Journal of Botany* 10, 707–709.

Sanderson, F.R., Scharen, A.L., Eyal, Z. and King, A.C. (1986) A study of genetic segregation using single ascospore isolates of *Mycosphaerella graminicola. Plant Breeding Symposium DSIR 1986. Agricultural Society of New Zealand Special Publication* 3, 195–201.
Schaller, A. and Ryan, C. (1996) Molecular cloning of a tomato leaf cDNA encoding an aspartic protease, a systemic wound response protein. *Plant Molecular Biology* 19, 1073–1077.
Schuh, W. (1990) Influence of tillage systems on disease and spatial pattern of *Septoria* leaf blotch. *Phytopathology* 80, 1337–1340.
Segarra, C., Casalongué, C., Pinedo, M., Cordo, C.A. and Conde, R. (2002) Changes in wheat leaf extracellular proteolytic activity after infection with *Septoria tritici. Journal of Phytopatology* 150, 105–111.
Segarra, C., Casalongué, C., Pinedo, M., Ronchi, V. and Conde, R. (2003) A germin-like protein of wheat leaf apoplast inhibits serine protease. *Journal of Experimental Botany* 54, 1335–1341.
Shaw, M.W. and Royle, D.J. (1989) Airborne inoculum as a major source of *Septoria tritici* (*Mycosphaerella graminicola)* infections in winter wheat crop in the UK. *Plant Pathology* 38, 35–43.
Simón, M.R., Perelló, A.E., Cordo, C.A., Larrán, S., van der Putten, P. and Struik, P.C. (2005) Association between *Septoria tritici* blotch, plant height and heading date, in wheat. *Agronomy Journal* 97, 1072–1081.
Tornero, P., Conejero, V. and Vera, P. (1996). Primary structure and expression of a pathogen induced protease (PR-P69) in tomato plants: Similarity of functional domains to subtisilin-like endoproteases. *Proceedings of the National Academy of Science USA*, 93, 6332–6337.
Tronsmo, A. (1986) Use of *Trichoderma* spp. in biological control of necrotrophic pathogens. In: Fokkema, N. and Heuvel, J. der (eds) *Microbiology of the Phyllosphere.* Cambridge University Press, Cambridge, UK.
Vera, P. and Conejero, V. (1988) Pathogenesis-related proteins in tomato. P-69 as an alkaline endoproteinase. *Plant Physiology* 87, 58–63.
Viterbo, A., Ramot, O., Chernin, L. and Chet, I. (2002) Significance of lytic enzymes from *Trichoderma* spp. in the biocontrol of fungal plant pathogen. *Journal of Antonie van Leeuwenhoek* 81, 549–556.
Witt, P. de, Hoffman, A., Velthuis, G. and Kuc, J. (1985) Isolation and characterization of an elicitor of necrosis isolated from intercellular fluids of compatible interaction of *Cladosporium fulveum* (syn. *Fulvia fulva*) tomato. *Plant Physiology* 77, 642–647.
Zadoks, J.C., Chang, T.T. and Konzak, C.F. (1974) A decimal code for the growth stages of cereals. *Weed Research, Oxford* 14, 415–421.
Zhan, J., Pettway, R.E. and McDonald, B.A. (2003) The global genetic structure of the wheat pathogen *Mycosphaerella graminicola* is characterized by high nuclear diversity, low mitochondrial diversity, regular recombination, and gene flow. *Fungal Genetics and Biology* 38, 286–297.

Part VI

Alternative Control Strategies

23 Review of *Thecaphora amaranthicola* M. Piepenbr., Casual Agent of Smut on *Amaranthus mantegazzianus* Pass.

M.C.I. Noelting,[1] M.C. Sandoval,[2] M.M.A. Gassó[1] and M.C. Molina[1,3]

[1]*Instituto Fitotécnico de Santa Catalina, Facultad de Ciencas Agrarias y Forestales, UNLP, Llavallol, Buenos Aires, Argentina;* [2]*Facultad de Ciencas Agrarias, UNLZ, Llavallol, Buenos Aires, Argentina;* [3]*CONICET (Consejo Nacional de Investigaciones Cientificas Tecnicas)*

Abstract

The amaranth (*Amaranthus* spp.) is becoming a socially and economically important crop due to the high level of quality proteins in its seeds and leaves. Among the factors that could limit the expansion of this crop are the smuts (Ustilaginales) that prevent normal development of the seeds. The objectives of the project were to: (i) characterize the pathogen which is responsible for smut on *A. mantegazzianus*, taking into account the cultural and morphobiometrical characters, and the germination of its teliospores; (ii) assess the incidence of smut in two amaranth cultivars; (iii) analyse fast techniques for the detection of the inoculum in seeds and plants cultivated in the field; and (iv) identify any possible reservoirs of the pathogen on wild amaranths. The results obtained allowed the authors to determine that: (i) *Thecaphora amaranthicola* is the causal agent of smut on *A. mantegazzianus*; (ii) the cultivar Don Manuel was affected most by the smut (36% incidence); (iii) the residue and plastic card techniques are fast and efficient for the detection of the pathogen inoculum; and (iv) the wild species *A. hybridus* and *A. retroflexus* are hosts of the pathogen. This is the first report in the world of *Thecaphora amaranthicola* as a pathogen of the *A. mantegazzianus* cultivated species.

Introduction

The amaranth is an ancestral precolumbine crop that was cultivated by the Aztecs, Mayas and Incas and which remained relegated for a long time after being banned by the Spanish conquistadors. However, in the past few decades it has been subjected to numerous investigations with the objective of studying its nutritional value, improvement and adaptation to new areas of cultivation (Afolabi *et al.*, 1981; Kulakow, 1987; Bressani, 1989; Espitia, 1991).

The seeds and leaves of this plant have a very high level of quality proteins. This property makes the amaranth a valuable resource, especially appropriate for a population which lives in areas that are considered marginal for the cultivation of traditional cereals.

The amaranth can be affected by pests and diseases. Among the diseases of fungal etiology which affect the normal development of its seeds are two species of smut, *T. amaranthi* (Hirschh) Vanky (syn. *Glomosporium amaranthi*) (Vánky, 1994) and

 Management of Fungal Plant Pathogens (eds A. Arya and A.E. Perelló)

T. amaranthicola M. Piepenbr. (Piepembring, 2000). The spores of both types of smut infect the ovaries, preventing normal development of the seeds. On the inside of the affected ovaries, hypertrophied sorus development takes place and these contain a mass of spores (teliospores). The species *T. amaranthi* was described and reported in wild species of amaranth (Vánky, 1985; Hirschhorn, 1986) and in cultivated species (Alcalde, 1995; Noelting and Sandoval, 2003). On the other hand, *T. amaranthicola* was described in only one wild species of amaranth collected in Ecuador (Piepembring, 2000). In Argentina, the data on this species are in a report where the symptoms caused by this smut on a cultivated species of amaranth are described (Noelting *et al.*, 2005a). There are no records of *T. amaranthicola* on cultivated species of amaranth in the rest of the world. Owing to the little information available on the biological and epidemiological aspects of this smut, and that its spreading could have a negative effect on the propagation of the amaranth crop, it was decided to start the present study, the objectives of which were:

1. To characterize the pathogen which is responsible for smut on *A. mantegazzianus*, taking into account the cultural and morphobiometrical characters, and the germination of its teliospores;
2. To assess the incidence of smut in two amaranth cultivars;
3. To analyse fast techniques for the detection of the inoculum in seeds and plants; and
4. To identify any possible reservoirs of the pathogen on wild amaranths.

Materials and Methods

Morphological and cultural characterization of pathogen

The taxonomic identification of the fungus was carried out by observing the teliospores, the evolution of the germination process (Piepembring, 2000) and the characteristics of the cultivated colonies *in vitro* which developed in potato dextrose agar (PDA 2%). Samples of teliospores obtained from infected seeds of amaranth (*A. mantegazzianus*) were used in order to study these characters. The seeds were harvested in an experimental field of the Instituto Fitotécnico de Santa Catalina, situated in the locality of Llavallol, in the south of the Buenos Aires Province, Argentina, in 2003. The teliospores were observed in an optical microscope and measured with a micrometric objective (Carl Zeiss Jena 7×) ($n = 50$).

In order to observe the germination, the teliospores were disinfected with a solution of sodium hypochlorite (2%) for 5 min, rinsed three times with sterile distillate water and dried between sterile paper filters. To inoculate Petri plates, teliospores were streaked on the surface of the agar media to dilute the inoculum. The plates were incubated at 25 ± 2°C with a 16 h photoperiod. To assess the colony characteristics, sections 6 mm in diameter, obtained from a colony in active growth, were transferred to dishes with PDA. The material was incubated in a chamber at 25 ± 2°C with a 16 h photoperiod. After 14 days of incubation, the cultural and morphological characteristics (colour, edge, diameter) of the developed colonies were described.

Incidence assessment

The incidence assessment was carried out in seeds of two cultivars of *A. mantegazzianus* (cvs. Don Juan and Don Manuel) which had been harvested in 2004 and presented spontaneous infections with smut. Each panicle was harvested and threshed by hand. Once the material had been threshed, samples of 200 seeds/panicle were taken, determining the number of seeds infected with *T. amaranthicola*. The incidence was calculated by using the formula:

$$\text{Incidence (\%)} = \frac{\text{Number of infected seeds}}{\text{Total number of analysed seeds}} \times 100$$

Techniques for Detection of Inoculum

In seeds

Samples of four different cultivated species of amaranth were used: *A. mantegazzianus*, *A. caudatus*, *A. hypochondriacus* and *A. cruentus* from Argentina, Bolivia and Mexico. The technique applied consisted in depositing 10 g of seed samples from the different countries between pieces of reticulated cellulose paper and then submitting them to three cycles of manual pressure. Next, the seeds were taken out of the paper and the residue contained in the reticules was observed with a stereoscopic magnifying glass (10×). The observations that tested positive for the presence of teliospores were confirmed by means of a preparation, with a solution of lactophenol and cotton blue, and observed with an optical microscope (450×).

In plants cultivated in fields

In order to determine the infection in the crop, 4 cm × 6 cm plastic cards were used, with lithium grease as an adhesive on one side of the cards. Eighteen cards were distributed randomly, 9 in each of the cultivars of *A. mantezzagianus* (cvs. Don Juan and Don Manuel), hanging from the plants for a month. Each card was later analysed with a stereoscopic magnifying glass to detect the teliospores.

Wild hosts of *T. amaranthicola*

To detect natural reservoirs of the inoculum of this smut, a sampling that involved three wild species of amaranth, *A. retroflexus* L., *A. hybridus* L. and *A. viridis* L., was carried out in 2006. The plants of these species were located around a crop of *A. mantegazzianus*, as well as areas further away at a distance of 1.5 km. The panicles of the collected material were taken to the laboratory and observed with a stereoscopic magnifying glass and an optical microscope.

Results and Discussion

Morphological and cultural characterization of pathogen

Morphobiometrical characteristics of teliospores

The observations carried out in samples of infected *A. mantegazzianus* seeds allowed determination of the presence of spore balls with the following characteristics: ochre colour, globose to subglobose (40.78 μm × 34.23 μm) (Fig. 23.1a). Each spore ball was formed by 8–23 teliospores, polyhedral to cuneiform: spore walls were deeply corrugated on the central parts of the teliospores, as seen by SEM (Fig. 23.1b). No individual teliospores were observed in any of the samples analysed. This phenomenon coincides with the one observed in smuts that have grouped spores and which develop on different kinds of plants (Barrus and Muller, 1943; Andrade *et al.*, 2004).

The germination of the teliospores plated on PDA was initiated after 24 h of incubation generating phragmobasidia, followed by the development of lateral and terminal basidiospores. Variations in the number of germinated teliospores in each spore ball were observed (Fig. 23.2a,b,c). In addition to this, multiple germinations occurred simultaneously (Fig. 23.2b,c).

The cultural characteristics of the colonies grown in laboratory conditions were as follows: velvety surface at the expense of the development of the mycelium, light beige colour, slightly serrated edges and softly lobated outline of the colony (Fig.23.2f). The growth of the colonies was slow, reaching a maximum diameter of 34 mm after 14 days of incubation. In the colonies analysed, no yeasty type of development characteristic of this type of fungus was found.

According to analysis, *T. amaranthicola*, belonging to the *Basidiomycota* Subkingdom, *Ustilaginomycetes* Class, *Ustilaginomycetidae* Subclass, *Ustilaginales* Order, *Glomosporiaceae* Family, was identified as the causal agent of smut in *A. mantegazzianus* (Fig. 23.3). The morphological and cultural data shown complement the

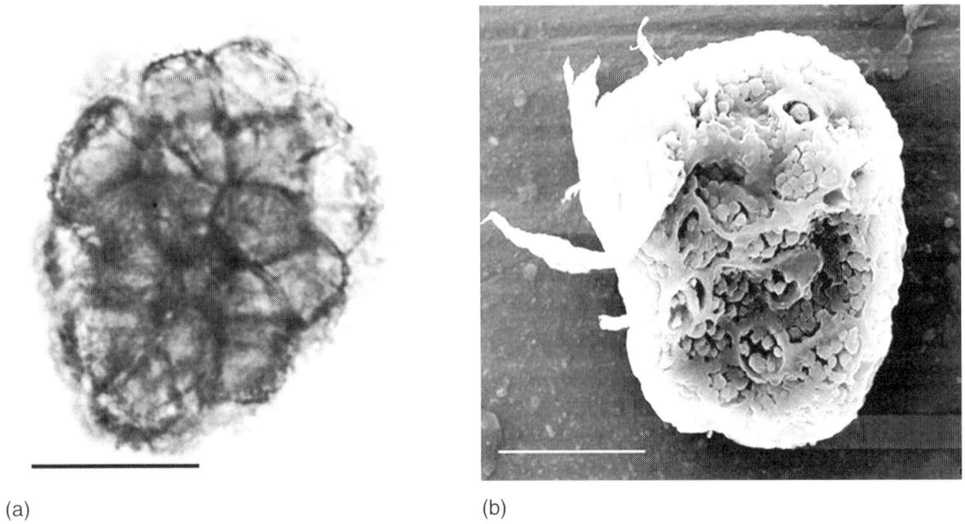

Fig. 23.1. Teliospore balls of *Thecaphora amaranthicola*: (a) under the light microscope (scale bar 10 μm); (b) SEM (scale bar 10 μm).

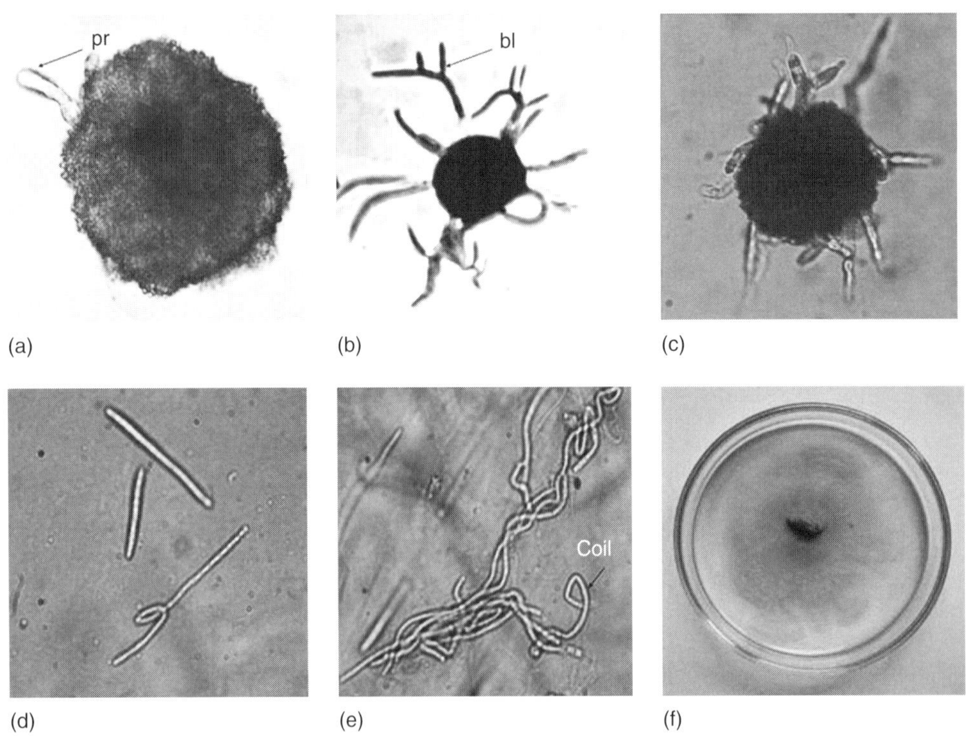

Fig. 23.2. Culture of *Thecaphora amaranthicola* on PDA: (a) single germination of a teliospore after 24 h; pr = probasidium; (b) and (c) multiple germination of teliospores, bl = lateral basidiospores; (d) hyaline basidiospores; (e) mycelia formation, coiling of hyphae (coil); (f) colony of *T. amaranthicola* developed on PDA after 17 days of incubation.

Fig. 23.3. (a) Panicle of *Amaranthus mantegazzianus*; (b) healthy seeds (12×); (c) seeds of *A. mantegazzianus* infected with *T. amaranthicola* (12×), seed coats show irregularities (cb).

preliminary information about the pathogen (Noelting *et al.*, 2005a).

Incidence

The percentage of incidence varied between 10 and 36.66% (average rates) for the Don Juan and Don Manuel cultivars, respectively. These results apparently indicate the existence of resistance mechanisms, especially in the Don Juan cultivar. More studies would have to be carried out in the future in order to learn more about said mechanisms. Nevertheless, it cannot be discarded that the relatively high rates of incidence which were detected may be so because the pathogen could have been introduced by contaminated germplasms in the region. With respect to this, it can be stated that the growing interest in amaranth cultivation crop has originated the incorporation of seeds from several countries and, since this pathology had not been reported previously in cultivated species of amaranth, there are no controls for it in Argentina.

Inoculum detection in seed samples

Teliospores of *T. amaranthicola* were found in 50% of the analysed samples (Table 23.1). The technique applied is effective, fast and simple (Noelting *et al.*, 2005b) compared with the test which consists of washing and filtering seeds and is used for the detection of smuts in the seeds of many crops (ISTA, 1985). The inoculum (teliospores) was detected on and among the seeds; therefore, individualization of the teliospores with this technique offered information about infection as well as contamination in amaranth seeds. Furthermore, the retrospective character of the analysis led to the conclusion that *T. amaranthicola* was already present in cultivated species of amaranth in Argentina prior to its first report (Noelting *et al.*, 2005a).

In plants cultivated in the field

The employment of cards to detect aerial inoculum in plants cultivated in the field allowed the detection of teliospores (*T. amaranthicola*) in 55% of the cards analysed, as well as other fungal propagules. This technique is a sampling method by deposition or capture similar to those which use slides covered by an adhesive substance (Bugiani and Govoni, 1991). The results obtained from both *A. mantegazzianus* cultivars indicate that the *T. amaranthicola* teliospores are spread by the wind.

Wild host of *T. amaranthicola*

Seeds infected by *T. amaranthicola* from panicles of *A. hybridus* and *A. retroflexus*

Table 23.1. Results of the analysis of amaranth seeds samples.

Seed sample	Locality	Province	Country	Year	*Thecaphora amaranthicola*
A. cruentus cv. Don Armando	Anguil	La Pampa	Argentina	1995	x
A. cruentus cv. Don Guiem	Anguil	La Pampa	Argentina	1997	x
A. hypochondriacus cv. G. Covas	Anguil	La Pampa	Argentina	1999	x
A. hypochondriacus Se143	Anguil	La Pampa	Argentina	1999	x
A. caudatus	Llavallol	Buenos Aires	Argentina	2003	–
A. hypochondriacus cv. G. Covas	Luis Guillon	Buenos Aires	Argentina	2003	–
A. hypochondriacus	–	–	Mexico	2004	–
A. caudatus	–	–	Bolivia	2004	–
A. hypochondriacus	Llavallol	Buenos Aires	Argentina	2004	–
A. cruentus cv. Se1MC	Luis Guillon	Buenos Aires	Argentina	2004	x
A. mantegazzianus cv. Don Manuel	Santa Rosa	La Pampa	Argentina	2005	–
A. mantegazzianus cv. Don Manuel	Colonia 25 de Mayo	La Pampa	Argentina	2005	–
A. mantegazzianus cv. Don Juan	Llavallol	Buenos Aires	Argentina	2005	x
A. mantegazzianus cv. Don Juan	Llavallol	Buenos Aires	Argentina	2005	x

plants were detected (Noelting *et al.*, 2006). The infections of a spontaneous nature which were found in the two species of wild amaranth that affect many crops and which grow in a vast region of Argentina are thought to be of epidemiological interest. This is due to the fact that they may turn into 'bridge' species for the entrance of the inoculum and the spreading of smut to cultivated amaranth species.

Conclusions

The presence of *T. amaranthicola* in *A. mantegazzianus* (cultivated crop) is the first report, not only in Argentina but also in the world, of smut as a pathogen. The inoculum detection techniques applied to samples of seeds and plants cultivated in the field are appropriate as a fast method for identifying the presence of smut. Two species of wild amaranth, *A. hybridus* and *A. retroflexus*, are hosts of *T. amaranthicola*. The interest in the amaranth crop has originated an intense interchange of germplasms among several countries of America, Asia and Europe. This situation suggests the need to undertake a major study of the biological and epidemiological characteristics of the seedborne pathogens that as the smuts (*T. amaranthi* and *T. amaranthicola*) have negative incidence in the crop.

References

Afolabi, A.O., Oke, O.L. and Umoh, I.B. (1981) Preliminary studies on the nutritive value of some cereal-like grains. *Nutrition Reports International* 24, 389–394.

Alcalde de L, M.A. (1995) Patógenos del amaranto (*Amaranthus* sp.) en el sur de la provincia de Córdoba, Argentina. *Resúmenes IX Jornadas Fitosanitarias Argentinas* 95.

Andrade, O., Muñoz, G., Galdames, R., Durán, P. and Honorato, R. (2004) Characterization, in vitro culture, and molecular analysis of *Thecaphora solani*, the causal agent of potato smut. *Phytopathology* 94, 875–882.

Barrus, M.F. and Muller, S.S. (1943) An Andean disease of potato tubers. *Phytopathology* 33, 1086–1089.

Bressani, R. (1989) The proteins of grain amaranth. *Foods Reviews International* 51, 1338.

Bugiani, R. and Govoni, P. (1991) Aerobiología e difusa delle pianta. *Informatore Fitopatologico* 11, 9–15.

Espitia, R.E. (1991) Revancha: variedad mejorada de amaranto para los valles altos de México. In: *Primer Congreso Internacional del Amaranto.* INIFAP, Chapingo, Mexico, 64 pp.

Hirschhorn, E. (1986) *Las Ustilaginales de la flora argentina*. Edition CIC, La Plata, Buenos Aires, 530 pp.

International Seed Testing Association (ISTA) (1985) International rules for seed testing. *Seed Science and Technology* 13, 299–513.

Kulakow, P.A. (1987) Genetics of grain amaranths. *Journal of Heredity* 78, 293–297.

Noelting, M.C. and Sandoval, M.C. (2003) Patógenos fúngicos en amarantos cultivados. *Boletin Sociedade Argentine de Botanica* 38 (Suppl), 273.

Noelting, M.C., Sandoval, M.C. and Astiz Gassó, M.M. (2005) Primer reporte en Argentina de *Thecaphora amaranthicola* como agente responsable del carbón en *Amaranthus mantegazzianus* (Res). *Fitopatología* 40, 76.

Noolting, M.C., Sandoval, M.C. and Astiz Gassó, M.M. (2005) Técnica rápida para detección de carbones (Ustilaginales) en semillas de amaranto (*Amaranthus* spp.) (Res). *Boletin Sociedade Argentine de Botanica* 40 (Suppl), 164.

Noelting, M.C., Astiz Gassó, M.M. and Sandoval, M.C. (2006) Especies silvestres de amarantos (*Amaranthus* spp.) hospedantes de *Thecaphora amaranthi* y *T. amaranthicola* (Ustilaginales) en la pcia de Buenos Aires (Res) *XII Jornadas Fitosanitarias Argentinas*. Catamarca, Argentina, pp. 312–313.

Piepembring, M. (2000) New neotropical smut fungi. *Mycological Research* 105, 762–763.

Vánky, K. (1985) Carpathian Ustilaginales. *Symbolae Botanicae Upsalienses* 24(2), 1–309.

Vánky, K. (1994) New combinations proposed. *Mycotaxon* 51, 170

24 Population Biology and Management Strategies of *Phytophthora sojae* Causing *Phytophthora* Root and Stem Rots of Soybean

Shuzhen Zhang[1] and Allen G. Xue[2]

[1]*Soybean Research Institute, Key Laboratory of Soybean Biology of Chinese Education Ministry, Northeast Agricultural University, Harbin, Heilongjiang, China;* [2]*Eastern Cereal and Oilseed Research Centre, Agriculture and Agri-Food Canada, Ottawa, Ontario, Canada*

Abstract

Soybean is an important oilseed crop; it is also the richest source of protein. Root and stem rot pathogen *Phytophthora sojae* can reduce yield by up to 40%. Symptoms, disease cycle and genetic diversity of *P. sojae* are described. The paper is a valuable document listing molecular markers and the role of 14 resistance genes located at eight genomic loci in the development of disease-resistant soybean varieties. There have been only a few fungicides available for the control of *P. sojae* and their effects are limited. The development of pathogen resistance to these fungicides is not known. Little information is available on cultural and biological controls of *P. sojae*. More effective management of *P. sojae* will require integration of all available strategies to address all stages of the disease cycle. The integrated approach may prove a boon to growers using a variety of susceptible cultivars. Suitable cultural, chemical and biological methods are recommended as alternative control strategies.

Introduction

Phytophthora root and stem rot of soybean (*Glycine max* (L.) Merr.), caused by *P. sojae* Kaufmann and Gerdemann, is a destructive disease throughout the soybean-planting regions of the world (Schmitthenner, 1985). The symptoms were first discovered as an unknown etiology in the state of Indiana in the USA in 1948, and subsequently in Ohio in 1951, but the causal agent was not described until 1958 (Kaufmann and Gerdemann, 1958). The disease has since been reported from many countries, including Canada (Hildebrand, 1959), Australia (Pegg *et al.*, 1980; Ryley *et al.*, 1998), Argentina and Brazil (Wrather *et al.*, 1997), China (Shen and Su, 1991) and the Republic of Korea (Jee *et al.*, 1998). The disease is more prevalent when soil is saturated for prolonged periods of time and susceptible cultivars are planted (Grau *et al.*, 2004). Diseased plants reduce yield by 10–40%, or a total crop loss when infection is severe (Anderson and Tenuta, 2003).

Pathogen and Disease Symptoms

P. sojae traditionally has been classified as a fungus due to its outward resemblance of growth habits and nutritional requirements. In fact, it is very distant from 'true' fungi evolutionarily and falls within the Kingdom *Stramenopila* (Förster *et al.*, 1990; Harper *et al.*, 2005), which constitutes a distinct branch of the eukaryotic evolutionary tree (Tyler, 2007).

The pathogen may infect soybean at any stage of plant development and cause seed rot, seedling pre- and post-emergence damping-off and root and stem rots of soybean (Kittle and Gray, 1979; Athow, 1987). The seeds could be rotted by infection of *P. sojae* in both heavy and light sandy soils, after periods of cool and rainy weather. The damping-off, stem and root rot symptoms often appear shortly after emergence and during early flowering when plants are under stress (Anderson and Tenuta, 2003). The infected seedlings have dull grey leaves and reddish, water-soaked lesions that occur from the base of the stem and slowly advance up the plant, and may collapse if the infection is severe. Symptoms on older plants are characterized by chocolate-brown discoloration extending from the soil line to the third or fourth node and into lower branches at any time from the first trifoliate to late R9 stage (Grau *et al.*, 2004). Severely infected plants have few lateral roots, with almost no nodules and only a short portion of taproots left, and brown and drooping leaves remaining attached to the stem, even though the plants die.

Disease Cycle

P. sojae has a narrow host range and is restricted primarily to soybean, but there are reports that lupin, lucerne, bean and sweetclover could be infected using artificial inoculation in controlled environments (Erwin and Ribeiro, 1996).

P. sojae has both asexual and sexual stages in the life cycle and produces sporangia, zoospore and chlamydospore in the asexual stage and oospore in the sexual stage, as shown in Fig. 24.1 (Tyler, 2007).

Oospores are produced by the fusion of a female organ, called oogonium, and a male organ, antheridium. Chlamydospores are thick-walled spores that protect the organisms surviving through periods of abiotic stress. *P. sojae* can survive for many years in soil, mainly as oospores that are formed in the roots and stems of infected soybeans

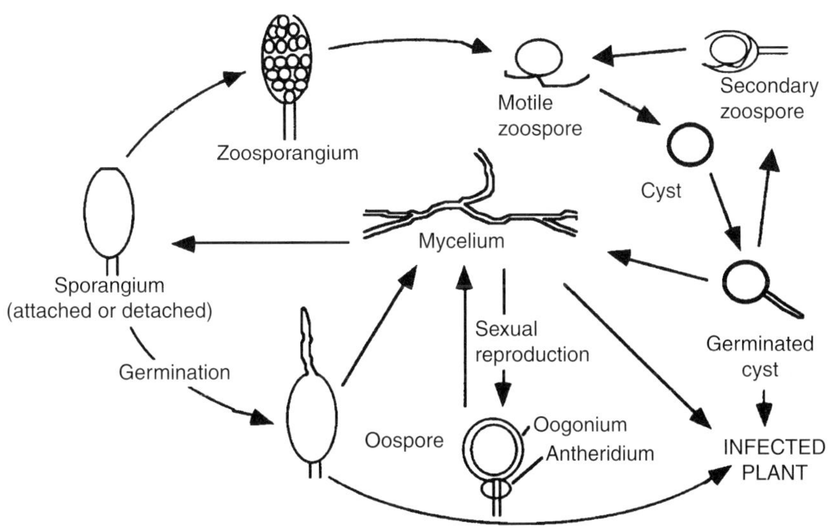

Fig. 24.1. *Phytophthora sojae* life cycle.

in large quantity and are released into the soil when these tissues decompose (Anderson and Tenuta, 2003).

Oospores serve as the primary inoculum and germinate to produce sporangia under flooded conditions or infective hyphae (Anderson and Tenuta, 2003). Zoospores do not have a cell wall and each has two flagella and are released by flooding. They can swim a short distance (1.0 cm or less) in saturated soil, but are disseminated primarily by moving flood water. At the end of the motile period, which may last up to several days, zoospore movement becomes sluggish and jerky and encystment occurs (Schmitthenner, 2000). Zoospores can be attracted towards the compounds excreted by soybean root tips (Morris and Ward, 1992; Tyler *et al.*, 1996). On reaching the root surface, the zoospores begin to encyst and germinate, and the hyphae penetrate directly between the cell walls of the epidermis (Beagle-Ristaino and Rissler, 1983). The infection process can be completed in 30 min in optimum conditions. On resistant cultivars, hypersensitive response (HR) may occur, the pathogen is contained in numerous necrotic or dead cells and there is no development of haustoria in the resistance interaction. However, there is no early HR reaction occurring in susceptible cultivar; the hyphae initially grow intracellularly and then form many haustoria in root cells, which remain alive in direct contact with the pathogen after infection for around 10 h (Enkerli *et al.*, 1997) or approximately 12 h (Ward, 1990), when *P. sojae* is able to colonize host cells in an initial biotrophic phase of growth without triggering any response from the plant. After the initial infection, the pathogen begins to enter a necrotrophic growth mode and causes many host cells to die. The hypha penetrates from the epidermal cells of the root into the deep layers and vascular tissues.

Isolation and Identification of Physiologic Races of *P. sojae*

P. sojae is known for the difficulty in isolation due to its slow growth. As a result, the pathogen is often covered by bacteria and saprophytic fungi. The commonly used methods of isolating *P. sojae* are the plant stem lesion and soil methods described by Dorrance *et al.* (2008). The protocol for plant stem lesion isolation is by surface-sterilizing cut-off tissues of symptomatic stems first and incubating on the selective medium, e.g. PBNIC agar, to control bacteria and other fungi, such as *Pythium*. *P. sojae* has a distinct growth pattern on PBNIC agar, showing white mycelium 2–3 days after incubation. The mycelium is coenocytic and have branches almost at right angles, with curved tips. The asexual sporangium looks like an inverted pear and the round oospores on solid culture media can be seen 8–10 days later. Soil isolation is usually done by grinding the soil to fine particles, flooding it for 24 h, then draining and air-drying the soil until it cracks or pulls away from the side of the container, although it is still damp. The process is to break dormancy and induce germination of oospores in the infested soil. The soil is then used for planting a susceptible cultivar and *P. sojae* can be isolated readily from collapsed hypocotyls of emerging seedlings 5–6 days later using the same procedure described for plant stem lesion isolation.

The common way to store *P. sojae* is to grow the pathogen on V8 juice agar slants for 2 weeks, then cover the culture with 2 ml of sterile deionized water and store the culture at 15°C. The fungus may be stored in such conditions for up to 3 years without losing its virulence and aggressiveness. For a longer-term preservation, *P. sojae* can be stored in liquid nitrogen for at least 4 years (Dorrance *et al.*, 2008).

Pathogenic and genetic diversity of *P. sojae*

The pathogenetic variation of *P. sojae* was first reported in 1958 (Kaufmann and Gerdemann, 1958). During the interaction process with the soybean varieties, the pathogenicity and virulence of *P. sojae* evolved rapidly and 55 physiologic races were identified based on their differential reaction on a set of 8 or 13 differentials with a single resistance gene. Of the 55 races, races 1 to 45 were identified using a set of 8 differentials (Bernard *et al.*, 1957; Morgan and Hartwig, 1965; Schmitthenner, 1972; Schwenk and Sim, 1974; Haas and Buzzel, 1976; Laviolette and Athow, 1977; Keeling, 1979, 1982; Laviolette, 1983; White, 1983; Layton, 1986; Wagner and Wilkinson, 1992; Henry and Kirkpatrick, 1995; Abney *et al.*, 1997) and races 46 to 55 on a set of 13 differentials (Ryley *et al.*, 1998; Leitz *et al.*, 2000).

Several DNA-based molecular markers such as SSR (simple sequence repeat), rDNA-ITS (rDNA-internal transcribed spacer), RFLP (restriction fragment length polymorphism) and RAPD (random amplified polymorphism DNA) have been used successfully to identify the genetic variation and diversity of *P. sojae*. Whisson *et al.* (1992) used RFLP to confirm sexual recombination of *P. sojae in vitro* and studied the segregation of avirulence genes. Föster *et al.* (1994) proposed that occasional outcrosses had been a major contributor to the origin of new physiological races of *P. sojae*, in addition to clonal evolution. Meng *et al.* (1999) used the RAPD method to study populations of *P. sojae* from Indiana, Iowa and Minnesota (USA) and found no correlation of populations with a geographic origin. Wang *et al.* (2003) analysed genetic diversity of 75 *P. sojae* isolates from China using the RAPD method and distinguished 12 genetic groups, but most of the isolates were clustered into one group and no relationship between clustering and geographic origin was found. Wang *et al.* (2006) studied the genetic variation among *P. sojae* in the USA and China and found that there existed higher genetic variations in populations in the USA compared to the Chinese populations based on the RAPD analysis. Gally *et al.* (2007) examined by RAPD analysis the diversity of 32 *P. sojae* isolates of different geographic origins from Argentina and detected intraspecific variability even among isolates of the same geographic origin. Xu *et al.* (2007) detected 17 *P. sojae* isolates from three locations in Heilongjiang Province, China, and demonstrated by sequence analysis the difference between the base constitution of ITS1 and ITS2 among isolates. The 17 isolates were classified into three groups based on the ITS sequence and those isolated from the same region belonged to the same group, which showed the variation in geography. These studies demonstrated that molecular tools could be used to disclose the intraspecific diversity of *P. sojae* isolates both within and among geographic origins.

Management Strategies

The management strategies for prevention against *Phytophthora* root and stem rot at present are mainly by deployments of cultivars with race-specific or race non-specific resistance, or a combination of the two. Chemical methods and cultural practices like crop rotation and tillage and integrated management are used to a lesser extent.

Screening for resources of race-specific and race non-specific resistance

Race-specific resistance (*Rps*) genes in soybean have been used extensively to manage *P. sojae* (Dorrance *et al.*, 2003). New sources of resistance to *P. sojae* have been reported, mainly from soybean varieties and germplasm in China, where soybean was originated. Lohnes *et al.* (1996) reported that the *Rps*1d gene was common in accessions from Anhui and Jiangsu Provinces after they evaluated 517 soybean germplasms collected from several provinces in central China. Kyle *et al.* (1998) investigated soybean accessions from southern China in response to several races of *P. sojae* and demonstrated that germplasm from Hubei, Jiangsu and Sichuan Provinces appeared to be valuable multi-gene resistance sources. Lv *et al.* (2001) screened 956 soybean accessions from north-east China (Heilongjiang, Jilin and Liaoning Provinces) and identified 23 varieties with resistance to both race 1 and race 25, the predominant and the most virulent races of *P. sojae* in the region, respectively. Zhang *et al.* (2007) evaluated

530 soybean germplasms including 280 native soybean accessions and 250 commercial cultivars and found that the percentage of resistance in native soybean varieties was higher than that of the commercial cultivars. Similarly, Zhu *et al.* (2000, 2004), Li *et al.* (2001), Huo *et al.* (2005) and Jin and Zhang (2007) reported the identification of *P. sojae* resistant germplasm with a number of *Rps* genes from wild soybean accessions and soybean varieties from China. In addition, Dorrance and Schmitthenner (2000) identified several soybean accessions with multi-gene resistance after evaluating 1015 plant introductions originated from the Republic of Korea. These single *Rps* genes, however, have often been short-lived, with an effective 'life' of 8–15 years, due to the emergence of new virulent races in response to selection pressure exerted by the continuous use of specific resistant cultivars (Schmitthenner and Van Doren, 1985; Ferguson, 1987; Schmitthenner *et al.*, 1994; Abney *et al.*, 1997; Ryley *et al.*, 1998).

With the known *Rps* genes defeated, race non-specific resistance or partial resistance, described as the ability of plants to survive root infection without displaying severe disease symptoms such as death, stunting or yield loss, is of great interest and gains more and more attention from soybean breeders (Buzzell and Anderson, 1982; Tooley and Grau, 1984; Schmitthenner and Van Doren, 1985). The strategy of the combination of race non-specific resistance with race-specific resistance is brought forward to provide long-term management of *Phytophthora* root and stem rot, as well as to avoid the boom-and-bust cycle of single gene deployment, since it reduces the severity of root rot and slows the rate of disease development (Buzzell and Anderson, 1982; Burnham *et al.*, 2003b).

Race non-specific resistance commonly had been evaluated under natural infection in the field until recently, owing to the unavailability of a suitable laboratory procedure. Jimenez and Lockwood (1980) first described a laboratory procedure for screening race non-specific resistance by growing soybean seedlings in cups that were placed in plastic trays containing a specified number of zoospores of *P. sojae*. Irwin *et al.* (1982) reported a laboratory assay by inoculating soybean seedlings with dry *P. sojae* mycelium for rapid determination of relative levels of race non-specific resistance. Dorrance *et al.* (2008) described a layer test and a tray test for screening soybeans for race non-specific resistance to *P. sojae*. The layer test is done by placing inoculum of a 14-day-old *P. sojae* culture 5 cm below the seeds in cups containing coarse vermiculite and the amount of root rot and seedling death is rated 3 weeks after planting. The tray test is assessed by wound inoculation of a mycelial slurry on the root of 7-day-old seedlings and root rot is rated after 7 days. Using the layer test, Jia and James (2008) identified several accessions with high levels of race non-specific resistance compared with Conrade, the common known race non-specific resistant soybean variety to *P. sojae*.

Resistance genes and marker-assisted selection

A single dominant resistance gene has been widely explored since the first resistance gene (*Rps*1a) was identified by Bernard *et al.* (1957). With the *Rps*1a defeated and the emergence of new races in response to selection pressure exerted by the continuous use of *Rps*1a, new *Rps* genes are identified. A total of 14 *Rps* genes including *Rps*1a, *Rps*1b, *Rps*1c, *Rps*1d, *Rps*1k, *Rps*2, *Rps*3a, *Rps*3b, *Rps*3c, *Rps*4, *Rps*5, *Rps*6, *Rps*7 and *Rps*8 at eight genomic loci have been reported so far (Bernard *et al.*, 1957; Kilen *et al.*, 1974; Laviolette and Athow, 1977; Mueller *et al.*, 1978; Athow *et al.*, 1980; Bernard and Cremeens, 1981; Athow and Laviolette, 1982; Ploper *et al.*, 1985; Anderson and Buzzell, 1992; Burnham *et al.*, 2003a). All of these loci have been placed on the soybean genetic map. *Rps*1 and *Rps*3 are mapped on molecular linkage groups (MLG) N and F, respectively (Diers *et al.*, 1992; Demirbas *et al.*, 2001; Burnham *et al.*, 2003a). *Rps*2 (MLG J), *Rps*4 (MLG G), *Rps*5 (MLG G), *Rps*6 (MLG G), *Rps*7 (MLG N) and *Rps*8 (MLG A2) have also been mapped (Diers *et al.*, 1992;

Lohnes and Schmitthenner, 1997; Demirbas *et al.*, 2001; Burnham *et al.*, 2003a).

Molecular markers have been used to facilitate selection for both single and multi-resistance genes (Bent and Yu, 1999; Kumar, 1999). SSR and RFLP markers have been identified for *Rps*1 (Diers *et al.*, 1992), *Rps*1a (Weng *et al.*, 2001), *Rps*1b and *Rps*1c (Demirbas *et al.*, 2001), *Rps*1d (Sugimoto *et al.*, 2008), *Rps*1k (Kasuga *et al.*, 1997; Bhattacharyya *et al.*, 2005), *Rps*2, *Rps*3, *Rps*4, *Rps*5, *Rps*6 (Diers *et al.*, 1992; Cregan *et al.*, 1999) and *Rps*7 (Lohnes and Schmitthenner, 1997). With the development of the molecular biotechnique, more and more markers closer to the resistance genes will be used in marker-assisted selection (MAS), which has been complementary to conventional breeding programmes and to shortening the breeding period.

In some cases, genes for complete race-specific resistance that have already been defeated by new races of *P. sojae* may contribute to race non-specific resistance (Gebhardt and Valkonen, 2001), which appears to be controlled by several genes (Walker and Schmitthenner, 1984; Glover and Scott, 1998) and is more durable (Tooley and Grau, 1984). Several QTLs have been mapped to linkage groups for race non-specific resistance to *P. sojae*. Burnham *et al.* (2003b) used three recombinant inbred line (RIL) populations with the cultivar Conrad as the race non-specific resistance parent and identified two putative QTLs on MLG F and D1b + W from Conrade in all three populations. Han *et al.* (2008) used the RILs population of Conrade and OX760-6-1 as the race non-specific resistance and susceptible parent, respectively, and detected three QTLs, i.e. QGP1, QGP2 and QGP3, for *Phytophthora* root and stem rot tolerance. They further confirmed that QGP1 was located on linkage group F and QGP2 in a different interval on linkage group F and QGP3 on linkage group D1b + W. Furthermore, an RIL population of a cross between Conrade and Hefeng 25 was constructed and four markers on three linkage groups, MLG D1b + W, MLG F and MLG A2, were identified as being associated significantly with race non-specific resistance (Li *et al.*, 2008).

Chemical, cultural, biological and integrated control

The most commonly used chemical in the prevention of *P. sojae* is metalaxyl, which is an acylalanine fungicide specific to oomycetes. The fungicide is commonly applied as seed treatment (apron fungicide) and in-furrow, spray or granule (ridomil fungicide) to reduce plant emergence loss and increase yields of susceptible varieties (Anderson and Buzzell, 1982; Guy *et al.*, 1989).

Since soybean is the only host of *P. sojae* in the field, crop rotation is an effective means of reducing the severity of the disease. Although a short-term crop rotation may not allow for reduction of inoculum, it does prevent the immediate build-up of *P. sojae* populations (Schmitthenner, 1985). Because saturated soil favours the occurrence of *P. sojae*, tillage that could promote soil drainage is proven to be effective in reducing the infection period (Grau *et al.*, 2004). Oospores could also be buried deeper in the soil by tillage (Workneh *et al.*, 1998).

Biological control has been considered a more natural and environmentally acceptable alternative to the existing chemical treatment methods (Cook and Baker, 1983; Baker and Paulitz, 1996). Several bacteria and fungi have been identified as potential bioagents in dual-culture and greenhouse experiments.

There are inevitable shortcomings for each of the prevention measurements, that is, the *Rps* genes in cultivars can be defeated by new races of *P. sojae*, and the varieties with race non-specific resistance, seed treatment, rotation and tillage cannot provide an effective control when disease pressure is high. A combination of two or more strategies to prevent the infection of *P. sojae* is very essential. The disease could be best managed with integrated strategies in a combination of deployment of cultivars incorporated into race-specific and race non-specific resistance genes, fungicide treatments, improved soil drainage and biocontrol. These management tactics can reduce inoculum in fields and limit the amount of water available for the pathogen to germinate and infect, therefore minimizing disease damage and increasing soybean production efficacy.

Conclusions

Owing to the shift of *P. sojae* races, use of race-specific resistance may quickly become inefficient in management of the disease. More efforts are needed in the identification and incorporation of multi-gene resistance and race non-specific or partial resistance into new soybean cultivars in the future for prevention against *Phytophthora* root rot and stem rot. Soybean cultivars with a combination of two types of resistance would be long-lived and more desirable by the soybean industry. Conventional breeding will still be the main method of resistance breeding in the foreseeable future but, with the rapid development of molecular technology, there will be more and more molecular markers closely linked to the *Rps* genes mapped, which could be useful in MAS to hasten the breeding and cultivar development process.

To our knowledge, limited research has been carried out on the expression of genes which are possibly involved in soybean resistance to *P. sojae*, notwithstanding that 14 dominant *Rps* genes at 8 loci have been identified and resistance gene mapping and quantitative trait loci have been explored. Moy *et al.* (2004) reported the patterns of gene expression on infection of soybean plants by *P. sojae* race 2 and demonstrated that genes identified as strongly upregulated during infection included those encoding enzymes of phytoalexin biosynthesis and defence and pathogenesis-related proteins. A better understanding of the resistance mechanism in the *P. sojae*–soybean interaction at the molecular level is needed for effective gene deployments and resistance breeding. Research into these new disease management strategies is required with transgenic soybean cultivars, in order to provide solutions for future needs when transgenic technologies will be more acceptable.

References

Abney, T.S., Melgar, J.C., Richards, T.L., Scott, D.H., Grogan, J. and Young, J. (1997) New races of *Phytophthora sojae* with *Rps*1-d virulence. *Plant Disease* 81, 653–655.

Anderson, T.R. and Buzzell, R.I. (1982) Efficacy of metalaxyl in controlling *Phytophthora* root and stalk rot of soybean cultivars differing in field tolerance. *Plant Disease* 66, 1144–1145.

Anderson, T.R. and Buzzell, R.I. (1992) Inheritance and linkage of the *Rps7* gene for resistance to *Phytophthora* rot of soybean. *Plant Disease* 76, 958–959.

Anderson, T.R. and Tenuta, A. (2003) *Phytophthora* rot. In: Bailey, K.L., Gossen, B.D., Gugel, R.K. and Morrall, R.A.A. (eds) *Diseases of Field Crops in Canada.* The Canadian Phytopathological Society, University Extension Press, University of Saskatchewan, Saskatoon, Canada, pp. 155–156.

Athow, K.L. (1987) Fungal diseases. In: Wilcox, J.R. (ed) *Soybeans: Improvement, Production, and Uses.* 2nd edn. Agronomy Monograph No.16. ASA, CSSA and SSSA, Madison, Wisconsin, pp. 687–727.

Athow, K.L. and Laviolette, F.A. (1982) *Rps*6, a major gene for resistance to *Phytophthora megasperma* f. sp. *glycinea* in soybean. *Phytopathology* 72, 1564–1567.

Athow, K.L., Laviolette, F.A., Mueller, E.H. and Wilcox, J.R. (1980) A new major gene for resistance to *Phytophthora megasperma* var. *sojae* in soybean. *Phytopathology* 70, 977–980.

Baker, R. and Paulitz, T.C. (1996) Theoretical basis for microbial interactions leading to biological control of soilborne plant pathogens. In: Hall, R. (ed.) *Principles and Practice of Managing Soilborne Plant Pathogens.* APS Press, St Paul, Minnesota, pp. 50–79.

Beagle-Ristaino, J.E. and Rissler, J.F. (1983) Histopathology of susceptible and resistant soybean roots inoculated with zoospores of *Phytophthora megasperma* f. sp. *glycinea*. *Phytopathology* 72, 727–733.

Bent, A.F. and Yu, I. (1999) Applications of molecular biology to plant disease and insect resistance. *Advances in Agronomy* 66, 251–298.

Bernard, R.L. and Cremeens, C.R. (1981) An allele at the *Rps*1 locus from the variety ‘Kingwa’. *Soybean Genetics Newsletter* 8, 40–42.

Bernard, R.L., Smith, P.E., Kaufmann, M.J. and Schmitthenner, A.F. (1957) Inheritance of resistance to *Phytophthora* root and stem rot in soybean. *Agronomy Journal* 49, 391.

Bhattacharyya, M.K., Narayanan, N.N., Gao, H., Santra, D.K., Salimath, S.S., Kasuga, T., Liu, Y., Espinosa, B., Ellison, L., Marek, L., Shoemaker, R., Gijzen, M. and Buzzell, R.I. (2005) Identification of a large cluster of coiled coil-nucleotide binding site-leucine rich repeat-type genes from the *Rps*1 region containing *Phytophthora* resistance genes in soybean. *Theoretical and Applied Genetics* 111, 75–86.

Burnham, K.D., Dorrance, A.E., Francis, D.M., Fioritto, R.J. and St Martin, S.K. (2003a) *Rps*8, a new locus in soybean for resistance to *Phytophthora sojae*. *Crop Science* 43, 101–105.

Burnham, K.D., Dorrance, A.E., VanToai, T.T. and St Martin, S.K. (2003b) Quantitative trait loci for partial resistance to *Phytophthora sojae* in soybean. *Crop Science* 43, 1610–1617.

Buzzell, R.I. and Anderson, T.R. (1982) Plant loss response of soybean cultivars to *Phytophthora megasperma* f. sp. *glycinea* under field conditions. *Plant Disease* 66, 1146–1148.

Cook, R.J. and Baker, K.F. (1983) *The Nature and Practice of Biological Control of Plant Pathogens*. APS, St Paul, Minnesota, 539 pp.

Cregan, P.B., Jarvik, T., Bush, A.L., Shoemaker, R.C., Lark, K.G., Kahler, A.L., Kaya, N., VanToai, T.T., Lohnes, D.G. and Chung, J. (1999) An integrated genetic linkage map of the soybean genome. *Crop Science* 39, 1464–1490.

Demirbas, A., Rector, B.G., Lohnes, D.G., Fioritto, R.J., Graef, G.L., Cregan, P.B., Shoemaker, R.C. and Specht, J.E. (2001) Simple sequence repeat markers linked to the soybean *Rps* genes for *Phytophthora* resistance. *Crop Science* 41, 1220–1227.

Diers, B.W., Mansur, L., Imsande, J. and Shoemaker, R.C. (1992) Mapping *Phytophthora* resistance loci in soybean with restriction fragment length polymorphism markers. *Crop Science* 32, 377–383.

Dorrance, A.E. and Schmitthenner, A.F. (2000) New sources of resistance to *Phytophthora sojae* in soybean plant introductions. *Plant Disease* 84, 1303–1308.

Dorrance, A.E., McClure, S.A. and Martin, S.K. (2003) Effect of partial resistance on *Phytophthora* stem rot incidence and yield of soybean in Ohio. *Plant Disease* 87, 308–312.

Dorrance, A.E., Berry, S.A., Anderson, T.R. and Meharg, C. (2008) Isolation, storage, pathotype characterization and evaluation of resistance for *Phytophthora sojae* in soybean. *Plant Health Progress Online* doi: 10.1094/PHP-2008-0118-01-DG.

Enkerli, K., Mims, C.W. and Hahn, M.G. (1997) Ultrastructure of compatible and incompatible interactions of soybean roots infected with the plant pathogenic oomycete *Phytophthora sojae*. *Canadian Journal of Botany* 75, 1493–1508.

Erwin, D.C. and Ribeiro, O.K. (1996) *Phytophthora Diseases Worldwide*. APS Press, St Paul, Minnesota, 592 pp.

Ferguson, M.W. (1987) Epidemic of *Phytophthora* root rot of soybeans in South Dakota. *Plant Disease* 71, 850.

Förster, H., Coffey, M.D., Elwood, H. and Sogin, M.L. (1990) Sequence analysis of the small subunit ribosomal RNAs of three zoosporic fungi and implications for fungal evolution. *Mycologia* 82, 306–312.

Föster, H., Tyler, B.M. and Coffey, M.D. (1994) *Phytophthora sojae* races have arisen by clonal evolution and by rare outcrosses. *Molecular Plant–Microbe Interactions* 7, 780–791.

Gally, M., Ramos, A.M., Dokmetzian, D. and Lopez, S.E. (2007) Genetic variability of *Phytophthora sojae* isolates from Argentina. *Mycologia* 6, 877–883.

Gebhardt, C. and Valkonen, J. (2001) Organization of genes controlling disease resistance in the potato genome. *Annual Review of Phytopathology* 39, 79–102.

Glover, K.D. and Scott, R.A. (1998) Heritability and phenotypic variation of tolerance to *Phytophthora* root rot of soybean. *Crop Science* 38, 1495–1500.

Grau, C.R., Dorrance, A.E., Bond, J. and Russin, J.S. (2004) Fungal diseases. In: Boerma, H.R. and Specht, J.E. (eds) *Soybeans: Improvement, Production, and Uses*, 3rd edn. Agronomy Monograph No. 16. ASA, CSSA and SSSA, Madison, Wisconsin, pp. 679–763.

Guy, S.O., Oplinger, E.S. and Grau, C.R. (1989) Soybean cultivar response to metalaxyl applied in furrow and as a seed treatment. *Agronomy Journal* 81, 529–532.

Haas, J.H. and Buzzell, R.I. (1976) New races 5 and 6 of *Phytophthora megasperma* var. *sojae* and differential reactions of soybean cultivars for races 1 and 6. *Phytopathology* 66, 1361–1362.

Han, Y.P., Teng, W.L., Yu, K.F., Poysa, V., Anderson, T., Qiu, L.J. and Li, W.B. (2008) Mapping QTL tolerance to *Phytophthora* root rot in soybean using microsatellite and RAPD/SCAR derived markers. *Euphytica* 162, 231–239.

Harper, J.T., Waanders, E. and Keeling, P.J. (2005) On the monophyly of chromalveolates using a six-protein phylogeny of eukaryotes. *International Journal of Systematic and Evolutionary Microbiology* 55, 487–496.

Henry, R.H. and Kirkpatrick, T.L. (1995) Two new races of *Phytophthora sojae*, causal agent of *Phytophthora* root and stem rot of soybean, identified from Arkansas soybean field. *Plant Disease* 79, 1074.

Hildebrand, A.A. (1959) A root and stalk rot of soybeans caused by *Phytophthora megasperma* Drechsler var. *sojae*. *Canadian Journal of Botany* 37, 927–957.

Huo, Y.L., Zhu, Z.D., Li, X.H., Huang, J.B. and Wu, X.F. (2005) Preliminary screening for *Phytophthora* root rot resistance in wild soybean. *Journal of Plant Genetic Resources* 6(2), 182–185.

Irwin, J.A.G., Langdon, P.W. and Aust, J. (1982) A laboratory procedure for determining relative levels of field resistance in soybeans to *Phytophthora megasperma* f. sp. *glycinea*. *Agricultural Research* 33, 33–39.

Jee, H., Kim, W. and Cho, W. (1998) Occurrence of *Phytophthora* root rot on soybean (*Glycine max*) and identification of the causal fungus. *Crop Protection* 40, 16–22.

Jia, H.Y. and James, K.J.E. (2008) Resistance and partial resistance to *Phytophthora sojae* in early maturity group soybean plant introductions. *Euphytica* 159, 27–34.

Jimenez, B. and Lockwood, J.L. (1980) Laboratory method for assessing field tolerance of soybean seedlings to *Phytophthora megasperma* var. *sojae*. *Plant Disease* 64, 775–778.

Jin, L.M. and Zhang, S.Z. (2007) Identification the resistance of wild soybean germplasm to *Phytophthora sojae*. *Soybean Science* 26(3), 300–304.

Kasuga, T., Salimath, S.S., Shi, J., Gijzen, M., Buzzell, R.I. and Bhattacharyya, M.K. (1997) High resolution genetic and physical mapping of molecular markers linked to the *Phytophthora* resistance gene *Rps*1-k in soybean. *Molecular Plant–Microbe Interacttions* 10, 1035–1044.

Kaufmann, M.J. and Gerdemann, J.W. (1958) Root and stem rot of soybean caused by *Phytophthora sojae*. *Phytopathology* 48, 201–208.

Keeling, B.L. (1979) Research on *Phytophthora* root and stem rot. In: Corbin, F.T. (ed.) *Proceedings of the World Soybean Research Conference II*. Westview Press, Boulder, Colorado, pp. 367–370.

Keeling, B.L. (1982) Four new physiologic races of *Phytophthora megasperma* f. sp. *glycinea*. *Plant Disease* 66, 334–335.

Kilen, T.C., Hartwig, E.E. and Keeling, B.L. (1974) Inheritance of a second major gene for resistance to *Phytophthora* root rot in soybeans. *Crop Science* 14, 260–262.

Kittle, D.R. and Gray, L.E. (1979) The influence of soil temperature, moisture, porosity, and bulk density on the pathogenicity of *Phytophthora megasperma* var. *sojae*. *Plant Disease Reporter* 63, 231–234.

Kumar, L.S. (1999) DNA markers in plant improvement: an overview. *Biotechnology Advances* 17, 143–182.

Kyle, D.E., Nickell, C.D., Nelson, R.L. and Pederson, W.L. (1998) Response of soybean accessions from provinces in Southern China to *Phytophthora sojae*. *Plant Disease* 82, 555–559.

Laviolette, F.A. (1983) Two new physiologic races of *P. megasperma*. f. sp. *glycinea*. *Plant Disease* 67, 497–498.

Laviolette, F.A. and Athow, K.L. (1977) Three new physiologic races of *Phytophthora megasperma* var. *sojae*. *Phytopathology* 67, 267–268.

Layton, A.C. (1986) New physiologic race of *P. megasperma*. f. sp. *glycinea*. *Plant Disease* 70, 333–338.

Leitz, R.A., Hartman, G.L., Pedersen, W.L. and Nickell, C.D. (2000) Races of *Phytophthora sojae* on soybean in Illinois. *Plant Disease* 84, 487.

Li, C.S., Lu, X.B., Liu, T.J., Li, L., Qi, J.S. and Xu, Z.T. (2001) Identification race of *Phytopthora sojae* and screening of soybean cultivar resistance. *Chinese Journal of Oil Crop Science* 23(2), 60–62.

Li, X.P., Han, Y.P., Ding, J.J., Zhang, S.Z., Ma, W.D. and Li, W.B. (2008) Mapping quantitative loci controlling tolerance to *Phytophthora* root rot in soybean. *Soybean Science* 27(4), 572–575.

Lohnes, D.G. and Schmitthenner, A.F. (1997) Position of the *Phytophthora* resistance gene *Rps7* on the soybean molecular map. *Crop Science* 37, 555–556.

Lohnes, D.G., Nickell, C.D. and Schmitthenner, A.F. (1996) Origin of soybean alleles for *Phytophthora* resistance in China. *Crop Science* 36, 1689–1692.

Lv, H.Y., Kong, F.J., Xu, X.H., Yuan, X.L. and Yang, Q.K. (2001) Response of soybean germplasm from the Northeast China to *Phytophthora* root rot caused by *Phytophthora sojae*. *Chinese Journal of Oil Crop Science* 23(4), 70–72.

Meng, X.Q., Shoemaker, R.C. and Yang, X.B. (1999) Analysis of pathogenicity and genetic variation among *Phytophthora sojae* isolates using RAPD. *Mycological Research* 103, 173–178.

Morgan, F.L. and Hartwig, E.E. (1965) Physiologic specialization in *Phytophthora megasperma* var. *sojae*. *Phytopathology* 55, 1277–1279.

Morris, P.F. and Ward, E.W.B. (1992) Chemoattraction of zoospores of the soybean pathogen, *Phytophthora sojae*, by isoflavones. *Physiological and Molecular Plant Pathology* 40, 17–22.

Moy, P., Qutob, D., Chapman B.P., Atkinson, I. and Gijzen, M. (2004) Patterns of gene expression upon infection of soybean plants by *Phytophthora sojae. Molecular Plant–Microbe Interactions* 17(10), 1051–1062.

Mueller, E.H., Athow, K.L. and Laviolette, F.A. (1978) Inheritance of resistance to four physiologic races of *Phytophthora megasperma* var. *sojae. Phytopathology* 68, 1318–1322.

Pegg, K.G., Kochman, J.K. and Vock, N.T. (1980) Root and stem rot of soybean caused by *Phytophthora megasperma* var. *sojae* in Queensland. *Australasian Plant Pathology* 9, 15.

Ploper, L.D., Athow, K.L. and Laviolette, F.A. (1985) A new allele at the *Rps*3 locus for resistance to *Phytophthora megasperma* f. sp. *glycinea* in soybean. *Phytopathology* 75, 690–694.

Ryley, M.J., Obst, N.R., Irwin, J.A.G. and Drenth, A. (1998) Changes in the racial composition of *Phytophthora sojae* in Australia between 1979 and 1996. *Plant Disease* 82, 1048–1054.

Schmitthenner, A.F. (1972) Evidence for a new race of *Phytophthora megasperma* var. *sojae* pathogenic to soybean. *Plant Disease Reporter* 56, 536–539.

Schmitthenner, A.F. (1985) Problems and progress in control of *Phytophthora* root rot of soybean. *Plant Disease* 69, 362–368.

Schmitthenner, A.F. (2000) *Phytophthora* rot of soybean. *Plant Health Progress. Online* doi: 10. 1094/PHP-2000-0601-01-HM.

Schmitthenner, A.F. and Van Doren, D.M. (1985) Integrated control of root rot of soybean caused by *Phytophthora megasperma* f. sp. *glycinea*. In: Parker, C.A.A., Rovira, D., Morre, K.J. and Wong, P.T.W. (eds) *Ecology and Management of Soilborne Plant Pathogens.* APS, St Paul, Minnesota, pp. 263–266.

Schmitthenner, A.F., Hobe, M. and Bhat, R.G. (1994) *Phytophthora sojae* races in Ohio over a 10-year interval. *Plant Disease* 78, 269–276.

Schwenk, F.W. and Sim, T. (1974) Race 4 of *Phytophthora megasperma* var. *sojae* from soybeans proposed. *Plant Disease Reporter* 58, 353–354.

Shen, C.Y. and Su, Y.C. (1991) Discovery and preliminary studies of *Phytophthora megasperma* on soybean in China. *Acta Phytopathology Sinica* 21, 198.

Sugimoto, T., Yoshida, S., Watanabe, K., Aino, M., Kanto, T., Maekawa, K. and Irie, K. (2008) Identification of SSR markers linked to the *Phytophthora* resistance gene *Rps*1-d in soybean. *Plant Breeding* 27, 154–159.

Tooley, P.W. and Grau, C.R. (1984) Field characterization of rate-reducing resistance to *Phytophthora megasperma* f. sp. *glycinea* in soybean. *Phytopathology* 74, 1201–1208.

Tyler, B.M. (2007) *Phytophthora sojae*: root rot pathogen of soybean and model oomycete. *Molecular Plant Pathology* 8(1), 1–8.

Tyler, B.M., Wu, M., Wang, J., Cheung, W. and Morris, P.F. (1996) Chemotactic preferences and strain variation in response of *Phytophthora sojae* zoospores to host isoflavones. *Applied and Environmental Microbiology* 62, 2811–2817.

Wagner, R.E. and Wilkinson, H.T. (1992) A new physiological race of *Phytophthora sojae* on soybean. *Plant Disease* 76, 212.

Walker, A.K. and Schmitthenner, A.F. (1984) Heritability of tolerance to *Phytophthora* rot in soybean. *Crop Science* 24, 490–491.

Wang, H.B., Wang, X.M. and Zhu, Z.D. (2003) Analysis of genetic diversity of *Phytophthora sojae* isolates in China using RAPD. *Mycosystema* 22, 219–227.

Wang, Z.Y., Wang, Y.C., Zhang, Z.G. and Zheng, X.B. (2006) Genetic relationships among Chinese and American isolates of *Phytophthora sojae* assessed by RAPD markers. *Chinese Science Bulletin* 51 (17), 2095–2102.

Ward, E.W.B. (1990) The interaction of soybeans with *Phytophthora megasperma* f. sp. *glycinea*: pathogenicity. In: Hornby, B. (ed.) *Biological Control of Soilborne Plant Pathogens.* CAB International, Wallingford, U K, pp. 311–327.

Weng, C., Yu, K., Anderson, T. and Poysa, V. (2001) Mapping genes conferring resistance to *Phytophthora* root rot of soybean, *Rps*1a and *Rps*7. *Heredity* 92(5), 442–446.

Whisson, S.C., Maclean, D.J. and Manners, J.M. (1992) Genetic relationships among Australian and North American isolates of *Phytophthora megasperma* f. sp. *glycinea* assessed by multicopy DNA probes. *Phytopathology* 82, 863–868.

White, D.M. (1983) Race of *P. me.* f. sp. *glycinea* on soybean in Eastern Nebraska. *Plant Disease* 67, 1281–1284.

Workneh, F., Yang, X.B. and Tylka, G.L. (1998) Effect of tillage practices on vertical distribution of *Phytophthora sojae. Plant Disease* 82, 1258–1263.

Wrather, J.A., Anderson, T.R., Arsyad, D.M., Gai, J., Ploper, L.D., Porta-Puglia, A., Ram, H.H. and Yorinori, J.T. (1997) Soybean disease loss estimates for the top 10 soybean producing countries in 1994. *Plant Disease* 81, 107–110.

Xu, P.F., Zhang, S.Z. and Li, W.B. (2007) Phylogenetic analysis of the sequences of rDNA Internal Transcribed Spacer (ITS) of *Phytophthora sojae*. *Journal of Genetics and Genomics* 34, 180–188.

Zhang, S.Z., Xu, P.F., Wu, J.J. and Li, W.B. (2007) Identification of resistance to *Phytophtora sojae* in soybean germplasm. *Soybean Science* 26(6), 914–917.

Zhu, Z.D., Wang, X.M., Chang, R.Z., Ma, S.M., Wu, X.F. and Tian, Y.L. (2000) Identification of race of *Phytophthora sojae* and reaction of soybean germplasm resources in Heilongjiang province. *Scientia Agricultura Sinica* 33(1), 62–67.

Zhu, Z.D., Wang, H.B., Wang, X.M., Chang, R.Z. and Wu, X.F. (2004) Response of soybean cultivars or lines developed in Heilongjiang province to five strains of *Phytophthora sojae*. *Journal of Plant Genetic Resources* 5(1), 22–25.

25 Management of Fungal Pathogens – A Prerequisite for Maintenance of Seed Quality During Storage

Anuja Gupta
Indian Agricultural Research Institute, Regional Station, Karnal, Haryana, India

Abstract
Storage of seed is essential for any seed programme to sow the crop in the next season, to maintain buffer stock as an insurance against crop failure in times of drought, excessive rainfall or other natural calamities, to maintain parental lines for the production of hybrid seed, to conserve germplasm for breeding purposes and for seed trade at national and international levels. Availability of good quality seed at the right time and place is a basic prerequisite for sustaining agriculture. While maintenance of seed germination is of utmost importance to any seed person, preservation of seed quality in terms of its health status is equally important. A quality seed should have high genetic purity, physical purity, seed germination, seed vigour and good health status.

Introduction

Seed health is being recognized as one of the important criteria in evaluating seed quality. Seed health refers primarily to the presence or absence of disease-causing organisms such as fungi, bacteria and viruses, or animal pests such as nematodes and insects, or physiological disorders due to deficiency of trace elements. One of the major problems associated with crop production in India is the maintenance of the prescribed level of seed vigour and viability from seed harvest till the next sowing season. In our plans for attaining self-sufficiency in food grains, preventing their loss in storage is as important as the various measures to increase production. About 60–70% of the annual output is retained by the farmer. Storage facilities with the farmer, as well as with the traders, are far from satisfactory. Farmers and traders are not fully aware of the large savings that can be obtained by proper storage and preservation techniques. The important factors that determine the longevity of seeds are seed moisture, the type of storage container and storage environment. These factors generally interact, leading to a number of physiological and biochemical changes in the stored seeds, which result in deterioration of seed both in quality and quantity, especially in tropical and subtropical countries. According to a current estimate, 10% of food grain is lost in storage due to microbial spoilage and insect attack. The damage caused by rodents and insects is visible and therefore remedial measures are adopted for their control, but microbial spoilage of seeds/grains cannot be seen easily.

Microbial Spoilage of Seed during Storage

Seeds are the end product of a series of steps that include sowing, growing, harvesting and threshing, wherein it becomes vulnerable to various pathogens/saprophytes. Among the different microbes, fungi form a major group of organisms that infest seeds. Nearly 150 species of fungi have been found associated with grains and seed in storage (Dharam Vir, 1974). Mechanical damage in the seeds, cracks, breaks or scratches in the pericarp or seed coat developed during threshing and processing substantially facilitate invasion by fungi, which find their way to the storage warehouses.

The fungi found associated with seeds during storage are known as storage fungi. They can grow without free water, on media with high osmotic pressure, at RH = 70–90%. Some common storage fungi include species of *Aspergillus*, *Penicillium*, *Rhizopus*, *Fusarium*, *Cladosporium*, *Alternaria*, *Mucor*, *Chaetomium*, *Epicoccum*, etc. Discoloration and distortion of seeds is a major degrading factor because of seedborne infection. Other common manifestations are reduction in seed size, seed rots, shrivelling of seeds, seedling decay and pre- and post-emergence mortality and abnormalities. The fungi primarily invade the embryo and in the early stages of infection, the seed may appear normal but, due to well-established infections, these embryos are killed and the seeds appear darkened. These fungi are responsible for a decrease in market value, germinability and nutrition of the produce, making the grains unfit for human consumption and reducing the viability of the seed. Excessive fungal growth may also result in heating, caking and decay. The seed may thus become totally spoiled, darkened or charred by prolonged exposure to the heat generated during storage, which brings about biochemical changes leading to the production of toxins and loss in seed weight.

The invasion by fungi leads to physical and chemical changes in the seeds. Prasad *et al.* (1990) observed changes in the amount of fatty acids, glycerol, sugar and amino acids in radish seeds infected with *Aspergillus flavus*. Dube *et al.* (1988) reported changes in starch, fatty acids and sugars in wheat grains infected with *A. flavus* and *A. niger*. Mishra and Dharam Vir (1991) observed higher milling losses ranging from 34.0 to 58.6% in discoloured rice grains. Joshi *et al.* (1988) reported 73% reduction in starch content in stored pearl millet seeds infected with storage fungi and an increase in the amount of reducing sugars and phenolic contents. Bilgrami and Sinha (1983) have reported aflatoxin contamination in maize, groundnut and a variety of agricultural foods and feeds. Vaidehi (1997) showed that storage fungi lowered the quality of maize grains due to the biochemical changes they brought about. Storage fungi may be present as dormant spores or mycelium on the seed surface or below the pericarp, which activate and multiply at a phenomenal rate under favourable storage conditions.

Seed Mycoflora

The initial mycoflora of the seeds can give an idea of the type of fungi that can initiate the process of deterioration in storage. With an increase in the storage period, there is an increase in the incidence of storage fungi and a decrease in seed germination (Gupta and Singh, 1990, 1993). Loss in germination due to storage fungi may be attributed to several factors. A toxin produced by *A. ruber* kills the tissues in the embryonic axes of pea seeds in advance of infection (Harman and Nash, 1972). Wheat seed infected with *Aspergillus* spp. and imbibed with water becomes a jelly-like mass, suggesting that cell wall degrading enzymes may be involved. In contrast, pea and squash embryos are killed without physical invasion by fungi, indicating the involvement of diffusible toxins (Harman and Pfleger, 1974). Mitochondria isolated from the embryonic axes of *A. ruber*-infected pea seeds were less active than those from non-infected seeds, suggesting that mitochondria damaged by fungi play a role in seed deterioration (Harman and Drury, 1973). Gupta *et al.* (1989) observed a decrease in the amount of volatile aldehyde compounds with increased levels of fungi on the seed and treatment with benomyl increased these compounds, indicating control of fungi and increased germination.

An increase in seed mycoflora is correlated directly to an increase in FFA content and leaching of solutes (especially electrolytes and water-soluble sugars) with advancing storage period (Gupta, 2003). Agarwal (1980) demonstrated that seed deterioration in okra, carrot and onion seeds was accompanied with leakage of sugars. Analysis of exudates from the seeds showed that the permeability of the membrane increased with the deterioration of seeds during ageing (Dadlani and Agarwal, 1983). According to Chen *et al.* (1998), with an increase in the fatty acid contents of different seeds like wheat and brassicas, storage potential decreased. Ramamoorthy and Karivaratharaju (1986) also found that with an increase in the storage period, the oil and protein content in groundnut seeds decreased gradually, while free fatty acid content increased, accompanied by a loss in seed viability under ambient storage conditions. Thus, during storage, especially under an ambient environment, seeds produce changes due to fungal activity, resulting in deterioration of their quality (Zagrebenyer and Bern, 1998; Gupta and Aneja, 2004).

Seed Mycoflora and Seed Viability

The incidence of fungal flora associated with different seeds is low initially, but it increases with an increase in the duration of storage and subsequently there is a decrease in seed viability. With the advancing storage period, the field fungi become limited and the produce becomes infested with storage fungi. A significant negative correlation ($r = -0.793$) between seed viability and seed mycoflora has been observed with advancing storage period (Table 25.1), and consequently a decrease in seed viability.

Treatments and seed mycoflora

Seed treatments, especially with fungicides like captan, thiram or mancozeb, restrict the growth of mycoflora on the seeds and maintain better seed viability (Gupta, 2003). Moreno *et al.* (1985) suggested the use of fungicides to protect the viability of corn seeds. In another study, Moreno and Ramirez (1983) recorded that after 330 days of storage at 26°C with 75% RH, the germination of untreated corn seed was only 61%, while it ranged from 68 to 90% in seeds treated with different fungicides either singly or in combination. The incidence of storage fungi was also very low in treated seeds.

Kushwaha and Raut (1994) reported that seeds treated with thiram and stored in poly-lined bags suppressed most of the fungi. Asalmol and Zade (1998) also observed

Table 25.1. Influence of seed mycoflora on seed viability and seed moisture during storage.

Storage period (months after seed treatment)	Seed germination* (%)	Seed mycoflora* (%)	Seed moisture* (%)
0	92.0	1.05	7.8
2	86.5	2.4	5.9
4	92.8	3.4	6.6
6	85.3	2.7	9.9
8	50.8	5.9	8.6
10	44.3	4.6	8.9
12	38.3	5.0	8.8
Correlation coefficient (r)	−0.872		0.35

Note: *Average of 20 treatments.

that pre-storage seed treatment helped to improve the shelf life of seeds and checked seed mycoflora during storage. Fungicide seed treatments were found to restrict the growth of mycoflora on different vegetable seeds (Gupta and Singh, 1993).

The incidence of *Colletotrichum dematium* associated with chilli seed at the time of storage was 5%. Seed treatment with captan controlled the pathogen just after its application. Other fungicides like thiride and carbendazim controlled the pathogen after 5 months of storage, whereas in the untreated control the pathogen persisted up to 7 months in a cloth bag and up to 15 months in an airtight container (Gupta *et al.*, 1992).

Thiram, bavistin and captan could control more than 96, 93 and 90% of the fungi associated with paddy seed as against 72% and 65% in hinosan and emisan + streptocycline treatments, respectively, after 17 months of storage under ambient conditions (Fig. 25.1). Mancozeb (78.6%) was most effective in the control of seed mycoflora on soybean seed during storage, followed by thiram (65.1%), bleaching powder (13.1%) and nimbecidine (10.1%) during storage (Fig. 25.2).

Onion seed variety Phule Safed dried at 6–7% moisture content and treated with 0.2% carbendazim could be stored safely in 700 gauge polyethylene bags for 32 months, as against 24 months in untreated seed (Mahajan *et al.*, 2001). Pumpkin seed treated with iodine-based halogen mixture at 3 g/kg seed and stored in 700 gauge polythene bags maintained seed quality. The disease, ginger yellow, was controlled effectively by seed treatment with 0.1% carbendazim (Rana and Sharma, 1995).

Sandhu (1989) reported seed dressing in the form of slurry with benlate at 1 g/kg and dry seed treatment with brassicol at 2.5 g/kg resulted in 100% and 93.9% inhibition of germination of pea seeds after their storage for 1 year. However, captan (0.25%) and bayletan (0.1%) treatments enhanced germination by 5.21 and 11.88%, respectively. He also found that in steam-sterilized soil, germination of poor quality seeds of pea variety Punjab-87 was enhanced from 17 to 56%. They were further enhanced up to 69% in seeds treated with captan.

Van Toai *et al.* (1986) observed that only reduced quality seed of soybean responded to fungicide seed treatment under prolonged storage. The accelerated ageing germination (AAG) results of the 24-month-old methanol-washed seeds were lower than the AAG results of the unwashed seeds (the fungicides were removed from the treated seeds by methanol prior to the AAG test), but significantly higher for all cultivars than the AAG values of the untreated seeds. The fungicidal seed treatments, in addition to protecting the seeds and seedlings during imbibition and germination, also helped to maintain the seed quality of soybean during storage.

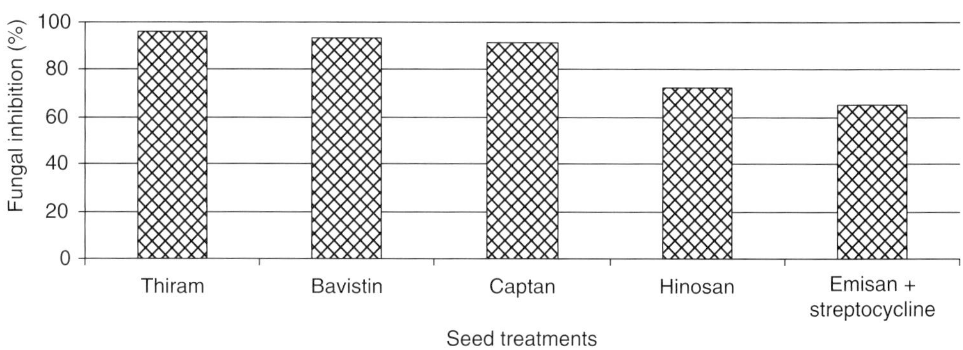

Fig. 25.1. Effect of different seed dressings on the control of seed mycoflora on paddy seed during storage.

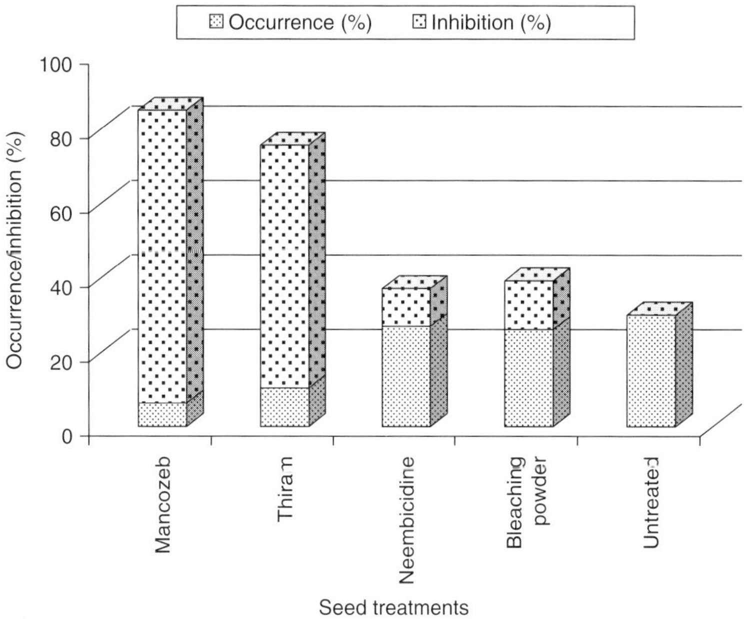

Fig. 25.2. Influence of seed dressings on the occurrence and inhibition of seed mycoflora on soybean seed during storage for 15 months.

Treatments and seed viability

The influence of seed treatments on seed viability and vigour is not apparent during the early period of storage, but becomes significant on prolonged storage. In some crops like mung bean, mustard, muskmelon, etc., the effect of seed treatments on seed germination is insignificant during storage, but in other crops like cowpea, sorghum, chillies, fenugreek, spinach and soybean, etc., seed treatments have a significant effect on maintaining seed viability for longer duration under ambient storage conditions.

Mung bean cv. PS-16 retained seed dormancy for up to 6 months of storage and viability for more than 36 months of storage (Table 25.2). Chickpea cv. P-256 treated with fungicides like thiram and ABC dust retained more than 85% germination for up to 24 months of storage. Thiram-treated seed had better germination compared to ABC dust treatment, and re-treatment with thiram further enhanced seed germination.

The germination of soybean seeds, cv. P-16, treated with mancozeb or thiram stored in poly-lined cloth bags (polythene bags of 400 gauge kept inside a cloth bag) remained above the prescribed standard of certification (70%), even on the 15th month of storage after seed harvest, whereas germination of both treated and untreated seeds stored in cloth bags fell below the certification standards on the 11th month of storage. Seed treatment with mancozeb and thiram resulted in significantly better root and shoot lengths of seedlings as compared to other treatments. Seedling vigour measured in terms of seedling dry weights and/or seedling lengths closely follow the pattern of seed germination. The results of treating mustard seeds with different seed dressings are shown in Table 25.3.

The seed germination of wheat variety HD-2329 remained above the prescribed standard of certification (85%) for up to 20 months of storage under ambient conditions when treated with carboxin, as against 16 months in untreated seeds. The stacking of seed bags (8 bags of 40 kg each stacked one above the other) in the seed warehouse had an insignificant effect on the viability

Table 25.2 Treatments to enhance storability of legume seeds under ambient storage.

Crop seeds	Effective seed treatment at 2 g/kg seed	Storability in months after seed harvest**	References
Cowpea	Captan/thiram*/mancozeb	28	Gupta and Singh, 1990
Mung bean	***	> 36	Gupta and Singh, 1990
Chickpea	Thiram*		Gupta and Singh, 1990

Note: *Seed treatment at 2.5 g/kg seed; **period that seed viability remained above the prescribed standards of certification; ***germination in treated and untreated seeds on a par.

Table 25.3. Treatments to enhance storability of oilseeds under ambient storage.

	Effective treatments		
Crop seeds	Seed treatment at 2 g/kg seed	Storage container	Storability in months after seed harvest**
Soybean	Mancozeb/thiram*	Poly-lined bag	15
Mustard	***	Poly-lined bag	24

Note: *Seed treatment at 2.5 g/kg seed; **period that seed viability remained above the prescribed standards of certification; ***germination in treated and untreated seeds on a par.

of both treated and untreated seeds. However, varietal differences were observed with respect to the storability of the seeds. Wheat seed of the HD-1553 variety maintained more than 85% seed germination for up to 15 months of storage, irrespective of seed treatments. Germination in untreated seeds of the HD-2285 variety fell below the certification standard on 9 months of storage as against seeds treated with captan, where germination remained above 85% for up to 15 months of storage. In the HD-2009 variety, the germination of seeds treated with carbendazim, captan or thiram was above the prescribed standard, even after 21 months of storage as against 15 months in untreated seeds. The decline in germination started at a faster rate after 21 months of storage and reached zero level after 39 months, irrespective of variety or fungicide seed treatment when stored under ambient conditions.

In paddy, seed germination remained above the prescribed standard of certification (80%) for up to 20 months after seed harvest and both the seed treatments and storage containers had insignificant effect on the viability of the stored seeds. The results of fungicidal seed treatments of sorghum cv. PC-9 are shown in Table 25.4.

Among vegetable seeds, fungicidal seed treatments also influenced germination significantly in spinach cultivar Pusa jyothi and fenugreek (Pusa kasuri) seeds (Table 25.5). However, the effect of seed treatments was insignificant in brinjal (Pusa kranti) and muskmelon (Pusa madhuras) seeds. The germination of brinjal and palak seeds remained above the certification standards (70% and 60%, respectively) for up to 18 months of storage after seed harvest. Spinach seeds treated with fungicides like captan or mancozeb had higher seed germination. Muskmelon seeds were stored for 42 months after seed harvest without any substantial loss in seed viability under ambient storage conditions. Fenugreek seeds retained viability for up to 30 months and treatment improved seed germination as against untreated seeds.

Fungicide treatments improved seed germination by about 5–7% on the 30th month of storage after seed harvest, but thereafter their influence on seed germination was negligible. The germination of chilli seeds

Table 25.4. Treatments to enhance storability of cereal seeds under ambient storage.

Crop seeds	Effective seed treatment at 2 g/kg seed	Storability in months after seed harvest**
Wheat	Carboxin*	20
Sorghum	Captan/thiram*/brassicol/carbendazim/mancozeb	21
Paddy	***	20

Note: *Seed treatment at 2.5 g/kg seed; **period that seed viability remained above the prescribed standards of certification; ***germination in treated and untreated seeds on a par.

Table 25.5. Treatments to enhance storability of vegetable seeds under ambient storage.

Crop seeds	Effective treatments		Storability in months after seed harvest**
	Seed treatment at 2 g/kg seed	Storage container	
Spinach	Mancozeb/brassicol		18
Brinjal	***		18
Chilli	Thiram*	Airtight	19
Muskmelon	***		42
Fenugreek	Thiram*/captan/carbendazim		30

Note: *Seed treatment at 2.5 g/kg seed; **period that seed viability remained above the prescribed standards of certification; ***germination in treated and untreated seeds on a par.

remained above the minimum prescribed standard (60%) for 19 months in airtight containers as against 10 months when stored in cloth bags, irrespective of fungicide treatments. However, thiram gave higher seed germination.

Adverse effects of copper-oxychloride fungicide (CuO) have been reported on vegetable seeds during storage (Gupta *et al.*, 1996). Fenugreek, brinjal and muskmelon seeds treated with CuO recorded 37, 50 and 36% germination as against 69, 83 and 80% in untreated seeds after 18, 30 and 24 months of storage, respectively. However, in spinach seed, the CuO treatment retained germination on a par with other seed dressings.

Paddy seeds of parental lines of paddy IR58025A and IR58025B retained seed longevity above the prescribed standards (80%) for up to 5 years after seed harvest when stored under controlled conditions (temperature = 15°C; RH = 30%), as against 2 years under ambient storage conditions in both the parental lines (Fig. 25.3). The germination of paddy seeds of both the parental lines stored in poly-lined bags (76.62%) was significantly higher than seeds stored in cloth bags (72.05%). Seed treatment with thiram and captan also improved seed germination as against the untreated control under both storage conditions (Fig. 25. 4).

The different treatments also influenced the germination of parental lines of pearl millet (MS841A, MS841B and D23) during storage (Gupta, 2007). Seeds of MS841A, MS841B and D23 retained germination above minimum seed certification standards (MSCS) (75%) for up to 16, 16 and 20 months after seed harvest, respectively. As in paddy, the germination of pearl millet seeds (Fig. 25.3) stored under controlled conditions (82.25%) was significantly higher than seeds stored under ambient conditions (66.43%). Thus, storage under low temperature can prolong the longevity of the precious seeds of the inbred parental lines of paddy and pearl millet.

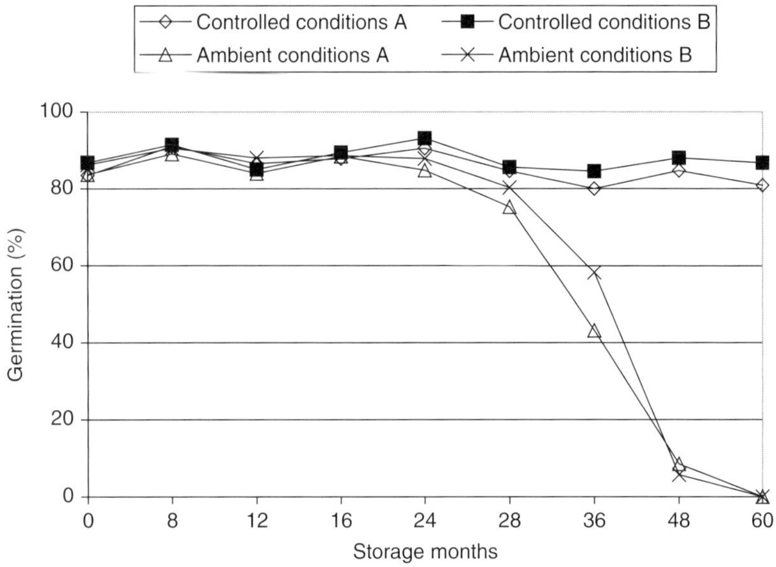

Fig. 25.3. Effect of storage conditions on seed germination in parental lines of paddy during storage.

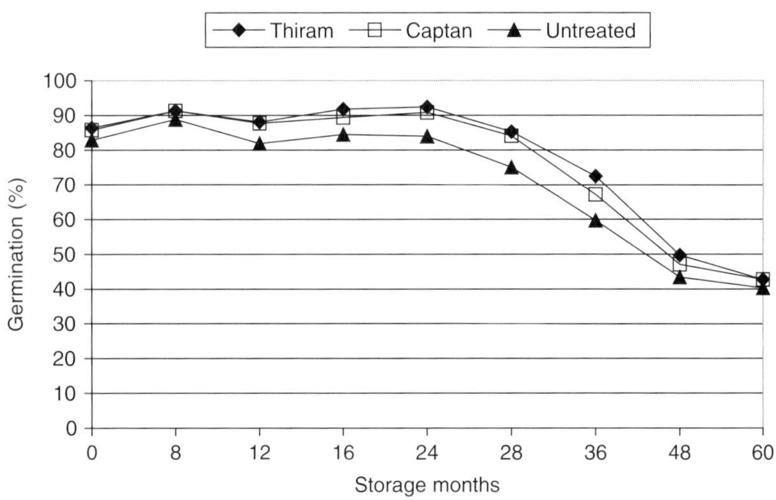

Fig. 25.4. Effect of seed treatments on seed germination in parental lines of paddy during storage.

Treatment of pearl millet seeds of different parental lines with bioagent *T. viride* maintained 78.30% germination during storage that was on a par with thiram (78.34%), captan (78.16%) and carbendazim (77.37%), as against 74.27% in the untreated control and 59.58% in *P. fluorescence* treatment (Fig. 25.6). Also, the seeds stored in polythene-lined cloth bags maintained higher seed germination in both paddy and pearl millet than the seeds stored in cloth bags (Fig. 25.7).

Role of Storage Environment in the Perpetuation of Fungi

The lifespan of seed is highly influenced by storage conditions, especially temperature

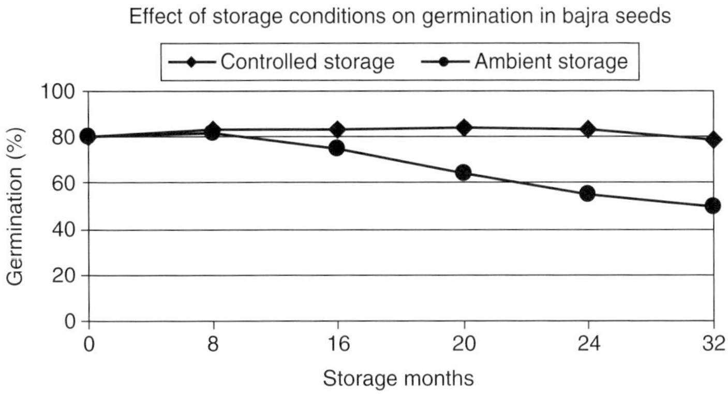

Fig. 25.5. Effect of storage conditions on seed germination in pearl millet.

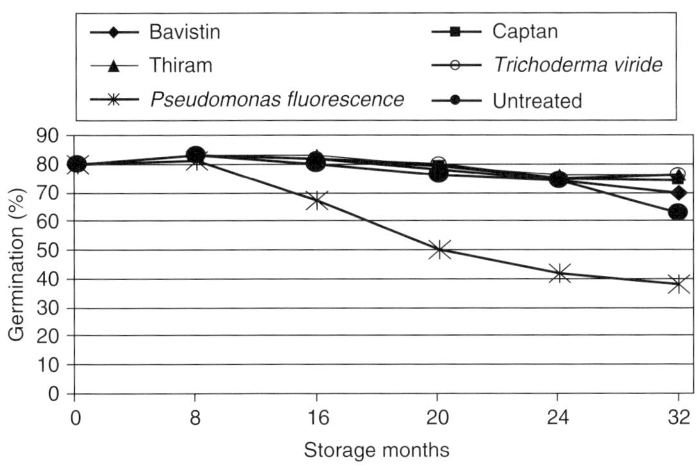

Fig. 25.6. Effect of seed dressings on seed germination in pearl millet.

and relative humidity (RH). The effects of temperature and RH (and its subsequent effect on seed moisture) of the storage environment are highly interdependent. Most crop seeds lose their viability when RH is about 80% and the temperature varies from 25 to 30°C. This hot and humid environment is congenial for the activity and growth of microorganisms, which leads to deterioration in seed quality. The expression of the mycoflora depends essentially on the temperature and humidity conditions of seed warehouses and also on the intergranular atmosphere of the seed. The moisture in the seed is present either on the seed surface or in the internal tissues, from where it moves to the surface and evaporates and is dependent on the RH of the atmosphere. At different RH levels, the equilibrium moisture content (MC) varies with different seeds. The MC of seeds ranges from 3 to 7, 6 to 10 and 9 to 14% at 20, 45 and 75% RH, respectively. According to Harrington's rule of thumb (Harrington and Douglas, 1970):

- For each 1% decrease in seed moisture content, the storage life of the seed is doubled.
- For each 10°F (5.6°C) decrease in seed storage temperature, the storage life of the seed is doubled.

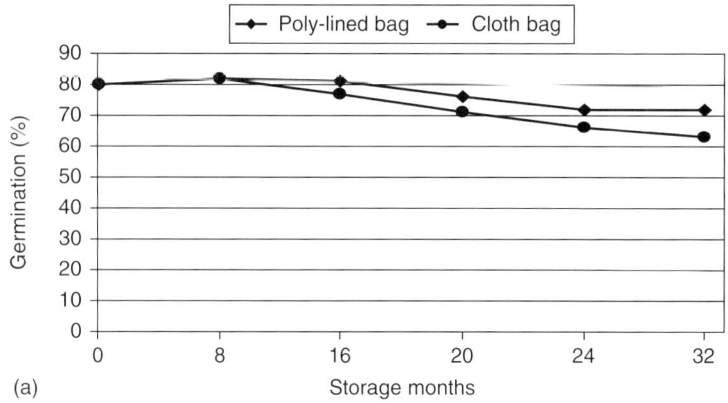

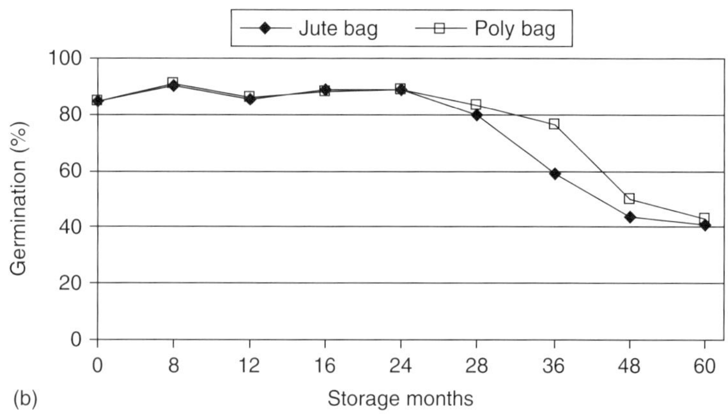

Fig. 25.7. Effect of storage containers on seed germination in pearl millet (a) and paddy (b) seed during storage.

- The arithmetic sum of the storage temperature in degrees F and the per cent RH should not exceed 100, with no more than half the sum contributed by the temperature.

However, this rule is valid only when the seed moisture varies from 5 to 14%.

Storage fungi are unable to grow and multiply if the MC of the stored produce is 12% or less. The optimum moisture for many types of seed is 6–8%, at which even damage by insects reduces (Dharam Vir, 1996). It was observed that the per cent of seed viability was highest at low temperatures and RH and short storage periods, but it decreased with increased storage period. Temperatures above 35°C are reported to cause rapid deterioration of soybean seeds in 6 months of storage (Nkang and Umoh, 1996).

Under normal storage conditions, the temperatures and relative humidity of the seed warehouse fluctuates with the environment. The temperature range in a seed warehouse located in Karnal varies from 13 to 33°C and 18 to 38°C (Fig. 25.8). The relative humidity of the seed warehouse varies from 51 to 75%. Since seed moisture is a function of RH, so it changes with variations in the RH of the seed warehouse. Seed dressings do not affect the MC of seeds appreciably, but storage containers do seem to affect it. It is higher in seeds stored in cloth bags as against those in poly-lined cloth bags because cloth bags, being pervious, allow a free flow of air from the surrounding atmosphere.

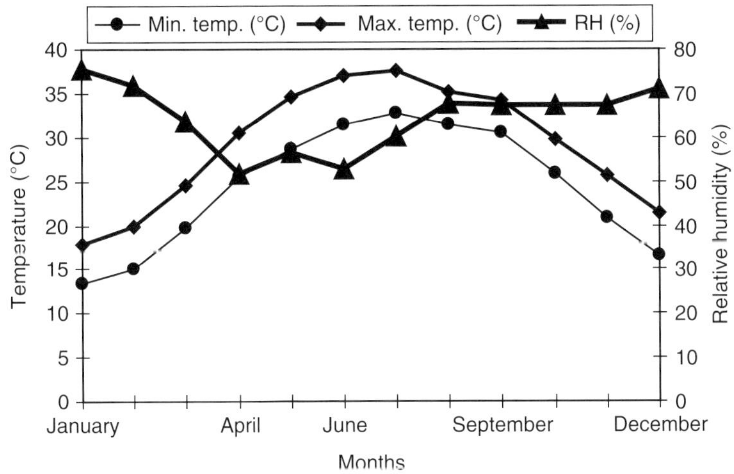

Fig. 25.8. Ambient conditions in a seed warehouse at Karnal during the year.

Although seed MC, type of storage container and storage temperature are interrelated, high temperatures hasten the deterioration of high-moisture seeds by increasing the metabolic activity of hydrolysed substrates and enzymes. Hence, maintaining these factors at low levels in the seed warehouses can improve the longevity of the seeds.

Persistence of seed dressings during storage

Suitable dressing at proper dosage and its uniform distribution on the seed is equally important for proper effect of seed treatments. This also goes a long way in enhancing the storage life of seeds, especially under ambient conditions in the seed warehouses. Dharam Vir (1977) observed that organomercurials retained their bioefficacy for a longer period as compared to antibiotics and dithiocarbamates, which degrade and become biologically ineffective after storage of treated paddy seeds for 1 year.

The persistence of fungicides on the seed is of paramount importance as it determines the longevity of effective seed treatment during storage. Certain fungicides, irrespective of the amount of dressing present on the seed, inhibit the fungi strongly, while others exhibit weak or almost nil inhibition. The storage container also seems to influence the residual activity of the chemical on treated seeds during storage. Vyas and Nene (1971) reported minimum loss of thiram on the seeds of cowpea, maize, paddy and soybean in tin boxes as compared to polyethylene bags, polyethylene-lined cotton bags or hessian bags. Gupta and Chatrath (1983) found that the quantity of thiram on soybean seeds decreased gradually with increase in the storage period and fungicide degradation was maximum when the seeds were stored in cloth bags, followed by paper and alkathene-lined jute bags. Sastry and Chatrath (1984) correlated the persistence of carbendazim on wheat seeds with storage conditions and type of container. The quantity of fungicide on seed decreased with the increased storage period, but the least loss was of seed stored in polythene-lined jute bags followed by polypropylene polyethylene, cloth and jute bags, and storage at 30°C resulted in more degradation as compared with lower temperatures. Lakshmi and Gupta (1997) reported a significant reduction in the quantity of thiophanate methyl on soybean seed with an extended period of storage. Maximum persistence of fungicide was found on seeds stored in polythene-lined jute bags, followed by polypropylene polyethylene bags. Gupta (2002) observed a loss of 10–100% in the activity of different chemicals during storage. The loss was more in treated seeds stored in cloth bag packaging compared to treated seed stored in a polythene bag inside a cloth bag.

The loss in the activity of the chemicals apparently may be due partly to evaporation of the active compound of the chemical and partly to diffusion of the compound into the seed. Raju and Chatrath (1978) reported that the process of the degradation of fungicides was influenced by storage conditions. Besides, other factors such as environmental changes or physiological changes may also be responsible for the depletion of the activity of the chemicals. However, this loss in the activity of the chemical can also be correlated to the presence of seed mycoflora during storage. The loss of activity is lower, the incidence of fungal flora is lower and seed germination is higher.

Management of Storage Fungi to Preserve Seed Longevity

Seed longevity can be maintained either by reducing or preventing the fungal inoculum, or by creating unfavourable conditions for their growth. Management strategies should include practices both at preharvest and postharvest stages. Proper storage and application of safe chemicals as postharvest treatments can control seed mycoflora effectively and reduce losses due to storage fungi considerably. Seed treatment is one of the most effective, safe and economic technologies which protects the seed from microbial deterioration, and thus improves its health status during storage and also ensures better field emergence and seed yield.

Preharvest management strategies

For seed to remain healthy during storage, management strategies need to be followed from the time the crop is in the field, i.e. before the crop is harvested. It is essential to keep the crop healthy and disease free. Use of pathogen-free/certified seed material is the most effective method of disease-free seed production. Disease-free seed production should be planned in safe areas and seasons where disease development is restricted or absent. Preharvest sprayings with suitable chemicals, namely fungicide or biocontrol agents, and harvesting the crop at proper maturity also help to maintain seed quality during storage. Stress conditions during plant growth also influence seed longevity.

Discoloration of seeds by fungi occurs when the crop is in the field. Govindrajan and Kannaiyan (1982) observed reduction in grain discoloration of rice through preharvest spraying with copper oxychloride. Seed discoloration in paddy increased with higher levels of nitrogen and phosphorus and decreased with larger spacing in the field (Misra and Dharam Vir, 1992). According to Deka *et al.* (1996), application of maneb at boot leaf stage, followed by spraying with common salt, was highly effective in reducing discoloration in paddy grains.

The association of fungi is likely to be greater in regions where the produce is harvested in the wet season. Indira and Rao (1968) observed higher association of storage fungi in samples obtained from areas with high humidity. Misra and Kanaujia (1973) considered the presence of antifungal substances in the seed coat of some oilseeds to be the reason for less storage fungi. Nair (1982) reported fewer fungi on seeds of *Luffa acutangula* because of their thick and hard seed coat, which has a low moisture-holding capacity. Varietal differences with regard to their susceptibility to fungal attack during storage has been observed by Sheeba and Ahmed (1994), who found higher fungal incidence on seeds of high-yielding varieties of paddy as compared to local cultivars.

Postharvest management strategies

Initial seed quality, seed moisture, storage temperature and RH play an important role in determining seed longevity. It is essential to avoid mechanical injuries to seed during harvesting/threshing. The produce needs to be dried properly to safe moisture levels before storage. The maximum drying temperature recommended for vegetable crop seed is 35°C. Sun drying of seeds can be practised at the farmer's level. The seeds are usually packed in gunny bags or cloth bags. Moisture-proof containers, hermetically sealed cans, polyethylene pouches or polylined aluminium foil packets are usually used for high-value, low-volume seeds, but it is essential to dry the seed to 5–6% moisture level before packing in these containers because moist seeds tend to deteriorate faster in sealed containers in comparison to ordinary containers.

Pre-storage seed treatment improves the shelf life of the seed, protects it from microbial deterioration and ensures better seed germination and better field stand. Many horticultural crops are propagated by stems, roots, leaves, tubers, corms, rhizomes, suckers, grafts and other vegetative stocks besides seed. This propagative material may carry several pathogens which cause different diseases, thereby affecting their field establishment. The pathogens present in the soil may also hamper field establishment of these propagules. Adopting proper seed treatment technology can reduce most of these problems. This technology is beneficial as it involves less wastage of chemicals, greater control over application, less environmental pollution, low risk for operators, minimum man power, is independent of weather conditions and has less deleterious effects on the treated material.

Lal (1975) reported propionic acid and potassium metabisulphite as effective against *A. niger*, *A. flavus*, *P. oxalicum* and *A. alternata* on wheat and maize grains. Acetic acid and propionic acid has proved effective against *A. flavus* and *C. lunata* on groundnut kernels. According to Vaidya and Dharam Vir (1986, 1987), sodium metabisulphite and propionic acid checked the growth of *Aspergillus* and *Penicillium* sp. on groundnut kernels. After ensuring the quality of seed material meant for storage, it is also essential to ensure that the seed warehouses where the material has to be kept is clean, dry, cool and properly aerated. The seed material should be packed in clean and, if possible, new containers. If old containers are being used, they should be disinfected or fumigated properly to avoid any carry-over pathogens.

Storage structures should not permit entry of water by seepage from the ground or walls. Low temperature retards the development of storage fungi on seeds and so is advisable, especially for low-volume, high-value seeds. It is also essential to ensure that the seed material meant for storage should be of high quality. The material should be stacked on wooden pallets, maintaining a proper distance from the walls and ceilings. The material should be checked regularly for the development of any pests and efficient remedial measures must be employed immediately to keep them under control. Thus, disease management of stored grains requires optimum storage conditions and deployment of treatments that do not pose any health hazards to consumers.

Conclusions

Seed storage is highly influenced by several intrinsic and extrinsic factors. Among them, genotypes, seed treatment and storage containers assume a prime role in successful seed storage. Studies are required to identify disease-free and disease-prone areas for seed production, as healthy seed produces healthy crops. Certification standards for many seed-borne diseases need to be developed. The development of suitable cost-effective packaging material for safe and prolonged storage of seeds is also needed. Identification of suitable pre-storage treatments with suitable agrochemicals and botanicals for the seeds is another important field which needs further work.

Some research has been initiated on an eco-friendly approach against diseases in the field, but their efficacy during storage needs elucidation. The methods available for the proper application of seed/soil dressings also need further refinement. The application of microorganisms as agents for the biocontrol of plant diseases in agriculture is an important alternative to chemical fungicides. Halogenation of seeds has also proved a better seed storage treatment for prolonging seed viability (Dharmalingam *et al.*, 2000). Invigoration treatments, namely the effect of hydration–dehydration or invigoration with different salt solutions on substandard seed lots for microbial growth, need elucidation. Mid storage hydration–dehydration treatment helps to maintain vigour, viability and productivity of crop seeds (Basu, 1994; Mandal *et al.*, 2000). Priming improves biological seed treatment, where the primed seeds of pea and French bean show superiority over other treatments (Rawat and Kumar, 2003). Pelleting of seeds is also advantageous, as the application of pesticides, micronutrients, biofertilizers or plant leaf powders can be incorporated into the seed for improvement in germination.

Some biotechnological approaches may also be exploited by incorporating genes for better storability of seeds. The production of artificial seeds has unravelled new vistas in plant biotechnology. These synthetic seeds are artificially encapsulated propagules used for sowing as a seed and they possess the ability to convert into a plant under *in vitro* or *in vivo* conditions. The preservation of viability and vigour of somatic embryos and synthetic seeds is one of the problems which has to be solved prior to applying synthetic seed technology practically. One of the future uses of synthetic seeds would be in germplasm conservation through cryopreservation. With the introduction of transgenic crops, it has become all the more important to ensure the quality of seeds that are traded across borders. The genetically modified seeds need to be assessed very carefully for any contamination with seed-borne pathogens, which can be accomplished by using the modern tools of biotechnology. Although Bt genes have proved to be quite effective in short-term protection against insect damage, there are concerns that widespread use of Bt varieties will accelerate the development of resistance to Bt in target pests. Thus, to facilitate seed trade, strict quarantine measures at national and international level are necessary in restricting high-risk diseases.

References

Agarwal, P.K. (1980) Relative storability of seeds of ten species under ambient conditions. *Seed Research* 8, 94–99.

Asalmol, M.N. and Zade, V.R. (1998) Effect of seed treatment on storability of seeds of different crops. *Seed Research* 26, 53–56.

Basu, R.N. (1994) An appraisal of research on wet and dry physiological seed treatment and their applicability with special reference to tropical and subtropical countries. *Seed Science and Technology* 22, 107–126.

Bilgrami, K.S. and Sinha, K.K. (1983) Aflatoxin contamination in agricultural commodities with reference to maize and groundnut. *Phytopathology* 73, 253–263.

Chen, X.L., Chen, S.P. and Lu, X.X. (1998) Studies on the seed storability of different types within the crop. *Scientia Agricultura Sinica* 31, 89–91.

Dadlani, M. and Agarwal, P.K. (1983) Mechanism of soybean seed deterioration. *Plant Physiology and Biochemistry* 10, 23–30.

Deka, B., Ali, M.S. and Chandra, K.C. (1996) Management of grain discoloration of rice. *Indian Journal of Mycology and Plant Pathology* 26, 105–106.

Dharam Vir (1974) Study of some problems associated with post-harvest spoilage of seeds and grains. In: Raychaudhri, S.P. and Verma, J.P. (eds) *Current Trends in Plant Pathology*. Botany Department, Lucknow University, Lucknow, India, pp. 221–226.

Dharam Vir (1977) Efficacy of fungicides. xx. Studies on biological activities of some antibiotics and fungicides during storage. *Pesticides* 11, 39–40.

Dharam Vir (1996) Seed treatment with fungicides for control of diseases in field and storage. In: Agnihotri, V.P., Prakash, O., Kishen, R. and Mishra, A.K. (eds) *Diseases Scenario in Crop Plants*, Volume II. International Books and Periodicals Supply Services, New Delhi, pp. 199–216.

Dharmalingam, C., Vasantha, R., Malarkodi, K. and Lakshmi, S. (2000) Halogenation treatment to safeguard pulse seeds in storage under ambient conditions. *Seed Research* 28, 42–46.
Dube, S., Shukla, H.S. and Tripathi, S.C. (1988) Changes in sugar and protein content of wheat due to *Aspergili. Indian Phytopathology* 41, 633–635.
Govindrajan, K. and Kannaiyan, S. (1982) Fungicidal control of grain infection. *International Rice Research Newsletter* 7, 1.
Gupta, A. (2002) Residual efficacy of seed dressings during storage using bioassay technique. *Indian Phytopathology* 55(1), 99–101.
Gupta, A. (2003) Role of seed mycoflora in deterioration of soybean seed during storage under ambient conditions. In: Singh, D.P. (ed.) *Implications of Plant Diseases on Produce Quality.* Kalyani Publishers, Ludhiana, India, pp. 85–96.
Gupta, A. (2007) Integrated management of seed health for prolonging seed storability in parental lines of pearlmillet (*Pennisetum glaucum*). *Indian Journal of Agricultural Science* 77(4), 212–219.
Gupta, A. and Aneja, K.R. (2004) Seed deterioration in soybean [*Glycine max* (L.) Merrill] cultivars during storage – physiological attributes. *Seed Research* 32(1), 26–32.
Gupta, A. and Singh, D. (1990) Viability of fungicide treated seeds of mungbean and cowpea in storage. *Seed Research* 18, 70–76.
Gupta, A. and Singh, D. (1993) Effoct of fungicidal treatment on seed viability and mycoflora of stored vegetable seeds. *Seed Research Special Volume* 1, 270–276.
Gupta, A., Singh, D. and Maheshwari, V.K. (1992) Effect of containers on the viability of fungicide treated chilli seeds. *Seed Research* 20, 160–161.
Gupta, A., Singh, D. and Maheshwari, V.K. (1996) Effect of copper oxychloride treatment on sorghum in different containers. *Agricultural Science Digest* 16, 39–41.
Gupta, J.P. and Chatrath, M.S. (1983) Persistence of Thiram on soybean seed in storage. *Indian Phytopathology* 36, 263–266.
Gupta, R.P., Mehra, U. and Pandey, U.B. (1989) Effect of various chemicals on viability of onion seed in storage. *Seed Research* 17, 99–101.
Harman, G.E. and Drury, R.E. (1973) Respiration of pea seeds (*Pisum sativum*) infected with *Aspergillus ruber. Phytopathology* 63, 1040–1044.
Harman, G.E. and Nash, G. (1972) Deterioration of stored pea seed by *Aspergillus ruber*: evidence for involvement of a toxin. *Phytopathology* 62, 209–212.
Harman, G.E. and Pfleger, F.L. (1974) Pathogenicity and infection sites of *Aspergillus* species in stored seeds. *Phytopathology* 64, 1339–1344.
Harrington, J.F. and Douglas, J.E. (1970) *Seed Storage and Packing. Applications for India.* NSC Ltd., New Delhi, 222 pp.
Indira, K. and Rao, J.G. (1968) Storage fungi in rice in India. *Kavaka* 14, 67–76.
Joshi, S., Williamson, D., Sharma, M. and Iyer, S.R. (1988) Combined effect of fungi on hexoses, pentoses and phenolic contents of *Pennisetum typhoides* (Bajra) grain during storage: a biochemical estimation. *Proceedings of the Indian National Science Academy* 58 B, 149–153.
Kushwaha, L.S. and. Raut, N.D. (1994) Effect of seed treatment, containers and storage period on seed health of soybean (*Glycine max*). *Seed Tech News* 24, 69–70.
Lakshmi Shantha, B. and Gupta, J.P. (1997) Persistence of thiophanate-methyl on stored soybean seed. *Indian Phytopathology* 50, 358–361.
Lal, S.P. (1975) Studies on storage fungi of wheat and maize. PhD thesis, IARI, New Delhi.
Mahajan, P.D., Padule, D.N., Perane, P.R and. Mate, S.N (2001) Improvement of storage life of seeds of onion (*Allium cepa* l.,) var Phule Safed. In: *Proceedings of 2001 National Seminar on Seed Science and Technology in New Millennium – Vistas and Vision.* Department of Studies in Applied Botany, Seed Pathology and Biotechnology, Manasagangotri, Mysore, India, 86 pp.
Mandal, A.K., De, B.K., Saha, R. and Basu, R.N. (2000) Seed invigoration treatment for improved storability, field emergence and productivity of soybean (*Glycine max* L.) *Seed Science and Technology* 28(2), 349–355.
Mishra, A.K. and Dharam Vir (1991) Assessment of losses due to discoloration of paddy grains. I. Loss during milling. *Indian Journal of Mycology and Plant Pathology* 21, 277–278.
Misra, A.K. and Dharam Vir (1992) Effect of different agronomic practices in incidence of seed discolouration of paddy. *Indian Journal of Plant Pathology* 22, 44–48.
Misra, R.R. and Kanaujia, R.S. (1973) Studies on certain aspects of seed borne fungi II. Seed borne fungi of certain oilseeds. *Indian Phytopathology* 26, 284–294.

Moreno, M.E. and Ramirez, J. (1983) Protective effect of fungicides on corn seed stored with low and high moisture contents. *Seed Science and Technology* 13, 285–290.

Moreno, M.E., Mandogaro, L., Mandoze, M. and Valevcin, G. (1985) Use of fungicides for corn seed viability preservation. *Seed Science and Technology* 13, 235–241.

Nair, L.N. (1982) Studies on mycoflora of seeds of some cucurbitaceous vegetables. *Journal of the Indian Botanical Society* 61, 343–345.

Nkang, A. and Umoh, E.O. (1996) Six month storability of five soybean cultivars as influenced by stage of harvest, storage temperature and relative humidity. *Seed Science and Technology* 25, 93–99.

Prasad, B.K., Rao, R.N., Narayan, N., Singh, R.N., Rahman, A., Singh, S.P., Daya, S. and Shankar, U. (1990) Changes in sugar and aminoacid contents in stored radish seeds due to *Aspergillus flavus*. *Indian Phytopathology* 42, 457–460.

Raju, T.N. and Chatrath, M.S. (1978) Persistence of thiram on stored wheat seed. *Indian Phytopathology* 31, 323–326.

Ramamoorthy, K. and Karivaratharaju, T.V. (1986) Storability and biochemical composition of groundnut seed as influenced by packaging and seed treatment. *Indian Agriculturist* 30, 101–106.

Rana, K.S. and. Sharma, B.K. (1995) Management of ginger yellows in storage and field. In: Gupta, V.K. and Sharma, R.C. (eds) *Integrated Disease Management and Plant Health*. Scientific Publishers, Jodhpur, India, pp. 203–206.

Rawat, A. and Kumar, J. (2003) Improving biological seed treatment through solid matrix priming. In: *Sovenir and Abstracts of 5th National Symposium on Biocontrol Agents for Sustainable Management of Pests*. GBPUAT, Pantnagar, Uttaranchal, India, 18–20 December 2003, 169 pp.

Sandhu, K.S. (1989) Effect of different seed and soil treatments on the viability of pea seed. *Plant Disease Research* 4, 43–47.

Sastry, R.K. and Chatrath, M.S. (1984) Persistence of carbendazin on stored wheat seeds. *Indian Phytopathology* 37, 637–640.

Sheeba, T. and Ahmed, R. (1994) Variation in seed borne mycoflora of paddy in Mangalore Taluk of Karnataka. In: Singh, T. and Trivedi, P.C. (eds) *Vistas in Seed Biology*. Printwell, Jaipur, India, pp. 220–228.

Vaidehi, B.K. (1997) Seed mycoflora in maize – an appraisal. In: Tewari, J.P., Saxena, G., Mittal, N., Tewari, I. and Chamola, B.P. (eds) *New Approaches in Microbial Ecology*. Aditya Books Pvt. Ltd., New Delhi, pp. 337–357.

Vaidya, A. and Dharam Vir (1986) Efficacy of fungicides XXXIV. Evaluation of antifungal chemicals for control of post-harvest spoilage of groundnut caused by *Aspergillus niger* and *A. flavus*. *International Journal of Tropical Plant Diseases* 4, 173–176.

Vaidya, A. and Dharam Vir (1987) Efficacy of fungicides XLVII. Evaluation of antifungal chemicals for control of post-harvest spoilage of groundnut caused by *Penicillium* spp. *Indian Journal of Mycology and Plant Pathology* 17, 241–244.

Van Toai, T.T., McDonald, M.B. Jr and Staby, G.L. (1986) Cultivar, fungicides seed treatment and storage environment interactions on carry over soybean seed quality. *Seed Science and Technology* 14, 310–312.

Vyas, S.C. and Nene, Y.L. (1971) Degradation of thiram on treated seed in storage. *JNKVV Research* 7, 121.

Zagrebenyer, D. and Bern, C.J. (1998) *Variety Effect on Carbon Dioxide Production and Free Fatty Acid Development in Soybeans*. American Society of Agricultural Engineers, No. 98182, Department of Agricultural and Biosystems Engineering, Davidson Hall, Iowa State University, Ames, Iowa, 17 pp.

26 Controlling Root and Butt Rot Diseases in Alpine European Forests

Paolo Gonthier
Department of Exploitation and Protection of Agricultural and Forestry Resources (DIVAPRA), Plant and Forest Pathology, University of Torino, Grugliasco, Italy

Abstract
Alpine European forests comprising of conifers and broadleaf trees at lower altitudes are facing major problems of poor regeneration and occurrence of numerous fungal diseases. Fungal organisms like *Armillaria mellea* and *Heterobasidion annosum* are taking their toll on a large number of conifers. These two pathogens are responsible for most of the root and butt rot diseases in natural forest stands. Diagnosis of disease can be done by macro/micromorphology of basidiomata. The wood-inhabiting fungi can also be identified by taxon-specific primers using PCR. This chapter deals with various biological and cultural control strategies and the promotion of disease-tolerant plants, which can reduce the occurrence of diseases. Integrated disease management plans are suggested for different species as found suitable in the Aosta Valley of the western Italian Alps.

Introduction

Mountains and uplands cover approximately one-fifth of the earth's surface and about one-tenth of the world's population lives in mountain regions (Ives *et al.*, 1997). Mountain forests have drawn growing attention in the past few decades in both North America and Europe. It has been hypothesized that if no forests existed in the Alps, humans would not inhabit most of the valleys (Motta and Haudemand, 2000). Forests protect cities and villages against avalanches, landslides, debris flows and rockfalls. They fix surface soil, prevent erosion and play an essential role in water resource management. They influence climate and air quality. At present, the main functions afforded by alpine forests are protection, tourism and recreation, wood production, landscape and nature conservation. Substantial economic and social changes in mountain areas over the past few decades have modified forest use drastically. Traditional forest functions (i.e. wood production) have been abandoned, while the importance of other functions has grown. Human activity has transformed mountain forests radically in various ways: large forest areas have been destroyed and the natural composition of forests has been modified through logging and thinning (Motta and Haudemand, 2000).

European alpine forests comprise mostly of conifers, while broadleaves are generally widespread at lower altitudes, where they may form significant stands. Only eight native coniferous tree species are present in the alpine region (Ozenda, 1985): *Abies alba* Miller (silver fir), *Picea abies* (L.)

Karsten (Norway spruce), *Larix decidua* Miller (European larch), *Pinus cembra* L. (Swiss stone pine), *P. sylvestris* L. (Scots pine), *P. uncinata* Miller (mountain pine), *P. mugo* Turra (dwarf mountain pine) and *P. nigra* Arnold (Austrian pine). These conifers can grow in pure or mixed stands, depending on the site.

Most forests in the Alps are naturally regenerated and current guidelines of forest management are aimed at maintaining adequate levels of natural regeneration. Another trait differentiating alpine forests from other forests located in flat or even in mountain areas is that in alpine forests, given their prominent protective function, clear cut followed by artificial planting is generally forbidden. Furthermore, mixed, irregular or uneven-aged stands are supported locally and maintained through appropriate silvicultural practices (e.g. selective cutting, 'forêt jardinée') where such features fulfil particular functions better (i.e. protection, landscape). Several forests in the Alps, most of which are protected forests, are affected by major problems (Mayer, 1982), including a lack of regeneration, a scarcity of medium-aged trees, insufficient stability and increasing vulnerability to natural disturbances.

Plant pathogens, including those causing root diseases or butt rots, may behave as natural disturbances. In addition to causing severe economic losses, they are reported to influence patterns and processes in forest ecosystems and to be affected by forest development and landscape characteristics, as well as by human activities (Castello *et al.*, 1995; Hansen and Goheen, 2000). This chapter reviews the significance and the epidemiology of the most important and widespread root and butt rot diseases of alpine forests, as well as the most effective and promising control strategies to fight them.

Root and Butt Rot Pathogens, Their Significance, Ecology and Infection Biology

Root and butt rot fungi are important components of forest ecosystems worldwide. Although a large number of lignicolous fungal species are reported on conifers in European mountain areas, including the Alps (Breitenbach and Kränzlin, 1986, 1991, 1995; Bernicchia, 2005), only a few of them are aggressive organisms having a significant impact on forests (Table 26.1). All these fungi belong to the *Basidiomycota* and are necrotrophic tree pathogens causing wood decay. They are facultative pathogens, being able to survive saprotrophically on dead wood. They may be classified either as white rot or brown rot agents depending on the component of the plant cell wall they are able to utilize, i.e. lignin or cellulose, respectively.

The infection biology of wood decay fungi in living trees has been reviewed previously (Rayner and Boddy, 1986). In general, primary infections occur by means of airborne meiospores, which allow for the infestation of new forest areas. Some of these fungi may also operate a secondary, vegetative spread, allowing for the expansion of individuals established through primary infection. Depending on the tree pathogen, this expansion may occur vegetatively through root grafts or contacts, leading to a tree-to-tree contagion, or by free growth of the fungus in the soil through rhizomorphs or mycelial cords. In some pathosystems, insect vectors are essential for the transmission of wood decay fungi (Slippers *et al.*, 2002).

The relative importance of primary and secondary infection is significant not only for our understanding of the epidemiology and population biology of these fungi, but also for control and management purposes. Pathogens like *Armillaria* spp. and *Heterobasidion* spp. are able to spread secondarily. When this happens, there is a carry-over of the pathogen into new generations; in this case, novel attacks are not necessarily caused by new primary infections, but by the inoculum established at that site previously.

Most root and butt rot agents are wound pathogens able to gain entry into the trees through wounds or lesions. Some of them are obligate wound pathogens (e.g. *Stereum sanguinolentum*), while others are facultative wound pathogens (e.g. *Heterobasidion* spp.).

Table 26.1. Summary of characteristics of the most important root and butt rot fungi present in alpine forests.

Fungi	Main hosts[1]	Rot type	Disease/symptoms	Infections	Reference[2]
Armillaria mellea (Vahl:Fries) P. Kummer *sensu lato*	Several conifer and broadleaf species	White, fibrous, wet, often with black zone lines	Root rot and mortality; occasionally heart decay: butt rot	Root contacts; active pathogenesis through rhizomorphs	Guillaumin and Legrand, 2005
Climacocystis borealis Kotl. & Pouzar	Norway spruce, silver fir, larch, pines	White, mottled rot	Heart decay: butt rot	Through wounds[3] on roots and bole	Solheim, 2006
Fomitopsis pinicola (Sw.) P. Karst.	Silver fir, Norway spruce, pines	Brown, cubic	Decay: butt and stem rot	By spores through stem wounds	Butin, 1995; Bernicchia, 2005
Heterobasidion annosum (Fr.) Bref. *sensu lato*	Conifers; mostly pines, Norway spruce, silver fir	White, pocket rot	Root rot and mortality; heart decay in roots, butt and stem	By spores through fresh stump surfaces or wounds; root contacts and grafts	Asiegbu *et al.*, 2005
Laetiporus sulphureus (Bull.) Murrill *sensu lato*	Larch, Norway spruce; oaks and chestnut among broadleaves	Brown, cubic	Heart decay: butt and stem rot; root rot and mortality when decay reaches the root system involving sapwood	By spores through wounds	Bernicchia, 2005; Butin, 2005
Onnia tomentosa (Fr.) P. Karst (= *Inonotus tomentosus*)	Norway spruce, Scots pine	White, pocket rot	Root rot and mortality	By spores through deep root wounds; vegetatively through root contacts and grafts	Tainter and Baker, 1996
Phaeolus schweinitzii (Fr.) Pat.	Norway spruce, silver fir, pines	Brown, cubic	Heart decay: butt and stem rot	Vegetatively in the roots	Barrett and Greig, 1985; Butin, 1995
Stereum sanguinolentum (Alb. & Schwein.) Fr.	Norway spruce	Pale brown, stringy rot	Butt and stem rot	By spores through wounds	Vasiliauskas *et al.*, 1996; Solheim, 2006

Note: [1]Species are listed based on their in-field susceptibility, the most susceptible listed first; [2]reference for the infection biology; [3]wounds are considered of mechanical origin.

Only a few root and butt rot disease agents do not need wounds to gain entry into the tree (e.g. *Armillaria* spp.). At the same time, most root and butt rot fungi are weak, secondary pathogens, being unable to attack vigorous trees. However, some of them are not (i.e. *Heterobasidion* spp., *Onnia tomentosa*) and behave as primary pathogens, which may cause significant disease with or without pre-existing tree stresses. Weakened physiological conditions caused either by primary or by secondary pathogens may then trigger off attacks by other, secondary parasites (i.e. bark beetles) (Tainter and Baker, 1996; Jakuš, 2001).

Wood decay fungi may rot standing trees in two ways, either starting from the cambium and then proceeding inward (sapwood decay), or by decaying the central portion of roots, bole and stem (heart decay) (Rayner and Boddy, 1986). When the cambium, functional xylem or outer sapwood are involved, several physiological functions in trees may be altered. This is particularly true for decays affecting the root system or the collar, which generally lead to a relatively rapid death of the host. In the second type of decay, only the smaller woody roots are killed, whereas the larger ones, the bole or the stem may remain physiologically functional for a long time (Rayner and Boddy, 1986; Gonthier *et al.*, 2003). Very few root and butt rot diseases have been studied in detail, for instance those caused by *A. mellea* or *H. annosum* species complexes (reviewed in Shaw and Kile, 1991; Woodward *et al.*, 1998a; Fox, 2000; Asiegbu *et al.*, 2005; Guillaumin, 2005). Current knowledge on the other pathosystems here described is still very limited.

For instance, only scanty information is available on *Climacocystis borealis*. This fungus is reported as a saprophyte and secondary pathogen (Bernicchia, 2005), being able to cause a typical heartwood rot in the roots and the bole, which seldom reaches more than 2–3 m in height (Solheim, 2006). Sometimes, the sapwood is also colonized. The borealis rot is a characteristic white mottle rot (Bernicchia, 2005) which, on a closer look, is cubic with white mycelium in between (Solheim, 2006). Cubes are easily distinguishable from those of typical cubical brown rots in that they are much finer (1–2 mm). The fungus is rarely lethal, but has been reported to cause significant economic losses locally and to amplify the mechanical instability of trees during storms, especially in mountain, mature Norway spruce stands (Rigling *et al.*, 2005).

Fomitopsis pinicola and *Laetiporus sulphureus sensu lato* are two powerful wood-destroying fungi, responsible for brown, cubic rots. In the alpine region, the former is associated mostly with severely damaged silver fir and Norway spruce trees. The second species, which recently was investigated phylogenetically (Vasaitis *et al.*, 2009), attacks mostly larch trees in alpine forests (Butin, 1995). They are wound pathogens, the second being able to progress towards the root system and here producing root rot and sapwood decay. *Phaeolus schweinitzii* is another widespread brown rot agent. It is reported on all coniferous tree species growing in the Alps and it causes a heart decay of the roots and the bole, spreading up to 1–2 m into the stem (Bernicchia, 2005). The infection biology of this fungus is still largely unknown. The pathogen is believed to infect the roots through the mycelium (Barrett and Greig, 1985; Bernicchia, 2005), which is unable to extend freely in the soil over long distances (Barrett and Greig, 1985). Thus, spores do not play any primary role in the infection. However, they are an important source of soil infestation (Barrett, 1985). It should be noted that other ways of infection, such as mechanical butt wounds or root contacts with diseased trees, have also been suggested for this pathogen (Tainter and Baker, 1996).

There is very scanty information on the significance of *O. tomentosa* in alpine forests. Its presence is probably overlooked. In fact, despite differences in the rot type, the fungus can be confused easily with *P. schweinitzii* since the basidiomata of the two species display similar macroscopic traits (Butin, 1995). *O. tomentosa* was found on Norway spruce and Scots pine trees. This fungus infects trees by spores through deep root wounds and is also capable of secondary spreading through root contacts and grafts (Tainter and Baker, 1996).

S. sanguinolentum is incapable of secondary spreading but is a very strong wound colonizer, especially on Norway spruce. Every wound, from root to top, is vulnerable to infection, even the oldest ones (Vasiliauskas *et al.*, 1996). Very important factors for infection are wound size and depth (Solheim, 2006). Wound rot is initiated by injuries caused by bark-stripping red deer, as well as by harvest-induced injuries (Cermák *et al.*, 2004). In Norway spruce, the potential economic losses caused by cut-off waste wood or low-quality logs are of considerable magnitude and wound rot affects the trees' stability negatively (Cermák *et al.*, 2004).

Several taxa in the *A. mellea* and *H. annosum* species complexes are responsible for most of the root and butt rot diseases of conifers in natural forest stands and plantations throughout the northern temperate regions of the world (Kile *et al.*, 1991; Asiegbu *et al.*, 2005). *A. mellea sensu lato* encompasses about 40 biological species of varying geographic distributions, host ranges and virulence (Pegler, 2000), seven of which are present in Europe (Marxmüller and Guillaumin, 2005): *A. mellea* (Vahl: Fries) P. Kummer *sensu stricto*, *A. ostoyae* (Romagnesi) Herink, *A. borealis* Marxmüller and Korhonen, *A. gallica* Marxmüller and Romagnesi, *A. cepistipes* Velenovský, *A. tabescens* (Scopoli) Emel and *A. ectypa* (Fries) Lamoure. This last species is only marginally important since it is a non-lignicolous, non-parasitic species. All the European species may be found in the alpine area, although some of them (i.e. *A. tabescens*, *A. mellea*), being thermophilic, are more common at low elevations and in the Mediterranean region (Marxmüller and Guillaumin, 2005). There is general agreement on the fact that basidiospores play a marginal role in wood colonization and infection (Guillaumin and Legrand, 2005). *Armillaria* root disease may spread either through root contacts or rhizomorphs, depending on the *Armillaria* species. Root contacts are essential for the spread of *A. tabescens* and *A. mellea*, which is characterized by fragile and short-lived rhizomorphs, while the less pathogenic *A. gallica* and *A. cepistipes* generally infect through rhizomorphs. The ability of *A. ostoyae* and *A. borealis* to produce rhizomorphs and to infect trees in this way is variable (Guillaumin and Legrand, 2005).

In Europe, *H. annosum sensu lato* comprises three species, responsible for losses estimated at more than 800m/year (Woodward *et al.*, 1998b). *H. parviporum* Niemelä & Korhonen primarily causes butt rots in Norway spruce, but it has also been reported to kill Scots pine saplings and attack exotics. *H. abietinum* Niemelä & Korhonen is commonly associated with root or butt rots in trees of the genus *Abies*, while *H. annosum sensu stricto* is associated typically with root rot and mortality of trees in the genus *Pinus*, but it can also be found on *Picea*, *Juniperus* and even on deciduous trees (Korhonen *et al.*, 1998a). All the three species of the fungus are widespread in alpine coniferous forests (Korhonen *et al.*, 1998a; Gonthier *et al.*, 2001) and they are extremely pervasive locally. For instance, levels of disease incidence of up to 95% were reported in some subalpine Norway spruce stands in the western Alps (Gonthier *et al.*, 2003). In a recent study, it was discovered that a large majority of gaps and mortality centres in mountain pine forests of the Swiss Alps was caused by a *Heterobasidion* species rather than by pathogenic *Armillaria* species (i.e. *A. ostoyae*) or other factors (Bendel *et al.*, 2006). *Heterobasidion* primarily infects its hosts by means of airborne meiospores, normally through freshly cut stumps or wounds, and is capable of secondarily spreading from tree to tree through root grafts and contacts (Asiegbu *et al.*, 2005). Airborne infection through thinning stumps not only may result in a rapid and heavy infection of a healthy stand in areas where *Heterobasidion* is common (Pratt and Greig, 1988; Swedjemark and Stenlid, 1993), but also may aid the spread of the fungus into new areas (Berry and Dooling, 1962). Thus, stumps play a crucial role in the epidemiology of this forest pathogen, as confirmed indirectly by the positive relationship between the incidence of disease in residual trees and the intensity of earlier thinnings, as well as the proportion of thinning stumps infected (Rishbeth 1957; Vollbrecht and Agestam, 1995).

Effects of Silviculture and Land Management

With very few exceptions, root and butt rot fungi are opportunistic pathogens being able to take advantage of habitat modifications for their establishment and spread. Thinning and logging, as well as other forest management activities, appear to increase the damage caused by root and butt rots (Garbelotto, 2004). It has been suggested that the current high incidence of *H. annosum* in unmanaged mountain pine forests of the central Alps is due to intense logging in the past (Bendel *et al.*, 2006). Large amounts of timber and fuel were needed to support mining activities, which were thriving in the area between the 14th and 17th centuries. Similarly, the high disease severity recorded in spruce forest stands of the western Alps (Gonthier *et al.*, 2003) could also be associated with mining activities in the past. In some areas of the western Alps, intensive cutting occurred during the 17th and 18th centuries and this led to the creation of stumps over large surfaces (Nicco, 1997). At the same time, the current high level of infestation of subalpine forests (Gonthier *et al.*, 2003) might also have a human origin. Cuttings have been performed regularly at the upper edge of forests to conserve alpine grasslands for summer grazing.

As previously stated, the creation of stumps is particularly important for *H. annosum*, as stumps behave as main infection courts for primary infections. It has been reported that thinnings also promote the tree-to-tree vegetative spread of the pathogen in infected Norway spruce stands (Bendz-Hellgren *et al.*, 1999; Piri and Korhonen, 2008). The growth rate of the fungus in roots increases after the felling of infected trees (Bendz-Hellgren *et al.*, 1999). In living spruce roots, *Heterobasidion* is confined typically to dead heartwood. Hence, the transfer of the fungus between living trees may be limited to functional root grafts, which enable the fungus to grow from the xylem of infected roots into the xylem of healthy roots. After the tree is cut, *Heterobasidion* begins to expand outwards from the centre of the root and it may be able to spread into the surrounding trees more easily through non-grafted root contacts. Such an increase in spreading ability after cuttings could also occur with other heart rot fungi.

Logging operations are likely to increase the probability of attack by most root and butt rot fungi, since new infection courts, i.e. wounds, are created. As an example, the wound rot caused by *S. sanguinolentum*, which is a relatively recent problem, seems to be the result of increasing mechanization of forestry (Butin, 1995). With the use of heavy machinery for the extraction of thinnings, large bark wounds occur much more frequently. This may be crucial for the establishment of fungi, like *S. sanguinolentum*, that need wounds larger than 10 × 10 cm to infect trees (Butin, 1995). In general, wounds play an important or fundamental role in the infection biology of fungi that are able to produce infective airborne inoculum. Nevertheless, it has been proposed that wounds could also trigger attacks by fungi unable to infect by spores, i.e. *Armillaria* spp., because they can induce a lowering of tree defences (Popoola and Fox, 1996).

Diagnosis and General Control Strategies

Management options are based on our knowledge of the ecology, epidemiology and infection biology of the causal agent. Hence, before planning control, there is good reason to perform an accurate diagnosis and identification of the causal agent.

The type of rot may aid in the diagnosis but in general it is not an exhaustive trait for the identification of wood decay fungi (Bernicchia, 2005; Solheim, 2006). Traditionally, diagnosis is based on the macro- and/or micromorphology of basidiomata and it may be achieved through the use of mycological keys (Eriksson *et al.*, 1984; Breitenbach and Kränzlin, 1986, 1991, 1995; Bernicchia, 2005; Gonthier and Nicolotti, 2007). When performing field diagnosis, a pathologist should consider that basidiomata of wood decay fungi usually emerge at advanced stages of the fungal infection and they may be rarely or sporadically visible. Furthermore, basidiomata of some species (e.g. *P. schweinitzii*) are short-lived (Butin, 1995) or may be found only during particular periods of the year (i.e. *Armillaria* spp.); thus, the timing of diagnosis is also important. Most root and butt rot fungi can be cultured easily from decayed wood. A few of them (i.e. *Heterobasidion* spp., *L. sulphureus*) develop a fast-growing conidial stage in culture or when colonized wood is incubated in a damp room. Usually, asexual mitospores of these fungi do not play any significant role in the infection biology. Nevertheless, they may have a diagnostic value. Identification of wood-inhabiting fungi, non-sporulating in pure culture is also possible by using appropriate keys (Nobles, 1965; Stalpers, 1978). Pure culture analysis, however, is difficult and time-consuming.

A number of molecular techniques have been developed and are now available for the identification of the most important root and butt rot fungi. A summary of them is given in Table 26.2. Some techniques allow for the identification of rots directly from wood. For instance, polymerase chain reaction (PCR) with taxon-specific primers provides reliable fungal diagnostics from both pure culture and environmental samples. Theoretically, a pathogen can be controlled during all stages of its life cycle, starting from primary infection and early establishment, through spreading inside the host, to formation, spread and survival of its propagules (Holdenrieder and Greig, 1998). Unfortunately, root and butt rot diseases are virtually impossible to eradicate once they are established. They may be controlled successfully only when pathogens have a small biomass and are therefore weakly competitive.

In general, when dealing with root and butt rot diseases characterized by abundant primary infection events (i.e. *H. annosum*, *S. sanguinolentum*, etc.), forest management

Table 26.2. Taxon-specific primers developed for the identification of some of the most important root and butt rot fungi present in alpine forests.

Fungi	Forward primer	Reverse primer	Amplicon size	Reference
A. mellea sensu lato[1]	ARM-1 (agggta tgtgcacgttcgac)	ARM-2 (ggaaagctaa gctcgcgcta)	660 bp	Schulze *et al.*, 1997
	ITS3 (gcatcgat gaagaacgcagc)	Armi2R (aaacccccat aatccaatcc)	184 bp	Guglielmo *et al.*, 2007
H. annosum sensu lato	HET-7 (cttctcac aaactcttcg)	HET-8 (caggtccccca caatcg)	400 bp	Bahnweg *et al.*, 2002
H. annosum sensu lato	MJ-F (ggtcctgtc tggctttgc)	MJ-R (ctgaagcacac cttgcca)	100 bp	Hantula and Vainio, 2003
H. parviporum	KJ-F (ccattaac ggaaccgacgtg)	KJ-R (gtgcggctcattc tacgctatc)	350 bp	
H. annosum sensu stricto	MLF (taaaaatttaa attagccataa)	Mito7 (gccaatttatttt gctacc)	230 bp	Garbelotto *et al.*, 1998; Gonthier *et al.*, 2001, 2003
H. abietinum		Mito5 (taagaccgctata	195 bp	
H. parviporum	MLS (aaattagcca tattttaaaag)	waccagac)	185 bp	
L. sulphureus sensu lato	25sF (tggcgaga gaccgatagc)	LaetR (ccgagcaaac gaatgcaa)	146 bp	Guglielmo *et al.*, 2007
O. tomentosa	It-ITS-209-f (gcta aatccactcttaacac)	It-ITS-700-rc (agga gccgaccacaaaagat)	491 bp	Germain *et al.*, 2002
Stereum spp.	ITS3 (gcatcgatg aagaacgcagc)	Ste2R (gtcgcaacaa gacgcactaa)	234–240 bp	Guglielmo *et al.*, 2007

Note: [1]Some *Armillaria* species may be distinguished through PCR-restriction fragment length polymorphism (RFLP) (Harrington and Wingfield, 1995; Schulze *et al.*, 1997; Sierra *et al.*, 1999).

should focus on minimizing those activities likely to create good primary infection courts, e.g. wounds on roots, stems and stumps. Care should be taken in order to avoid harvest-induced injuries. Containment of wildlife populations, especially of bark-stripping deers, may also have some effects on wound rot severity (Cermák *et al.*, 2004).

Lowering stand density may regulate secondary infection events, especially for pathogens spreading through root grafts and contacts. In addition, a regulation of stand density aimed at reducing tree competition may prevent or reduce infections by a wide range of weakness pathogens, regardless of their mode of transmission (i.e. *Armillaria* spp., *F. pinicola*, etc.).

The management of mature alpine forests that are diseased or under the risk of root and butt rots could be achieved through several other and more specific methods, which are reviewed below. Some of them are hardly applicable in mountain areas or may be justified only locally. Some others are currently used over large areas, including the alpine region. An integrated disease management approach could be more efficient than single methods in controlling these diseases.

Trenching

Digging isolation trenches around diseased trees to prevent the secondary, vegetative spread of root and butt rot fungi is one of the most traditional control methods recommended against several pathogens including *H. annosum sensu lato* and *A. mellea sensu lato* (Korhonen *et al.*, 1998b; Kliejunas *et al.*, 2005; Legrand *et al.*, 2005; Eyles *et al.*, 2008). The trench should be at least 70–100 cm deep. Instead of an open trench, it could be more practical to bury a plastic sheet in a vertical position into the soil (Korhonen *et al.*, 1998b; Legrand *et al.*, 2005). As suggested by Eyles *et al.* (2008), the effectiveness of this method depends on the regular maintenance of trenches, to prevent the re-establishment of root contacts, and on the proper siting of the trenches to ensure all infected trees are isolated. It is usually quite difficult to determine whether or not trees are infected; colonization of the root systems may go undetected for long periods of time. Trenching is a very impractical control method and its use is advisable only when pathogen inoculum is very localized.

It should be noted that in the case of pathogens that are also able to spread aerially, trenching would offer limited protection. In the case of *H. annosum*, for instance, observational data suggest that instead of controlling the disease, trenching could actually promote the spread of the fungus by breaking and injuring roots (Korhonen *et al.*, 1998b).

Tree and stump removal

Secondary mycelia of most root and butt rot fungi will survive and produce basidiomata for a long period of time on colonized wood. Thus, infected trees should be removed from the stand promptly in order to reduce the airborne inoculum of fungi spreading through spores. As most of these fungi can attack timber, asymptomatic felled trees should also be removed. In general, timber is unselective or less selective than standing trees to infection by wood decay fungi (Rayner and Boddy, 1986). As an example, *S. sanguinolentum*, which commonly attacks Norway spruce trees, is reported to colonize felled wood of spruce but also of pine and silver fir, in which it causes a red streaking (Butin, 1995). Thus, attention should be given not only to preferential hosts, but also to other tree species that can become saprophytically colonized.

Removing stumps and roots from the soil has been recommended for controlling both *H. annosum sensu lato* and *A. mellea sensu lato* (Korhonen *et al.*, 1998b; Legrand *et al.*, 2005). Benefits of this method have been reviewed recently by Vasaitis *et al.* (2008). It should be noted that de-stumping can show effectiveness not only against diseases spreading from tree to tree through root grafts and contacts, but rather it may be effective against a wide range of wood decay agents. In fact, for instance, basidiomata of *P. schweinitzii* growing on stumps are reported as sources of soil infestation for long periods of time (Barrett, 1985). Stump removal is an expensive, time-consuming control method (Korhonen *et al.*, 1998b; Legrand *et al.*, 2005) that requires the use of machines (Omdal *et al.*, 2001). This method can be adopted optionally in certain artificial plantations after clear-felling. It is rarely used in mountain forests or after selective cuttings. Furthermore, de-stumping contrasts with current trends in forest management since it may have negative effects on biodiversity.

Promoting tolerant species

If a forest is heavily infested by a root or butt rot agent with restricted or defined host range, control may be achieved by increasing the proportion of trees more resistant to the pathogen. In principle, a rotation of a resistant tree species can clean the site of the pathogen inoculum (Korhonen *et al.*, 1998b). The concept of forest rotation has been widely advocated for the management of root rots of forest trees, with contrasting results (Korhonen *et al.*, 1998b; Lygis *et al.*, 2004). The pathogen inoculum usually persists in stumps and roots for decades after felling, and this was reported not only for *H. annosum* or *A. mellea* species complexes (Korhonen and Stenlid, 1998; Guillaumin and Legrand, 2005) but also for other root and butt rot fungus, i.e. *P. schweinitzii* (Barrett, 1985).

Changes in tree species composition, turning a susceptible forest into a more tolerant one, can be achieved in a relatively short period of time through clear-felling only, followed by artificial plantation. In naturally regenerated, uneven-aged or irregular alpine forests, shifts in tree species composition are more difficult to obtain and they should be driven by appropriate silvicultural practices. Obviously, in such conditions, for a reduction of the pathogen inoculum to occur, several decades, or even centuries, may be necessary. As for most plant diseases, it could be advisable to avoid the complete removal of the susceptible species, since the resulting selection pressure on the pathogen might result in the infection of plants normally considered as non-host (Garbelotto, 2004).

This method of control might show some effectiveness in controlling *Heterobasidion* root and butt rots. In a recent unpublished study conducted in the western Italian Alps, and based on the analysis of about 2300 recently felled trees, it was found that disease incidence was significantly higher on Norway spruce (43% of average incidence) than on other native tree species (Table 26.3). Silver fir was also rather susceptible, while larch and, especially, Scots pine trees were more tolerant. Moreover, with the exception of *H. annosum sensu stricto*, a strict host preference of *H. annosum* species has been reported in the Alps (Gonthier *et al.*, 2001). Thus, based on these data and on other observational data (Table 26.3), it is likely, for instance, that the establishment of deciduous tree species would have beneficial effects in most *Heterobasidion*-infested forests, but not necessarily in *H. annosum sensu stricto* infected stands (Table 26.3). Also, the control of *H. abietinum* in heavily infected silver fir forests could be achieved either by promoting Scots pine trees or Norway spruce trees. The disease in severely damaged Norway spruce stands can be lowered by favouring the more resistant larch.

A. mellea sensu lato species display a lower degree of host preference with respect to *H. annosum sensu lato* species. Nevertheless, variation of tree species composition could have some beneficial effects for this pathosystem also. For instance, Swiss stone pine trees have been reported as very susceptible to *A. ostoyae* in subalpine forest (Anselmi and Lanata, 1989). Cuttings promoting the regeneration and establishment of the most tolerant larch could be effective in the management of Armillaria root rots in these high elevation forests.

Timing of thinning and cutting

The timing of logging and harvesting may have a strong impact on the incidence of

Table 26.3. Susceptibility of native alpine forest trees to *H. annosum* species based on the author's experience and on previously published (Gonthier *et al.*, 2002, 2003) and unpublished data. The list does not include susceptibility and symptoms of seedlings.

Tree species	Average *Heterobasidion* incidence[1]	*H. parviporum*		*H. abietinum*		*H. annosum sensu stricto*	
Abies alba	17%			H	+++		
Picea abies	43%	H	++++			H	++
Larix decidua	12%	H	++			H	+
Pinus cembra	15%	H	++			H	++
P. sylvestris	4%					M	++
P. uncinata	?					M	++?[2]
Broadleaves	?					–	+

Note: [1]Based on the analysis of about 2300 recently felled trees from 22 forest stands in the western Italian Alps; [2]symptoms and tentative susceptibility according to Bendel *et al.*, 2006. Symbols: +, recorded; ++, occasionally diseased; +++, susceptible; ++++, very susceptible; M, root rot and mortality; H, heart rot; –, saprophyte.

airborne infectious diseases. As a rule, operations should be done preferably when environmental conditions are unfavourable to the pathogen. For *H. annosum sensu lato*, winter thinning and logging operations have been used in Fennoscandia to take advantage of the low inoculum pressure during the cold season (Brandtberg *et al.*, 1996; Piri and Korhonen, 2008). In that area, infections follow a bell-shaped curve with very low spore deposition rates in winter; an average of only 2% of Norway spruce stumps was infected following thinning in November–February compared with 34% in June–July (Brandtberg *et al.*, 1996).

Seasonal patterns of spore deposition of *Heterobasidion* species have been studied in the alpine region recently with the aid of woody traps (Gonthier *et al.*, 2005). Here, the airborne inoculum of this pathogen, although present starting in February at most sites, is higher in August–October, reaching a peak in September. A relative peak, lower than the late summer one, appears in late spring (Fig. 26.1). Thus, despite the development of perennial basidiomata, in the Alps the inoculum production of *Heterobasidion* spp. is concentrated largely in a period of 2–3 months. Spore inoculum in winter, spring and early summer is generally low (Gonthier *et al.*, 2005). Although the hazard of stump infection is not always described accurately by spore loads on woody traps (Driver and Ginns, 1969), recent unpublished data confirm the above seasonal patterns of spore deposition and indicates that the highest risk of stump infection occurs in autumn (Gonthier and Nicolotti, unpublished).

Winter operations may not be feasible at all sites in the Alps. However, *Heterobasidion* primary infections would be controlled successfully by planning logging and thinning in most of spring and in early summer. Although the above timing may be somewhat impractical, its advantage in Norway spruce includes limiting infection through wounds not only by *H. annosum sensu lato* but also by *S. sanguinolentum*, whose annual basidiomata are produced in autumn (Solheim, 2006).

Biological and Chemical Control

Several biological and chemical methods have been tested for the control of root and butt rot fungi, especially of *H. annosum sensu lato* and *A. mellea sensu lato* (reviewed in Holdenrieder and Greig, 1998; Pratt *et al.*, 1998; Guillaumin *et al.*, 2005a,b). Most

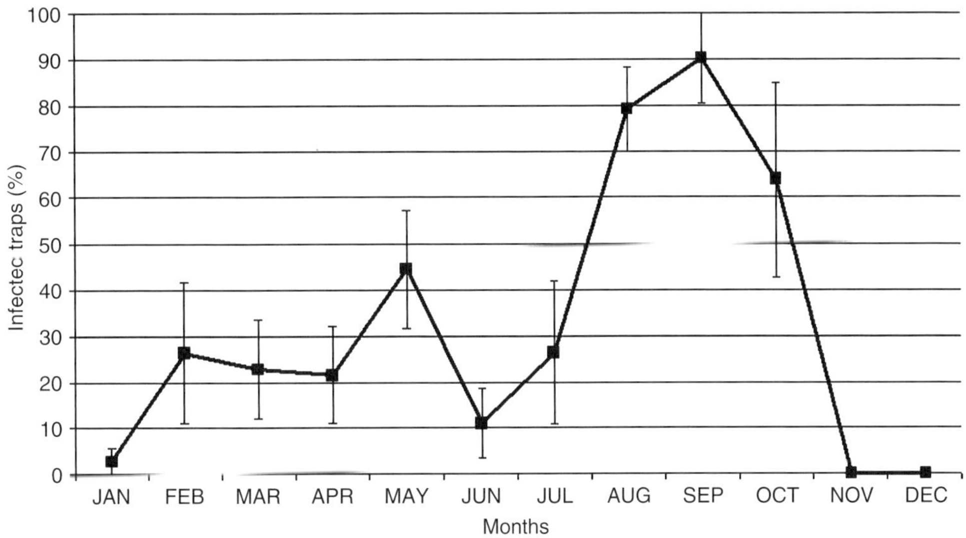

Fig. 26.1. Average percentage of woody traps infected monthly by *Heterobasidion* spores in four forests of the western Alps from 1998 to 2000. Re-elaborated from data published by Gonthier *et al.* (2005). Bars show standard errors.

experiments were conducted *in vitro*. Currently, only a very few control approaches are recommended in practical forestry and they are all devoted to *H. annosum sensu lato*. Stump treatment with appropriate biological or chemical products immediately after felling may prevent *H. annosum* primary infections, therefore reducing timber losses.

A number of fungi have been tested on stumps as competitors or antagonists throughout North America and Europe (Holdenrieder and Greig, 1998), including in the alpine forests (Nicolotti *et al.*, 1999). Only *Phlebiopsis gigantea* (Fr.) Jül is used currently, with good results over large areas (Holdenrieder and Greig, 1998; Thor, 2003; Berglund and Rönnberg, 2004; Thor and Stenlid, 2005). Three distinct products based on this saprotrophic fungus have been developed: PG Suspension® in the UK, PG IBL® in Poland and Rotstop® in Fennoscandia. Rotstop® showed a very good effectiveness in alpine Norway spruce stands heavily infected by *Heterobasidion* (Nicolotti *et al.*, 1999; Nicolotti and Gonthier, 2005). In a recent comparative study performed in Austrian alpine protection forests, strains of *Phlebiopsis gigantea* from Poland resulted in a higher colonization frequency of spruce stumps with respect to the Rotstop® strain (Cech *et al.*, 2008).

Several chemicals proved to be effective as stump protectants against *Heterobasidion* airborne infections (Pratt *et al.*, 1998), the most known of which are urea in Europe and borax in North America. Both compounds, as well as other chemicals, were tested in the western Alps on spruce stumps and they showed very good results, comparable to those obtained with the Rotstop® treatment (Nicolotti *et al.*, 1999; Nicolotti and Gonthier, 2005). The effectiveness of urea was dependent on the concentration of the water solution: the best results were obtained with a 30% concentration. The rise of pH of stump surfaces, which occurs commonly during hydrolysis after treatment, rather than a toxicity of urea or urea-derivate compounds per se (e.g. ammonia and ammonium ions), is responsible for the inhibition of *Heterobasidion* germination and growth (Johansson *et al.*, 2002). Such a high urea concentration allows high pH values on stumps to be maintained for at least the length of time these remain susceptible to infection, i.e. approximately 1 month.

Because the costs of registration are very high, it is unlikely that any biological or chemical product will be registered in the near future for stump treatments in the alpine area. However, urea and borax are currently classified as fertilizers and their use is mostly unregulated for forestry purposes (Nicolotti and Gonthier, 2005). Urea could be preferred for its long history in stump treatment in Europe (Nicolotti and Gonthier, 2005; Oliva *et al.*, 2008) and for its moderate effects on non-target organisms inhabiting stumps (Table 26.4). Furthermore, urea is effective on stumps of several native alpine tree species (Gonthier and Nicolotti, unpublished).

Integrated Disease Management and Forest Protection: A Concluding Example

It is generally agreed that systems combining cultural, biological, chemical or other methodologies to reduce parasites are more effective and even cheaper than single control methods. In the field of forest trees, integrated pest management (IPM) systems have been developed especially for the protection of forests against insects or nurseries against diseases (Volney and Mallett, 1998; South and Enebak, 2006). Appropriate IPM systems may be developed only with a good understanding of the pathogen biology and disease epidemiology. Except for a few pathosystems (i.e. *H. annosum sensu lato*, *A. mellea sensu lato*), our current understanding of the epidemiology of root and butt rot diseases is still limited to allow the development of efficient IPM systems. However, the biology and epidemiology of *H. annosum sensu lato* are well known and some weak points exist in its life cycle (i.e. stage of infection by spores).

An integrated management system was designed to control *H. annosum* root and butt rots in the Aosta Valley, western Italian Alps. The system is based on stump treatment

Table 26.4. A summary of the effectiveness and impact on non-target fungi of biological and chemical treatments against *Heterobasidion* airborne infections on Norway spruce stumps in the western Alps.

Antagonist/competitor or active ingredient	Application method or commercial product	Effectiveness against *Heterobasidion*[1]	Impact on non-target fungi[2]
Biological			
Hypholoma fasciculare	Wheat mash	Low	Low (after 2 years)
Phanerochaete velutina	Wheat mash	High	Low (after 2 years)
Phlebiopsis gigantea	Rotstop®	High	High
Vuilleminia comedens	Wheat mash	Low	Low (after 2 years)
Verticillium bulbillosum	Culture filtrate	Medium	Very low
V. bulbillosum	Conidial and mycelial suspension	Medium	Very low
Trichoderma harzianum	Conidial and mycelial suspension	Low	Very high
Chemical			
Copper oxychloride	Azuram®	High	Low (after 2 years)
Propiconazole	TILT (25% emulsion)	High	Low (after 2 years)
Sodium tetraborate decahydrate	Borax powder	High	Very high
Urea	Water solution 10% conc.	Low	Very low
Urea	Water solution 20% conc.	High	Low (after 2 years)
Urea	Water solution 30% conc.	High	Low (after 2 years)

Note: [1]Categories (low, medium, high) were designed based on previously reported results (Nicolotti *et al.*, 1999; Nicolotti and Gonthier, 2005); [2]categories (very low, low, high, very high) were designed based on previously reported results (Varese *et al.*, 1999; 2003a,b).

with urea at 30% concentration, combined with an appropriate timing of thinning and logging operations, and with practices aimed at promoting tolerant species (Fig. 26.2). In the Aosta Valley, forest management activities are planned yearly by the Regional Forest Administration, who decides the stands that need to be thinned each year. Every forest harvesting team is in charge of thinning a variable number of stands (3–8). While planning the timing of thinnings, priority is given to the most susceptible (see Table 26.3) and heavily infected stands, which are thinned preferably in spring and early summer, when the risk of stump infection is still limited. The average minimum air temperature of a 4-week period has been identified as a suitable predictor for modelling *Heterobasidion* primary infections in the Alps (Gonthier *et al.*, 2005) and may be used for an accurate estimation of the seasonal risk of stump infection for each forest stand. Remaining, less susceptible and uninfected stands are thinned in summer or autumn. Stump treatment is necessary during summer and autumn thinnings and is strongly recommended whenever dealing with uninfected or susceptible stands, regardless of their location and distance from an infection source. In fact, despite a general limited potential dispersal range of *Heterobasidion* spores (Gonthier *et al.*, 2001), with spore densities undergoing huge dilution after the first metres (Stenlid, 1994), the migration of even a few spores may be significant for areas still not colonized by the pathogen (Garbelotto, 2004).

Winter operations may be possible locally, in low elevation stands. Sanitation fellings are advisable in Scots pine forests to reduce bark beetle attacks. Depending on the forest function and on the economic injury level, practices such as de-stumping and, especially, the transformation of heavily infected susceptible forests into more tolerant ones can be arranged locally and they may be suited. For instance, increasing the larch component in subalpine spruce forests would have positive effects not only in reducing *Heterobasidion* incidence, but also in improving general forest stability (Motta and Haudemand, 2000). At lower elevations, the regeneration of diseased spruce forests with silver fir or Scots pine could be advisable since *H. parviporum* very seldom attacks mature firs or pines (Korhonen *et al.*, 1998b).

Obviously, any integrated disease management system to fight root and butt rots of forest trees should fit and meet the requirements of the general forest management

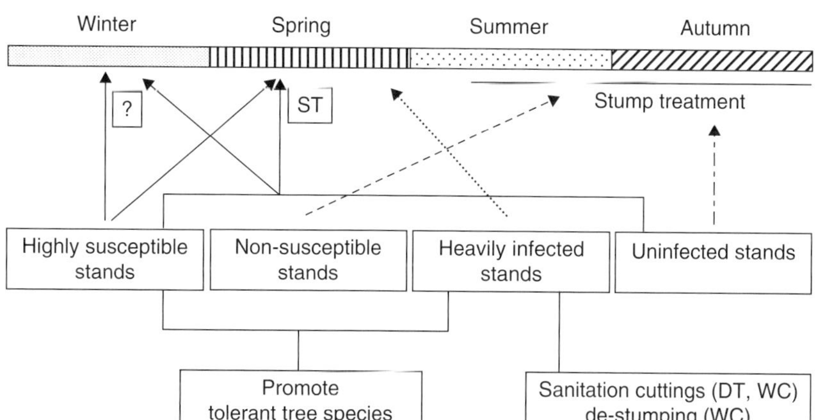

Fig. 26.2. Diagram of the integrated disease management system developed to fight *Heterobasidion* root and butt rots in the Aosta Valley, western Italian Alps. Arrows indicate the appropriate timing of thinning. Stump treatment is performed with urea at 30% concentration. Symbols: ?, where possible; ST, stump treatment; DT, dead trees; WC, where convenient.

system of the area (Tainter and Baker, 1996). Within an integrated forest protection approach, the integrated disease management system here described could combine other pest management systems, for instance those designed for the control of *Ips typographus* L. or other bark beetles threatening alpine forests.

References

Anselmi, N. and Lanata, F. (1989) Distribuzione e comportamento delle specie di *Armillaria* riscontrate nell'Italia settentrionale. *Micologia Italiana* 3, 57–70.

Asiegbu, F.O., Adomas, A. and Stenlid, J. (2005) Conifer root and butt rot caused by *Heterobasidion annosum* (Fr.) Bref. s.l. *Molecular Plant Pathology* 6, 395–409.

Bahnweg, G., Möller, E.M., Anegg, S., Langebartels, C., Wienhaus, O. and Sandermann, H. Jr (2002) Detection of *Heterobasidion annosum* s.l. [(Fr.) Bref.] in Norway spruce by polymerase chain reaction. *Journal of Phytopathology* 150, 382–389.

Barrett, D.K. (1985) Basidiospores of *Phaeolus schweinitzii*: a source of soil infestation. *European Journal of Forest Pathology* 15, 417–425.

Barrett, D.K. and Greig, B.H.W. (1985) The occurence of *Phaeolus schweinitzii* in the soils of Sitka spruce plantations with broadleaved or non-woodland histories. *European Journal of Forest Pathology* 15, 412–417.

Bendel, M., Kienast, F., Bugmann, H. and Rigling, D. (2006) Incidence and distribution of *Heterobasidion* and *Armillaria* and their influence on canopy gap formation in unmanaged mountain pine forests in the Swiss Alps. *European Journal of Plant Pathology* 116, 85–93.

Bendz-Hellgren, M., Brandtberg, P.O., Johansson, M., Swedgemark, G. and Stenlid, J. (1999) Growth rate of *Heterobasidion annosum* in *Picea abies* established on forest land and arable land. *Scandinavian Journal of Forest Research* 14, 402–407.

Berglund, M. and Rönnberg, J. (2004) Effectiveness of treatment of Norway spruce stumps with *Phlebiopsis gigantea* at different rates of coverage for the control of *Heterobasidion*. *Forest Pathology* 34, 233–243.

Bernicchia, A. (2005) *Polyporaceae* s.l. Candusso, Alassio, Italy, 808 pp.

Berry, F.H. and Dooling, O.J. (1962) *Fomes annosus* on shortleaf pine in Missouri. *Plant Disease Reporter* 46, 886–887.

Brandtberg, P.O., Johansson, M. and Seeger, P. (1996) Effects of season and urea treatment on infection of stumps of *Picea abies* by *Heterobasidion annosum* in stands on former arable land. *Scandinavian Journal of Forest Research* 11, 261–268.

Breitenbach, J. and Kränzlin, F. (1986) *Champignons de Suisse, champignons sans lames*, Tome II. Mykologia, Lucerne, Switzerland, 412 pp.

Breitenbach, J. and Kränzlin, F. (1991) *Champignons de Suisse, bolets et champignons à lames*, Tome III. Mykologia, Lucerne, Switzerland, 364 pp.

Breitenbach J. and Kränzlin, F. (1995) *Champignons de Suisse, champignons à lames*, Tome IV. Mykologia, Lucerne, Switzerland, 371 pp.

Butin, H. (1995) *Tree Diseases and Disorders*. Oxford University Press Inc., New York, 252 pp.

Castello, J.D., Leopold, D.J. and Smallidge, P.J. (1995) Pathogens, patterns, and processes in forest ecosystems. *Bioscience* 45, 16–24.

Cech, T.L., Steyrer, G. and Łakomy, P. (2008) Preliminary results of Norway spruce stump treatment with *Hypholoma fasciculare* and *Phlebiopsis gigantea* in an Austrian Alpine protection forest. In: Garbelotto, M. and Gonthier, P. (eds) *Proceedings of the 12th International Conference on Root and Butt Rots of Forest Trees, Berkeley, California–Medford, Oregon, USA, 12–19 August 2007.* The University of California, Berkeley, California, pp. 129–132.

Cermák, P., Jankovský, L. and Glogar, J. (2004) Progress of spreading *Stereum sanguinolentum* (Alb. et Schw: Fr.) Fr. wound rot and its impact on the stability of spruce stands. *Journal of Forest Science* 50, 360–365.

Driver, C.H. and Ginns, J.H. (1969) Ecology of slash pine stumps: fungal colonization and infection by *Fomes annosus*. *Forest Science* 15, 2–10.

Eriksson, J., Hjortstam, K. and Ryvarden, L. (1984) *The Corticiaceae of North Europe*, Volume 7. Fungiflora, Oslo.

Eyles, A., Beadle, C., Barry, K., Francis, A., Glen, M. and Mohammed, C. (2008) Management of fungal root-rot pathogens in tropical *Acacia mangium* plantations. *Forest Pathology* 38, 332–355.
Fox, R.T.V. (2000) *Armillaria Root Rot: Biology and Control of Honey Fungus.* Intercept, Andover, UK, 222 pp.
Garbelotto, M. (2004) Root and butt rot diseases. In: Burley, J., Evans, J. and Youngquist, J.A. (eds) *Encyclopedia of Forest Sciences*, Volume 2. Elsevier, Oxford, UK, pp. 750–758.
Garbelotto, M., Otrosina, W., Cobb, F.W. and Bruns, T.D. (1998) The European S and F intersterility groups of *Heterobasidion annosum* may represent sympatric protospecies. *Canadian Journal of Botany* 76, 397–409.
Germain, H., Laflamme, G., Bernier, L., Boulet, B. and Hamelin, R.C. (2002) DNA polymorphism and molecular diagnosis in *Inonotus* spp. *Canadian Journal of Plant Pathology* 24, 194–199.
Gonthier, P. and Nicolotti, G. (2007) A field key to identify common wood decay fungal species on standing trees. *Arboriculture and Urban Forestry* 33, 410–420.
Gonthier, P., Garbelotto, M., Varese, G.C. and Nicolotti, G. (2001) Relative abundance and potential dispersal range of intersterility groups of *Heterobasidion annosum* in pure and mixed forests. *Canadian Journal of Botany* 79, 1057–1065.
Gonthier, P., Garbelotto, M. and Nicolotti, G. (2002) European pines may be simultaneously infected by more than one species of *Heterobasidion*. *Plant Disease* 86, 814.
Gonthier, P., Garbelotto, M. and Nicolotti, G. (2003) Swiss stone pine trees and spruce stumps represent an important habitat for *Heterobasidion* spp. in subalpine forests. *Forest Pathology* 33, 191–203.
Gonthier, P., Garbelotto, M. and Nicolotti, G. (2005) Seasonal patterns of spore deposition of *Heterobasidion* species in four forests of the western Alps. *Phytopathology* 95, 759–767.
Guglielmo, F., Bergemann, S., Gonthier, P., Nicolotti, G. and Garbelotto, M. (2007) A multiplex PCR-based method for the detection and early identification of wood rotting fungi in standing trees. *Journal of Applied Microbiology* 103, 1490–1507.
Guillaumin, J.J. (2005) *L'armillaire et le pourridié-agaric des végétaux ligneux*. INRA, Paris, 485 pp.
Guillaumin, J.J. and Legrand, P. (2005) Cycle infectieux de l'armillaire – stratégie biologique des espèces Européennes. In: Guillaumin, J.J. (Coord.) *L'armillaire et le pourridié-agaric des végétaux ligneux.* INRA, Paris, pp. 177–201.
Guillaumin, J.J., Lung-Escarmant, B. and Dubos, B. (2005a) Lutte chimique. In: Guillaumin, J.J. (Coord.) *L'armillaire et le pourridié-agaric des végétaux ligneux.* INRA, Paris, pp. 365–375.
Guillaumin, J.J., Legrand, P. and Lung-Escarmant, B. (2005b) L'environnement microbiologique de l'armillaire et la lutte biologique. In: Guillaumin, J.J. (Coord.) *L'armillaire et le pourridié-agaric des végétaux ligneux.* INRA, Paris, pp. 389–406.
Hansen, E.M. and Goheen, E.M. (2000) *Phellinus weirii* and other native root pathogens as determinants of forest structure and process in western North America. *Annual Review of Phytopathology* 38, 515–539.
Hantula, J. and Vainio, E. (2003) Specific primers for the differentiation of *Heterobasidion annosum* (s.str.) and *H. parviporum* infected stumps in Northern Europe. *Silva Fennica* 37, 181–187.
Harrington, T.C. and Wingfield, B.D. (1995) A PCR-based identification method for species of *Armillaria. Mycologia* 87, 280–288.
Holdenrieder, O. and Greig, B.J.W. (1998) Biological methods of control. In: Woodward, S., Stenlid, J., Karjalainen, R. and Hüttermann, A. (eds) *Heterobasidion annosum, Biology, Ecology, Impact and Control.* CAB International, New York, pp. 235–258.
Ives, J.D., Messerli, B. and Spiess, E. (1997) Mountains of the world: a global priority. In: Ives, J.D. and Messerli, B. (eds) *Mountains of the World: A Global Priority.* Parthenon, Carnforth, UK, pp. 1–15.
Jakuš, R. (2001) Bark beetle (*Coleoptera, Scolytidae*) outbreak and system of IPM measures in an area affected by intensive forest decline connected with honey fungus (*Armillaria* sp.). *Journal of Pest Science* 74, 46–51.
Johansson, S.M., Pratt, J.E. and Asiegbu, F.O. (2002). Treatment of Norway spruce and Scots pine stumps with urea against the root and butt rot fungus *Heterobasidion annosum* – possible modes of action. *Forest Ecology and Management* 157, 87–100.
Kile, G.A., McDonald, G.I. and Byler, J.W. (1991) Ecology and disease in natural forests. In: Shaw, C.G. and Kile, G.A. (eds) *Armillaria Root Disease.* USDA Forest Service, Washington, DC, pp. 102–121.
Kliejunas, J.T., Otrosina, W.J. and Allison, J.R. (2005) Uprooting and trenching to control annosus root disease in a developed recreation site: 12-year results. *Western Journal of Applied Forestry* 20, 154–159.

Korhonen, K. and Stenlid, J. (1998) Biology of *Heterobasidion annosum*. In: Woodward, S., Stenlid, J., Karjalainen, R. and Hüttermann, A. (eds) *Heterobasidion annosum, Biology, Ecology, Impact and Control.* CAB International, New York, pp. 43–70.

Korhonen, K., Capretti, P., Karjalainen, R. and Stenlid, J. (1998a) Distribution of intersterility groups in Europe. In: Woodward, S., Stenlid, J., Karjalainen, R. and Hüttermann, A. (eds) *Heterobasidion annosum, Biology, Ecology, Impact and Control.* CAB International, New York, pp. 93–104.

Korhonen, K., Delatour, C., Greig, B.J.W. and Schönhar, S. (1998b) Silvicultural control. In: Woodward, S., Stenlid, J., Karjalainen, R. and Hüttermann, A. (eds) *Heterobasidion annosum, Biology, Ecology, Impact and Control.* CAB International, New York, pp. 283–313.

Legrand, P., Lung-Escarmant, B. and Guillaumin, J.J. (2005) Lutte contre l'armillaire en forêts: méthodes sylvicoles et cultureales. In: Guillaumin, J.J. (Coord.) *L'armillaire et le pourridié-agaric des végétaux ligneux.* INRA, Paris, pp. 349–363.

Lygis, V., Vasiliauskas, R. and Stenlid, J. (2004) Planting *Betula pendula* on pine sites infested by *Heterobasidion annosum*: disease transfer, silvicultural evaluation, and community of wood inhabiting fungi. *Canadian Journal of Forest Research* 34, 120–130.

Marxmüller, H. and Guillaumin, J.J. (2005) Description et distribution des armillaires européennes. In: Guillaumin, J.J. (Coord.) *L'armillaire et le pourridié-agaric des végétaux ligneux.* INRA, Paris, pp. 63–84.

Mayer, H. (1982) Waldbauliche Zukunftsperspektiven für den Gebirgswald. *Schweizersche Zeitschrift für Forstwesen* 133, 759–780.

Motta, R. and Haudemand, J.C. (2000) Protective forests and silvicultural stability: an example of planning in the Aosta Valley. *Mountain Research and Development* 20, 180–187.

Nicco, R. (1997) I boschi tra settecento ed ottocento. In: Regione Autonoma Valle d'Aosta (eds) *Uomini e boschi in Valle d'Aosta.* Tipografia Valdostana, Aosta, Italy, pp. 98–135.

Nicolotti, G. and Gonthier, P. (2005) Stump treatment against *Heterobasidion* with *Phlebiopsis gigantea* and some chemicals in *Picea abies* stands in the western Alps. *Forest Pathology* 35, 365–374.

Nicolotti, G., Gonthier, P. and Varese, G.C. (1999) Effectiveness of some biological and chemical treatments against *Heterobasidion annosum* on Norway spruce stumps. *European Journal of Forest Pathology* 29, 339–346.

Nobles, M.K. (1965) Identification of cultures of wood-inhabiting hymenomycetes. *Canadian Journal of Botany* 43, 1097–1139.

Oliva, J., Samils, N., Johansson, U., Bendz-Hellgren, A. and Stenlid, J. (2008) Urea treatment reduced *Heterobasidion annosum s.l.* root rot in *Picea abies* after 15 years. *Forest Ecology and Management* 255, 2876–2882.

Omdal, D.W., Shaw, G.C. and Jacobi, W.R. (2001) Evaluation of three machines to remove Armillaria- and annosum-infected stumps. *Western Journal of Applied Forestry* 16, 22–25.

Ozenda, P. (1985) *La végétation de la chaîne alpine.* Masson, Paris, 330 pp.

Pegler, D.N. (2000) Taxonomy, nomenclature and description of *Armillaria*. In: Fox, R.T.V. (ed.) *Armillaria Root Rot: Biology and Control of Honey Fungus.* Intercept, Andover, UK, pp. 81–93.

Piri, T. and Korhonen, K. (2008) The effect of winter thinning on the spread of *Heterobasidion parviporum* in Norway spruce stands. *Canadian Journal of Forest Research* 38, 2589–2595.

Popoola, T.O.S. and Fox, R.T.V. (1996) Effects of root damage on honey fungus. *Arboricultural Journal* 20, 329–337.

Pratt, J.E. and Greig, B.J.W. (1988) *Heterobasidion annosum*: development of butt rot following thinning in two young first rotation stands of Norway spruce. *Forestry* 61, 339–347.

Pratt, J.E., Johansson, M. and Hüttermann, A. (1998) Chemical control of *Heterobasidion annosum*. In: Woodward, S., Stenlid, J., Karjalainen, R. and Hüttermann, A. (eds) *Heterobasidion annosum, Biology, Ecology, Impact and Control.* CAB International, New York, pp. 259–282.

Rayner, A.D.M. and Boddy, L. (1986) Population structure and the infection biology of wood-decay fungi in living trees. *Advances in Plant Pathology* 5, 119–160.

Rigling, D., Blauenstein, H. and Dobbertin, M. (2005) Incidence of root and butt rot in storm-damaged versus non-damaged Norway spruce. In: Małgorzata, M. and Łakomy, P. (eds) *Proceedings of the 11th International Conference on Roots, Root and Butt Rots of Forest Trees, Working Party 7.02.01, Poznan-Bialowieza, Poland, 16–22 August 2004.* The August Cieszkowski Agricultural University, Poznań, Poland, p. 274.

Rishbeth, J. (1957) Some further observations on *Fomes annosus* Fr. *Forestry* 30, 69–89.

Schulze, S., Bahnweg, G., Möller, E.M. and Sandermann, H. Jr (1997) Identification of the genus *Armillaria* by specific amplification of an rDNA-ITS fragment and evaluation of genetic variation within *A. ostoyae* by rDNA-RFLP and RAPD analysis. *European Journal of Forest Pathology* 27, 225–239.

Shaw, C.G. and Kile, G.A. (1991) *Armillaria Root Disease.* USDA Forest Service, Washington, DC, 233 pp.

Sierra, A.P., Whitehead, D.S. and Whitehead, M.P. (1999) Investigation of a PCR-based method for the routine identification of British Armillaria species. *Mycological Research* 103, 1631–1636.

Slippers, B., Wingfield, B.D., Coutinho, T.A. and Wingfield, M.J. (2002) DNA sequence and RFLP data reflect relationships between *Amylostereum* species and their associated wood wasp vectors. *Molecular Ecology* 11, 1845–1854.

Solheim, H. (2006) White rot fungi in living Norway spruce trees at high elevation in southern Norway with notes on gross characteristics of the rot. *Aktuelt fra skogforskningen.* 1, 5–12.

South, D.B. and Enebak, S.A. (2006) Integrated pest management practices in southern pine nurseries. *New Forests* 31, 253–271.

Stalpers, J.A. (1978) *Identification of Wood-Inhabiting Fungi in Pure Culture.* Studies in Mycology 16. Centraalbureau Schimmelcultures, Baarn, The Netherlands, 248 pp.

Stenlid, J. (1994) Regional differentiation in *Heterobasidion annosum.* In: Johansson, M. and Stenlid, J. (eds) *Proceedings of the 8th IUFRO Conference on Root and Butt Rots,* Wik, *Sweden and Haikko, Finland, 9–16 August 1993.* Swedish University of Agricultural Sciences, Uppsala, Sweden, pp. 243–248.

Swedjemark, G. and Stenlid, J. (1993) Population dynamics of the root rot fungus *Heterobasidion annosum* following thinning of *Picea abies. Oikos* 66, 247–254.

Tainter, F.H. and Baker, F.A. (1996) *Principles of Forest Pathology.* John Wiley and Sons, Inc, New York, 805 pp.

Thor, M. (2003) Operational stump treatment against *Heterobasidion annosum* in European forestry – current situation. In: Laflamme, G., Bérubé, J.A. and Bussières, G. (eds) *Proceedings of the 10th International Conference on Root and Butt Rots, Québec City, Canada, 16–22 September 2001.* Information Report LAU-X-126. Laurentian Forestry Centre, Sainte-Foy, Québec, Canada, pp. 170–175.

Thor, M. and Stenlid, J. (2005) *Heterobasidion annosum* infection of *Picea abies* following manual or mechanized stump treatment. *Scandinavian Journal of Forest Research* 20, 154–164.

Varese, G.C., Buffa, G., Luppi, A.M., Gonthier, P., Nicolotti, G. and Cellerino, G.P. (1999) Effects of biological and chemical treatments against *Heterobasidion annosum* on the microfungal communities of *Picea abies* stumps. *Mycologia* 91, 747–755.

Varese, G.C., Gonthier, P. and Nicolotti, G. (2003a) Long-term effects on other fungi are studied in biological and chemical stump treatments in the fight against *Heterobasidion annosum* coll. *Mycologia* 95, 379–387.

Varese, G.C., Gonthier, P. and Nicolotti, G. (2003b) Impact of biological and chemical treatments against *Heterobasidion annosum* on non-target micro-organisms. In: Laflamme, G., Bérubé, J.A. and Bussières, G. (eds) *Proceedings of the 10th International Conference on Root and Butt Rots, Québec City, Canada, 16–22 September 2001.* Information Report LAU-X-126. Laurentian Forestry Centre, Sainte-Foy, Québec, Canada, pp. 145–154.

Vasaitis, R., Stenlid, J., Thomsen, I.M., Barklund, P. and Dahlberg, A. (2008) Stump removal to control root rot in forest stands. A literature study. *Silva Fennica* 42, 457–483.

Vasaitis, R., Menkis, A., Woon, Lim, Y., Seok, S., Iomsovsky, M., Jankovsky, L., Lygis, V., Sippers, B. and Stenlid, J. (2009) Genetic variation and relationships in *Laetiporus sulphureus s. lat.*, as determined by ITS rDNA sequences and *in vitro* growth rate. *Mycological Research* 113, 326–336.

Vasiliauskas, R., Stenlid, J. and Johansson, M. (1996) Fungi in bark peeling wounds of *Picea abies* in central Sweden. *European Journal of Forest Pathology* 26, 285–296.

Vollbrecht, G. and Agestam, E. (1995) Modeling incidence of root rot in *Picea abies* plantations in southern Sweden. *Scandinavian Journal of Forest Research* 10, 74–81.

Volney, W.J.A. and Mallett, K.I. (1998) Integrated pest management in western Canadian boreal forests. *Forestry Chronicle* 74, 597–605.

Woodward, S., Stenlid, J., Karjalainen, R. and Hüttermann, A. (1998a) *Heterobasidion annosum, Biology, Ecology, Impact and Control.* CAB International, New York, 589 pp.

Woodward, S., Stenlid, J., Karjalainen, R. and Hüttermann, A. (1998b) Preface. In: Woodward, S., Stenlid, J., Karjalainen, R. and Hüttermann, A. (eds) *Heterobasidion annosum, Biology, Ecology, Impact and Control.* CAB International, New York, pp. xi–xii.

27 Some Important Fungal Diseases and Their Impact on Wheat Production

Aakash Goyal and Rajib Prasad
Agriculture and Agri-Food Canada, Lethbridge Research Center, Lethbridge, Canada

Abstract

Wheat, an important cereal crop, is cultivated worldwide and is second highest in production, just after maize. Due to the increasing world population, there is need for a 40% increase in wheat production to meet global food requirements. Wheat production is diminished mainly by biotic and abiotic stresses all over the world. Of these, pathological diseases are the most important limiting factor of wheat production as different pathogens infect wheat plants, causing severe losses in yield and quality. Wheat can be infected by biotrophic fungi, necrophytic species and nematodes, as well as viruses and bacteria. Among these, different fungal diseases are the most prominent and pose a great challenge to wheat production. Development of resistant varieties is the only solution to overcome this problem and to attain the required wheat production. The development of resistant varieties has benefited immensely from the use of molecular markers, genetic maps, physical maps, QTL analysis and marker-assisted selection (MAS). However, we have to develop multidisease-resistant varieties to fulfil the demand for wheat globally. This review highlights some major fungal diseases of wheat in different parts of the world and the associated problems.

Introduction

Common wheat (*Triticum aestivum* L. em. Thell) is an important staple food crop and ranks first among the three major crops (wheat, maize and rice), which together constitute about half of the total world food production. Wheat feeds about 40% of the world population and provides 20% of the total food calories and protein in human nutrition (Varshney *et al.*, 2006). It is not only used for bread making but is also used for making biscuits, cakes, breakfast cereals, pasta and fermented products like beer, alcohol, vodka, etc. It is also becoming popular as a forage crop. Wheat straw has been used as a fodder material and for ethanol production for the past few years. Unlike rice and maize, which prefer tropical environments, wheat is best adapted to temperate regions, occupying 17% (one-sixth) of the total crop acreage worldwide (Gupta *et al.*, 2008). According to the FAO (2007), wheat occupies 20% of the cultivated crop area (in 2007, 213m ha versus 150m for rice and 143m for maize) and its annual production is 619 Mt of grain. Over the past 20 years, there has only been a small increase in the area of land on which wheat is cultivated worldwide, but the tonnage of wheat grain produced on this land has tripled as a result of improved farming practices and

 Management of Fungal Plant Pathogens (eds A. Arya and A.E. Perelló)

the development of better wheat varieties (Marshall *et al.*, 2001). A significant increase in wheat production has been observed in the past four decades; however, a slowing down has been witnessed during the past few years (Gupta *et al.*, 2008). Due to a consistent increase in world population, there is need for a 40% increase in wheat production to meet this requirement. Despite the enormous progress that has taken place around the world, there is less hope in achieving this goal. Resistance to both biotic and abiotic stresses will be critical for reaching this target. Abiotic stresses include drought, untimely or excess heat, untimely or excess rain, water logging of soils, wind, extreme cold, frost, acid soils and salinity, nutrient imbalances and/or shortages, as well as micronutrient deficiencies.

The impact of biotic stress on wheat production and quality is highly devastating. Diseases in wheat, most caused by fungal pathogens and a few by viruses and bacteria, are important production constraints in almost all wheat-growing environments (Rajaram and van Ginkel, 1996). Wiese (1987) identified over 40 fungal, 32 viral and 81 bacterial diseases that attack wheat plants at different growth stages. Although it is difficult to obtain accurate estimates of crop losses to different fungal diseases, the British Agrochemicals Association (1993) suggests that, under farm conditions where crop rotations, good husbandry and the application of pesticides are practised, losses to diseases can still be around 13%, while under conditions where crop protection measures are not taken, losses can be as high as 50%. It is the goal of wheat breeders to introduce genetic resistance into their varieties to minimize chemical protection measures and losses due to diseases. Under different environments, breeders face problems of different spectra of locally prevalent diseases caused by specific biotypes, serotypes and strains. For many diseases, genes for resistance segregating in a simple 'Mendelian' fashion have been identified; while for other diseases, resistance genes still remain to be detected, due to either a complex mode of inheritance or imprecise disease-screening procedures (Gowda *et al.*, 2006). Recently, available molecular markers and functional genomics tools have helped the breeder to manipulate the wheat genome to develop disease-resistant cultivars and achieve the target of wheat production.

Fungal Diseases of Wheat

Most of the important diseases of wheat are caused by fungal pathogens, while only a few are caused by viruses and bacteria (McIntosh *et al.*, 1995; Rajaram and van Ginkel, 1996). Infection of fungal diseases in wheat depends on the availability of free water on the host plant surface, susceptibility of the host, the density of inoculum, temperature and other environmental factors. Moreover, host–parasite interaction plays a significant role in the development of disease and subsequent symptoms on the wheat plant. In this chapter, some of the commonly reported fungal diseases of wheat are described. Table 27.1 lists major fungal diseases of wheat reported by different pathologists around the world.

Fusarium head blight (FHB)

Several species of *Fusarium* can cause *Fusarium* head blight (FHB), also known as scab of cereal crops. Among these, *F. graminearum*, found mainly in the USA, Canada, China and the EU, is accountable for severe losses in yield and quality of wheat production (Parry *et al.*, 1995). An epidemic of FHB in the USA and Canada in 1993 was a result of changes in crop management practices (minimum or reduced tillage), changes in rainfall patterns and a low resistance in the cultivars against FHB (Dill-Macky and Jones, 1997). In the case of wheat, *Fusarium* spp. attacks different plant organs but mainly targets the ear, which leads to great loss in seed quality. On the ear, *Fusarium* enters through the stomata to the palea and lemma and destroys these tissues completely. The first symptom of FHB is a tan or brown discoloration at the base of a floret within the spikelets of the head. The infection may be

Table 27.1. The major fungal diseases of wheat reported across the world.

Name of the disease	Pathogenic fungal species	Tolerant varieties/ genotypes	References
Black point/kernel smudge	*Alternaria alternata*	Sunco, Cascades	Lehmensiek *et al.*, 2004
Common bunt	*Tilletia caries*, *T. foetida*	AC Domain	Fofana *et al.*, 2008
Common root rot	*Cochliobolus sativus*	ND 652	Mergoum *et al.*, 2005
Ergot	*Claviceps purpurea*	Carleton, Kenya farmer	Platford and Bernier, 1970
Fusarium head blight (scab)	*Fusarium graminearum*	Bizel, Sumai 3	Bourdoncle and Ohm, 2003
Leaf rust	*Puccinia recondita* (*P. triticina*)	Tangmai 4, ND 652	Li *et al.*, 2004; Mergoum *et al.*, 2005; Kolmer *et al.*, 2007
Loose smut	*Ustilago tritici*	DT676	Knox *et al.*, 2008
Powdery mildew	*Erysiphe graminis*	Tangmai 4	Li *et al.*, 2004
Speckled leaf blotch	*Septoria avenae* f. sp. *tritici*	Arina and Riband; Courtot and Tonic	Chartrain *et al.*, 2009
Glume blotch	*Stagonospora nodurum*	Red Chief	Laubscger *et al.*, 2008
Spot blotch	*Cochliobolus sativus*	Ning 8201, K8027	Sharma *et al.*, 2007
Stem rust	*P. graminis* f. sp. *tritici*	ND 652, Tangmai 4	Li *et al.*, 2004; Mergoum *et al.*, 2005
Stripe rust	*P. striiformis*	Tangmai 4	Li *et al.*, 2004
Take-all	*Gaeumannomyces graminis* var. *tritici*	Xinong 1376, Xinong 918, R859	Xiaoning *et al.*, 2004

limited to one spikelet, but if the fungus invades the rachis, the entire head may develop symptoms of the disease. Discoloration of the head starts due to the production of mycotoxins [zearalenones and deoxynivalenol (DON)] by the *Fusarium*. The mycotoxins affect seed quality adversely, producing toxic dust and thus making the seeds unsuitable for human and livestock consumption (Eudes and Laroche, 2003). The mycotoxin DON, even in low doses of 1–3 ppm, can cause reduced feed intake and less weight gain in animals, while a high dose up to 10 ppm can cause vomiting and refusal to feed. DON is also very harmful to humans; therefore, different countries have established laws to protect consumers. For example, the EU Member States allow a maximum of 1.25 ppm DON in unprocessed bread, 0.5 ppm in bread and bakery products and only below 0.2 ppm in baby foods (Buerstmayr *et al.*, 2009). The USA Food and Drug Administration recommend only 1 ppm DON in finished wheat products, while Health Canada have established guidelines of 2 ppm and 1 ppm DON in soft wheat in non-stable and baby foods, respectively.

Control of this disease has been difficult, because of the complex nature of the host/pathogen interaction. Cultural practices, such as rotation with non-host crops and management of crop residues, in combination reduce primary infection. A mixed fungicide composed of carbendazim and triadimefon was reported to have a significant synergistic action (Wang, 1997). Under high disease pressure, Bravo or Folicur were reported to reduce levels of FHB, though these are not cost-effective under low disease pressure (Agrios, 1997). Host resistance is a promising and effective management solution, but resistance has not been easy to achieve in the adapted cultivars.

Wheat rust

Wheat rust pathogens belong to genus *Puccinia*, family *Pucciniaceae*, order *Uredinales* and class *Basidiomycetes*. Rust disease is capable of causing considerable economic loss throughout the world (FAO, 2008). Rust in cereals, found back in the late 17th century, was caused by a fungal parasite which was named later as Persoon's *P. graminis* (Chester, 1946). In the beginning of the 20th century, different fungal species were identified for different rusts with contrasting host ranges. In wheat, rust diseases are so important that in 2007, the CSIRO, Australia, published a special issue on wheat rust in the *Australian Journal of Agricultural Research*.

Stem rust

Stem or black rust of wheat is a major disease problem, caused by the fungus, *P. graminis* Pers. f. sp. *tritici*. It has been a major disease on wheat since the rise of agriculture and the Romans even prayed to a stem rust god, 'Robigus'. The Italians, Fontana and Tozzetti, independently provided the first report on stem rust in wheat in 1767. In the early to mid 1950s, stem rust epidemics caused approximately 50% yield losses of wheat in North America (Leonard, 2001). During the 1950s, Norman Borlaug and other scientists started developing high-yielding wheat varieties that were resistant to stem rust and other diseases in North America and throughout the world. The rust-resistant, high-yielding wheat variety banished chronic hunger in much of the world, ended stem rust outbreaks and won Borlaug the Nobel peace prize in 1970 (Singh *et al.*, 2006). In most areas of the world, the life cycle of *P. graminis* consists of continual uredinial generations. The disease spreads either via airborne spores or occasionally locally from wild susceptible barberry (*Berberis* sp.) plants (Eversmeyer, 2000).

Ug99, so called as it was first seen in Uganda in 1999, is a new devastating race of 'stem rust' which has already travelled from Africa to Iran and can proceed to India, Pakistan and Bangladesh (Pretorius *et al.*, 2000). It is particularly dismaying because of its ability to infect crops in just a few hours and its vast cloud of invisible spores can be carried by the wind for hundreds of miles (Singh *et al.*, 2006).

Cultural control provides at least partial control of wheat rust epidemics. Planting early maturing varieties is an efficient way to avoid losses due to stem rust infection. Propiconazole (Tilt) and triadimefon (Bayleton) are found to be effective against stem rust (Agrios, 1997), though these chemicals are cost-prohibitive. To save the world from the wheat epidemic, CIMMYT and ICARDA started the Global Rust Initiative (GRI) to coordinate efforts to track and study Ug99 and develop resistant varieties of wheat (Stokstad, 2007). Later in 2008, it was taken over by the Borlaug Global Rust Initiative (BGRI), chaired by Dr N.E. Borlaug, who said he was optimistic that the fungus would be beaten again (Stokstad, 2007). Efforts were also taken to understand the rust's epidemiology and evolution, which led to the barberry eradication programme in North America and Europe (Singh *et al.*, 2006).

Leaf rust

Wheat leaf rust, also known as brown rust, is caused by the rust fungus, *P. triticina* Rob. Ex Desm. f. sp. *tritici* Eriks (syn. *P. recondita*). De Candole (1815) reported for the first time that leaf rust was caused by fungus and named it *Uredo rubigovera*. Later in the 19th century, the name was changed to *P. recondita* (Cummins and Caldwell, 1956). However, the present name, *P. triticina*, was suggested by Savile (1984) and Anikster *et al.* (1997). Up to 2007, more than 50 races of leaf rust were detected all over the world (Kolmer *et al.*, 2007; Mebrate *et al.*, 2008). Leaf rust is the most prevalent of all the wheat rust diseases, occurring in nearly all areas where wheat is grown. Depending on the severity and duration of infection, losses in wheat can vary by up to 50% (Nagarajan and Joshi, 1975; McIntosh *et al.*, 1995). The disease has caused serious epidemics in North America, Mexico, South America and some other countries. This fungus can infect wheat plants with a 3 h dew period at temperatures near 20°C. However, more infections occur with longer dew periods. The fungus initially starts covering leaves with orange pustules of urediniospores (uredinia). The urediniospores are reddish-brown, elliptical to egg-shaped, echinulate structures. In the later stage, the postules eventually darken due to the formation of black teliospores (Roberson and Luttrell, 1987). Infections can result in a 1–20% yield loss since infected leaves die earlier and all the nutrients are directed to the growing fungi. Infection can also cause grains to shrivel. The loss in yield depends on several factors that include time of initial infection, crop development stages, relative resistance or susceptibility of the wheat cultivars. Higher yield losses result when the initial infection occurs early in the growing season before tillering. Infection occurring after heading when grain filling is in progress will cause lesser crop loss (Agrios, 1997).

Chemical control with trizole fungicides has been reported as useful in controlling infections up to ear emergence, but is difficult to justify economically in attacks after this stage. Varietal control is again the best control for leaf rust. Resistant varieties possess one or more special leaf rust resistance genes called *Lr* genes. Currently, there are more than 58 different *Lr* genes available in wheat (Bansal *et al.*, 2008; Chhuneja *et al.*, 2008; McIntosh *et al.*, 2008), but most varieties have only a few *Lr* genes. So, there is a need to develop multi *Lr* gene-carrying varieties to defeat leaf rust disease.

Yellow rust or stripe rust

Another rust of wheat, stripe or yellow rust which is caused by *P. striiformis* f. sp. *tritici*, can be as damaging as other rusts. Due to a requirement for a very low optimum temperature for its development, stripe rust is not found in many areas of the world. However, a total area of 9.4m ha (> 35%) under wheat cultivation is affected by stripe rust (Singh *et al.*, 2004). On the world level, stripe rust is found predominantly in northern Europe, the Middle East, East Africa, China, India and the continents of South America, Australia and New Zealand (Saari and Prescott, 1985). In the USA, it was first reported in 1915 (Carleton, 1915) and serious outbreaks were reported in the western states in the 1960s (Line, 2002; Boyd, 2005).

For this disease, generally no cultural control measures are applicable, but in the USA, where the disease occurs commonly, the removal of the alternate host is an established method of cultural control. Identification and use of the resistant gene in resistant varities is the only way to reduce the impact of the disease on wheat production. Many yellow rust resistance genes have been identified in wheat by different wheat workers and to date, 41 of these (*Yr1* to *Yr41*) have been designated (McIntosh *et al.*, 2008). Most of the identified yellow rust resistance genes have proven to be race-specific, with resistance being effective only against isolates of *P. striiformis* f. sp. *tritici* carrying the corresponding avirulence gene. Different wild wheat varieties were also used to transfer the resistance gene to hexaploid wheat for stripe rust resistance (Kuraparthy *et al.*, 2007a,b; Singh *et al.*, 2007; Chhuneja *et al.*, 2008). More recently, a highly resistant gene with broad spectrum on strip rust races, namely *Yr36*, from wild emmer wheat was used for positional cloning (Fu *et al.*, 2009).

Karnal bunt

Karnal bunt (partial bunt) of wheat has become a disease of serious concern in some parts of the world as it causes direct yield losses and also has significance as an export problem because many believe the pathogen to be a quarantine pest. Consequently, stringent quarantine measures have been adopted in several countries, which may affect not only the wheat grain trade but also germplasm exchange (Royer and Rytter, 1988). Karnal bunt caused by the smut fungus *Tilletia indica* Mitra *Neovossia indica* (Mitra, 1931), a *Basidiomycetes* fungus, is a serious floral-infecting disease of wheat in the major wheat-growing areas of India (Gill, 1990) and some other wheat-growing countries of the world (Nath *et al.*, 1981). The pathogen is known to infect bread wheat, durum wheat and triticale (Agarwal *et al.*, 1977). The disease was first reported in 1931 in experimental wheat crop at the Botanical Station at Karnal, India (Mitra, 1931), and was for many years known only in the plains of India and Pakistan (Ahmad and Attaudin, 1991). Currently, it occurs in Afghanistan, India, Iran, Iraq, Mexico, Nepal and Pakistan and in limited areas of the USA (Durán, 1972; Munjal, 1975; Singh *et al.*, 1989; Ykema *et al.*, 1996). Recognition of fungal structures (teliospores) on grain samples from Lebanon and Syria suggest that the disease is established in these countries as well (Locke and Watson, 1955).

Karnal bunt requires free water in the soil for teliospores, the overwintering life stage of Karnal bunt, to germinate. Teliospores are brown to dark brown, spherical or subspherical, or oval, 22–42 × 25–40 µm in diameter, occasionally having an apiculus (Roberson and Luttrell, 1987), papilla (Mitra, 1931) or a vestige of attached mycelium (Durán and Fischer, 1961). The disease cycle (Fig. 27.1) starts with the introduction of teliospores on to a field. Contaminated seeds are considered to be the major source of teliospores, while other sources include wind, animals, contaminated equipment or contaminated vehicles. Teliospores may remain dormant but viable for several years (Ottman, 2002). Although planting infected seed is the primary means of getting the spores into the soil, this may or may not produce infected plants directly in the first year. The greater threat of disease occurs the following year as the soil is turned over, bringing these teliospores back to the surface. At the flowering stage of host plants, the teliospores produce sporidia that infect the plant florets and fungal hyphae enter the ovary (Aujla *et al.*, 1977; Singh and Prasad, 1978; Khetarpal *et al.*, 1980; Krishna and

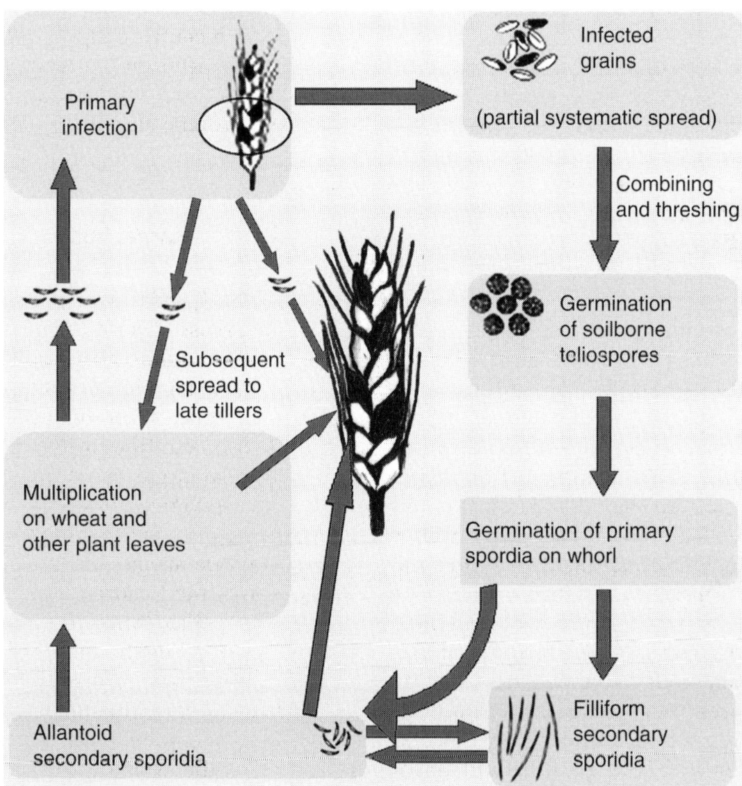

Fig. 27.1. The life cycle of Karnal bunt caused by *Tilletia indica* Mitra.

Singh, 1982). Subsequent disease development in the embryo end of the kernel results in the formation of new teliospores, which are deposited back in the soil at harvest, adding further to soil inoculum. Cool, cloudy and very humid conditions or rainfall between awn emergence and the end of flowering are required for sporidia production, infection and for the disease to flourish (Dhaliwal *et al.*, 1983; Goates, 1988). The incidence of Karnal bunt is usually very low and rarely seen if the environmental requirements are not met. Karnal bunt affects the heads of wheat plants. The disease is not easily detected in the field because few florets are typically infected and the area of the kernel affected might be small and facing inwards. A mass of black teliospores is found at the embryo end of the kernel and, at higher levels of infection, along the crease or in the entire kernel (Goel *et al.*, 1977; Dhaliwal *et al.*, 1983). A fishy odour is emitted from infected seeds due to the presence of trimethylamine (Mehdi *et al.*, 1973).

Conventional approaches to control this disease consist of the adoption of various cultural practices such as crop rotation for longer periods, sowing of disease-free seeds, adjustment of the nitrogen balance in the soil and adjustment of the time of irrigation to minimize disease incidence (Mitra, 1937; Munjal, 1974; Goel *et al.*, 1977; Singh and Prasad, 1978; Aujla *et al.*, 1981, Singh and Singh, 1985; Gill *et al.*, 1993). Control through fungicides is not completely effective as the disease is seed- and soilborne (Singh *et al.*, 1985). However, application of Tilt at heading and 1 week later can reduce disease incidence by 90% when environmental conditions are conducive to disease development (Ottman, 2002). Hence, the most economical, eco-friendly and effective approach to control the disease is the cultivation of resistant varieties. The main sources of resistance against Karnal bunt have been the Indian, Chinese and Brazilian wheats (Fuentes-Davila and Rajaram, 1994). A new range of genetic variability for resistance to Karnal bunt has been observed in synthetic hexaploid wheat derived from *T. turgidum* × *T. tauschii* crosses (Villareal *et al.*, 1996).

Powdery mildew

Powdery mildew of wheat, a wind-dispersed disease, is an important and most common disease worldwide, particularly in humid regions (Oerke *et al.*, 1994). It is of special interest in epidemiology because it results in reduced kernel size and seed weight, and ultimately lower yield. The fungal pathogen, *Blumeria graminis* f. sp. *tritici* (an *Ascomycete*), causing powdery mildew on wheat, is a biotrophic obligate parasite (Cooke *et al.*, 2006), which is highly sensitive to the environment and its presence can vary from season to season (Jenkyn and Bainbridge, 1978; Jorgensen, 1988; Wolfe and McDermott, 1994). The powdery mildew fungus is made up of different races and forms that are highly specialized. Wheat cultivars might be resistant to a certain race of the mildew fungus, but susceptible to another race. Some of the special features of powdery mildew, such as wide distribution, rapid development within or on host tissue, massive production of spores, the ability to remain viable after long-distance dispersal and a high capacity to become virulent on previously resistant cultivars, make it a devastating disease of wheat (Boshoff *et al.*, 2002).

Powdery mildew oversummers on volunteer crops in the asexual stage, infects the autumn-sown crop and, eventually, overwinters on the volunteers to infect the crops in spring (Zadoks, 1961). In mild areas, volunteer wheat plants are abundant because of the relatively frequent rainfall in summer, while in dry regions, oversummering can depend on grass species (Boshoff *et al.*, 2002). For mildew, the asexual cycle is the production of haploid conidia, while occasionally the ascospores, which are the result of sexual cycle, can initiate epidemics (Cooke *et al.*, 2006). Mildew also differentiates a sexual stage, which contributes to oversummering. In early summer, *B. graminis* f. sp. *tritici* initiates the formation of generative mycelium and cleistothecia starting on the lower leaves. In the cleistothecia, 15–20 asci develop, each containing eight haploid ascospores which are dispersed by wind, even under high humidity after rain (Gotz *et al.*, 1996). Ascospores can develop at any time during the last half of the year; therefore, sexual reproduction is more important for powdery mildew on wheat. Apart from ascospores, conidia from the summer crop can also infect volunteer plants; thus, a mixture of ascospores and conidia forms the inoculum for the winter crop. However, the mildew population grown during autumn on the winter crop can survive the cold period in vegetative stage on overwintering green plants (Cooke *et al.*, 2006). Mildew is more severe in dense stands of heavily fertilized wheat. Plants are most susceptible during periods of rapid growth, especially from stem elongation through heading growth stages.

Powdery mildew on wheat is recognized by small, effuse patches (colonies) of cottony mycelia on the upper and lower surfaces of the leaves. As these patches sporulate and age, they become a mass of dull tan colour. Chlorotic (yellow) patches may later surround the mildew colonies (Purdy, 1967; Kingsland, 1982). Powdery mildew attacks the leaves, but stems and heads are also affected. The fungus grows primarily on the surface of the host and feeds on the living green cells of the plant. Damage occurs from reduced photosynthetic ability when green surfaces are shaded and the host is robbed of moisture and food by fungal growth. Yields may be reduced by 20% or more. Spring wheat, other than soft white wheat, are seldom affected at economic levels on the prairies, while winter wheat is affected to a greater degree. The disease will reduce yields seriously if the flag and second leaves are affected (Gotz *et al.*, 1996; Boshoff *et al.*, 2002).

Incorporating wheat residues into the soil, destroying volunteer wheat and crop rotation can lessen the amount of overwintering inoculum in the field. Powdery mildew thrives where high rates of nitrogen have been used. Therefore, use of a correct and balanced fertilization programme with proper levels of N, P and K is advised. It is important to keep the top two leaves of the plant as disease free as possible so that the plant can use its full potential to fill the grain. Fungicides can be applied based on the level of disease in the field, the known susceptibility of the variety and the selling price of the grain (Agrios, 1997). Growing mildew-resistant cultivars is the most economical way to control powdery mildew, though wheat varieties vary in their resistance to powdery mildew and new races of the fungus can attack previously resistant varieties.

Conclusions

Wheat is a most important cereal crop and is becoming more in demand due to the significant increase in the world population. To protect the world from the upcoming threat of hunger, food and nutrition, wheat production must be doubled in time. Major concern about wheat quality and production is related to biotic stress. Different methods, such as chemical control, cultural methods and eradication of alternate hosts, are used to prevent the disease, but the most important and effective one is the development of resistant varieties. For durability of resistance against fungal diseases, breeders should focus on new sources of race-specific resistance genes from either adapted cultivars or wild varieties. However, extensive knowledge of the pathogen population is a vital criterion in assessing resistance and guidelines for breeders to incorporate useful resistance genes into the desired background. Recent studies have proved the usefulness of different marker systems and association mapping of genes/QTLs controlling resistance against different fungal diseases (Crossa *et al.*, 2007). Similarly, MAS was also employed successfully to improve quality and resistance against disease (see Dubcovsky, 2004; Anderson, 2007; Sorrells, 2007). In future, new molecular marker systems (e.g. ESTs, SNPs and DArTs) and functional genomics approaches (e.g. TILLING, RNAi and epigenetics) can be used to facilitate the development of resistant varieties in bread wheat.

References

Agarwal, V.K., Verma, H.S. and. Khetarpal, R.K. (1977) Occurrence of partial bunt on a triticale. *FAO Plant Protection Bulletin* 25, 210–211.

Agrios, G.N. (1997) *Plant Pathology*, 4th edn. Academic Press, London, 635 pp.

Ahmad, S. and Attaudin (1991) Appraisal of Karnal bunt of wheat in the North West Frontier Province of Pakistan. *Pakistan Journal of Phytopathology* 3, 35–37.

Anderson, J.A. (2007) Marker-assisted selection for *Fusarium* head blight resistance in wheat. *International Journal of Food Microbiology* 119, 51–53.

Anikster, Y., Bushnell, W.R., Eilam, T., Manisterski, J. and. Roelfs, A.P. (1997) *Puccinia recondita* causing leaf rust on cultivated wheats, wild wheats and rye. *Canadian Journal of Botany* 75, 2082–2096.

Aujla, S.S., Sharma, Y.R., Chand, K. and. Sawney, S.A. (1977) Influence of weather factors on the incidence and epidemiology of Karnal bunt disease of wheat in the Punjab. *Indian Journal of Ecology* 4, 71–74.

Aujla, S.S., Gill, K.S., Sharma, Y.R., Singh, D. and Nanda, G.S. (1981) Effect of date of sowing and level of nitrogen on the incidence of Karnal bunt of wheat. *Indian Journal of Ecology* 8, 175–179.

Bansal, U.K., Hayden, M.J., Venkata, B.P., Khanna, R., Saini, R.G. and Bariana, H.S. (2008) Genetic mapping of adult plant leaf rust resistance genes *Lr48* and *Lr49* in common wheat. *Theoretical and Applied Genetics* 117, 307–312.

Boshoff, W.H.P., Pretorius, Z.A. and Van Niekerk, B.D. (2002) Establishment, distribution and pathogenicity of *Puccinia striiformis* f. sp. *tritici* in South Africa. *Plant Disease* 86, 485–492.

Bourdoncle, W. and Ohm, H.W. (2003) Fusarium head blight-resistant wheat line 'Bizel' does not contain rye chromatin. *Plant Breeding* 122, 281–282.

Boyd, L.A. (2005) Can Robigus defeat an old enemy? – Yellow rust of wheat. *Journal of Agricultural Science* 143, 233–243.

British Agrochemicals Association (1993) Focus on wheat (*Triticum aestivum*). *Grapevine* No. 37.

Buerstmayr, H., Ban, T. and Anderson, A. (2009) QTL mapping and marker-assisted selection for Fusarium head blight resistance in wheat a review. *Plant Breeding* 128, 1–26.

Carleton, M.A. (1915) A serious new wheat rust in this country. *Science* 42, 58–59.

Chartrain, L., Sourdille, P., Bernard, M. and Brown, J.K.M. (2009) Identification and location of *Stb9*, a gene for resistance to *septoria tritici* blotch in wheat cultivars Courtot and Tonic. *Plant Pathology* 58, 547–555.

Chester, K.S. (1946) *The Nature and Prevention of the Cereal Rusts as Exemplified in the Leaf Rust of Wheat.* Chronica Botanica Co., Waltham, Massachusetts.

Chhuneja, P., Kaur, S., Garg, T., Ghai, M., Kaur, S., Prashar, M., Bains, N.S., Goel, R.K., Keller, B., Dhaliwal, H.S. and Singh, K. (2008) Mapping of adult plant stripe rust resistance genes in diploid A genome wheat species and their transfer to bread wheat. *Theoretical and Applied Genetics* 116, 313–324.

Cooke, B.M., Jones, D.G. and Keye, B. (2006) *The Epidemiology of Plant Diseases.* 2nd edn. Springer, Dordrecht, The Netherlands, 576 pp.

Crossa, J., Burgueño, J., Dreisigacker, S., Vargas, M., Herrera-Foessel, S.A., Lillemo, M., Singh, R.P., Trethowan, R., Warburton, M., Franco, J., Reynolds, M., Crouch, J.H. and Ortiz, R. (2007) Association analysis of historical bread wheat germplasm using additive genetic covariance of relatives and population structure. *Genetics* 177, 1889–1913.

Cummins, G.B. and. Caldwell, R.M. (1956) The validity of binomials in the leaf rust fungus complex of cereals and grasses. *Phytopathology* 46, 81–82.

De Candolle, A. (1815) Uredo rouille des cereales. In: *Flora francaise, famille des champignons*, 83 pp.

Dhaliwal, H.S., Randhawa, A.S., Chand, K. and Singh, D. (1983) Primary infection and further development of Karnal bunt of wheat. *Indian Journal of Agricultural Science* 53, 239–244.

Dill-Macky, R. and Jones, R.K. (1997) The effect of previous crops and tillage on Fusarium head blight of wheat. *Cereal Research Communications* 25, 711–712.

Dubcovsky, J. (2004) Marker-assisted selection in public breeding programs: the wheat experience. *Crop Science* 44, 1895–1898.

Durán, R. (1972) Further aspects of teliospore germination in North American smut fungi. *Canadian Journal of Botany* 50, 2569–2573.

Durán, R. and Fischer, G.W. (1961) *The genus Tilletia.* Washington State University Press, Pullman, Washington, 138 pp.

Eversmeyer, M.G. (2000) Epidemology of wheat leaf and stem rust in the central great plains of the USA. *Annual Review of Phytopathology* 38, 491–513.

Eudes, F. and Laroche, A. (2003) Fusarium head blight resistance and reduced levels of mycotoxin in small grain cereals: current status and prospects. In: Huang, H.-C. and Acharya, S.N. (eds) *Advances in Plant Disease Management.* Research Signpost, Kerala, India, pp. 1–24.
FAO (Food and Agriculture Organization of United States) (2007 and 2008) http://www.fao.org/.
Fofana, B., Humphreys, D.G., Cloutier, S., McCartney, C.A. and Somers, D.J. (2008) Mapping quantitative trait loci controlling common bunt resistance in a doubled haploid population derived from the spring wheat cross RL4452 x AC Domain. *Molecular Breeding* 21, 317–325.
Fontana, F. (1767) *Observations on the Rust of Grain.* Pirone, P.P., translator. Classics No. 2. American Phytopathological Society, Washington, DC 1932 (originally published in 1767).
Fu, D., Uauy, C., Distelfeld, A., Blechl, A., Epstein, L., Chen, X., Sela, H., Fahima, T. and Dubcovsky, J. (2009) A kinase-start gene confers temperature-dependent resistance to wheat stripe rust. *Science* 323, 1360–1363.
Fuentes-Davila, G. and Rajaram, S. (1994) Sources of resistance to *Tilletia indica* in wheat (*Triticum aestivum*). *Crop Protection* 13, 20–24.
Gill, K.S. (1990) *Genetic evaluation and utilization of wheat germplasm for breeding varieties resistant to Karnal bunt. Third Annual Research Report.* Department of Plant Breeding, Punjab Agricultural University, Ludhiana, Punjab, India.
Gill, K.S., Sharma, I. and Aujla, S.S. (1993) *Karnal Bunt and Wheat Production.* Punjab Agricultural University, Ludhiana, India, 153.
Goates, B.J. (1988) Histology of infection of wheat by *Tilletia indica,* the Karnal bunt pathogen. *Phytopathology* 78, 1434–1441.
Goel, L.B., Sing, D.V., Srivastava, K.D., Joshi, L.M. and Nagarajan, S. (1977) *Smuts and Bunts of Wheat in India.* Bulletin. Division of Mycology and Plant Pathology, IARI, New Delhi, 38 pp.
Götz, M., Friedrich, S. and Boyle, C. (1996) Cleistothecia as a source of infection of wheat during summer. *Zeitschrift fuer Pflanzenkrankheiten und Pflanzenschutz* 103, 134–141.
Gowda, M., Roy-Barman, S. and Chattoo, B.B. (2006) Molecular mapping of a novel blast resistance gene *Pi38* in rice using SSLP and AFLP markers. *Plant Breeding* 125, 596–599.
Gupta, P.K., Mir, R.R., Mohan, A. and Kumar, J. (2008) Wheat genomics: present status and future prospects. *International Journal of Plant Genomics* Article ID 896451, 36 pp.
Jenkyn, J.F. and Bainbridge, A. (1978) Biology and pathology of cereal powdery mildews. In: Spencer, D.M. (ed.) *The Powdery Mildews.* Academic Press, London, pp. 284–321.
Jorgensen, J.H. (1988) *Erysiphe graminis*, powdery mildew of cereals and grasses. *Advances in Plant Pathology* 6, 137–157.
Khetarpal, R.K., Agarwal, V.K. and Chauhan, K.P.S. (1980) Studies on the influence of weather conditions on the incidence of black-point and karnal bunt of triticale. *Seed Research* 8, 108–110.
Kingsland, G.C. (1982) Triadimefon for control of powdery mildew of wheat. *Plant Disease* 66, 139–141.
Knox, R.E., Campbell, H.L., Clarke, J.M., Fernandez, M.R., McLeod, J.G. and DePauw, R.M. (2008) Registration of DT676 loose smut resistant durum wheat germplasm. *Journal of Plant Registrations* 2, 156–161.
Kolmer, J.A., Oelker, L.M. and Liu, J.Q. (2007) Genetics of leaf rust resistance in three Americano landrace-derived wheat cultivars from Uruguay. *Plant Breeding* 126, 152–157.
Krishna, A. and Singh, R.A. (1982) Effect of physical factors and chemicals on the teliospore germination of *Neovossia indica. Indian Phytopathology* 35, 448–455.
Kuraparthy, V., Chhuneja, P., Dhaliwal, H.S., Kaur, S., Bowden, R.L. and Gill, B.S. (2007a) Characterization and mapping of cryptic alien introgression from *Aegilops geniculata* with novel leaf rust and stripe rust resistance genes *Lr57* and *Yr40* in wheat. *Theoretical and Applied Genetics* 114, 1379–1389.
Kuraparthy, V., Sood, S., Chhuneja, H., Dhaliwal, S., Kaur, S., Bowden, R.L. and. Gill, B.S. (2007b) A cryptic wheat–*Aegilops triuncialis* translocation with leaf rust resistance gene *Lr58. Crop Science* 47, 1995–2003.
Laubscher, F.X., von Wechmar, B. and van Schalkwyk, D. (2008) Heritable resistance of wheat varieties to glume blotch (*Septoria nodorum* Berk.). *Journal of Phytopathology* 56, 260–264.
Lehmensiek, A., Campbell, W., Williamson, P.M., Michalowitz, M., Sutherland, M.W. and Daggard, G.E. (2004) QTLs for black-point resistance in wheat and the identification of potential markers for use in breeding programmes. *Plant Breeding* 123, 410–416.
Leonard, K.J. (2001) Stem rust–future enemy? In: Peterson, P.D. (ed.) *Stem Rust of Wheat: From Ancient Enemy to Modern Foe.* APS Press, St Paul, Minnesota, pp. 119–146.
Li, H.J., Connor, R.L., Mccallum, B.D, Chen, X., Su, H., Wen, Z.Y., Chen, Q. and Jia, X. (2004) Resistance of 'Tangmai 4' wheat to powdery mildew, stem rust, leaf rust, and stripe rust and its chromosomal composition. *Canadian Journal of Plant Science* 84, 1015–1023.

Line, R.F. (2002) Stripe rust of wheat and barley in North America: a retrospective historical review. *Annual Review of Phytopathology* 40, 75–119.
Locke, C.M. and Watson, A.J. (1955) Foreign plant diseases intercepted in quarantine inspections. *Plant Disease Reporter* 39, 518.
McIntosh, R.A., Wellings, C.R. and Park, R.F. (1995) *Wheat Rusts: An Atlas of Resistance Genes.* Kluwer Academic Publishers, Dordrecht, The Netherlands.
McIntosh, R.A., Yamazaki, Y., Dubcovsky, J., Rogers, J., Morris, C., Somers, D.J., Appels, R. and Devos, K.M. (2008) *Catalogue of Gene Symbols for Wheat. Proceeding 11th International Wheat Genetics Symposium.* University of Sydney Press, Australia (http://wheat.pw.usda.gov/GG2/Triticum/wgc/2008/).
Marshall, D.R., Langridge, P. and Appels, R. (2001) Wheat breeding in the new century: preface. *Australian Journal of Agricultural Research* 52, 1–4.
Mebrate, S.A., Dehne, H.-W., Pillen, K. and Oerke, E.-C. (2008) Postulation of seedling leaf rust resistance genes in selected Ethiopian and German bread wheat cultivars. *Crop Science* 48, 507–516.
Mehdi, V., Joshi, L.M. and. Abrol, Y.P. (1973) Studies on chapati making quality: (vi) effect of wheat grains with bunts on the quality of chapaties. *Bulletin of Grain Technology* 11, 195–197.
Mergoum, M., Frohberga, R.C., Stackb, R.W., Rivelandc, N., Olsona, T. and Millerd, J.D. (2005) Registration of spring wheat germplasm ND 652 resistant to common root rot, leaf, and stem rusts. *Crop Science* 45, 2667–2668.
Mitra, M. (1931) A new bunt of wheat in India. *Annals of Applied Biology* 18, 178–179.
Mitra, M. (1937) Studies on the stinking smut or bunt of wheat in India. *Indian Journal of Agricultural Science* 7, 459–478.
Munjal, R.L. (1974) Technique for keeping the cultures of *Neovossia indica* in sporulating condition. *Indian Phytopathology* 27, 248–249.
Munjal, R.L. (1975) Status of Karnal bunt (*Neovossia indica*) of wheat in Northern India during 1968–69 and 1969–70. *Indian Journal of Mycology and Plant Pathology* 5, 185–187.
Nagarajan, S. and Joshi, L.M. (1975) A historical account of wheat rust epidemics in India and their significance. *Cereal Rusts Bulletin* 2, 29–33.
Nath, R., Lambat, A.K., Mukewar, P.M. and Rani, L. (1981) Interceptions of pathogenic fungi on imported seed and planting material. *Indian Phytopathology* 34, 282–286.
Oerke, E.C., Dehne, H.W., Schonbeek, F. and Weber, A. (1994) *Crop Production and Crop Protection. Estimated Losses in Major Food and Cash Crops.* Elsevier, Amsterdam.
Ottman, M. (2002) Cultural practices for karnal bunt control. Cooperative extension, The University of Arizona, Tucson, Arizona (http//ag.arizona.edu/pubs/crops/az1287.pdf, accessed 22 March 2009).
Parry, D.W., Jenkinson, P. and McLeod, L. (1995) Fusarium ear blight (scab) in small grain cereals – a review. *Plant Pathology* 44, 207–238.
Platford, R.G. and Bernier, C.C. (1970) Resistance to *Claviceps purpurea* in spring and durum wheat. *Nature* 226, 770.
Pretorius, Z.A., Singh, R.P., Wagoire, W.W. and Payne, T.S. (2000) Detection of virulence to wheat stem rust resistance gene *Sr31* in *Puccinia graminis* f. sp. *tritici* in Uganda. *Plant Disease* 84, 2003.
Purdy, L.H. (1967) Wheat powdery mildew, a 1966 epiphytotic in eastern Washington. *Plant Disease Reporter* 51, 94–95.
Rajaram, S. and van Ginkel, M. (1996) A guide to the CIMMYT bread wheat section. In: *Wheat Special Report No. 5.* CIMMYT, Mexico, DF.
Roberson, R.W. and Luttrell, E.S. (1987) Ultrastructure of teliospore ontogeny in *Tilletia indica. Mycologia* 79, 753–763.
Royer, M.H. and Rytter, J. (1988) Comparison of host ranges of *Tilletia indica* and *T. barclayana. Plant Disease* 72, 133–136.
Saari, E.E. and Prescott, J.M. (1985) World distribution in relation to economic losses. In: Roelfs, A.P. and Bushnell, W.R. (eds) *The Cereal Rusts*, Volume II. Academic Press, Orlando, Florida, pp. 259–298.
Savile, D.B.O. (1984) Taxonomy of the cereal rust fungi. In: Bushnell, W.R. and Roelfs, A.P. (eds) *The Cereal Rusts, Origins, Specificity, Structures, and Physiology*, Volume 1. Academic Press, Orlando, Florida, pp. 79–112.
Sharma, R.C., Duveillerb, E. and Ortiz-Ferrara, G. (2007) Progress and challenge towards reducing wheat spot blotch threat in the Eastern Gangetic Plains of South Asia: is climate change already taking its toll? *Field Crop Research* 103, 109–118.
Singh, A. and Prasad, R. (1978) Date of sowing and meteorological factors in relation to occurrence of Karnal bunt of wheat in U.P. Tarai. *Indian Journal of Mycology and Plant Pathology* 8, 2–10.

Singh, D.V., Srivastava, K.D., Joshi, L.M. and Verma, B.R. (1985) Evaluation of some fungicides for the control of Karnal bunt of wheat. *Indian Phytopathology* 38, 571–573.

Singh, D.V., Agarwal, R., Shrestha, J.K., Thapa, B.R. and Dubin, H.J. (1989) First report of *Tilletia indica* on wheat in Nepal. *Plant Disease* 73, 273.

Singh, K., Chhuneja, P., Ghai, M., Kaur, S., Goel, R.K., Bains, N.S., Keller, B. and Dhaliwal, H.S. (2007) Molecular mapping of leaf and stripe rust resistance genes in *Triticum monococcum* and their transfer to hexaploid wheat. In: Buck, H., Nisi, J.E. and Solomon, N. (eds) *Wheat Production in Stressed Environments*. Springer, The Netherlands, pp. 779–786.

Singh, R.P., William, H.M., Huerta-Espino, J. and Rosewarne, G. (2004) Wheat rust in Asia: meeting the challenges with old and new technologies. In: *New Directions for a Diverse Planet. Proceedings of the 4th International Crop Science Congress, Brisbane, Australia, 26 September–1 October 2004*. Regional Institute, Gosford, NSW, Australia (http://www.cropscience.org.au/icsc2004/symposia/3/7/141_singhrp.htm).

Singh, R.P., Hodson, D.P., Jin, Y., Huerta-Espino, J., Kinyua, M.G., Wanyera, R., Njau, P. and Ward, R.W. (2006) Current status, likely migration and strategies to mitigate the threat to wheat production from race Ug99 (TTKS) of stem rust pathogen. *CAB Reviews* 54, 1–14.

Singh, S.L. and Singh, P.P. (1985) Effect of some fungicide applications against Karnal bunt (*Neovossia indica*, Mitra) Mundkur of wheat. *Indian Phytopathology* 38, 593.

Sorrells, M.E. (2007) Application of new knowledge, technologies, and strategies to wheat improvement. *Euphytica* 157, 299–306.

Stokstad, E. (2007) Deadly wheat fungus threatens world's breadbaskets. *Science* 315, 1786–1787.

Tozetti, G.T. (1767) True nature, causes and sad effects of the rusts, the bunts, the smuts, and other maladies of wheat and oats in the field. In: Tehon, L.R. (ed.) *Translated. Phytopathological Classics No. 9.* American Phytopathological Society, St Paul, Minnesota, 1952, 139 pp. (originally published 1767).

Varshney, R.K., Balyan, H.S. and. Langridge, P. (2006) Wheat. In: Kole, C. (ed.) *Genome Mapping and Molecular Breeding in Plants, Cereals and Millets*, Volume I. Springer, Germany, pp. 79–134.

Villareal, R.L., Mujeeb-Kazi, A., Fuentes-Davila, G. and Rajaram, S. (1996) Registration of four synthetic hexaploid wheat germplasm lines derived from *Triticum turgidum* × *T. tauschii* crosses and resistant to karnal bunt. *Crop Science* 36, 218.

Wang, Y.Z. (1997) Epidemiology and management of wheat scab in China. In: Dubin, H.J., Gilchrist, L., Reeves, J. and McNab, A. (eds) *Fusarium Head Scab Global Status and Future Prospects*. CIMMYT, Mexico, DF, pp. 97–105.

Wiese, M.V. (1987) *Compendium of Wheat Diseases*, 2nd edn. APS, St Paul, Minnesota.

Wolfe, M.S. and McDermott, J.M. (1994) Population genetics of plant pathogen interactions: the example of the *Erysiphe graminis–Hordeum vulgare* pathosystem. *Annual Review of Phytopathology* 32, 89–113.

Xiaoning, G., Qili, L., Lili, H., Guorong, W. and Zhensheng, K. (2004) Resistance assessments of high yield wheat cultivars to take-all disease. *Journal of Yunnan Agricultural University* 19, 384–386.

Ykema, R.E., Floyd, J.P., Palm, M.E. and Peterson, G.L. (1996) First report of Karnal bunt of wheat in the United States. *Plant Disease* 80, 1207.

Zadoks, J.C. (1961) Yellow rust on wheat, studies in epidemiology and physiologic specialization. *Tijdschrift over Planteziekten* 67, 69–256.

Index

Note: Page numbers in *italic* refer to tables and figures in the text